HEALTH AND MEDICAL GEOGRAPHY

HEALTH and MEDICAL GEOGRAPHY

FOURTH EDITION

Michael Emch
Elisabeth Dowling Root
Margaret Carrel

THE GUILFORD PRESS
New York   London

Library of Congress Cataloging-in-Publication Data

Names: Emch, Michael, author. | Root, Elisabeth D., author. | Carrel,
    Margaret, author.
Title: Health and medical geography / Michael Emch, Elisabeth Dowling Root,
    Margaret Carrel.
Other titles: Medical geography (Guilford Press)
Description: Fourth edition. | New York : Guilford Press, [2017] | Revision
    of: Medical geography / Melinda S. Meade, Michael Emch. | Includes
    bibliographical references and index.
Identifiers: LCCN 2016037367 | ISBN 9781462520060 (hardcover)
Subjects: LCSH: Medical geography.
Classification: LCC RA792 .M42 2017 | DDC 614.4/2--dc23
LC record available at *https://lccn.loc.gov/2016037367*

Preface

Health and Medical Geography is the fourth edition of the book formerly titled *Medical Geography*. Our hope is that this text will be a sound foundation for the future development and practice of health and medical geography, and that it will inspire geographers and others to bring their own special subdisciplinary knowledge and theoretical approaches to enrich and advance this growing course of study. It provides a historical review of the field and includes the latest advances in health and medical geography. It is intended for both undergraduate- and graduate-level courses, and it should be an invaluable resource for all health and medical geographers. It can also provide an overview of the field to students and scholars in related fields, including public health, population health, and social and biophysical sciences that contribute to the study of health.

ORGANIZATION

The book has 14 chapters organized into five sections. We recommend that it be taught in sequence over a semester. The parts are organized into groups of chapters with common themes. Part I, "Introduction and Big Ideas," introduces health and medical geography, explores the historical roots of the subdiscipline, and provides background in order to understand the diverse and holistic approaches that are used today by health and medical geographers. Part II, "Maps and Methods," focuses on the methodological approaches of the field, including mapping sciences, methods that measure the spread of diseases, and genetic methods that inform the field. Part III, "What We Eat and Where We Live," explores how our changing food behaviors and living environments impact human health. Part IV, "Environments and Climates," focuses on the environmental, weather, and climate impacts on human health. Part V, "Health Care and Final Thoughts," focuses on geographic understandings of the provision of health services and access to health care, followed by a concluding chapter.

We have attempted to write a general textbook that can be read without the constant interruption of strings of citations, most of which are publications themselves citing the common source of an idea/term or general knowledge in a field ("It is hot in the tropics [x, 1958]" or "The malaria plasmodia (schistosome, tick, etc.) has several life stages [y, 1999]"). Where possible, we have referenced the seminal ideas and influential discussions, not all of which represent the most recent applications of those ideas. Our references are not meant to be comprehensive bibliographies of works in health and medical geography. The approaches, information, or examples we have consulted (i.e., actually used in writing the chapter) are listed under "References." Students can usually identify relevant materials simply by their titles. "Further Reading" contains relevant geographic studies or foundations that have not actually been used in writing the chapter. Some of the "Further Reading" entries were cited in past editions or are important, somewhat parallel, works. Many of them are suitable for term papers and further study. Finally, original ideas, quotes, or individual studies used for data are specifically cited in the body of the text.

PEDAGOGICAL FEATURES

The book includes the following pedagogical tools to help guide students and instructors:

- *Part introductions*: For each of the five parts, introductions provide a description of the theme for the part and introductions to the chapters within them.
- *Key terms*: Specialized terminology is highlighted in boldfaced text.
- *Glossary*: Short definitions of all boldfaced key terms are included in a consolidated glossary at the end of the book.
- *Quick reviews*: Each chapter includes one or more review sections of key concepts that are covered in the chapter.
- *End-of-chapter review questions*: Review questions are provided for students to consider and apply the material and concepts presented in each chapter.
- *Reference sections*: Specific references for work that is cited in the chapter are provided at the end of each chapter.
- *Further Reading sections*: A list of other suggested scholarly work that is not specifically cited in each chapter but that is related to the topic is provided.
- *PowerPoints*: PowerPoint presentation files for each chapter are provided on

the website of The Guilford Press to help instructors present the material to their classes.

NEW COVERAGE FOR THIS EDITION

With a new title and two new authors comes a book very different from the third edition. This new edition includes five wholly new chapters, and all other chapters were substantially revised and extended with new and updated material. The new chapters and topics are as follows. Chapter 3, "Expanding Disease Ecology: Politics, Economics, and Gender," expands the theory of disease ecology to include upstream political and economic forces that change these ecologies and create patterns of unequal mortality. Chapter 7, "Emerging Infectious Diseases and Landscape Genetics," describes the forces that drive emergence and reemergence of infectious diseases, such as globalization and migration, and introduces the field of landscape genetics and its application to questions from health and medical geography. Chapter 8, "Food, Diet, and the Nutrition Transition," discusses the emergence of obesity, diabetes, and other diet-related diseases that come with changes in food production, marketing, and distribution. The context for these discussions is the nutrition transition. Chapter 10, "Urban Health," focuses on how the movement of the global population into urban areas is impacting human health. Chapter 12, "Climate and Health," focuses on how weather and climate affect human health. It also offers a discussion of climate change science and reviews how future climate change scenarios are being used to predict how climate change will impact health and disease distributions.

Much has changed in the world and in the discipline of geography since the third edition of this text was published in 2010. Globalization has extended and intensified the processes that drive spatial patterns of disease, including the demographic transition, migration and circulation, income inequality, and cultural/economic/technological connectivity. The power of the geographic perspective of health is that these and other processes can be integrated by place, by region, and by geography. This text is about both geography as integration in place and geography as spatial analysis. Specific post-2010 changes worthy of attention in this edition of the book include the following:

 • World population has grown to about 7.4 billion, more than half of whom now live in urban areas; the global fertility rate has dropped to 2.5 children per family, although many countries in sub-Saharan Africa still have rates of 6 or 7 children per family; and the population growth rate has decreased to 1.2% a year. The richest European and northeast Asian countries have raised life expectancy to over 80 years, but their birth rates have dropped so low that they have negative population growth today. Life expectancy in the United States is slightly lower at 79 years because of large socioeconomic disparities in health. Sub-Saharan Africa has a life expectancy of only 57, with some countries in southern Africa in the 40s.

In our discussion of the demographic transition, we examine how the population trajectories of countries have changed in the last 5 years and some of the drivers of these changes.

- Extreme poverty rates have fallen in many countries, as can be seen through progress toward the Millennium Development Goals (MDG), which the United Nations established in 2000 to be met by 2015. MDG 1, to halve the proportion of people living in extreme poverty, was achieved several years early. In 2012, 22% of the world's population lived on $1.25 per day, compared with 47% in 1990. While MDG 4, to reduce the global child mortality rate by two-thirds, was not achieved, much progress was made. The global rate in 2012 was about half of what it was in 1990. Southern Asia and sub-Saharan Africa lagged behind the rest of the world, although there are exceptions such as Bangladesh, which reduced child mortality from 144 to 41 child deaths per thousand from 1990 to 2015. Some progress was made on MDG 5, to reduce maternal mortality by three-quarters, but the target was not met and progress was highly variable by region and country. From 1990 to 2013, the rate dropped from 380 to 210 maternal deaths per 100,000 live births. Notable success stories include Cambodia, Laos, Eritrea, Rwanda, and Nepal, all of which had maternal mortality reductions greater than 75%. In contrast, several sub-Saharan African countries had reductions of less than 33%, including the Democratic Republic of Congo, Cameroon, Kenya, and Côte d'Ivoire.

- There has been great progress in treatment and prevention of the leading causes of death in low-income countries, especially in sub-Saharan Africa. The three leading causes of death in low-income countries are lower respiratory infection, HIV/AIDS, and diarrheal diseases, in that order. In sub-Saharan Africa, the order of causes of death is HIV/AIDS, respiratory infections, and diarrheal diseases, with malaria a fourth important cause. Much progress has been made during the past decade in malaria and HIV/AIDS prevention and control, especially in sub-Saharan Africa.

  - Malaria mortality decreased by almost half from 2000 to 2012 in this region, which is home to approximately 90% of the world's cases. In 2013, there were an estimated 584,000 malaria deaths worldwide, but increased prevention and control measures have led to a reduction in malaria mortality rates by 47% globally since 2000 and by 54% in sub-Saharan Africa. The main control method has been the mass distribution of insecticide-treated bed nets.

  - The HIV/AIDS epidemic has slowed since the last edition of this book, especially in sub-Saharan Africa, which comprises three-quarters of the world's cases. Increased antiretroviral therapy (ART) access in sub-Saharan Africa has led to a steep decline in HIV/AIDS incidence. From 2009 to 2013, the number of people dying from HIV/AIDS decreased by 22% and the number of children dying from HIV/AIDS decreased by 31%. While there has been progress, in 2013 there were still an estimated 1.5 million people who died from HIV/AIDS globally. ART programs prevented an estimated 7.6 million deaths between 1995 and 2013. During the past few years, there has been growing evidence that HIV treatment can be used to prevent transmission. A landmark study, HPTN 052, showed that ART can reduce transmission rates by 96%. Therefore, it might be possible for the world to treat itself out of this epidemic with mass distribution of ART. Although there has been great progress in developing treatments for this disease—enough that some misguided people in the richest countries are starting to consider HIV/AIDS a chronic disease—there are still disparities in who gets ART. This is changing, however, with multibillion dollar programs, such as the United States–funded President's Emergency Plan for AIDS Relief, that distribute ART drugs in sub-Saharan Africa. HIV is discussed throughout the book, but most notably in Chapter 3, "Expanding Disease Ecology."

● China has developed faster than any other society in history. It is now the world's second biggest economy, and is on its way to becoming the first. Its urban areas are almost unrecognizable from 30 years ago, and there has been enormous environmental change. It has completed many massive projects of environmental engineering, including the Three Gorges Dam, which redistributes the flow of water from the south to the arid north. There has been devastating environmental degradation associated with this alteration, as well as with industrial growth and uncontrolled coal burning. The unique case of China is discussed in multiple chapters, notably Chapter 10, "Urban Health," and Chapter 11, "Environment and Health."

● New diseases have emerged, such as MERS, Middle East respiratory syndrome; old diseases have spread and been redistributed, such as Ebola and Zika; and the aging world population means that we may see epidemics of cancer and chronic diseases associated with older people. A new chapter is included that specifically addresses emerging and reemerging infectious diseases, including the processes that underlie their emergence.

● What may be the first impacts of climate warming on human health have been expressed in increasing drought conditions in arid lands and increasing flooding and cyclone severity, which have already destroyed crops and villages and have displaced tens of millions in sub-Saharan Africa and South Asia. Chapter 12, "Climate and Health," is updated with the most recent research on how global climate change will impact human health.

A genetic component has been discovered for many diseases; however, the environmental context for the expressions of those genes is still largely unknown and must be explored in order to understand the complex etiology of many diseases. Social science has developed an interdisciplinary theoretical perspective and a new vocabulary for addressing old processes. Geographic information systems (GIS), with their ability to manage and portray spatial data, have become the dominant tools in geography and have transformed the structuring of public data and health analyses. Health and medical geography as a subdiscipline has become less concerned with the optimization of health service delivery or a dichotomy between health service and disease ecology (etiology). Instead, it has become increasingly concerned with health and medical geography as a behavioral and social construction, and with disease ecology as an interface between the natural (physical world) and cultural dimensions of existence.

In this text, we endeavor to provide a broad-based, comprehensive survey of the rich diversity of health and medical geography, while also serving as a sound reference for the complexities of classifications, processes, and systems. Our perspective is holistic and global in scope. We hope to provide the necessary biological background for geographers to understand disease processes, as well as the necessary geographic background for health researchers to understand spatial processes. Students who have used the text in the past decade have included medical doctors pursuing doctorates in epidemiology; graduate students working on doctorates in geography; graduate students from such public health disciplines as epidemiology, biostatistics, health behavior, nutrition, health administration, and public policy; undergraduate geography majors; and premedical undergraduates with majors in chemistry or biology, but little background in the social sciences or geography.

ACKNOWLEDGMENTS

We would like to thank two students at the University of North Carolina at Chapel Hill. Cory Keeler, a PhD candidate who has wonderful cartographic skills, created several maps and graphics. Katharine Mather, an undergraduate student, helped with various tasks, including formatting and organizing the final draft. We would also like to thank the staff at The Guilford Press and the reviewers of this text, including Sue C. Grady, Michigan State University; Christopher K. Uejio, Florida State University; Tony Dzik, Shawnee State University; and L. Benjamin Zhan, Texas State University. Finally, we wish to acknowledge and appreciate the examples and even words in this text that belong to the four coauthors of the first three editions, Melinda Meade, John Florin, Wilbert Gesler, and Robert Earickson, whose work is a seamless part of the text.

This book is dedicated to Melinda Meade, who passed away in 2013. It would not have been possible without her wisdom and guidance of us all. Melinda trained many graduate students who went on to shape this field. She also informally mentored many other faculty and students, particularly women scholars. Melinda was a well-respected and well-loved mentor. Her intellect and compassion were always on hand when working through research or contemplating life as a student. She gave unselfishly of her time to her students and approached research with an enthusiasm and excitement that was contagious. She encouraged people to think deeply and to strive not just to "do research" on a population, but to truly understand and appreciate the different ways in which diverse populations live their lives. Melinda engaged her students in scholarly dialogue that demanded a maturity of thinking that naturally led to intellectual development. She read voraciously and widely. Her academic legacy is a far-reaching network of scholars who focus on people and places and health outcomes. We encourage you to go to *swraex.wistia.com/medias/ plaluvak07* to view a beautiful video tribute to Melinda's life entitled *Atlas Hands*.

MICHAEL EMCH
ELISABETH DOWLING ROOT
MARGARET CARREL

Contents

Purchasers of this book can download PowerPoint slides
of the figures at *www.guilford.com/emch-materials*
for personal and classroom use.

On Airs, Waters, and Places

Whoever wishes to investigate medicine properly, should proceed thus: in the first place to consider the seasons of the year, and what effects each of them produces (for they are not at all alike, but differ much from themselves in regard to their changes). Then the winds, the hot and the cold, especially such as are common to all countries, and then such as are peculiar to each locality. We must also consider the qualities of the waters, for as they differ from one another in taste and weight, so also do they differ much in their qualities. In the same manner, when one comes into a city to which he is a stranger, he ought to consider its situation, how it lies as to the winds and the rising of the sun; for its influence is not the same whether it lies to the north or the south, to the rising or setting sun. These things one ought to consider most attentively, and concerning the waters which the inhabitants use, whether they be marshy and soft, or hard and running from elevated and rocky situations, and then if saltish and unfit for cooking; and the ground, whether it be naked and deficient in water, or wooded and well watered, and whether it lies in a hollow, confined situation, or is elevated and cold; and the mode in which the inhabitants live, and what are their pursuits, whether they are fond of drinking and eating to excess and given to indolence, or are fond of exercise and labor, and not given to excess in eating and drinking.

From these things he must proceed to investigate everything else.

—HIPPOCRATES (c. 400 B.C.E.)

PART I

INTRODUCTION and BIG IDEAS

Health and medical geography is a subdiscipline of geography that uses the theory and methods of the field to investigate health. This part introduces health and medical geography, explores the historical roots of the subdiscipline, and provides background in order to understand the diverse and holistic approaches that are used today by health and medical geographers.

Chapter 1 introduces and describes the subfield, including a discussion of debates over health versus medical geography as a name for the subfield. It summarizes the roots and history of the field from the time of Hippocrates to the present, with a special emphasis on its development in the United States during the 20th and 21st centuries. The chapter also defines health and disease as well as several epidemiological terms that are used throughout the book. The most recent and exciting areas of expansion of the field of health and medical geography and areas that need further development are described. The chapter ends with a challenge to students and scholars of the field that was offered more than 40 years ago but that still holds true today.

Chapter 2 explores how geographers use the concept of "ecology" to investigate health and disease. It introduces a model called the "triangle of human ecology," which is an approach for studying health and disease that has three mutually interacting vertices: population, habitat, and behavior. *Population* refers to biological attributes of humans such as their genetic susceptibility, *habitat* to the environment in which people live, and *behavior* to the observable aspects of culture that protect or put people at risk of disease. Another approach called *landscape epidemiology,* which focuses mainly on vector-borne infectious diseases, is defined. The chapter explores how the triangle of human ecology and landscape epidemiology can be applied as frameworks for understanding spatial patterns of disease through several case studies, including colorectal cancer, cholera, Rocky Mountain spotted fever, and Lyme disease.

Chapter 3 expands on the ecological concepts discussed in Chapter 2 by focusing specifically on more upstream determinants of health and disease, namely, politics, economics, and gender. These large-scale structural forces shape many of the environmental and behavioral risk factors that more directly affect health. At the same time, political, economic, and social forces often set research agendas or discourage the exploration of links between gender and health or environmental conditions and health.

1

The study of genetics has changed dramatically during the last decades and as such health and medical geography studies increasingly consider genetic factors. The chapter explores an expanded approach to the ecology of health and disease through several case studies including the genetics and political ecology of cancer, HIV/AIDS, and women's health.

Chapter 4 focuses on the demographic transformation of the human population from premodern times to the present. The population has grown dramatically since premodern times and life expectancy has reached an all-time high. With fertility rates dropping rapidly, and sometimes stabilizing at below-replacement levels, this demographic transformation translates to an aging population. In this chapter we examine demographic transition theory and the near universal trends in fertility, mortality, and population growth seen around the world. These population changes are related to significant changes in the major causes of death, which is often called the "epidemiological transition." The related mobility and nutrition transitions are also introduced. We frame these transitions within a larger discussion of globalization and how this process has contributed to the transformation of world populations.

CHAPTER 1

What Is Health and Medical Geography?

Health and medical geography uses concepts and methodologies from the discipline of geography to investigate health-related topics. The viewpoint is holistic, dealing with a variety of cultural systems and a diverse biosphere. Drawing freely from the concepts, theories, and techniques of other social, physical, and biological sciences, it approaches health and disease through its own core questions and perspectives, and uses its own techniques of spatial analysis. Health and medical geography is an integrative, multistranded subdiscipline that has room within its broad scope for a wide range of specialist contributions.

Health and medical geography is both an ancient perspective and a new specialization. As illustrated by the quotation preceding this chapter, Hippocrates (c. 460–377 B.C.E.) was familiar with the importance of cultural–environmental interactions more than 2,000 years ago. The study of these interactions, which are important to disease etiology, health promotion, and health service provision alike, continues to this day as health and medical geography. As old diseases, almost forgotten, are reemerging amid new risks; as the majority of the world's population has become urban and moves toward stabilization of numbers; as biotechnology transforms medicine, agriculture, and our understanding of the nature of life; and as climate change, air and water pollution, metastasizing consumption, and overwhelming inequalities transform the ecology of disease, so the ancient study of how people, their cultures, and their societies in different environments create and spread disease, promote health, and provide care has never been more relevant. It is important to understand the interactive processes of cultural and environmental change, as well as the importance of distance and location. By doing this work, health and medical geography not only advances knowledge, but also has real-world applications for improving human health.

This first chapter undertakes three topics. First, it introduces health and medical geography. To do so, it presents a framework for the complexity of this specialization and explains something of the divergence of "ecological" paradigms between the social sciences and the biological sciences, which health and medical geography straddles. It also takes a look at the overall questions and nature of the discipline of geography. Second, this chapter addresses the history of health and medical geography, especially in the United States, and its evolution in a changing world. Third, it briefly introduces some of the most basic terminology for talking about the subject. It closes by revisiting a challenge to geographers made in the early 1970s by John Hunter that helped define the field and still rings true today.

WHAT'S IN A NAME?

Over the past two decades *health geography* has become the more common term for the field and the authors of this book believe it to be a more appropriate and current name for the field than *medical geography*. That said, there is still some disagreement and debate about what the field should be called. A brief history of the field is offered in this chapter including a discussion of the history of the nomenclature. The title of this new edition is *Health and Medical Geography* but we usually refer to the field as "health geography" for the sake of brevity.

The term *health geography* is relatively new, with its roots in the early 1990s. The term *medical geography* has more than half a century of use in the United States. A range of opinions and definitions are offered in several chapters of a 2009 book entitled *Companion to Health and Medical Geography* edited by Brown, McLafferty, and Moon in Part 1: Debates in Health and Medical Geography. The three chapters in this section debate the meanings of the terms *health geography* and *medical geography* and whether there should be separate subfields with separate names. The last chapter of the section of that book is called "Doubting Dualisms," and contends that "health" and "medical" are arbitrary divisions than cannot be separated into different subfields. In practice, many scholars who might have called themselves medical geographers in the past now call themselves health geographers, including the authors of this book. Whatever one's opinion about the nomenclature it is also clear that there has been great progress in all parts of the field during the last three decades. The field has always been holistic but has expanded its breadth in recent years. As the field has evolved, it has become more inclusive of concepts, methodologies, and subject matter related to the geography of health and disease. It arches over biological, environmental, and social sciences; it uses both the quantitative and qualitative methodologies of cartography and statistical analysis, as well as participatory fieldwork, interview techniques, and perceptions best addressed through the humanities. The subdiscipline of health geography is a dynamic, evolving, vibrant area of scholarship. In this book, after this initial chapter that includes descriptions of the debates and history of the field, the term *health geography* will be used, with the view that there is one unified, integrated field that welcomes the contributions of all geographers to further the understanding of various geographic dimensions of health, healthcare, and disease.

The accelerating rush of **globalization** is spreading and connecting political/social/economic structures and behaviors around the world in ways both deleterious and advantageous to health. Individual social behaviors and priorities of society that increase well-being act to promote health and prevent disease. Those that expose people to more risk or increase hazards in their environment promote diseases of body, mind, or society. By whatever name, the subdiscipline is really public health geography—a rather awkward but more descriptive name. In the spring of 2009 the Medical Geography Specialty Group of the Association of American Geographers (AAG) recognized the contributions and research perspectives of all its members by renaming itself the Health and Medical Geography Specialty Group. Others have kept the historic name; indeed, the largest international conference in the field is called the International Medical Geography Symposium (IMGS), which hosted its 16th biannual conference in Vancouver in 2015 and whose 17th biannual conference will be held in Angers, France, in 2017. The conferences include the entire breadth of the field including scholars who call themselves medical or health geographers.

Some Complexities of Geography and of Ecology

Geography is a complex subject, as the above-described debates about a name for our subdiscipline suggest. Some of the points of contention among geographers and certainly among epidemiologists arise out of the layers of complexity that the term *ecology* has developed. This section explains more about the nature of geographic study. It also looks at some theoretical differences between health geographers, epidemiologists, and sociologists.

One way of viewing the complexities of the discipline of geography is portrayed in the matrix of Figure 1.1. The matrix shows what are known as the topical or systematic specializations of study within the discipline (i.e., the subdisciplines), each of which usually has a cognate discipline of study, which is *aspatial*. Health geography's aspatial cognate is public health. Studying one or more topical aspects over space, across multiple scales, addresses the core questions of geography: "Where?" and "Why there?" Why are things—rainfall, carcinogens, primary care physicians—distributed unevenly over space? Why are they located where they are? The question "Therefore, what?" (i.e., what the interactions and consequences of that distribution in space are) is addressed through the holistic integrations of region. The scales in Figure 1.1 for regionalization may be those of analysis, or they may be those of integration of all relevant topical processes and variables, which occur together in space and create its reality—and its state of health. The distinctiveness—the uniqueness of place that can be so emotively powerful—is thus a microregionalization, not different in kind from those considered at different scales: as country (the United States), as region of country (the South), as city (home town), or as census block (the neighborhood).

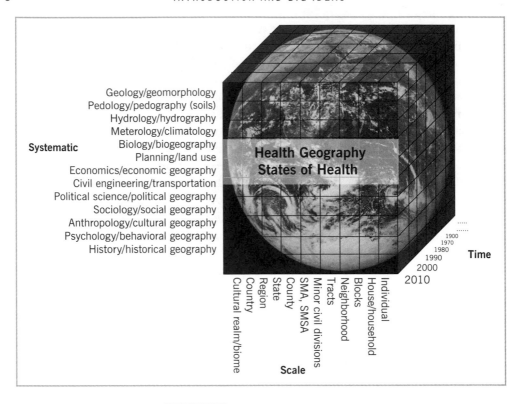

FIGURE 1.1 The matrix of geography.

Integrating the interactions in a place and a time is a difficult endeavor approached through a variety of methodologies. These include qualitative ones, such as fieldwork, participant observation, interviews, and focus groups; they also include more quantitative ones, such as multivariate statistical analysis of patterns, mathematical analysis of diffusion, or remote sensing (satellite images) of land use change. Many kinds of study and realms of knowledge can be bridged and connected in geography because they all occur in space and can be integrated in place or region. As the cube of Figure 1.1 shows, the study of air, water, soil, and their processes and expression in the physical environment; the study of biological systems, from molecular to organism to ecosystem; the study of society in varied culture systems, with different religions and governmental forms; the study of individual people and their perceptions and beliefs, as well as genetic constitution—all these and more are connected by their overlay in space. Not surprisingly, academic divisions between the biophysical sciences and social ones, and of both types of sciences with meaning and perception in the humanities, are all mirrored in arguments and divisions within geography. There are so many ways to classify and to interrelate things. True integration even in space is so difficult!

Various efforts have been made to conceptualize the complexity of the discipline of geography discussed above in terms relevant to the developing theories and structures of the broader science community. Previous editions of this text have emphasized Pattison's (1964) "four traditions" of geography: earth science; spatial (and locational) analysis; man–land (today, society–environment, human–environment, nature–society, or [as in this text] cultural ecology); and regionalization (integration). They, and all geography, address questions about why things are located and distributed in space the way they are (where); why they are moving here and not there (migration, circulation, diffusion); what importance distance has; how and why people vary in perceptions of these things; and how objects, landscapes, ideas, and people interact and overlap in space to create place. There is today a greater emphasis on the larger picture in the context of society. Whether this is addressed through concepts of structure, agency, power, and dependency (in social theory); through identifying the overarching economic and political processes in globalization (in political economy); or through questions about decision, purpose, gain, and loss at different scales (in political ecology), often the individual, local, microscale, "where the rubber meets the road" interactions and effects are neglected.

What's in a name? There are so many disciplines and perspectives involved in the study of health and disease that scholars and professionals can often talk past each other by addressing totally different paradigms even while using the same words. Probably the most fundamental of these is the term—the process, the system, the fallacy, the methodology—*ecology*. Consider the definitions and disciplinary descriptions in Figure 1.2.

The concept of **ecology** changed within biology as it matured from an emphasis on organisms to an emphasis on relations and functions of the system. The foundational biological concepts and processes, however, entered in the 1960s into cultural ecology in both anthropology and geography, and thus into the emerging subdisciplines of medical anthropology and medical geography. The concepts of cultural–environmental interactions in layers (scales) of systems seemed especially useful for understanding the basis of infectious and parasitic diseases in the newly independent tropical countries. As the ideas of systems, multiple layers, interactions, and complexity entered the social sciences as **social ecology,** Hawley's

---

Ecology (biology) is the study of the interactions between living organisms and their biotic and abiotic environments. (Odum, 1959)

Ecology (sociology, human ecology) is a descriptive term applied to complex relationships between organisms and their environment. (Hawley, 1950)

Social ecology (sociology) is the study of the relationships between individuals, social groups, and their environments.

Political ecology is concerned with how higher-level socioeconomic and political structures and processes create the context in which local cultural and individual behavior interacts with the environment.

FIGURE 1.2 Definitions and disciplinary descriptions of ecology.

ecology as a "descriptive term applied to complex relationships" was increasingly used to focus on relationships between individual people and their social groups. The connection with the natural environment and processes was progressively lost. The environment became mostly social, although some built environment of urban blight or noxious facility was occasionally included. What were measured and modeled were the socioeconomic and demographic variables. They were studied with new statistical methodologies, factor analysis and its relatives, which were able to model complex relationships. The dimensions of socioecological differentiation thus found could be used to develop theories of socioecological structure, beginning with macrolevel units based on sets of variables. Through sociology into political science and other social sciences, including social and urban geography, "ecological" came to mean complex, multivariate, quantitative (statistical, factorial) analysis of characteristics of people, their groups, and social/economic/political conditions (perhaps housing, crowding, or other "environmental" effects), conditions identified and measured as variables. The macroscale approach was not about individuals, families, streets, and neighborhoods, but about populations, such as teenagers or households below the poverty line in income.

Ecological studies thus became almost anathema to epidemiology, since they seemed to be full of confounding variables and spoke to population-level data and not individual-level causation or intervention. "Ecological" came to mean population studies: multivariate, multiple-scaled perhaps, complex, uncontrollable, and fuzzy—interesting, perhaps, but of limited scientific use. Data "averaged" for a group made causal connections between an exposure and disease results very murky. A group average could neither connect individuals of that group with exposure or results, nor connect larger populations or other groups with those average exposures. This is the **ecological fallacy**, the idea that associations statistically identified at one scale of analysis are valid at or can be generalized to either larger or smaller scales. (The ecological fallacy is discussed at several points later in this text, most notably in Chapter 5.) The real environment and its ecology largely disappeared from perspective. Epidemiology (and the U.S. surgeon general) could say by the 1980s that more than 80% of cancer was "environmentally caused," meaning caused not by pollution or radiation, but by behavioral inputs like smoking tobacco. This "environment" was that of a cell, not the biosphere of a human population. Health promotion became concerned primarily with the institutional and structural impediments to changing behavior and accessing health care. For a couple of decades, epidemiological questions were mostly limited to small controlled populations, and any generalizations (such as those about cigarette smoking and cancer risk) were sought through methodology from them. One of the most exciting things happening in public health today is that the power of **geographic information systems** (GIS) is allowing epidemiology to ask questions and consider relationships at different scales again. They facilitate the analysis of health at multiple levels in "nested" relationships of schools, neighborhoods, and communities within cities, states, and nations.

Medical geographers throughout the 1970s and 1980s could differentiate between two "schools" of ecology: disease ecology and health service delivery.

Social/urban/economic geographers drew on their models to study health service delivery. They used the term *ecology* to mean multivariate complexity as the sociologists did, although as geographers they recognized the importance of study at different scales. They sought to understand the roles of distance and accessibility in determining the optimal locations to provide health services and to locate physicians and other caregivers, facilities, and equipment. They also sought then to determine the obstacles and social/political/economic creation of need and obstacles to its provision. (The main body of their work is explained in Chapter 13.) In contrast, geographers who were especially concerned with human–environment interactions identified primarily with cultural ecology. Their study of disease ecology became inclusive of the influences in the biological and physical as well as the social environment as they modeled disease etiology. They sought to integrate these different aspects of environment within place in order to understand the creation and maintenance of health and disease systems. This would enable the vulnerable points and protective interventions to be identified, and so to permit illness to be reduced and health promoted. Over time, disease ecology became concerned not just with infectious diseases, but with the cultural ecology of chronic and degenerative diseases and other health conditions of people in economically and demographically advanced countries as well. Such studies have also needed to address higher social/economic/political structures, processes, and questions.

Many geographers called everything that was not about health service "disease ecology." This included studies of disease diffusion, methodologies of mapping disease, and social-factorial analyses of the pattern of disease. This dichotomous approach was and still is a source of confusion, however, as these studies seldom involved environmental interactions or ecological processes. Since most research specialists tend to read research literature relevant to their own paradigms and approaches, the cross-definitions and evolved usages explained above became and have remained major sources of misunderstanding. With the help of GIS (see Chapter 5), most of these ways of studying health and disease today seem to be converging—but some scholars are still speaking the language of different paradigms, as the following brief history of medical geography makes clear.

A BRIEF HISTORY OF HEALTH AND MEDICAL GEOGRAPHY

Place was important to medicine until the middle of the 19th century. For 2,000 years medicine was concerned with geographic variations in air, water, soil, and vegetation; insects and animals; diet, habit, and custom; clothing and house type; government; and economy. The late 19th century saw a paradigm shift—that is, a change in the great overarching idea of how disease occurred and what questions were worth asking. In the 20th century germ theory, otherwise known as the "doctrine of specific etiology," resulted in revolutionary advances. Specific etiology—one cause (germ) that is both necessary and sufficient for each disease—is less relevant to a society in which people die from heart disease, cancer, kidney failure, addiction, and violence. These are diseases of multiple, complex causes based as much in culture and society as in biology. Even for infectious diseases, germs are no longer considered to be the "sufficient" causes. The tubercle bacillus is necessary to cause tuberculosis, but the disease and its course depend on many factors: nutrition; genetics (resistance); treatment; bacillus Calmette–Guérin (BCG) vaccination; the presence of disease-promoting conditions; crowding, ventilation, and housing quality; the custom of spitting; and mental attitudes. The etiology and control of strokes or diabetes are even more complex.

The progress of specific etiology was paralleled by the progress of specialization in medical study and treatment, and by the increasing divorce of body from mind and environment (and even organ from body, and cell from organ). As the contradictions between the dominant biomedical orientation and the health needs of people increased, the social sciences became more involved. Besides health geography, the last few decades have seen the development of flourishing concentrations in medical anthropology, medical sociology, health economics, and health psychology. Historians also have reconsidered the significance of disease for major social and economic change and the consequences of the connection of empires.

Geographic variation in health and disease has long been studied under such interdisciplinary rubrics as geographic pathology, medical ecology, medical topography, geographical epidemiology, geomedicine, and so forth. The perspective that cultural–environmental interactions were important for disease etiology and for health continued to be philosophically important, even dominant, until the emergence of germ theory. Thus the 18th- to 19th-century physicians (e.g., Finke, 1792–1795) who first used the term *medical geography,* and who struggled in dozens of works to describe and organize the avalanche of new information about human diseases, cultures, and environments that came with world exploration, were continuing the holistic Hippocratic tradition. Their descriptions of where diseases were endemic and where they didn't occur, and their explanations for this and for epidemic spread, are being rediscovered and reevaluated by geographers and other scientists once again concerned with disease ecology (Barrett, 1980). Barrett (2000) established that "medical geographical and geographical medical" research reached its zenith in the decade of the 1850s with the publications of Drake (1854), Fuchs (1853), and others and was being increasingly challenged. It had been thought that understanding the geographic patterns might lead to understanding why there was disease, but the rise of bacteriology brought other answers. By the time Hirsch (1883–1886) published his masterful handbook, the "doctrine of specific etiology" of disease was finding one bacterium after another. The discovery that microbes invade human bodies and cause alterations that result in disease led to asepsis and sterilization, vaccination, antibiotics, chlorination of water, treatment of sewage, and over 30 more years of life for the average person in a developed country. The enormously successful paradigm of germ theory is probably responsible for the existence of half the people reading this book, as without it they would never have been born.

After World War II, with the development of widely available antibiotics and powerful insecticides such as DDT, people dreamed that infectious disease might become a thing of the past. At the same time, however, degenerative diseases such as heart disease, stroke, cancer, liver and kidney failure, respiratory obstruction, and such became the major killers and disablers. It was obvious that these diseases did not have a specific etiology; they had complex etiologies involving many factors—cultural, social, and environmental, as well as biological. New theories about society and cultural behavior were needed to explain their causes and to guide research and interventions.

The perspective and methodology of geography have now been applied to the study of health, disease, and health care for over 60 years. The emergence of a systematic (topical; see Figure 1.1) interest in health geography can be dated from the first *Report of the Commission on Medical Geography (Ecology) of Health and Disease to the International Geographic Union* in 1952 (Geddes, 1978). The next 15 years saw pioneering researchers and teachers in a dozen countries develop a substantive, international focus on health geography. The following sections follow the developments in the United States.

Jacques May

Jacques May was the "father" of medical geography in the United States. He initiated the subdiscipline (May, 1950, 1954) with the paper "Medical Geography: Its Methods and Objectives," developed the first maps of global distributions (May, 1950–1954), and wrote the book *The Ecology of Human Disease* (May, 1958). This represented the culmination of a long intellectual evolution on May's part. He began his career as a French surgeon in Siam (now Thailand) and spent more than a decade as head of surgery at the major French hospital in Hanoi, Vietnam. He started his intellectual journey by questioning why his patients in the tropics experienced and responded differently to disease and surgery than did the European patients described in his textbooks. May progressed to an interest in his patients' multiple, simultaneous infections and then to the conditions of their lives. Then, from his own travel and experience, he came to understand the cultural and environmental conditions that produced and limited their health and disease. May was influenced in his development of disease ecology by the French approach to geography, which emphasized distinctive local–regional expressions of interaction between human and physical domains, expressed as *genres de vie*. He referenced Sorre (1947) in his own seminal article (1950), including Sorre's concept of "pathogenic complexes": physical, biological, and anthropological factors that lead to disease (Barrett, 2000). World War II found May in New York City, where he conducted research at Rockefeller University; he then taught briefly at several major northeastern universities.

Jacques May became the founding director of the American Geographical Society's Medical Geography Department. By 1951 he was publishing the first maps of distribution of species of helminthes and microbes, and their tick and mosquito vectors; of the distribution of diseases such as leprosy; and of key variations in diet and nutritional deficiencies. Plate 1 ("Distribution of Malaria Vectors") and Plate 2 ("Study in Human Starvation") reproduce two of these groundbreaking maps, which were published as inserts in the *Geographical Review,* the journal of the American Geographical Society. The Plate 2 map of diet and nutritional deficiencies is especially notable because May spent the next decades publishing almost annually, from 1961 to 1974, books on the ecology of malnutrition in a world cultural region.

The Early Years

John Hunter (1966, 1974) taught medical geography at Michigan State University from the 1960s until the 1990s and seeded American colleges with geographers who had at least an initial exposure to the subject. One of your authors, Michael Emch, was Hunter's last PhD student before he retired in the late 1990s. In the 1960s the curriculum of human ecology at the University of Illinois strongly affected the founding of a medical anthropology curriculum at the University of California at Berkeley. Also in the 1960s, the first studies on locational analysis for health services—the ground-breaking regionalization of the Swedish health system by geographers at the University of Lund (Godlund, 1961)—occurred, and Richard Morrill brought both Gerald Pyle (1971) and Robert Earickson (Morrill & Earickson, 1968) into the geography of health services through the Chicago Regional Hospital studies. Soon afterward, Gary Shannon and Alan Dever (1974) wrote the widely read book that first introduced a generation of graduate students in geography to the subject: *Health Care Delivery*. Settlement forms, systems of cities (urban hierarchy), transportation networks, and population mobility had become objects of modeling in urban geography. As a doctoral student in urban geography at the University of Chicago, besides writing a dissertation on the future of heart disease and the spatial dimensions of the need for hospital care in that city, Pyle (1969) studied how the changing process of diffusion of cholera in the 19th-century United States reflected the maturing infrastructure and urban system of the country. In doing so, he wrote the seminal paper in spatial analysis of disease diffusion (see Chapter 6).

At its November 1974 meeting, the Council of the Association of American Geographers established a Committee on Medical Geography and Health Care. Pyle was appointed chair, with a committee composed of John Hunter, Alan Dever, Warwick Armstrong (a New Zealander), and J. L. Girt (a Canadian). By January 1975 this committee had proposed an atlas with chapters on the following topics: the "continental ecospatial framework," "biogeochemical environment and health," "man-made environmental hazards," "organizations and use of health services," "illness and health behavior," and "policy planning and implementation." A similar framework, though with much more emphasis on health services, could have been used for paper sessions at meetings 20 years later. In 1975, also, the Medical Geography "Special Interest Group" (as the AAG then considered it) elected its first steering committee, with Melinda Meade emerging with the most votes as chair. By organizing in 1975 two sections of paper presentations (one in "general" medical geography and one in disease perception) and offering a sold-out workshop ("The mini-course organized by Gary Shannon was designed to provide an organization scheme for health/disease research, a working knowledge of appropriate data collection and analytic techniques, and health planning program developments pertinent to geographers"), the Medical Geography Special Interest Group became a model for the reorganization of the AAG and its national meetings by specialty group. In January 1979, having gathered an abundance of signatures following the protocol for founding the new AAG organizations, Melinda Meade became the first chair of the new Medical Geography Specialty Group—the organizing body today is now known as the Health and Medical Geography Specialty Group.

The Middle Modeling Years

The numbers of medical geographers and of their college course offerings increased in the 1980s, but government funding was almost exclusively allotted to health service research. The geographic research agenda followed the funding. Annual meetings presented hours of papers on health service location and utilization; scores of medical geographers found jobs in planning and marketing. The U.S. health care system was an "un-system" that had grown rapidly without much spatial planning or organization at any level of government (Chapter 13). Cities and states did not know where hospitals were located, let alone doctors or specialists. Unschooled in geography, those who began to grapple with provision of services, construction of needed beds, location of expensive new scanners and other technology, and racial desegregation, among other things, had no idea about the friction (and its costs) of distance or how to optimize the locations of or accessibility to new facilities. Techniques and computer programs were developed to regionalize emergency rooms and route ambulances, as well as to estimate the accessibility consequences of hospital closings or retirements of physicians in rural areas. The classic "journey to work" studies and their locational analysis models of economic geography became health service demand and market area analyses.

As geographers began to focus on the spatial distribution and inequalities of health care provision, it became clear that the way countries organized their health care systems had a major effect on the dimensions of health inequality. Understandings of the political economy became increasingly important. Such topics as public policy and the structure of society moved out of sociology into geographic analysis (Eyles & Woods, 1983) and from Britain across the ocean to geographers involved with the national Canadian health promotion endeavor. Studies increasingly focused on topics such as territorial injustice, environmental racism, and the structural context, which influenced why people get sick, rather than simplicities of distance. Jones and Moon's (1987) influential book moved the social context of not only health care but health itself into the mainstream of geographic study. The deinstitutionalization of persons with mental illnesses in the United States, and the failure to develop the community service provision and shelter to care for those affected by mental illness, substance abuse, or domestic violence, influenced the focus and methodologies of geographic research and exposition (Dear & Wolch, 1987; Wolch & Dear, 1993).

During this time there were also developments in spatial analysis, which involved fewer geographers but had a large impact upon public health. Spatial analysis is so intrinsic to geographic thinking that many geographers would find it strange to discuss it separately. It received special attention in two areas: the study of disease diffusion and the analysis of diseases of unknown etiology. When the epidemiological and mathematical study of the spread of contagious disease within a population and its subgroups is considered spatial, some of the most authoritative models in geography come to bear on population movement, settlement hierarchy, and transportation network. Analytical models of the diffusion of influenza, measles, cholera, and hepatitis during this time became elaborate and powerful, as reviewed by Cliff and Haggett (1988) and Thomas (1992) (see Chapter 6). A different aspect of spatial analysis involved the study of the covariation in space of disease occurrence and possibly related factors of environmental and social conditions. As noted above, these studies have often been classified with disease ecology, but the "ecology" in them has been of the social–factorial kind and during this time involved little or no "real" environment (see Chapters 3 and 9).

The special and basic tool of geographers, in the service of all its traditions, is the map. **Cartography**, the construction and interpretation of maps, has held a central place in geography for more than 2,000 years. Most geographers have a profound love affair with maps. An old saying in the field is "If it can't be mapped, it's not geography." Although a wide variety of statistical techniques are used in geography, as in other social sciences, maps are unique, powerful, and flexible tools for the analysis of geographic phenomena. A map is a model of the world that, through the use of point, line, and area symbols, can integrate many dimensions of reality. Following on the great success of a world atlas produced at Heidelberg University in Germany (Rodenwaldt & Jusatz, 1952–1961), the studies in geomedicine at Heidelberg (Jusatz, 1968–1984) attempted to pull the multiple dimensions of the geography of health and disease together in place and region by mapping. In the 1980s the ancient dimension of cartography as an art as well as a science gave way to the accessible but mindless precision of computer cartography. In the 1990s this gave way in turn to the larger application of the computer to the methodologies of spatial data management GIS (see Chapter 5). GIS have not so much changed the questions that geographers ask as revolutionized their ability to answer them. The long history of mapping in geographic and public health analyses of prevalence, etiology, diffusion, and health care distribution is exhaustively related in two books by Tom Koch (2005, 2011) called *Cartographies of Disease* and *Disease Maps: Epidemics on the Ground*. Cliff, Haggett, and Smallman-Raynor (2004) have shown multiple forms of mapping as they have taken on global, regional, and local maps and interpretations of 50 of the major epidemic diseases in their eloquent *World Atlas of Epidemic Diseases*.

Recent Times

Over the last two decades, developments in availability of digital data; in theory in social science; in population mobility and world urbanization; and in globalization of pollution, trade, and other processes that could be listed to fill this page have changed everything—and changed nothing. In this section we look quickly at developments in social theory, place and therapeutic landscapes, and political ecology.

Social science has seen fundamental questioning of and turbid debates over **epistemology**. One result is a social-theoretical framework that is much more interdisciplinary than before, and that values qualitative methods and is more open to questions about meaning (previously the domain of the humanities). Geographic concepts of place, landscape, and global interconnectedness have become familiar and commonly expressed in other disciplines. They have also become more dominant concerns in geography. Social theories variously known as social-interactionist, structuralist, structurationist, poststructural, and critical have progressed from being positivist—empirically trying to verify data and test observations and models (disease etiology)—to concerns about interpreting the "meaning of illness" to an individual and, with or without human agency, describing the deep structures of economic and political domination and the inequalities they produce (see Gattrell, 2002).

Kearns and Moon (2002, p. 606) have identified three new themes in health geography: emergence of place as a framework for understanding health; application of social- and cultural-theoretical positions to health; and a quest to develop "critical" geographies of health. Their approach to health geography brings to bear the concerns underlying the development of social theories over the past decades. Individual people are persons, not just "observations"; place is a "living construct" that has meaning, not a container from which to draw observations or study participants (as when counties are used as units in mapping cases of influenza). Qualitative methods of interview, observation, focus groups, and text or discourse analysis are more appropriate for analysis. Well-being and justice are the objects of interest. These perspectives and concerns have been directed at physical disabilities, mental health problems, and old age—vulnerabilities that were seldom addressed before.

Gesler (2003) developed the constructs of landscape and place (Kearns & Gesler, 1998) in an original study of therapeutic places, such as the ancient healing grounds of the sanctuary of Asclepius at Epidauros or the healing waters of Lourdes. Gesler has noted (2005, p. 295) that following Williams (1999), others have taken the concept and used it to look at places that support well-being (such as children's camps), as well as landscapes of risk (such as refugee camps).

The development of **political ecology** is another response to the need to consider the political, social, and economic structures and processes that construct policy and form the context of local and individual exposure to health risks (conditions good and bad that people must cope with and react to) and medical treatment. Political ecology is quite analogous to political economy, but is concerned with more than the consequences of capitalist production or with power, domination, and blame. Political ecology is explicitly about causal interactions with local ecology. (Yes, it can be a two-way flow! Drought can change political and social institutions and structures just as well as result from them.) Political/economic/social policies, structural institutions, and processes are not ecologically neutral. The construction of a large dam and water impoundment to provide electricity for a major city, industrialized jobs for the urban migrants, and profits for the corporations can create a habitat for *Anopheles* mosquitoes and malaria where before there was none. Commercial logging and subsequent settler efforts at agricultural conversion change local ecological systems and reduce feedback, which, with the induced population mobility, makes species connections and promotes the emergence of new disease. Land tenure and land use policies can determine who in what place have no "entitlement" to food and so affect regional levels of malnutrition. And so on. Women are especially vulnerable to these changes in their societies generally. Political ecology is concerned with understanding how political, economic, and social forces—which interact and influence each other at multiple geographic scales—shape choices and actions about the natural environment and resource use.

Public health research is also struggling again with ecological studies, taken as nested systems of multiple interactive variables. As digital data become more available, the capacity to analyze context and to compare relative influence at different scales is a siren song. The neighborhoods-and-health framework for analyzing multiple nested layers of disease causation is described in Chapter 9. It develops out of the power of spatial data management and the integrative power of spatial organization. The use of GIS in public health is already well under way (Cromley & McLafferty, 2012).

Future Directions

The only thing certain about the future of health geography is that there will be new developments. The earth is warming, and rain very likely will be falling in different spatial patterns. The human population is slowing its growth; in fact, population growth may stop altogether in a few decades. The majority of the world now lives in an urban setting, and in developed countries and others (e.g., China), the population is getting much older. Global connectivity continues to increase exponentially, along with what must be considered revolutions in communication and in diffusion of ideas and information. Satellite imaging can monitor and show us the surface of our planet almost foot by foot. The availability of digital data also continues to increase exponentially, as does the need for means to interpret it and to understand what the almost infinite layers of spatial variation mean to real people leading lives in real places. The human genome has been "mapped," spelled out letter by letter. It turns out that we are all the same, with barely any adaptations of immune systems and other forms of direct interaction with the varied environments to which we have migrated to live. Our differences are our cultural creations—a project that seems to continue.

Infectious disease has not gone away. New diseases are emerging, and old ones are reemerging. We need to study infectious diseases again, this time in the context of higher-level political/economic/ecological processes. This can now be done with newly available genetic information. Chapter 7 describes geographic contributions to a field called "landscape genetics," which helps us understand how diseases spread and evolve differentially in space. The evolutionary biologist Paul Ewald (2000) controversially claims that these studies of infection should include many human degenerative diseases. Even as the physical and social sciences increasingly struggle with the principle of indeterminacy and the precautionary principle, he suggests that the germ theory standards of evidence may need to be changed. The influence of systems theory on ways of thinking in cultural ecology (Moran, 1990) has also had a kind of revival, developing from the way theorization on information and ecology has emerged from Prigogine's studies of self-organizing systems (Kellert, 1993; Prigogine & Stengers, 1984).

The intensive development of social-theoretical frameworks now enables the study of those social and economic organizations, belief systems, and technologies that constitute the behavioral vertex of the triangle of human ecology. The methodologies and perspectives of health geography deepen and enrich our ability to understand ourselves. Cultural ecology has as a great claim to study culture, but it needs to inform those spatial-analytical studies more. GIS are creating the opportunity to do that and, through community participation, to return the ever-proliferating microspecializations that fracture geography to the ancient integration of the map.

DEFINITIONS AND TERMINOLOGY

Definitions of Health and Disease

Everyone knows what "health" is, and yet a precise definition of health is difficult to come by. This problem is shared by researchers who, in studying health, ironically need to measure disease. **Health**, however, is more than the absence of disease. It is a positive thing—the actualization of the creative force of human life for individuals, communities, and societies. We know that greater health is usually equated with lower mortality and morbidity rates. The problem remains of how to define health without reference to disease.

The first major definition to present health as a positive entity—a presence to be promoted and not merely an absence to be regretted—occurs in the 1946 charter (preamble to the constitution) of the World Health Organization (WHO): "*Health is a state of complete physical, mental, and social well-being and not merely the absence of disease or infirmity.*" This influential statement was important for the philosophical position it stated and for the goals it set for government programs and research funding. It has not proved useful, however, for implementing any standards or research designs that require criteria. It is a utopian ideal. Researchers have for specific studies and purposes created various "indexes" of health, but they seem both too particular and too cumbersome for widespread use.

May's (1961, p. xv) definition of disease was for some years referred to by geographers. He stated that disease is "that alteration of living cells or tissues that jeopardizes survival in their environment." There are several important points in this definition. The organism has an environment to which it relates. The idea that disease jeopardizes survival implies that there may be different levels of health without there being disease. An office worker, for example, need not have the physique or eyesight of a hunter. One may be born with a physical disability and lead a productive life into old age, depending on the society within which one lives and its technology. May, however, was a physician before he became a geographer, and it shows; the present understanding of and interest in such dimensions as mental health, mind–body interaction, and sociocultural morbidity context are difficult to reconcile to this more biomedical definition.

An influential definition comes from Dubos (1965, p. xvii): "States of health or disease are the expressions of the success or failure experienced by the organism in its efforts to respond adaptively to environmental challenges." This definition implies a system whose parts can exist in different states of interaction. Health is not necessarily a condition of physical vigor, but a condition suited to reaching goals defined by the individual. The most important word here is "adaptively." Dinosaurs were highly adapted to their environment but could not cope with environmental changes. There is a dynamic quality to health. Dubos's definition, however, defines what health results from, not what it is.

In this book, we adhere mainly to J. Ralph Audy's (1971, p. 142) definition: *health* as a "continuing property" that can be measured by an "individual's ability to rally from a wide range and considerable amplitude of insults, the insults being chemical, physical, infectious, psychological, and social" (p. 140). One might prefer the term *stimuli* or *hazards* to *insults*. Such stimuli may be either negative or positive; the crucial thing is that the individual must respond to them.

Infectious insults consist of **pathogens,** agents that cause disease. Every person is infected at all times with many billions of viruses, bacteria, and protozoa that cause no harm, such as intestinal bacteria, and that usually enable our body to function as it is supposed to. This **microbiome,** the world of other species that live inside of each human, outnumbers our own cells by a ratio of at least 10 to one. Changes in health status, however, can cause a normally benign relationship to alter and become pathogenic.

We also constantly receive nonorganic physical insults, such as from electromagnetic radiation. The trauma of tissue damage and broken bones can result from falls and violence. We live in a chemical soup. Our bodies are chemical systems, quite literally composed of what we eat, and what our mothers ate while we were in utero or breastfeeding. Petroleum derivatives and nicotine are now part of our chemistry. The absence of an essential vitamin or an excess in a basic food component, such as cholesterol, can also be a chemical insult. Mental and social insults further influence physiological functioning.

Examples of Insults or Stimuli

Chemical	*Physical*
Carbon monoxide	Trauma
Drugs	Radiation
Benzene	Light
Formaldehyde	Noise
Calcium deficiency	Electricity
Oxygen deprivation	Air pressure
Infectious	*Psychosocial*
Prions	Danger
Viruses	Crowds

Rickettsiae	Isolation
Bacteria	Anxiety
Protozoa	Community
Helminths	Love

It is possible to map, at a variety of scales, every kind of insult. The areas of a town could be mapped based on noise, people's fear of walking down the street at night, air pollution, visual blight or beauty, mosquito density, or alcohol consumption. Such maps could be overlaid to show regions of health hazards. These regions of insults form the environments to which individuals are exposed at the microscale as they move through space.

At the microscale exist self-specific environments. Everyone is wrapped in an envelope of heat, humidity, bacteria, fungi, and mites, and may also host lice and fleas. The driver of a car encounters a set of insults that differs by sections of the road and by the number and types of other vehicles around. The infectious and other insults encountered on a bus are quite different from those in a car. Within buildings, one is insulted by microwaves from the walls, magnetism from electricity, light, changes in humidity, infections from other people, and various psychosocial challenges (e.g., from books, television, and conversation). The exact nature and range of insults to which an individual is exposed during a day are unique. Behavioral roles associated with age, sex, class, and occupation create some groupings of insults, however, and geographical location delimits other groupings. These differences in exposure to various health hazards can be modeled.

The idea that health is a "continuing property" and not a characteristic that is either present or absent involves recognition that health exists at various levels. The only absence of health is death. Health can exist at a threshold, marginally, or it can exist amply with great reserves. Audy points out that an insult can have a "training" impact. That is, after the body has successfully rallied from the insult, the body is better able to cope with future insults of that kind. One's first public talk, first date, or first exam in college is more difficult to cope with than the 20th. While a person is reacting to a stimulus, however, the level of health is decreased, and that person becomes less able to cope with another insult. Audy has suggested that the way insults affect the level of health can most easily be diagrammed for immunological health. In the example illustrated in Figure 1.3, two individuals are conceived, are born, and experience infectious insults.

The first becomes infected with a cold virus, and while she is coughing, sneezing, and slightly feverish, her health level declines a little. Soon, however, she is immune to that cold virus, and her health rebounds to a higher level. In this way she proceeds through a succession of infectious episodes. Through time, her level of health continues to increase until her early 20s, and then gradually declines over the next several decades. The second individual also survives the massive insults that attend birth, but poor maternal nutrition has given her a lower birthweight and level of health. Soon after she rebounds from the cold virus with increased health, she is infected with bacteria that give her diarrhea. She is removed from the food supplements believed to be the cause and even from water, in an effort

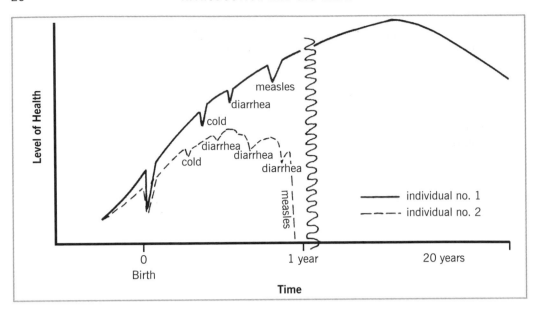

FIGURE 1.3 Health level and insult stress. The discontinuous solid line indicates health level over a lifetime, with peak of health at about age 20. The part before the wavy vertical line indicates levels of health for two individuals within their first year.

to stop the diarrhea. Her health level increases slightly as she masters the bacteria, but the episode has precipitated malnutrition because her diet had been marginal even before. With health lowered by malnutrition, before she can rally and restore herself, she is again assaulted by enteric (intestinal) bacteria. Now further malnourished and dehydrated, her health level is greatly decreased. As she struggles to rally and grow, she becomes infected with the measles virus. Depleted of health reserves, she cannot rally, and she dies.

In one study about nutrition and diarrhea, the graph of weight gain by babies and their bouts of infection and illness perfectly reproduced Audy's graph of level of health and stress from insults (Rhode, 1986). Weighing babies in a sling-scale every month is a method commonly used in well-baby clinics around the world to find early indication that a baby is failing to thrive. Healthy babies usually gain weight and grow in a regular manner, within population-specific norms well known and graphed. Against this curve of normal gain, the ups and downs of the weight of babies over their first 3 years of life were tracked and graphed along with demarcation of each bout of diarrhea, fever, measles, bronchitis, and other illnesses. On the chart of weight curves, one infant with three bouts of diarrhea, a high fever, and the measles at 18 months had reached only the normal weight of an 8-month-old; at the end of 3 years, after further attacks of bronchitis and fevers and repeated diarrhea, it had reached only the weight of a 9-month-old.

The scenario of multiple insults piling up and cumulatively lowering the health level even below the vital threshold occurs frequently in countries with high infant mortality. Mental and physical loads can also exact a long-term, cumulative penalty on individuals.

The stress of life events such as marriage and divorce, promotion and being fired, moving, losing a spouse, or having a baby can predict the likelihood of illness events (see the discussion of stress in Chapter 9). It has also been noticed that employee absences due to sickness tend to cluster in time. Both these examples of the timing of illness events can be addressed in the terms of the framework that is outlined below. Insults require adaptation. Health is lowered during that process, making a person temporarily less able to adapt to the next insult. As final examination time approaches, for example, many students get sore throats as the accumulated stresses of little sleep, poor diet, anxiety, and other insults lower their levels of health until throat bacteria that have been well controlled suddenly cause clinical illness.

Terminology

A familiarity with some terminology is necessary before the availability and limitations of data on health and disease can be appreciated. This section presents some of the most commonly used terms.

Diseases are referred to as **congenital** when they are present at birth. These may be of genetic origin, as hemophilia is; they may be acquired in the womb, as chemical-induced deformity is; or they may be acquired during the process of birth itself, as when severe inflammation of the eyes results from passage through a birth canal infected with the bacteria of gonorrhea. Diseases are referred to as **chronic** when they are present or recur over a long period of time, and as **acute** when their symptoms are severe and their course is short. **Degenerative diseases** are characterized by the deterioration or impairment of an organ or the structure of cells and the tissues of which they are a part. **Infectious diseases** result from the activities of living creatures, usually microorganisms, that invade the body. **Contagion,** transmission of infectious disease agents between people, may be direct through person-to-person contact or indirect through the bites of insect **vectors** or via **fomites** (vehicles) such as contaminated blankets, money, or water.

Figure 1.4 illustrates one way of looking at the continuum of health and disease. The term **clinical** refers to the appearance of symptoms that can be presented to a physician for observation and treatment. In a **subclinical** condition, an infectious agent may enter the body, multiply, stimulate the production of antibodies, and be eliminated from the body without the person's being consciously aware of any illness. Usually the only way that subclinical infections can be detected is through serology (the identification of antibodies and other immune reactions in the blood). Quite a few diseases produce acute reactions in only a small proportion of those infected. Other common diseases are mild infections when acquired in childhood and often pass unnoticed. In either case, public health officials are sometimes startled to find from serology that a "rare" disease has in fact infected the majority of the population.

Usually an infectious disease has a **latency period** between the time the infection occurs and the appearance of clinical symptoms. People are sometimes infectious before the disease is manifest. This is because the infectious agent needs time, known as **incubation,** to adapt to its host and multiply and become numerous

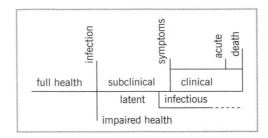

FIGURE 1.4 The health continuum, illustrating terminology and stages of ill health between full health and death.

in its mode of transmission (respiratory droplets, body fluids, etc.). Sometimes, however, it completes this incubation before it causes symptoms (i.e., while it is still latent), and thus can spread before the infected person (host) is treated or becomes bedridden or otherwise restricted in mobility from exposing others. The common cold, for example, usually has an incubation period of 1–3 days, but people may be infectious for 24 hours before their own symptoms appear. Measles has an incubation period of about 10 days until the onset of fever; people are then infectious for the 3–4 days until the rash appears, as well as for several days afterward. People infected with HIV are most infectious in the first months after incubation; the latency period until the symptoms of AIDS appear may be years, allowing untested and unaware people to spread the infection, even through childbirth. Because disease data usually are not produced until clinical symptoms are diagnosed, statistics at any given time usually underestimate the amount of disease in a population. Problems with time lag, apparent health, and time of diagnosis are accentuated when degenerative diseases are studied. The latency period between initial stimulus and the diagnosis of presented symptoms for cancer, for example, is commonly more than 20 years and may vary by several years for individuals, depending at what stage of the disease they are diagnosed.

A disease is **endemic** when it is constantly present in an area. It may occur at low levels, occasionally popping up here or there, as typhoid does in the United States (**hypoendemic**); or it may occur with intense transmission, as malaria does in parts of Africa (**hyperendemic**). Sometimes a hypoendemic disease can flare into rapid spread, perhaps in response to the dislocations of warfare or malnutrition from crop failure. Sometimes diseases that are not endemic are introduced and spread rapidly. A disease is said to be epidemic when it occurs at levels clearly beyond normal expectation and is derived from a common or propagated source. **Epidemic disease** may include an outbreak spreading through a population or widespread degenerative diseases, such as lung cancer.

The terms **incidence** and **prevalence** are often confused or used loosely and wrongly. Incidence refers to the number of cases of a disease diagnosed or reported for a population during a defined period of time, most commonly a year. It refers to new cases. Prevalence refers to the number of people in a population sick with a disease at a particular time, regardless of when the illness began. Thus the incidence of tuberculosis in Texas in 2016 refers to the number of new cases diagnosed there that year, whereas the prevalence of tuberculosis in Texas in 2016 includes the total number of Texans suffering from the disease that year, no matter what year they acquired it. The incidence and prevalence rates of chronic diseases, especially, can be very different.

THE CHALLENGE OF HEALTH GEOGRAPHY

"Geography is destiny." So declared the researchers who analyzed health care and the outcome of treatment by composing *The Dartmouth Atlas of Health Care* (Dartmouth Medical School, 1998) by health service areas across the United States. "Health and our geographies" are inextricably linked, declared Gattrell (2002, p. 3). "Where you live affects your risk of disease, access to basic resources of food, clean water, and housing, and concepts of health." In the words of Hippocrates, "airs, waters, and places" are still the things from which to start the study of everything else.

It has been said in the context of studies of world regions as syntheses and not just collections of events and facts that if geography did not exist, it would be necessary to invent it. In the face of global transformations of environment, population, economics/politics/society/religion, biotechnology, communication, and so on, scientists concerned with disease globalization and emergence and persistence have, in a way, tried to invent geography. Aron and Patz (2002) target public health graduate students with a global perspective on ecosystem change, including models of population movements, water cycles, and epidemiological design with GIS. Guest (2005) presents globalization, health, and the environment in chapters on poverty, medicinal plants, population dynamics, health ecology of the Nunavut, and urban land use in Baguio City. The best presentation is McMichael's (2001) look at past patterns and future uncertainties in human frontiers, environments, and disease. It states as its central theme that looking at humankind's long evolutionary and historical experience shows how the social and natural environments affect patterns of disease and survival. Appreciating the ecological perspective on human population health when such large-scale stresses are appearing in our world is a prerequisite, it says, to achieving a sustainable future.

Health geography was at its founding inclusive, open, aiming at synthesis and synergy. This text tries to survey the field and provide an introduction to the geographic concepts and methodologies, the cultural and environmental influences, and the ecological and socioeconomic structures that explain the patterns of health and disease. The integrative perspective of geography and all the questions and methodologies of geography's various traditions are needed. In the words of John Hunter (1974, pp. 3–4),

> The application of geographical concepts and techniques to health-related problems places medical geography, so defined, in the very heart or mainstream of the discipline of geography. I would suggest that there is no professional geographer,

whatever his or her systematic bent or regional interest, who cannot effectively apply a measure of his or her particular skills or regional insights towards the understanding, or at least partial understanding, of a health problem. This is the essential challenge of medical geography.

The challenge was to all geographers to contribute to health studies using diverse methods and using and building from the breadth of the field. The challenge is being met and the field of health geography will continue to prosper.

QUICK REVIEW

✓ Medical and health geography are both terms for the study of spatial or geographic components of health and disease. Some scholars believe they are separate fields but the authors of this book believe they encompass one integrated and inseparable field called health geography.

✓ Jacques May was the "father" of medical geography in the United States. He brought an ecological focus to the study of disease to this country after World War II.

✓ The focus of the field in the 1980s was on modelling including spatial aspects of health care utilization and provision and disease diffusion and the analysis of disease etiology.

✓ More recent themes in health geography include using a political ecology framework to understand patterns of health and disease, the emergence of place as a framework for understanding health, and the application of social- and cultural-theoretical positions to health.

✓ The challenge of health geography is the application of geographical concepts and techniques to health-related problems. All geographers in all subspecialties can contribute to answering the challenge.

REVIEW QUESTIONS

1. Chapter 1 emphasizes how the field of health geography involves multiple perspectives within the field of geography and other disciplines. Examine the course catalog at your university and list courses that are offered in geography that will provide systematic specializations of study within the discipline that would be relevant to health geography studies.

2. Figure 1.1 shows aspatial cognate fields that are associated with the different subdisciplines of geography. Examine the course catalog at your university and list courses that are offered in cognate fields that would be relevant to health geography studies.

3. Health geography is a field that is built upon and has borrowed from many different expressions of ecology. How have ecological concepts been incorporated into the field of health geography?

4. Some would argue that the history of health geography begins with Hippocrates but the field wasn't formalized in the United States until the 20th century. Make a timeline of important periods in the history of the field noting important scholars and their contributions during each.

5. Hippocrates theorized about human health around 400 B.C.E. What are some of the similarities between how Hippocrates and modern health geographers theorize human health and disease?

6. The authors of the text describe their views on the changing nomenclature of the subdiscipline of health geography and some controversies and debates within the subdiscipline. Some scholars believe there are two separate fields called medical geography and health geography but the authors argue that this is a false dichotomy and there is really only one field. Explain why you agree or disagree with their argument.

7. At your university library find copies of the following journals: *Health and Place, Social Science and Medicine,* and the *International Journal of Health Geographics.* Look at the titles and abstracts of some of the articles in recent editions and find some that you think should be categorized as health geography studies. Explain why you think the papers fit into the field of health geography.

8. Figure 1.4 describes the terminology used to describe a continuum of health from full health to death. Choose an example of an infectious disease that you know something about (e.g., influenza, HIV/AIDS) to describe the life cycle in a human from infection to death. Write a short narrative of your example using the words *infection, impaired health, subclinical, symptoms, infectious, clinical, acute,* and *death* in the description.

9. What is the challenge of health geography? In a discussion question at the end of this text you will be asked whether you think that the challenge is being met.

REFERENCES

Aron, J. L., & Patz, J. A. (Eds.). (2001). *Ecosystem change and public health: A global perspective.* Baltimore: Johns Hopkins University Press.

Audy, J. R. (1971). Measurement and diagnosis of health. In P. Shepard & D. McKinley (Eds.), *Environmental: Essays on the planet as a home* (pp. 14–162). Boston: Houghton Mifflin.

Barrett, F. A. (1980). Medical geography as a foster child. In M. S. Meade (Ed.), *Conceptual and methodological issues in medical geography* (pp. 1–15). Chapel Hill: University of North Carolina, Department of Geography.

Barrett, F. A. (2000). *Disease and geography: The history of an idea* (Geographic Monograph No. 23). Toronto: Atkinson College, York University.

Cliff, A. D., & Haggett, P. (1988). *Atlas of disease distributions: Analytic approaches to epidemiological data.* Oxford, UK: Blackwell.

Cliff, A. D., Haggett, P., & Smallman-Raynor, M. (2004). *World atlas of epidemic diseases.* London: Arnold.

Cromley, E. K., & McLafferty, S. L. (2012). *GIS and public health* (2nd ed.). New York: Guilford Press.

Dartmouth Medical School, Center for the Evaluative Clinical Sciences. (1998). *The Dartmouth atlas of health care 1998.* Chicago: American Hospital Association.

Dear, M. J., & Wolch, J. R. (1987). *Landscapes of despair.* Princeton, NJ: Princeton University Press.

Drake, D. (1854). *A systematic treatise, historical, etiological and practical, on the principal diseases of the interior valley of North America.* New York: Lenox Hill.

Dubos, R. (1965). *Man adapting.* New Haven, CT: Yale University Press.

Earls, F., & Carlson, M. (2001). The social ecology of child health and well-being. *Annual Review of Public Health, 22,* 143–166.

Ewald, P. W. (2000). *Plauge time: The new germ theory of disease.* New York: Free Press.

Eyles, J., & Woods, K. (1983). *The social geography of medicine and health.* New York: St. Martin's Press.

Finke, L. L. (1792–1795). *Versuch einer allgeminen medicinisch–pratkischen geographie.* 3 vols. Leipzig: Weldmannische Buchhandlung.

Fuchs, C. F. (1853). *Medizinische geographie.* Berlin: Duncker.

Gattrell, A. C. (2002). *Geographies of health.* Oxford, UK: Blackwell.

Geddes, A. (1978). Report to the Commission on Medical Geography. *Social Science and Medicine: Part D. Medical Geography, 12*(3–4), 227–237.

Gesler, W. M. (1991). *The cultural geography of health care.* Pittsburgh, PA: University of Pittsburgh Press.

Gesler, W. M. (2003). *Healing places.* Lanham, MD: Rowman & Littlefield.

Gesler, W. M. (2005). Therapeutic landscapes: An evolving theme. *Health and Place, 11,* 295–297.

Godlund, S. (1961). Population, regional hospitals, transportation facilities, and regions: The location of regional hospitals in Sweden. In *Lund Studies in Geography Series B: Human Geography No 21.* Lund, Sweden: Department of Geography, Royal University of Lund.

Guest, G. (Ed.). (2005). *Globalization, health, and the environment: An integrated perspective.* Lanham, MD: Rowman & Littlefield.

Hawley, A H. (1950). Human ecology. In *International encyclopedia of the social sciences* (pp. 328–336). New York: Macmillan.

Hippocrates. (1886). *The genuine works of Hippocrates* (F. Adams, Trans.). New York: Wood.

Hirsch, A. (1883–1886). *Handbook of geographical and historical pathology* (C. Creighton, Trans.). 3 vols. London: New Sydenham Society.

Hunter, J. M. (1966). River blindness in Nangodi, Northern Nigeria: A hypothesis of cyclical advance and retreat. *Geographical Review, 56,* 398–416.

Hunter, J. M. (1974). The challenge of medical geography. In J. M. Hunter (Ed.), *The geography of health and disease* (pp. 1–31). Chapel Hill: University of North Carolina, Department of Geography.

Jones, K., & Moon, G. (1987). *Health, disease and society.* London: Routledge & Kegan Paul.

Jusatz, H. J. (Ed.). (1968–1984). *Medizinische landerkunde.* 6 vols. *Geomedical Monograph Series.* Berlin: Springer Verlag.

Kearns, R., & Moon, G. (2002). From medical to health geography: Novelty, place and theory after a decade of change. *Progress in Human Geography, 25*(5), 605–625.

Kearns, R. A., & Gesler, W. M. 1998. *Putting health into place: Landscape, identity, and well-being.* Syracuse, NY: Syracuse University Press.

Kellert, S. H. (1993). *In the wake of chaos: Unpredictable order in dynamical systems.* Chicago: University of Chicago Press.

Koch, T. (2005). *Cartographies of disease: Maps, mapping, and medicine.* Redlands, CA: ERSI Press.

Koch, T. (2011). *Disease maps: Epidemics on the ground*: Chicago: University of Chicago Press.

Learmonth, A. T. A. (1975). *Patterns of disease and hunger.* North Pomfret, VT: David & Charles.

May, J. M. (1950). Medical geography: Its methods and objectives. *Geographical Review, 40,* 9–41.

May, J. M. (1950–1954). *Atlas of the distribution of diseases.* New York: American Geographical Society.

May, J. M. (1954). Medical geography. In P. SE. James & C. F. Jones (Eds.), *American geography: Inventory and prospect.* Syracuse, NY: Syracuse University Press.

May, J. M. (1958). *The ecology of human disease.* New York: MD Publications.

May, J. M. (1961). *Studies in disease ecology.* New York: Hafner.

McGlashan, N. D. (1972). *Medical geography: Techniques and field studies.* London: Methuen.

McMichael, A. J. (2001). *Human frontiers, environments and disease: Past patterns, uncertain futures.* Cambridge, UK: Cambridge University Press.

Moran, E. F. (Ed.). (1990). *The ecosystem approach in anthropology.* Ann Arbor: University of Michigan Press.

Morril, R. L., & Earickson, R. (1968). Variation in the character and use of Chicago area hospitals. *Health Services Research, 3,* 224–238.

Odum, E. P. (1959). *Fundamentals of ecology.* Philadelphia: Saunders.

Pattison, W. D. (1964). The four traditions of geography. *Journal of Geography, 63,* 211–216.

Phillips, D. R., & Verhasselt, Y. (Eds.). (1994). *Health and development.* London: Routledge.

Prigogine, I., & Stengers, I. (1984). *Order out of chaos: Man's new dialogue with nature.* New York: Bantam Books.

Pyle, G. F. (1969). The diffusion of cholera in the United States in the nineteenth century. *Geographical Analysis, 1,* 59–75.

Pyle, G. F. (1971). *Heart disease, cancer, and stroke in Chicago: A geographical analysis with facilities, plans for 1980.* Chicago: University of Chicago, Department of Geography.

Rhode, J. E. (1986). *Diarrhea is a nutritional disease.* Paper presented at the Second International Conference on Oral Rehydration Therapy, Washington, DC.

Rodenwaldt, E., & Jusatz, H. J. (Eds.). (1952–1961). *Welt-Seuchen atlas* (3 vols.). Hamburg, Germany: Falk.

Rothman, K. J. (1986). *Modern epidemiology.* Boston: Little, Brown.

Shannon, G. W., & Denver, G. E. A. (1974). *Health care deliver: Spatial perspectives.* New York: McGraw-Hill.

Shoshin, A. A. (1962). *Principles and methods of medical geography.* Moscow: Academy of Sciences.

Sorre, M. (1947). *Les fondements biologiques: Essai d'une écologie.* Vol. 1 of *Les fondements de géographie humaine.* Paris: Libraire Armand Colin.

Thomas, R. W. (1992). *Geomedical system: Intervention and control.* London: Routledge.

Williams, A. (Ed.). (1999). *Therapeutic landscapes: The dynamic between wellness and place.* Lanham, MD: University Press of America.

Wolch, J. R., & Dear, M. J. (1993). *Malign neglect.* San Francisco: Jossey-Bass.

FURTHER READING

Banks, A. L. (1959). The study of the geography of disease. *Geographical Journal, 125,* 199–216.

Brown, T., & Moon, G. (2004). From Siam to New York: Jacques May and the "foundation" of medical geography. *Journal of Historical Geography, 30*(4), 747–763.

Cassel, J. (1964). Social science theory as a source of hypotheses in epidemiological research. *American Journal of Public Health, 54,* 1482–1487.

Clark, J. (1998). A social ecology. In M. Zimmerman et al. (Eds.), *Environmental philosophy.* Upper Saddle River, NJ: Prentice Hall.

Cliff, A. D., & Haggett, P. (1989). Spatial aspects of epidemic control. *Progress in Human Geography, 13,* 315–337.

Council of National Research. (1997). *Rediscovering geography.* Washington, DC: National Academy Press.

Dubos, R. (1987). *Mirage of health: Utopias, progress, and biological change.* New Brunswick, NJ: Rutgers University Press. (Original work published 1959)

Farmer, P. (2003). *Pathologies of power: Health, human rights, and the new war on the poor.* Berkeley & Los Angeles: University of California Press.

Hawley, A. H. (1950). *Human ecology: A theory of community structure.* New York: Ronald Press.

Howe, G. M. (1972). *Man, environment and disease in Britain.* New York: Barnes & Noble.

Inhorn, M. C., & Brown, P. J. (Eds.). (1997). *The anthropology of infectious disease: International health perspectives.* Amsterdam, The Netherlands: Gordon & Breach.

Janson, C.-G. (1980). Factorial social ecology: An attempt at summary and evaluation. *Annual Review of Sociology, 6,* 433–456.

Joseph, A. E., & Phillips, D. R. (1984). *Accessibility and utilization: Geographical perspectives on health care delivery.* New York: Harper & Row.

Learmonth, A. T. A. (1998). *Disease ecology: An introduction.* Oxford, UK: Blackwell.

Mayer, J. D. (1996). The political ecology of disease as one new focus for medical geography. *Progress in Human Geography, 20.*

McGlashan, N. D., & Blunden, J. R. (1983). *Geographical aspects of health: Essays in honour of Andrew Learmonth.* New York: Academic Press.

Meade, M. S. (1986). Geographic analysis of disease and care. *Annual Review of Public Health, 7,* 313–335.

Paul, B. K. (1985). Approaches to medical geography: An historical perspective. *Social Science and Medicine, 20,* 399–407.

Pyle, G. F. (1979). *Applied medical geography.* New York: Wiley.

Stamp, L. D. (1964). *The geography of life and death.* Ithaca, NY: Cornell University Press.

Stokols, D. (1992). Establishing and maintaining healthy environments: Toward a social ecology of health promotion. *American Psychologist, 47*(4), 6–22.

Stokols, D. (1996). Translating social ecological theory into guidelines for community health promotion. *American Journal of Health Promotion, 10*(4), 282–293.

CHAPTER 2

Ecology of Health and Disease

The **human ecology of disease** is concerned with the ways by which human behavior, in its cultural and socioeconomic contexts, interacts with environmental conditions to produce or prevent disease among susceptible people. This constitutes the etiology, or causal evolution, of health and disease. Population genetics, physiology, immunological status, and nutritional status are important to disease processes and must be understood as prerequisites to sound research into these processes. Geography is also important, as its roots are firmly anchored in the study of cultural and environmental interactions.

To avoid confusion, let us state succinctly the difference between human ecology and a term we have used earlier, cultural ecology. Human ecology is a broad term used in anthropology and epidemiology, as well as in geography, to denote the patterns of human interaction with the physical environment, including not only behavior but genetic adaptation and physiological reaction to environmental stimuli (such as air pressure or trace elements in water) as well. Cultural ecology is more specific and refers to behaviors and belief systems within a particular culture, such as those regarding diet, house construction, or hygiene. The usage of ecological to mean multivariate studies of complex systems stems from the sociological tradition (see Chapter 1).

Geographers have traditionally studied the creation of landscape; the mobility and composition of population; the determinants of economic activity and its location; and the diffusion of things, ideas, and technology. All these are of consequence to health geography. The landscape is composed of insects, medicinal herbs, and hospitals, as well as topography, vegetation, animals, water sources, house types, and clothing. Mobility is important in exposure to and transmission of disease (discussed in Chapter 4). Elements of population composition include not only age structure, but also immunological and nutritional status and genetic susceptibility.

The main purpose of this chapter is to establish a conceptual framework for understanding why human disease and health vary over the surface of the earth. First we examine the causative agents of infectious disease and the modes by which they are transmitted. Then, the triangle of human ecology, with its three interacting dimensions of population, habitat (environment), and behavior, is presented as a model for understanding patterns of health and disease, using specific examples of childhood asthma, colorectal cancer, and cholera. The role of that natural environment in influencing patterns of disease is then examined more closely via the field of **landscape epidemiology**, which is of particular relevance for understanding diseases that are **vectored** by species such as flies and mosquitoes.

DISEASE AGENTS AND TRANSMISSION PROCESSES

An infectious disease's causative organism—variously known as a germ, microbe, pathogen, or parasite—is the **agent** of the disease. It may be a virus, a rickettsia, a bacterium, a protozoan, or a worm. The type of disease agent is important for understanding the possibilities of intervention in the disease cycle. At the smallest and simplest end of the infectious agent continuum are various bits and strings of protein; tiny short loops of DNA; and similar mysterious molecules called prions, plasmids, and new things seemingly every year. They do not seem to meet the historic definitions of life. They have no nucleus of control, no container of their being, apparently no being to contain as they drift among individual animals and species of animals and even between animals and plants. But they can be replicated. They can change the behavior of DNA and, for example, make a strain of bacteria more virulent, or cause the destruction of brain cells. Prions are the agent of Creutzfeldt–Jakob disease in people, and of bovine spongiform encephalopathy (mad cow disease) in cattle. Next are the viruses, simple loops of information that formerly were not considered living, either. Outside living tissue, viruses cannot reproduce. Pharmaceutical drugs have little effect on them, except for the potential of some of the new pharmacopeia developed through research on HIV/AIDS. The best weapon against viruses is a vaccine.

Rickettsiae cannot live outside cells, but in other respects they are similar to bacteria and have been regarded as degenerated bacteria that have lost their containers. They live very well within arthropods and get introduced to people during "blood meals." Rickettsiae and bacteria both are sensitive to antibiotics, which means that infected people can be treated. Mass treatment programs based on a single injection are sometimes possible. Protozoa have more DNA information and more than one life stage, generally defying vaccines and learning to resist drugs. The amoeba that causes dysentery, for example, exists as a trophozoite that can act commensally or invasively, causing massive pathology; or it can exist as a durable cyst, the infectious form that is resistant to chlorine, drying, or sun as it travels through water, food, or even dust on money. The protozoan that causes malaria, a plasmodium, has four life stages that have defied vaccine production; these include sexual reproduction inside an *Anopheles* mosquito. The more complicated organisms, protozoa and especially helminths (flukes, filaria, and intestinal worms), are difficult to treat on a mass basis because powerful drugs that have many adverse side effects must be used, requiring hospitalization. Hospitalizing half the population to treat schistosomiasis, for example, is not only too costly in agricultural work time and medical care, but can seem useless when people return home only to be reinfected. Helminths infecting various body organs are even more difficult to treat, especially when they exist in heavy loads of hundreds and thousands of organisms. When they die, the body attacks them. Only intestinal worms have a ready means of egress from the body; the others, dead and unable to disguise themselves, become masses of internal foreign matter (clogging the liver, lymph nodes, veins, etc.). Currently, there are no successful vaccines for any of the complicated organisms. They have enough genetic material that they seem able to adapt and even prevent complete immunity to them by their hosts.

Multiple terminologies are often used to refer to a specific insect or parasite. Usually this draws on genus or family classification to invoke known characteristics (of breeding, attaching, eating, etc.) that separate the groups. For example, an article might refer to "the trombiculid mite" or "the trematode," and, in the same paragraph, to "the fluke" or "the schistosome." People commonly say "insect bite" when they mean "arthropod bite" (e.g., ticks are not insects). A filarial worm is a long, thread-like worm that may live in lymph glands or under the skin, but that reproduces and disseminates itself by producing microfilaria (microscopic larvae) that arthropods can take up with a blood meal. Elephantiasis and river blindness are filarial diseases vectored by arthropods; roundworm and tapeworm infestations are not. To refer to a bacteria by its classification as a spirochete, or a vibrio, or a bacillus, is to comment on the bacteria's mobility, its ability to survive in the soil or water, its contagion.

The organism infected by a disease agent is called the **host**. When animals are the ordinary hosts, the disease is known as a **zoonosis**. A disease that often infects both people and animals is an anthropo-zoonosis. When animal hosts serve as a continuing source of possible infection for human beings, our anthropocentric name for them is a **reservoir**, and we humans regard ourselves as the hosts (although the animals are actually the primary hosts). A disease always present in, or endemic (literally "in the population") to, an animal population is said to be **enzootic**, and a disease epidemic (spreading in a population) among animals is said to be **epizootic**.

An arthropod that transmits a disease agent between hosts, and in which the agent multiplies and often goes through life cycle changes in form, is known as a biological **vector**. Biology distinguishes these vectors from flies or inanimate objects that may merely transport the agents mechanically (as vehicles or fomites). Many people speak of vectors as "insects," but ticks and mites are not, strictly speaking, insects. The more inclusive category is the phylum, Arthropoda. The term **arbovirus** includes all arthropod-borne viruses. The major vectors are mosquitoes, biting black flies, ticks, mites, sand flies, fleas, and lice, but gnats, midges, and other arthropods are occasionally involved. Most disease agents are strictly limited to transmission by a single species, or at most a genus, of vector. This establishes limits to their geographic distribution because vectors have specific habitat requirements.

Intermediate hosts are organisms that are necessary to some stage of an agent's life cycle. Hosts in which the agent attains maturity or its sexual life stage are the primary hosts or definitive hosts; hosts for the agent's asexual or larval stage are intermediate hosts. The fluke that causes schistosomiasis, for example, must alternately infect people and snails. The snails do not transmit the agent to people; they are not vectors, but intermediate hosts that are eaten and die themselves. Nevertheless, it is often convenient to treat intermediate hosts as though they were vectors because the same kinds of population dynamics and intervention strategies are involved. Sometimes this concept is even further enlarged, as when dogs are treated as "vectors" of rabies to people. Such careless usage may nevertheless be helpful for mathematical modeling of a disease system from our anthropocentric viewpoint.

Types of Transmission

Three types of transmission are frequently modeled, and understanding them is useful when trying to sketch out an ecology for a specific disease (Figure 2.1). The first type is direct transmission, either through physical contact or through the air (chain 1). Agents of yaws, ringworm, and trachoma are spread by physical contact, and sometimes cold and other viruses can be communicated by hand as well. Sexual transmission is a specialized form of physical contact. More explosively, direct transmission by means of respiratory aerosol droplets can send pandemics of influenza, whooping cough, measles, or (previously) smallpox whizzing around the world.

Some disease agents can also be spread from one person to another through contamination of food and/or water with the disease agents (chain 2). This second model is known as the fecal–oral route of transmission. It can involve contact (usually between mother and small child), but the primary danger is water-borne transmission of typhoid, cholera, hepatitis, and many other of our most serious infections. This form of transmission is addressed below.

Chain 3 represents transmission of disease from animals to humans through direct interaction, such as through the bite of a rabies-infected dog. The important differentiation in transmission for the third model is with vectored diseases, agents of which are transmitted into people when they are being eaten by arthropods. Chain 4 illustrates a vectored human disease that has no animal reservoir. The absence of a reservoir raises a possibility that does not exist for most anthropozoonoses, disease eradication. Although birds, reptiles, and monkeys have their own forms of malaria, for example, human malaria is solely a human disease vectored by Anopheles mosquitoes. There was therefore some hope in the 1950s that it could be eradicated. We consider later some of the reasons why it could not.

Most vectored diseases involve transmission of agents among animals (chain 5). Many of these cycles are so ancient that agent and host have mutually adapted, and no disease symptoms appear in the infected hosts. Sometimes people are

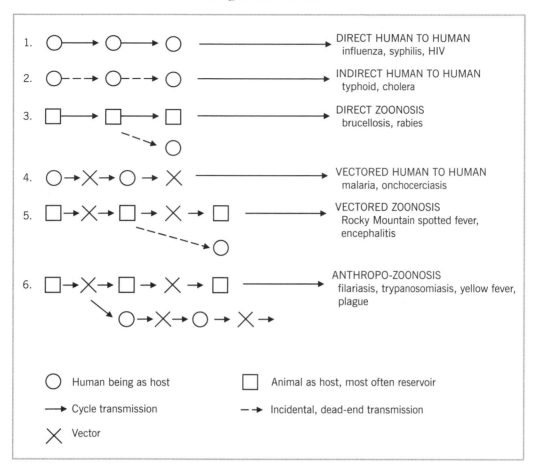

Chains of disease transmission.

accidentally infected when they intrude on the animal habitat. Usually they develop no disease at all, as they are not compatible hosts. Occasionally they are so susceptible to the rare infection that case mortality rates are high. Whether they don't even get sick or they get so sick they die from the infection, people are usually dead-end hosts for zoonotic agents. This means that they cannot give the agent to a vector for transmission to another host. The amount of virus (or bacteria or any agent) circulating in the blood is known as *viremia*. When people are incidentally infected with viral encephalitis by a mosquito that has previously fed on an infectious chicken, for example, they do not develop viremia sufficient to serve as a source of infection for another mosquito imbibing a little of their blood. Encephalitis cannot, therefore, be spread among people by mosquitoes. Even when clusters of cases occur, they result from exposure to the zoonotic chain.

The final chain in the diagram represents anthropo-zoonoses, which can be transmitted between animals or between people by vectors, and which sometimes cross between the two systems.

The Role of Water in Infectious Disease

Directly transmitted diseases between people, with the exception of those that are sexually transmitted, need to exist for a period of time in the environment. For respiratory droplets of tuberculosis bacilli, viruses for influenza, and even helminth eggs passed into soil, this means exposure to the ultraviolet rays of sunlight. Outside the built environment, extremely cold temperatures of winter or dry conditions of heat may kill germs. Soil surfaces get flooded; water evaporates. Microbes, larval helminths, and such themselves face parasitism and predation. Such environmental conditions can eliminate or control infectious agents, or at least cause seasonality. Environmental conditions, especially continued warmth and humidity, can also nurture and protect disease agents. In wet–dry climates, the coming of the first rains in months can cause animal and human feces on the ground to wash into wells. Alternatively, the dry season can close wells and concentrate microbial life of all kinds in the remaining wells.

During the WHO's decade of water development in the 1980s, water-related diseases were classified into three development-related categories that continue to be useful: **water-borne** diseases, which are ingested; **water-washed** (or, perhaps more accurately, unwashed) diseases, which are preventable by hand/hair/clothes/floor washing and other hygiene; and **water-based** diseases, which are vectored diseases requiring water for the vector to breed. As Table 2.1 shows, these include most of the major infectious diseases.

Water-unwashed diseases are those that are preventable by washing with water and soap. These include the common cold, which is often transmitted by contaminated hands; conditions caused by infectious agents that encyst when dry and blow as dust in the wind or contaminate money; and flea- or louse-vectored diseases, which hot soapy laundry and floor/rug washing can prevent. One of the most serious unwashed diseases today is trachoma. *Chlamydia trachomatis* now infects 70 million people, 5 million of them in the late stage and 2 million already blind. It infects the mucosa of the eyes, causing conjunctivitis. Flies, especially *Musca sorbens*, crave the eye secretions for food and spread the infection from child to child. The flies lay their eggs in human feces. Dirty cloths and fingers also spread the infection. Repeated infection irritates and scars the underside of the eyelid, causing the lid to turn inward and perpetually scrape the cornea with rigid lashes.

TABLE 2.1. Some Major Water-Related Diseases

Water-borne	Water-unwashed	Water-based
Typhoid	Intestinal worms	Malaria
Cholera	Amebic dysentery	Filariasis
Hepatitis A	Colds	Japanese encephalitis
Diphtheria	Typhus (louse-vectored)	River blindness
E. coli, salmonella	Plague (flea-vectored)	Schistosomiasis
Polio cryptosporidium	Pesticide residue	Dengue/dengue hemorrhagic fever

Water-borne diseases are "carried" in water, and often in contaminated food, and ingested (i.e., drunk or eaten) by people. Bacteria like diphtheria can spread in this way and kill whole armies. Some water-borne viruses, like hepatitis, continue to cause health consequences worldwide. Poliomyelitis is such a virus with a special ecology. This virus used to be so ubiquitous that it is thought almost all children ingested it very young, when they were still protected by their mothers' antibodies. The maternal antibodies provided a grace period in which children could develop their own antibodies to the polio virus. It wasn't until treated water and sanitation—life-saving public health improvements—prevented infants from getting infected with the polio virus that the horror of the paralyzing disease developed. Encountering the virus for the first time without antibody protection at 8, or 14, or 39—perhaps while swimming, like Franklin D. Roosevelt—a few were killed outright, but many became paralyzed. Infantile paralysis was a public terror in the United States into the 1950s, when the first effective Salk vaccine was tested. Vaccination has since eliminated the disease in the United States, but as "safe drinking water" has become increasingly available and saved the lives of millions of children in the developing world, so have epidemics of paralysis followed. At the end of the 20th century, the WHO, the Carter Center, Rotary International, and other organizations launched a massive effort to make polio the second disease eradicated from the earth. As of 2015 there remain only three countries in the world where polio is endemic: Nigeria, Afghanistan, and Pakistan. Of the ~360 cases of wild polio virus observed in 2014, 340 of them occurred in these three countries.

Although case mortality for diarrhea has been reduced in the past 20 years by the diffusion of a simple and affordable new treatment technology—**oral rehydration therapy (ORT)**, which uses an inexpensive and readily available mixture of salt, sugar, and boiled water to stop dehydration—diarrhea continues to be one of the leading killers in the developing world. About 2 million children a year die from it—more than 3% of total world mortality. The highest rates are in sub-Saharan Africa and South Asia, especially among the malnourished (in accordance with Audy's definition of health; see Chapter 1). Even in developed countries, moreover, there are increasing hazards from *Salmonella* species and *Escherichia coli* (*E. coli*), ubiquitous intestinal bacteria that are developing antibiotic-resistant strains. Protozoa, such as *Cryptosporidium* from the intestinal tracts of cattle and *Giardia*, can survive municipal chlorination unscathed and cause epidemics and mortality even in U.S. cities. In the following section, the ecology of cholera is developed as an example of a water-borne infection that takes the direct, fecal–oral route.

THE TRIANGLE OF HUMAN ECOLOGY

Population, habitat, and behavior form the vertices of a triangle that encloses the state of human health (Figure 2.2) (Meade, 1977). *Population* is concerned with humans as biological organisms—as the potential hosts of disease. The ability of a population to cope with insults of all kinds depends on its genetic susceptibility or resistance, its nutritional status, its immunological status, and its immediate

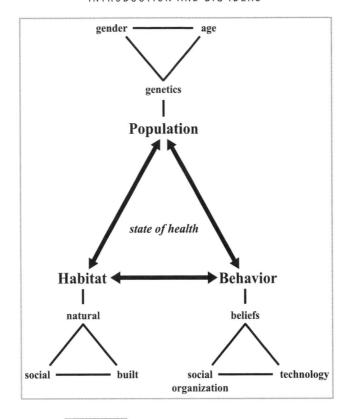

FIGURE 2.2 The triangle of human ecology.

physiological status with regard to time of day or year. The effects of age, gender, genetics, and other population components are pervasive but largely implicit in the remainder of this book.

Habitat is that part of the environment within which people live—that which directly affects them. Houses and workplaces, settlement patterns, naturally occurring biotic and physical phenomena, health care services, transportation systems, schools, and governments are parts of the habitat thus broadly conceived. The following chapters develop other aspects of the habitat; in this chapter the discussion is limited to the constructed part of the habitat, where humans live and work.

Behavior is the observable aspect of culture. It springs from cultural precepts, economic constraints, social norms, and individual psychology. It includes mobility, roles, cultural practices, and technological interventions. The triangular ecological model differs from sociological models in its separate consideration of behavior and population. Educational or socioeconomic status, for example, is an element of behavior rather than population. Education involves behavioral exposures to an opportunity in the habitat; educational status is behavior but the location of the school and its environment are under the habitat vertex of the triangle.

The three vertices of the triangle are mutually interacting, but dividing them is useful when sketching out an ecology of disease. Through their behavior, people create habitat conditions, expose themselves to or protect themselves from habitat conditions, and move elements of the habitat from place to place. The habitat presents opportunities and hazards to the population genetics, nutrition, and immunology. The status of the population affects the health outcome from the habitat stimuli and the energy and collective vigor needed to alter behavior and habitat.

It is possible to elaborate on this basic triangle in many ways. Subsystems can be created for economics, politics, or religion. The effects of global warming (itself affected by human behavior) on natural habitat and behavioral feedbacks can be analyzed. Motivation systems for behavioral alteration can be analyzed. The effects of changes in major cultural paradigms, such as the purpose and role of humans in the environment, can be isolated and spotlighted as insets. The triangle of human ecology is useful when brainstorming causal pathways of disease or trying to identify opportunities for intervention or prevention of disease and poor health.

Population

The nature of the population—that is, the characteristics, status, and conditions of individuals as biological organisms—does much to determine the health consequences of any stimulation or insult. Whether the stimulus is a bacterium, light, drug, sound, or thought, the reaction will differ according to the body's biochemical state. This physiology is in part inborn through the genetic code, but it is also influenced by nutrition, previous experience, age, and so on.

Genetics

Genes set the limits of our possible responses to social, biological, and physical experiences. Once it became possible to "read" the deoxyribonucleic acid (DNA) sequence of bases that encodes the structure and processes of life, the sciences of genetics and biochemistry expanded explosively. DNA, the sum total of all genetic information, is a long sequence of small molecules termed **nucleotides**. There are four of them, which are denoted by the letters A, C, T, and G. Their specific order and position code the genetic information. The two strands of DNA join in the famous structure of the double helix. There are approximately 6 billion nucleotides in the human genome. The DNA of all human beings on Earth is more than 99.9% the same. Now that the human genome has been read—all those billions of nucleotides have been spelled out—genetic research for marking, determining, intervening, and treating has become very important and well funded. Our knowledge of how genetic information is stored, transmitted, and activated at the appropriate time is undergoing almost daily revision. The DNA chain can be broken in some places but apparently not in others. Pieces can be switched, overridden, repeated, or deleted. Some genetic information has persisted from ancient times, with no known function today; apparently at least some of this information is that of ancient viruses. Moreover, some genetic information is activated to govern the production of enzymes, hormones, or other proteins only at certain times in life and then is "turned off." Cancer may turn some of these genes (e.g., for tissue growth) back on, and shut the inhibiting proteins off. The functioning of the genetic code is much more complex, and the encoded instructions much more varied, than had ever been imagined previously.

Several genes responsible for serious diseases have already been identified: those involved in cystic fibrosis, Huntington's chorea, familial breast cancer susceptibility, and familial early-onset Parkinson's disease. Recombinant genetic technology—using, for example, common viruses to "vector" normal genes into cells to replace damaged and dysfunctional ones—is already leading to new therapies. Even diseases influenced by several genes (e.g., some types of breast cancer) are yielding to the new technologies, and there is excitement about treatments for diabetes, schizophrenia, and severe forms of obesity, to name a few of the current targets.

Genetically *caused* disease is uncommon, however. Rather than being caused by dominant lethal genes, most genetic diseases are associated with rare recessive traits that, while terribly important to the individuals and families involved, are not a concern at the population level except under conditions of inbreeding. Hemophilia among the royal families of Europe in the 19th and early 20th centuries is the most famous example of such a disease. Harmful traits in a population that is not inbred tend to be eliminated rather than reproduced. When they are found to occur frequently in a population, one looks for some advantage that they bestow. The gene for sickle cell anemia in Africans was the classic conundrum. The sickle cell trait causes high mortality, but wasn't rare; in much of sub-Saharan Africa, where malaria was and still is most prevalent (see Figure 4.4 in Chapter 4), more than half the population carried the trait. Eventually it was found that though having two genes for cell sickling was deadly, having a single gene for the trait was protective against malaria.

Most diseases that tend to be transmitted in families or that occur more frequently in particular population groups result from genetic *susceptibility* rather than genetic causation. That is, the disease requires a stimulus or "cause" to occur, but another person who is not genetically susceptible will not respond to the same stimulus with disease. Often the ability of the liver or proteins in the blood to block or detoxify a specific insult is involved. Some types of cancer, probably Parkinson's disease, and the virulence of several diseases are thought to be related to genetic susceptibility or resistance. Great scientific resources are being expended to find and identify these genes for susceptibility, but (alas, in our views) very little for identifying the stimuli (e.g., agricultural chemicals) to which they are susceptible.

The field of **epigenetics** suggests that genetic susceptibility can be passed from one generation to another without changing the underlying DNA of an individual or his or her offspring. In epigenetics, the way genes are expressed or activated is changed due to some insult or stimulus in the environment. Studies in Sweden and the Netherlands have indicated that famine experienced by a grandparent, for instance, may impact the way his or her children and even his or her grandchildren respond to food and produce greater or lesser risks of cardiovascular disease and diabetes. Epigenetics is being intensely studied as a cause of many types of cancer and other conditions, such as schizophrenia.

ETHNICITY AND GENETIC VARIABILITY

Research in biochemistry and genetics has been overwhelmed by the recent findings on the variability and plasticity of human inheritance. The paired, rod-shaped chromosomes in the nucleus of each body cell contain the paired genes (composed of those DNA nucleotides) that control heredity. There are specific points, known as structural loci, on the chromosomes for genes governing each characteristic (trait). Genes that occupy the same locus on a specific pair of chromosomes and control the heredity of a particular characteristic, such as blood type, are known as **alleles**. When more than one version of the same trait is common (e.g., blue or brown eyes; Type A, Type B, Type AB, or Type O blood), the population is said to be polymorphic for that trait. Humans have long been recognized to be polymorphic for blood type, skin color, hair texture, stature, and other traits that used to be categorized by the concept of ethnicity. There are, at a conservative estimate, more than 50,000 structural loci. About one-third of these are polymorphic. Each individual has two different alleles for about one-third of these polymorphic structural loci, or about one-tenth of his or her entire genetic inheritance.

One way to illustrate the importance of polymorphism for human health is to consider the histocompatibility or human leukocyte antigen (HLA) system—in short, the genetic control of the body's immune system. Histo- refers to tissue; HLAs are white blood cell substances that induce production of antibodies. The HLA region of the chromosomes has at least four loci. More than 20 alleles may occur at one of these, 40 at another. In total, more than 80 alleles are involved at these four loci, and the possible number of reproductive combinations exceeds 20 million. It has long been known that people have four different types of blood—A, B, AB, or O—and that the Rh factor is positive or negative. Currently, more than 160 red blood cell antigens are identifiable. Most have been implicated in blood transfusion reactions and presumably are involved in mother–fetus exchanges. More than 100 variants of human hemoglobin are also known. This is the type of genetic variability involved in acceptance or rejection of organ transplants, defense against cancer, and resistance to diseases such as malaria or measles.

Studies of monozygotic (identical) twins have identified an astonishing amount of genetic propensity. Such twins, who have identical DNA, may be raised together under the same social influences or separately in different families and circumstances. Twin studies allow genetic effects and shared environmental effects to be separated. The range of genetic influences includes many things we would assume to be environmentally conditioned, such things as color preference and favorite foods, occupational choice, form of exercise, alcohol and tobacco usage, and "personality type" in risk of cardiovascular disease.

The manner in which each genetic distribution compares with environmental, cultural, or disease patterns is of interest. Both physical environmental factors (e.g., solar radiation) and cultural factors (e.g., method of maintaining livelihood, perceptions of beauty, marriage ideals, and migration histories) have affected genetic patterns. Geographers have joined the search for associations between blood type and disease susceptibility and between metabolic differences and aspects of cultural evolution, such as cropping systems. Any effort to categorize all distributions, however, has to involve rather arbitrary criteria and great simplification. The concept of ethnicity is still used occasionally as a convenient biological categorization for the relative frequencies of many alleles, in full recognition that group boundaries and the traits included are arbitrary.

When ethnicity is used as a grouping of alleles to indicate the overall genetic distance or closeness of populations, it sometimes establishes a useful research framework. The Japanese (historically considered Mongoloid) in Japan have different health problems—kinds of cancer, for example—than do white Europeans and Americans (historically considered Caucasian) also living in industrialized countries at the same latitudes. The disease experience of successive generations of Japanese in the United States has been studied to help untangle genetic (changed by intermarriage over generations), cultural (behavioral changes in diet, etc., over decades), and environmental (immediate changes by migration) factors in cancer etiology.

In most scientific usage, ethnicity has replaced the use of the designation of **race**, for it has been many decades since behavioral, linguistic, or mental characteristics were scientifically associated with the genetic inheritance of physical traits. Anyone conducting or reading about research that involves a racial classification needs always to question closely the nature, purpose, and appropriateness of the characteristic used; for example, is careless and lazy usage neglecting socioeconomic or cultural causal associations? Race has, of course, become a social construction—a category experienced or studied as a reality, with real consequences for social marginalization, group perception, and health care, and should often be included, or is the underlying cause of other items included, in the behavior vertex of our triangle.

GENETICALLY BASED DIFFERENCES IN METABOLISM

There is considerable interest today in investigating the genetic basis of differences in human metabolism (energy and material transformation within cells), because these differences interact with culture and health. Lactose intolerance and alcoholism are briefly discussed here.

Lactose intolerance is a classic geographic puzzle in human ecology. Lactose, the only carbohydrate in milk, is split in digestion by the enzyme lactase into glucose (a sugar) and galactose, which are absorbed into the bloodstream as nutrients. Lactase appears in the human fetus in the third trimester of pregnancy, reaches a peak at birth, and falls gradually in childhood. The condition of lactose intolerance, or being unable to digest milk fully because of an inability to produce lactase, is the usual condition of most adult mammals, including human beings. Among some populations, notably those of European origin, the gene for producing lactase usually does not shut off, and so most adults are able to digest dairy products. Scientists and others of European descent used to consider this the normal human condition.

Simoons (1969, 1970) proposed a geographic hypothesis of biological and cultural interaction to explain the distribution of lactose intolerance around the world. He estimated that among Asians, about 90% of the population is lactose-intolerant. The question raised by his research is whether, for example, the Chinese lost their ability to digest lactose because they defined cattle keeping as barbaric and consequently excluded from their diet milk and the meat of herd animals (eating the meat of only "dooryard scavengers" such as chickens and pigs), or whether they developed their cuisine because they were unable to digest milk. Similarly, did Europeans develop a high frequency of lactose tolerance because they became herders of cattle, with mixed farming systems to support their draft animals, or did they take to dairy foods and raise the animals that produced milk because they could digest lactose? In Africa the ethnic groups that herd cattle are generally much more lactose-tolerant than the farming groups that do not, but the genetic pattern is complex and highly variable among villages as a result of invasions, migrations, and intermarriage with Muslim traders (e.g., across the Sahara as Islam spread). Lactose intolerance has important ramifications for worldwide emergency relief (usually involving powdered milk) and protein supplementation, agricultural extension, and development aid of all kinds, as well as vitamin D and calcium supplementation.

Studies involving the metabolism of alcohol have shown that Chinese people, even when acculturated to U.S. drinking habits, tend to metabolize alcohol differently than members of European ethnic groups do. A single glass of wine may produce the dizziness, flushing, and nausea that usually characterize much higher levels of alcohol consumption among Europeans. Some ethnic groups, such as Jews, have low levels of alcoholism; others, such as Russians, have rather high levels. Most such differences have been explained in sociocultural terms, as alcohol plays different roles in different societies. But the question arises: Is the different role of alcohol in European and Chinese cultures a result of or a cause of differences in metabolism of alcohol? As the study of alcoholism progresses, it is being found that upbringing, life stress, familiarity, and social custom are important, but that alcoholism also clusters in families. Even when adopted in infancy, the children of alcoholic fathers are more likely to develop alcoholism, and identical twins reared apart and brought up differently seem to show similar susceptibility (these studies have involved small numbers, however). Alcohol has been produced for millennia in some cultures to store excess grain and to provide an alternative to polluted water; in other societies it has been little used in these ways. Today alcohol is an especially severe problem in ethnic groups as diverse, yet similar, as the Inupiat in Greenland, Australian aborigines, and Native Americans. As in the case of lactose intolerance, geographic associations of culture, alcohol, and population may help establish patterns and promote understanding of a serious human health problem.

Other Influences on Population Status

Age is a critical factor for health status. Those who study geriatrics (diseases of old age) are just beginning to understand the aging process, but many of the biochemical changes are obvious. Metabolism changes in response to different energy requirements when behavioral roles change with age and growth, physical maturation, and completion of reproduction. Hormone and enzyme production; organ function; and deposition and storage of fats and chemicals in the bones, liver, and blood vessels all change with age. Life experience also accumulates, and the immune system recognizes and copes with a greater variety of infectious agents—and is sometimes more sensitized from long exposure to a variety of allergenic substances. The regulation of some homeostatic systems, such as temperature control, tends to become less efficient with age; other systems gain in efficiency, from practice.

The age structure of a population in large part determines consequences as diverse as the spread of an infectious agent and the severity of the illnesses it causes; the effects of changes in the weather or of an air pollution episode; the expression of carcinogenic agents in clinical disease; and the need for health services. Because age affects so many dimensions of health status, it needs to be accounted for in virtually every study of disease ecology. The relevance of life stages, and techniques for adjusting data for age so that populations can be compared, are described later.

Immunological and nutritional status are two other aspects of population that impact health patterns in space and time. Both are conditioned or determined by aspects of habitat and behavior (Have you been previously exposed to a pathogen? Were you given a vaccine? Can you access appropriate types and amounts of food?) but are then biologically relevant to how well your body can adapt to insults and stimuli. Finally, there are some diseases or health conditions for which sex is an important factor to consider. Sex in this case denotes biological categorization as male or female, with attendant differences in sex hormones such as estrogen and testosterone. Gender, the social construction of roles and habits and expectations based upon sex, also influences health but belongs in the behavior vertex of the triangle. Female sex, for instance, is associated with higher rates of autoimmune diseases such as lupus, rheumatoid arthritis, and Crohn's diseases: some 80% of people with these autoimmune diseases are women. Exactly why this is the case remains unknown, but research suggests that women biologically produce a greater immune response and antibody production than men, and that estrogen increases inflammation during infections, increasing the possibility of a subsequent autoimmune response.

Habitat

Another leg of the triangle is habitat. Habitat (or environment) can roughly be divided into three different types: the natural, the built, and the social. The natural environment's effect on human health, the plant and animal species that surround us, the weather and climate that we experience, the availability of water or food, are discussed further in this chapter's section on Landscape Epidemiology and in Chapter 12: Climate and Health. Below we discuss the other two types of environment that influence where we fall on the continuum of health: the built and social environments.

The Built Environment

Asleep or awake, humans spend most of their lives inside their houses or other buildings. Dubos (1965) pointed out that evolutionary stimuli now come more from the environment we have constructed than from nature, a statement perhaps even more true today than in the mid-1960s.

What stimuli do you receive in your house? Is the house heated and cooled, so that the humidity is also affected? Is it well ventilated? Do any insects live in the woodwork or basement, or do you prefer the vapors of insect nerve poison? Does your dog or cat sleep in the house? Are there windows to let in light? Are they screened, or do you usually keep them closed for air conditioning, preventing mosquitoes or other animals from entering? Are there dark corners and rooms where roaches, spiders, or perhaps mice live? Is the concentration of dandruff, hair, dust, and allergenic materials higher inside or outside your house? Is there perhaps formaldehyde in the insulation or carpet backing, or lead solder in the pipe joints, or asbestos in the shingles? Do you have radiation sources, such as a television or a microwave oven? Or do you have radiation only from electric wiring in the walls? Could your stone foundation emit radon concentrated in the basement? What is the noise level? Do you feel crowded or isolated there?

The types of houses, the presence of domestic animals and their parasites, and the kinds of pens or outbuildings are all of consequence to health. Consider the house types presented in Plate 3, and add your own examples. It matters to insect ecology (and hence to disease transmission) whether roofs are made out of thatch, in which insects nest, or of corrugated iron, which excludes them. It matters what barriers there are to entry of mosquitoes. It matters for the survival and contagion of germs and vectors whether architecture is oriented toward private, shaded, inner courtyards or is open and almost continuous with the outdoors. It matters whether kitchens are inside or outside the dwellings (because of smoke), and whether inside there are piped water and flush toilets. In wealthy settings, "sick buildings" can result from sealing the windows in the interest of energy efficiency and thereby trapping fumes from cleaning fluids and paint, plastic emissions, and other noxious gases inside. Plate 4 presents several examples of wells, and Plate 5 presents examples of toilets and other sanitation facilities. Water-borne contagious human diseases have been major causes of death at least since the time of the first cities. What barriers protect water from fecal contamination? What means of access to protected water do people have? As you consider these pictures, think about changes in these conditions from wet to dry seasons or hot to cold. What facilitates disease transmission?

DETAILS MATTER IN THE ECOLOGY OF DISEASE

A certain kind of chimney can cause a room to be smoky. This may repel mosquitoes, but it may chronically irritate eyes and eventually cause blindness. Flooring has one common purpose, but many effects. Floorboards can be spaced so that food and dirt fall through to be scavenged by pigs and chickens below; or they can be placed tightly together or solidly covered, so that one needs to learn to use a dustpan for sanitation; or they can be carpeted, so that fleas can spend their entire life cycle in the living room. Houses can be built out of cold, damp stone and be full of drafts. Alternatively, they can be constructed from mud and straw and provide good nesting cracks for insects. Humans create much of their disease environment.

Consider the following description by Amato (2000) of the living conditions of poor European peasants (i.e., most of the population) from medieval times until the early 20th century. After explaining how peasants lived "mired in muck" in one-room houses (perhaps with a second room for storage/stable), with dirt floors, no chimneys, and dark, damp, soot-filled interiors, he quotes (p. 26) the Italian historian Camporesi's assertion that peasants were

> dirty, almost always barefooted, legs ulcerated, varicose and scarred, badly protected by meager and monotonous diets, living in humid and badly ventilated hovels, in continuous, promiscuous contact with pigs and goats, obstinate in their beliefs, with dung heaps beneath their windows, their clothes coarse, inadequate and rarely washed, parasites spread everywhere—on their skin, in their hair and in their beds—their crockery scarce or nonexistent, often attacked by boils, herpes, eczema, scabies, pustules, food poisoning from the flesh of diseased animals, malignant fevers, pneumonia, epidemic flus, malarial fevers . . . lethal diarrhea (not to mention the great epidemics, the diseases of vitamin deficiency like scurvy and pellagra, the convulsive fits, so frequent in the past, epilepsy, suicidal manias and endemic cretinism).
>
> Villagers carried around with them a whole fauna of fleas and lice. Not only did they scratch themselves, but friends and relations from all levels in the social scale deloused one another.

The thumb, Amato notes, "was called the louse-killer (*tue-poux*)."

Changes in the built environment can result in profound alterations of disease conditions. We do not know exactly why some diseases have disappeared from Europe and others have increased during the last few centuries. Leprosy, for example, used to be common in Europe but no longer is endemic. Dubos (1965) argued that changes in the built environment, such as cheaply produced window glass and architectural principles that allowed construction of multiple chimneys, flooded even the houses of the poor with light and warmth, drastically changing the habitat for disease agents. In the cities of early industrial Europe, and in New York City early in the 20th century as it received a million immigrants a year from peasant Europe, the construction of dark, unventilated, and crowded tenements provided an ideal habitat for tuberculosis bacteria (see Chapter 10 for more on this topic).

Settlement patterns—the ways people are clustered and distributed on the land—also influence health conditions. On a microscale, geographers look within settlements at the spatial arrangement of residences and land uses. Usually three general settlement forms are distinguished: **nuclear**, **dispersed**, and **linear**. At a macroscale, geographers look at how the settlements are distributed with regard to each other. A settlement system consists of various sizes of settlements (from dispersed farmhouses to large cities), together with the distances, directions, and connections among them that form the structure of a functional region of trade, ideas, and other interactions. The role played by this hierarchical system in disease diffusion is discussed in Chapter 6. It is also important for the accessibility and specialization of health care.

The most common settlement form is nuclear. Most rural people in the world live in houses clustered in a village from which they walk out to the surrounding agricultural fields, with forest and grassland lying beyond. The settlement land use buffers most households from any insect-transmitted diseases from woods and fields. The nuclear form, however, facilitates the fecal contamination of water sources and the spread of directly contagious diseases. Houses in a dispersed settlement form are located on the farmland of their owners, and neither air nor water provides much focus of contagion for the scattered population. Each household, however, is exposed to vectored diseases originating in the natural surroundings. A linear settlement, in which houses are lined up along both sides of a river, canal, or road, has an intermediary position and often is characterized by the worst conditions of the other two settlement forms. People are only partially buffered from insect-transmitted diseases because the rear of the dwelling is exposed; yet the clustering of houses provides a focus for contagion, especially for those households downstream from other dwellings. In the contemporary U.S. context, the inner city is like a nuclear settlement, exposed to contagion and concentrated pollution. Suburbs, fragmenting the landscapes into which they disperse, reap the hazards of increased tick vectors even as their reduced density retards contagion. Housing sprawled along the highways suffers from the air pollution and vehicular hazards of the city, as well as the mosquito and tick pests of the countryside.

Health services are an integral part of the human habitat. Whether a disease is diagnosed and how it is treated depend partly on whether facilities for urine analysis, X-rays, brain scans, and various blood tests are available to the physician. Some health facilities lack electricity and in many ways constitute a very different health habitat from that presented by a university research hospital. Factors as diverse as international economics and roads washed out by monsoon rains can affect the availability of antibiotics or blood for transfusion. The availability and physical accessibility of health facilities and health personnel constitute a critical part of the health habitat. For more on the distribution of and access to health care, see Chapter 13.

The Social Environment

The social environment consists of the groups, relations, and societies within which people live. Audy's (1971) conception of health at different levels and depths of reserves, and Dubos's (1965) and McKeown's (1988) observations—that (1) all people exposed to an infection or other risk did not get sick, and (2) once they were sick, people followed varied courses of illness and outcome—had come to seem obvious by the mid-1980s but were still mysterious. These authors had pointed out that the major declines in mortality resulting from many infectious diseases had occurred prior to effective therapy. Genetic predisposition, other population attributes, and even such things as public health improvements in sanitation and changes in year-round availability of good nutrition seemed to explain only part. The groundbreaking studies by Marmot and colleagues (Marmot, 1986; Marmot, Kogevinas, & Elston, 1987), known as the "Whitehall studies," followed 10,000 British civil servants over two decades. They established the importance to health of relative status and not simply economic deprivation.

In these studies, everyone had abundant food, good shelter, excellent water and sanitation, accessible good-quality health care, public health information, and the general benefits of living in an economically developed country. Yet the clerical/manager civil servants had three and a half times the mortality rates of senior administrator civil servants. The effects of status seemed to work especially through the response to stress. Everyone's blood pressure rose during the work week, for example, but at home over the weekend the blood pressure of senior administrators fell to healthy levels, whereas that of lower-level civil servants declined only slightly. They seemed to remain continually stressed. Social theorists were quick to realize the implications for race discrimination, gender stereotyping, access to higher education, and social status constructs of many kinds. Addressing the question of how apparently equivalent risks to health became manifested so differently, Robert Evans's (e.g., Evans, Barer, & Marmor, 1994) influential research and writing brought the issue to the forefront of Canadian efforts to promote health as a national priority. Many researchers have confirmed the importance of relative differences in status, and not simply deprivation. The old emphases on "unhealthy lifestyle choices" and economic deprivation are clearly incomplete and inadequate. There are larger structural and context processes involved in determining health.

Even "laboratory" studies at the Duke Primate Center (in North Carolina) and elsewhere have confirmed that moving a high-status primate to a group where its status is low has effects on heart disease, cancer, and the course of infections. But what are the plausible mechanisms? Evans and colleagues (1994) elucidated the internal biological responses—especially processes of the immune system and endocrine production—that link health status to the perception of the external environment (note that this is an interaction of the behavior vertex with the population vertex to cope with the habitat vertex). They pointed out that many cultural variables affect how distress and disease become manifest in a particular cultural context. There are implications for such things as how groups respond to the stress of culture change through coping strategies based on different traditional organization and key values, or how epidemiologists conduct community surveys of mental health. The links between health and the social environment are covered in more detail in Chapter 9, Neighborhoods and Health.

Behavior

Human beings are animals who are cultural beings. Culture creates social organization, structuring relationships of power, status, and control of resources. Culture creates belief systems, values, and perceptions. Culture develops technology. The study of this vertex comprehends most of social science and the humanities. Cultural behavior interacts with the triangle of human ecology in four ways:

1. Humans create many habitat conditions.
2. Behavior exposes individuals and populations to some hazards and protects them from others.

3. People move not only themselves from place to place, but also other elements of disease systems.

4. Behavior affects the health of the population by controlling genetics through marriage customs, nutritional status through food customs, and immunological status through the technology of vaccination and customs of deliberate childhood exposure.

Each of these aspects of behavior is discussed in this section.

Very little of the earth's surface is unaltered by human activity. Vegetation has been burned, removed, and planted. Water has been withdrawn from below the earth's surface and stored and distributed over hundreds of miles on the surface. The chemical composition of the air and water has been altered. The kinds of buildings people construct, the kinds of industries they work in, the kinds of vehicles and roads they build, and the kinds and scales of the settlement systems they live in all help form the hazards to which they are exposed. This cultural creation of disease is as true for malaria as for lung cancer. For example, in the U.S. Northeast over a century ago, people cleared most of the forest for agriculture and growing cities. They thereby eliminated many of the animals. Land has in recent decades reverted to old fields and new forests, and presently is being fragmented by expanding suburban settlement and abandonment of industrial sites. With the restoration of habitat and now without large predators to control them, the deer population has rebounded to the highest level in two centuries. This has promoted the spread of Lyme disease, vectored by the deer tick, and established it as a major public health concern throughout the region.

Besides creating hazards, cultural practices also function to protect people. Many protective customs have been developed. The European custom of using a handkerchief for blowing the nose is as protective as the Chinese custom of blowing the nose onto and spitting upon the street is endangering. Wearing shoes provides almost total protection against hookworm. Frequent washing of hands, especially after defecation or before handling food, is effective protection from diseases as diverse as the common cold and hydatidosis (invasion of tissue by tapeworms).

We have learned to construct cultural **buffers** against disease deliberately: consider the chlorination of water supplies, the use of seat belts, and the prevalence of the water-sealed toilet. There are cultural origins and diffusion paths for protective buffers, and sometimes diffusion and adoption take time. There are also socioeconomic barriers to their occurrence. For example, case mortality from breast cancer is higher among African American women than among European American women, in large part because fewer of the former receive mammograms for early detection. This is a case where "race" could be indicated as a risk factor for breast cancer while really it is the behavioral and cultural differences surrounding the constructs of race that matter more so than any biological difference.

Behavioral roles, varying by age and gender and often by ethnicity or class, largely determine who is exposed to what. There is great cultural variation in social norms. In one culture men are exposed to the infections of the crowded marketplace because women are thought not to understand money, whereas in another culture women are exposed because they are judged superior at bargaining and men are considered too unreliable to handle the family money. Who herds animals in pastures and gets hookworm? Who handles dangerous industrial chemicals and gets cancer? Who washes laundry in the morning when *Mansonia* mosquitoes are biting and gets filariasis (a worm infestation)? Who tends the orchard in the daytime when *Aedes* mosquitoes are biting and gets dengue fever? Who goes to school and gets influenza? Who drives a vehicle and suffers a concussion? Who fights a war and is maimed? Who moves to the city for wage labor and contracts HIV while separated from home? Who leaves the home village for marriage and faces abuse or depression without social support? Globalization may be affecting many socioeconomic structures and the behaviors they channel, but traditional cultural behavior and beliefs remain influential locally.

Mobility patterns are of critical importance for the diffusion and incidence of disease; they are discussed at some length in Chapter 4. Medicinal plants, as well as insects that transmit disease, have been spread around the world by humans. When people migrate, they carry customs that are adaptive in their place of origin but are often inappropriate in the new place, such as building a house in Hawaii with a steeply pitched roof originally designed to shed snow. Such "cultural baggage" may include ideas about poisonous or nutritious plants. Formerly protective behavior may even be harmful in the new place. A good example is **geophagy**, the practice of earth eating. In parts of Africa pregnant women eat earth formed into molds and sold in the markets (Hunter, 1973). The earth comes from special sites and is often high in calcium, iron, and other minerals. It may provide critical nutrients missing in the diet. It may be eaten to control diarrhea and eliminate parasites that could affect the fetus, just as the clay kaolin is eaten in Kaopectate. The practice persisted late into the 20th century in the United States among a few people of African descent. An old woman in rural Georgia would keep a jar of her favorite clay from a local stream bed in her pantry, although she did not need its minerals (inappropriate behavior); and her daughter who had migrated to a northern city instead consumed laundry starch as a substitute for earth (harmful behavior). With mobility, custom often becomes dissociated from purpose.

It is easy to understand how behavior can affect the genetics of a population. Marriage customs have evolved partly to control genetics. All societies appear to forbid incest, although it is defined in different ways. Some societies prohibit racial intermarriage, whereas others are far more concerned with class status. Some peoples in southern India and in Arabia favor cousin marriages to keep wealth or property in the family, and so promote genetic diseases. Technology can affect the genetics of the population. In one society people with poor vision make poor hunters and providers; in another eyeglasses have rendered eyesight an irrelevant criterion for marriage. Health services and technology enable people with genetic traits for diabetes, hemophilia, or deficient immune systems to live and reproduce. The immunological status of a population used to depend almost entirely on its age structure because experience with infectious agents accumulates over time. Now technology allows even newborns to be vaccinated for diphtheria, whooping cough, and other frequently fatal diseases. The degree of disease resistance of the population thus has become a cultural construct.

The Triangle of Human Ecology in Practice

The three vertices of the triangle of human ecology can be used to model and summarize what is known about the complex ecology of a disease and to raise or test new hypotheses. Here we outline how the triangle can be used to outline the ecology of two noninfectious diseases (childhood asthma and colorectal cancer) as well as an infectious disease (cholera).

The Human Ecology of Childhood Asthma

Childhood asthma has a political–ecological context of where noxious facilities are located or how health care is made accessible. Figure 2.3, however, presents the human ecology of genetics, behaviors (breastfeeding, smoking, and using various chemicals), and the built and social (siblings, with their germs) environment and interactions (Ward, 2007). The question being addressed is the reason for increasing rates of asthma in rural and suburban areas without roaches, old housing, and other known risks. The hypothesis being tested is the so-called **hygiene hypothesis**. It avers that some of our autoimmune and inflammatory diseases are increasing because we have so thoroughly sterilized our habitat and removed the worms that we evolved with, even from our intestinal tracts, that our immune response cells no longer get properly trained. Instead we bring their powerful armament to bear

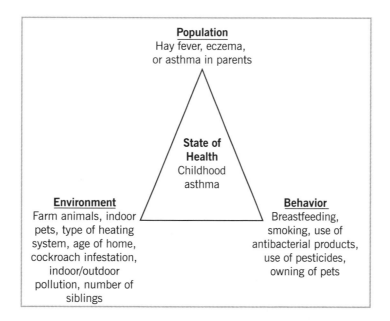

FIGURE 2.3 The triangle of human ecology and childhood asthma. Used by permission of Ashley Ward.

on inappropriate targets in our own bodies. The hygiene hypothesis is related to new research into the importance of the microbiome, the species of bacteria and viruses and fungi that live inside human bodies. When the microbiome is damaged, through antibiotic use, for instance, or when the microbiome is not as complex as it otherwise should be, due to Caesarean section birth or overly simplified diets or lack of exposure to the outdoors, humans are not as healthy. Restoring the microbiome, through fecal transplants, for instance, and decreasing the hygiene of the body, can sometimes reverse the course of autoimmune diseases or intractable bacterial infections. As regards the asthma example, do you understand why each variable in the diagram is in the vertex that it occupies?

The Human Ecology of Colorectal Cancer

Colorectal cancer (CRC) begins with a precancerous phase that, if left untreated, can progress to cancer. The triangle of human ecology is useful in considering not only this precancerous phase, wherein polyps grow and are detected, but also in whether or not the polyp becomes cancerous. Within the population vertex of the triangle, the primary risk factor for CRC is age. The role of age in CRC is so strong that population screenings (see Behavior and Environment) are recommended based upon attaining a certain age. Areas with higher numbers of old people may be places where CRC rates are higher. After accounting for age, the genetic composition of the individual plays a role in the age at onset of polyps and the speed with which the polyp develops. While the general recommendation is that all persons be screened for polyps starting at age 50, screening is recommended for younger people with known genetic predisposition, such as family history of the disease, or the presence of a comorbidity such as inflammatory bowel disease. Male sex is also associated with greater CRC risk.

The development of polyps may have an environmental component such as proximity to—or direct handling of—carcinogenic materials. Social environments, such as your information network and your knowledge of colonoscopy screening, the availability of healthy food options and recreation space, will directly influence behavior surrounding health care utilization and lifestyle risks associated with CRC. The spatial distribution of oncologists and colonoscopy provision sites will also influence uptake of screening and treatment.

Precancerous polyps on the colon and rectum are preventable. Behaviorally speaking, individuals who engage in healthy activities, such as regular exercise, healthful diets, or have care-seeking behaviors are less likely to develop polyps. Behavioral risk factors for CRC include smoking, excessive alcohol consumption, obesity, lack of exercise, and poor diet. Care-seeking behavior, along with other risky activities such as smoking and poor diet, are typically associated with socioeconomic status and race/ethnicity in the United States, and help explain why some race/ethnicity and socioeconomic groups have higher rates of late-stage CRC, cancer that is not diagnosed until the symptoms can no longer be ignored and treatment options are limited. The likelihood of death from CRC also increases with its extent and severity, so the behavioral factors that influence lack of screening behavior and late-stage detection also influence mortality patterns.

The Human Ecology of Cholera

Cholera is a disease of poverty, but the bacteria responsible for the disease are linked to the environment. This disease is caused by the colonization of the small intestine with *Vibrio cholerae* 01 or 0139. An infective dose, which is quite large at about a million bacteria, causes massive watery diarrhea within 2 days of infection. Its victims become dehydrated. If they go untreated, the disease will kill as many as half of the people who contract it. ORT (see above) can stop the dehydration and has been a miracle treatment for millions of people in resource-poor countries who are infected with cholera. Everyone is susceptible to infection, with the course of the disease largely depending on the individual level of health reserves (resilience) and nutritional, immunological, and hydration status. *V. cholerae* bacteria inhabit estuaries, brackish waters, rivers, and ponds of coastal areas of the tropical world, especially in South Asia and Africa. In the behavioral vertex, socioeconomics, demographic change, and sanitary practices/hygiene are key drivers of the human dimension of outbreaks.

Figure 2.4 summarizes the ecology of cholera. Humans become infected with the bacteria through contaminated water or food and interaction with the natural aquatic reservoir. Poor people are much more likely to drink contaminated water and to eat contaminated food, as well as to interact with their aquatic environment.

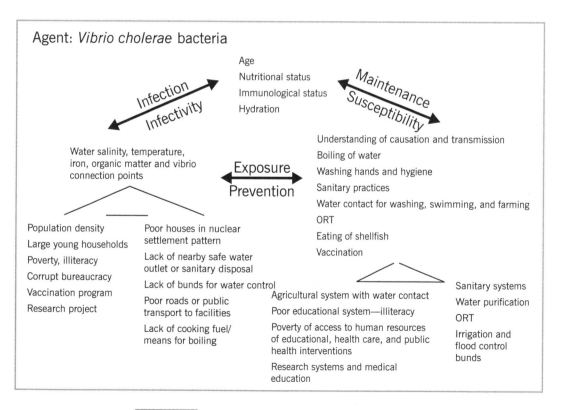

FIGURE 2.4 The triangle of human ecology and cholera.

The habitat of the disease begins with the aquatic flora and fauna, including algae, phytoplankton, zooplankton, and copepods, in which the bacteria flourish and which serve as reservoirs (Islam, Drasar, & Bradley, 1990). *V. cholerae* survival appears to be dependent also on abiotic characteristics, including alkalinity, salinity, iron concentration, temperature, nutrient concentrations, and the number of available attachment sites for the vibrio (plankton). The local sea surface temperature influences the growth of phytoplankton concentrations, which in turn encourage the subsequent multiplication of commensal copepods. Temporal fluctuations in cholera are likely related to variations in physical and nutritional aquatic parameters, including conditions in both ocean reservoirs and the brackish ponds and canals in endemic areas.

These factors are influenced by larger-scale climate variability (e.g., El Niño southern oscillation [ENSO] of the jetstream) and seasonal effects (i.e., sunlight, temperature, precipitation, monsoons), but the exact mechanisms are not well understood. Cholera has been shown to be related to upper troposphere humidity, cloud cover, and top-of-atmosphere absorbed solar radiation. Similarly, ENSO raises water temperature, bringing about increased zooplankton blooms that may influence longer cycles in cholera periodicity. Climate can affect the temperature of the sea surface as well as local ponds and rivers, possibly increasing the incidence of cholera through the faster growth rate of the pathogen in aquatic environments. The heating of surface water may also lead to an increase in the phytoplankton blooms that feed zooplankton, encouraging the subsequent multiplication of commensal copepods that house *V. cholerae*. Periodic climatic and temperature cycles such as ENSO have an effect on cholera variability (Koelle & Pascual, 2004). Warm El Niño events are linked to cholera outbreaks in Bangladesh, and also to the emergence of new harmful algal blooms throughout Asia. Cholera outbreaks in Bangladesh between 1966 and 2002 demonstrate a 9- to 14-month lag between Indian Ocean sea surface temperatures, on the one hand, and atmospheric temperature changes and subsequent cholera outbreaks, on the other (examples include the 1987–1988 El Niño, the 1988–1989 La Niña, and the 1997–1998 El Niño).

The built environment in Bangladesh is based on houses clustered on raised land in rice fields. Some people live on houseboats (see Plate 3). The population density of this land is very high, and the population is very young. Embankments called *bunds* have been built to control water flow, especially to keep floodwaters from the river's flow inundating the fields and villages. Two of your authors (Emch & Carrel) have done research about the alterations in concentration and persistence of the vibrio associated with this water control (Carrel et al., 2009; Carrel et al., 2010; Emch, Feldacker, Islam, & Ali, 2008; Emch et al., 2008). Water sources and sanitation systems are illustrated in Plates 3 and 4. Some wells are better protected than others, but there are few outlets people can access for safe drinking. Even many of the tube wells protected from surface contamination produce arsenic poisoning. Roads and transport facilities to get sick children to hospitals for intravenous rehydration are few and poor. The local fuel for heating in this land of few trees is fuelwood, which is thus expensive, relatively speaking. The social environment includes the population density and large young households (although the average fertility rate of Bangladesh's total population has been lowered below three children today).

Although the environment is important for whether cholera exists in a particular area of the world, the magnitude of epidemics is related to socioeconomic factors (Emch, 1999). Secondary transmission occurs when the infection is passed by the persons directly infected by the original source to other people. Perhaps the original persons contaminate a different water site, handle food or dishes, touch a child, or are helpfully touched in their own illness. Factors affecting secondary transmission control the extent to which the disease is present (i.e., whether it will reach epidemic proportions). The severity of secondary transmission depends primarily on water sources for household consumption. People who use contaminated surface water for drinking, cooking, and bathing are more likely to contract cholera than those who do not. In other words, there is an inverse relationship between cholera and access to improved water sources and sanitation. People living on less than a dollar a day, of course, cannot afford fuelwood for boiling water. We could virtually eliminate cholera throughout the world if drinking water and sanitation in the poor countries of the world could be improved to the standards of rich countries such as the United States. In the 1800s there were large cholera epidemics in the United States and Europe, but it has been virtually eliminated in these areas since then. In contrast, cholera has rapidly expanded throughout Africa during the past decade.

Other cultural buffers to cholera have been developed and practiced or utilized. The oldest, in a new promotion, is illustrated in Figure 2.5. It shows a cholera education billboard in Malawi, with a heading that can be roughly translated

FIGURE 2.5 Cholera education advertisement in Malawi.

as "Get Rid of Cholera in Malawi." The main message states that people should wash their hands every time they use the toilet, after they change a baby's diaper, before they prepare food, before they eat food, and before they breastfeed. There are other, newer cultural buffers as well. Water is sometimes filtered through multiple layers of old cloth in the household. New ceramic filters have been designed (by graduate students in public health at the University of North Carolina); in the future these may be made locally and provide employment as well as safer water. Efforts to design and distribute locally functional solar cookers may help boil water also.

QUICK REVIEW 1

✓ The human ecology of disease is concerned with the ways in which human behavior, in its cultural and socioeconomic contexts, interacts with environmental conditions to produce or prevent disease among susceptible people.

✓ The causative organism of infectious disease is an agent, agents infect hosts, and infection can be either direct or vectored through insects such as mosquitoes or ticks.

✓ Zoonotic diseases, diseases of animals, are important to human health.

✓ The triangle of human ecology consists of three vertices, population, behavior, and habitat, and is a framework for understanding patterns of health and disease.

LANDSCAPE EPIDEMIOLOGY AND VECTORED DISEASES

While the importance of the natural environment has been alluded to above in our discussion of directly transmitted or water-borne diseases, it plays a particularly important role in influencing the spatial patterns and processes of vectored diseases and zoonotic diseases. To try to understand patterns of a zoonotic disease newly emergent in humans without considering the disease in animal hosts would be useless: the biogeography of the arthropod vectors and their animal hosts must also be examined. The Soviet geographer Eugene N. Pavlovsky (1966) called this exploration of vectored diseases "landscape epidemiology." Today studies of the spatial patterns of vectored diseases, often using geographic information systems (GIS) and sometimes satellite imagery, are increasing internationally. Most of them use the following concepts, but few of them apply Pavlovsky's term. This section will explore the landscape epidemiology of several diseases such as malaria, Lyme disease, and Rocky Mountain spotted fever in the context of the triangle of human ecology.

Each of these diseases is of major importance not only to human health, but also to the advance of scientific understanding about the nature and occurrence of different types of transmission systems. There are many "models" available in various disciplines for predicting secondary infections and epidemic size, or for analyzing niche competition among species. The importance of human agency (i.e., cultural behavior) in creating the conditions with which people come in contact, modifying them, exposing susceptible individuals to them, and preventing and treating different disease outcomes is almost always ignored. We look at why these disease systems occur where they do; how they are relocated, contained, eradicated, or maintained; and how cultural buffers can be used for intervention or prevention. The use of landscape epidemiology allows us to integrate and examine the disease expression in place.

Regions

There are at least four types of regions involved in the study of landscape epidemiology. The first is biotic. Climate, altitude, and latitude combine to create broad biotic regions, **biomes**, with predictable locations. These are themselves compounded of other regions, of course—climatic, geomorphic, soils, and glaciers, to name a few—and include not only plants and animals, but also particular types of insects and microbes. If the biome's natural fauna and flora are displaced through land development, their replacements tend to be crops and domestic animals also distinctive to that biome, even when historically not that cultural region.

The second type of region is also biotic: **realms of evolution**. Because oceans, deserts, and great mountain ranges formed barriers to the exchange of genetic information as the continents drifted apart, complexes of plants and animals followed separate paths of evolution in South America, in Africa south of the Sahara, in Asia south of the Himalayas and the Yangzi (Yangtze) River, in Australia, and in the remainder of Eurasia (Palearctic) and North America (Nearctic). These realms of evolution include distinct microbes and insects, as well as the more visible mammals and birds (e.g., kangaroos or giraffes) that popularly demarcate the regions. As trade and political empires established contacts beyond their culture realms, the biotas of the evolutionary realms were exchanged. Deliberately, in the case of food crops, and unintentionally, in the case of disease agents, sweeping processes of global readaptation were set in motion. Historians have only relatively recently paid attention to the health consequences that followed upon connection of these realms (McNeill, 1976). When the maritime and land-based transportation and trade networks of the Roman Empire connected with those of the Chinese Empire, for example, terrible epidemics swept both worlds as they exchanged smallpox, measles, diphtheria, typhoid, and much more. The best-studied connection of evolutionary realms is that between the Old World of Eurasia and Africa and the New World of the Americas. The so-called Columbian exchange involved the two-way exchange of scores of crops, animals, and diseases (see Chapter 8).

The third type of region, **culture realm**, is a broad cultural area delimited by the extent of particular cultural practices and beliefs that largely originated in the primary culture hearths. China and the United States, for example, occupy vast territories with the same latitude and often similar climatic patterns, but because their inhabitants use the land differently and have different settlement forms, house types, and technologies, their landscapes are distinctive. The use of human waste for fertilizer in the Chinese culture realm (which includes Korea, Japan, and Vietnam) and the use of cows for milk in the European and North American cultural spheres are examples of regional patterns associated with health consequences: intestinal parasites and dysentery in the former, and tuberculosis of the bone and lymph system, Q fever, and brucellosis in the latter, as well as nutritional advantages and disadvantages in each.

The fourth type of region, **natural nidus** (focus or cluster), has been defined by Soviet and German geographers but is probably unknown to most of our readers. A natural nidus is a microscale region constituted of a living community, among the members of which a disease agent continually circulates, and the habitat conditions necessary to maintain that circulation in the disease system. The more familiar term "focus" is frequently used instead of the unfamiliar term, but it does not carry the connotation of microecology of soil, vegetation, and animal movement as well as *nidus* does. The life cycles and transmission chains of nidal diseases are complex. This chapter describes several examples of different types of systems.

Vector Ecology

Most disease agents require specific hosts in which they live. They are even more restrictive about the cold-blooded arthropods in which they reproduce. Because an agent usually goes through changes in life stages and multiplies within a vector, most arthropods cannot be biological vectors. An inappropriate arthropod finds its throat blocked, its abdomen too heavy for flight, and its life expectancy so shortened that both arthropod and agent die before transmission. Such poor adaptation of a vector to an agent implies that the relationship is of recent origin.

A full range of arthropods is involved in disease transmission almost everywhere, but one may identify broad patterns that generally coincide with biomes. At high altitudes and latitudes, ticks are especially important vectors (Table 2.2). This is because a tick is capable of transmitting rickettsia and viruses *transovarially* (into its eggs), so that the thousands of tick larvae that hatch from an infected tick's eggs are equally infected, through generations. In this way the disease agent is transmitted despite freezing temperatures, which otherwise stop transmission by killing, for example, adult mosquitoes and the agents they carry.

Mosquito-transmitted diseases are most important in warm, humid lands where these insects can live long and reproduce frequently (Table 2.3). Reproduction in mosquitoes is temperature-dependent and occurs more quickly when the temperature is warm. The female mosquito takes blood to get protein to lay her eggs (male mosquitoes do not bite people). If a mosquito transmitting malaria in Georgia were to become infected from her first blood meal after emerging as an adult, she might infect 11–13 people in her life. The same species in New York would be unlikely to infect more than four or five. As temperatures get colder, the metabolism of mosquitoes slows down. Below freezing it stops entirely, and death follows. A few mosquitoes, such as the vector of La Crosse encephalitis, sometimes pass the infection transovarially, but this is rare (see Table 2.3). Thus the disease cycle is interrupted annually in cold climates. A disease such as yellow fever can be introduced only as an epidemic in the middle latitudes and does not become endemic there, as it did in the Caribbean. In the wet–dry tropics, some mosquitoes have the ability to estivate

TABLE 2.2. Some Diseases Vectored by Ticks and Mites				
Disease	Agent	Major vector	Endemic locale	Reservoir/ comments
Viral				
Encephalitis, Central European	Togavirus, group B	Tick: *Ixodes* spp.	Europe	R: rodents
Encephalitis, Russian spring—summer	Togavirus, group B	Tick	Europe, North Asia	R: rodents; tick transovarial transmission
Louping ill	Togavirus, group B	Tick	Great Britain	R: sheep
Rickettsial				
Rocky Mountain spotted fever (similar to boutonneuse fever, South African tick typhus, India tick typhus, etc.)	*Rickettsia ricketsii*	Tick: *Ixodes* spp.	North America, especially Southeast piedmont	R: small mammals; tick transovarial transmission
Scrub typhus (tsutsugamushi disease)	*Rickettsia tsutsugamushi*	Larval trombiculid mites: *Leptotrombidum akamushi* and *deliens*	Asia: Siberia to Indonesia	R: rodents; mite transovarial transmission
Bacterial				
Tularemia	*Francisella tularensis*	Ticks: dog, rabbit, Lone Star	North America, continental Europe	Zoonosis of rabbits and rodents
Relapsing fever (endemic)	*Borrelia recurrentis*	Ticks: *Ornithrodoros* spp.	Tropical Africa, Mediterranean, Asia Minor, Americas	R: rodents; tick transovarial transmission
Protozoan				
Babesiosis	*Babesia microti*	Tick	Nantucket Island, United States	R: rodents; rare in humans

(or to hide and dehydrate during the dry season) and survive. The infected adult mosquitoes can renew the transmission chain when the rains come.

There is growing appreciation of yet another biological strategy. Many viruses transmitted by mosquitoes, including several kinds of encephalitis found in the United States, are actually zoonoses of birds. Winter kills mosquitoes and breaks transmission over most of the United States, yet early every summer there have been puzzling new cases of Eastern equine encephalitis in Virginia. The viruses, it seems, are annually reintroduced along the great bird flyways during avian migration and arrive just in time for the new spring mosquitoes to become infected. The virus simply winters in warmer climes. When West Nile virus (WNV), a bird zoonosis that causes encephalitis in people, was first introduced into New York City, health officials struggled to contain the arbovirus. They wanted to prevent it from migrating south to Florida for the winter, to survive, spread, and return the next year. They targeted the main breeding sites in the stagnant swimming pools that had been created by water conservation measures that summer. Cold winter appeared to have killed the mosquitoes and ended transmission. The arbovirus's reemergence early the following spring was baffling, until it was noticed that swarms of mosquitoes emerged when the cover of a sewer line was removed. The vector of WNV likes organic pollution, and found warmth underground in New York City. Perhaps aided by the vehicles of people migrating to keep warm in Florida, WNV did spread to the Gulf of Mexico, and managed to cross to the Mississippi and West Coast flyways also. Birds were hard hit. As is the case with the native Eastern equine encephalitis, however, infection of people, who are dead-end hosts, is incidental.

TABLE 2.3. Some Diseases Vectored by Mosquitoes

Disease	Agent	Major vector	Endemic locale	Reservoir/ comments
Viral				
Chikungunya	Togavirus, group A	*Aedes* spp.	Africa, Southeast Asia	R: monkeys
Encephalitis, Eastern equine	Togavirus, group A	*Aedes* spp.	Americas	R: birds
Encephalitis, Western equine	Togavirus, group A	*Culex tarsalis*	Americas	R: small mammals, reptiles?
Encephalitis, Japanese B	Flavivirus, group B	*Culex tritaeniorhynchus, C. gelidus*	East Asia, Pacific islands	Pig amplifying host
Encephalitis, St. Louis	Flavivirus, group B	*Culex* spp.	Americas	R: birds
Encephalitis, Murray Valley	Flavivirus, group B	*Culex annulirostris*	Australia, New Guinea	R: birds
Encephalitis, La Crosse	Bunyavirus, group C	*Anopheles triseriatus*	Central United States	Transovarial transmission
Dengue fever	Flavivirus, group B	*Aedes aegypti*	World tropics; hemorrhagic form, Southeast Asia	R: monkeys, Southeast Asia
Yellow fever	Flavivirus, group B	*Aedes aegypti*	Tropical Africa, tropical forest South America	R: monkeys
Protozoan				
Malaria	*Plasmodium: vivax* (tertian) *falciparum* (malignant) *malariae* (quartan) *ovale* (West African)	*Anopheles* spp.	Tropics and subtropics	Humans are source
Helminthic				
Filariasis (elephantiasis)	Nematode: *Wuchereria bancrofti*	*Culex fatigans, Aedes* spp.	Tropics and subtropics	No reservoir: infective microfilaria, night or day periodic, depending on local vector; zoonosis of wild and domestic animals in Malaysia
Filariasis (elephantiasis), Malayan	Nematode: *Brugia malayi*	*Mansonia* spp.	Southeast Asia, India, China	

Analyzing the efficacy of a vector requires detailed knowledge of its particular biology. The following characteristics, for example, affect mosquitoes' usefulness as vectors: flight range, altitude range, sex ratio, breeding habits, preferred breeding sites, life expectancy, age structure of local mosquito population, alternative (other than blood) food, activity time, host preference, resting habits, biting habits, and tolerance for specific agents (viruses, microfilariae, protozoa, and others). Working with entomologists, epidemiologists and geographers can attach probabilities to each step, and each contact, each stage of an epidemic, and can model its course or monitor an intervention. Even without such detailed information from costly field research, knowing the time of biting, flight range, and type of habitat for breeding helps geographers understand viable interventions and cultural buffers.

Some mosquito species require fresh water with no pollution for the larvae; others flourish on organic pollution such as sewage. Some breed in containers, tree holes, or leaf tendrils; others need open expanses of water. A few like brackish conditions. Some mate for life when they emerge as adults from the water; others hover around blood sources, hoping to find a mate. Some are weak fliers; others are easily carried long distances by wind; and yet others are strong fliers, even upwind. Most mosquitoes prefer to bite other animals, including snakes, frogs, and birds, rather than humans—but a few species prefer people, and others tolerate them when hungry enough. Some mosquitoes readily enter human dwellings and after feeding rest on the walls; others never enter dwellings and bite only outside. Some are active during morning or evening; others are active only at night. The detailed ecological requirements for specific mosquitoes are inherent in the landscape, and the distribution of relevant species determines where various disease agents can be spread.

Flies inhabit a wide range of biomes (see Table 2.4). In general, they are most important in arid places. The flies of Australia, of the wet–dry tropics, and of arid grazing lands are infamous. A few types of these flies are bloodsucking vectors of human disease. The most notorious of these is the tsetse fly, which transmits trypanosomiasis (African sleeping sickness). This disease mainly affects animals. In one form, *Trypanosoma rhodesiense,* it is an anthropo-zoonosis that has a reservoir in African wildlife and is virulent in humans. In another, presumably older, form, *T. gambiense,* it is milder, is slower to kill, and has its reservoir mainly among people, with animals being incidental. There are still other forms that affect only animals. The disease has had a major impact on human population distribution and nutrition by making the raising of livestock for protein and draft power virtually impossible across large areas of Africa. Even this fly, however, needs water to breed and becomes less active when humidity is very low. Another vector, the biting blackfly, *Simulium damnosum,* is the major vector of onchocerciasis (river blindness). It lays its eggs in oxygenated, fast-flowing water by anchoring them to underwater stones. Exposure to the disease cycle, as the name suggests, is associated with use of the fertile soils near the streams and rivers or with use of the rivers for laundry, fording, or fishing. Hunter (1980) first described the complexity of the disease system and the multisided, integrated attack that is needed to control it. The various forms of leishmaniasis are transmitted by sand flies. Although characteristic of rather arid terrains, they require some humidity and are usually found at specific altitudes or

TABLE 2.4. Some Diseases Vectored by Flies and Reduviid Bugs

Disease	Agent	Major vector	Endemic locale	Reservoir/comments
Viral				
Sand fly fever	Bunyavirus, group C	Midge, sand fly: *Phlebotomus papatasii*	Mediterranean climate areas, Asia Minor, Central and South America	Human—fly complex crucial
Bacterial				
Tularemia	*Francisella tularensis*	Deer fly	North America, Europe, Japan	Zoonosis of rabbits and rodents Humans are source
Bartonellosis (Oroya fever, Carrion's disease)	*Bartonella bacilliformis*	Sand fly: *Phlebotomus* spp.	Andean valleys in Peru, Ecuador, Colombia	Humans are source for *gambiense* R: wild game and cattle for *rhodesiense*
Protozoan				
African trypanosomiasis (sleeping sickness)	*Trypanosoma gambiense, T. rhodesiense*	Tsetse fly: *Glossina* spp.	Tropical Africa	R: wild and domestic animals
American trypanosomiasis (changas)	*Trypanosoma cruzi*	Reduviid bugs	Tropical America	R: dogs, cats, rodents in different places Fatal untreated
Leishmaniasis, visceral (kala-azar)	*Leishmania donovani*	Sand fly: *Phlebotomus* spp.	Tropics and subtropics, Mediterranean	R: dogs, rodents
Leishmaniasis, cutaneous (Oriental sore espundia, uta)	*Leishmania tropica, L. brasiliensis, L. mexicana*	Sand fly: *Phlebotomus* spp.	Arid margins of Asia, Africa, Mediterranean, South and Central America	Humans are source
Helminthic				
Onchocerciasis	Nematode: *Onchocerca volvulus*	Blackfly: *Simulium* spp.	Tropical Africa, Central America to Amazon	Humans are source
Loaiasis	Nematode: *Loa loa*	Mangrove fly: *Chrysops* spp.	West and Central Africa	

ecological niches where the relative humidity is high enough. It finds shelter from the sun and low humidity in the plaster and adobe walls of human dwellings and the burrows of rodents.

One of the lesser-known and most unusual vectors of deadly disease is the cone-nosed "kissing bug," or the reduviid bug. These bugs (that's their name!) transmit the protozoal agent, a trypanosome, of American trypanosomiasis. Much better known as Chagas disease, it infects people in most countries in South and Central America, as there are many species of reduviid adapted to the different habitats. There are hundreds of reservoir animals in the forest, from giant sloths to dogs, but the disease has been most intense among people living in thatch-roofed housing, especially on the forest edge or frontier. The bugs, which resemble cockroaches in appearance and living habits, differ in the important way that they suck blood: they like to suck blood from lips and exposed membranes. After eating they defecate, and a person's hand then rubs the bite, infecting it with thousands of trypanosomes in the bug's feces. After initial, usually mild, sickness, the trypanosomes multiply for years until eventually the person dies from complications of an enlarged heart (and related effects).

The Landscape Epidemiology Approach

Landscape epidemiology is a geographic delimitation of the territory of a transmitted disease in order to identify cultural pathways for disease control. Working with the American Geographical Society, Jacques May mapped mosquito vectors, and rickettsial and helminthic diseases, among others (see Chapter 1 and Plate 1). Several holistic approaches to the study of arthropod-borne or naturally occurring diseases have been developed in Europe. In Germany, for example, a holistic ecological approach was developed under the leadership of Helmut J. Jusatz. He produced the first world atlas of epidemic diseases (Rodenwaldt & Jusatz, 1952–1961), which not only portrayed disease distribution but also analyzed its association with climate, topography, hydrology, and flora/fauna. This work was expanded into a series of monographs on the health geography of such countries as Kenya, Kuwait, and Korea (Jusatz, 1968–1980). This holistic approach and focus on ecological associations (*landschaftokologische,* in German) includes settlement and cultural patterns. The modeling approach of Pavlovsky provides a framework to do this. It needs to be noted that in U.S. health geography in general, the study of the natural ecology of disease, even when enhanced with May's influential cultural processes, has been weak.

May's Multifactor Zones and Pavlovsky's Nidus

May (1958) described areas where transmissible diseases were present as regions where multiple factors had to occur and overlap for disease systems to occur. He wrote of *three-factor* and *four-factor* diseases. Each factor has its own spatial distribution, and all factors had to coincide in space to create the disease conditions. In the example that follows, plague is a four-factor disease: the agent is one factor; the rat host (reservoir) is a second; the flea vector is a third; and the human host is a fourth. May called a place where the three factors of agent, vector, and reservoir coincide, but disease is not known because no humans are present, a **silent zone** of disease. People build houses, roads, or campgrounds, or otherwise intrude into a silent zone, and are surprised to develop encephalitis, yellow fever, or plague. When people such as pastoralists in the grasslands or woodcutters in the forests penetrate into the silent zone, or farmers settle it, disease can result. In the former Soviet Union, this was a common occurrence in the settlement of Siberian mining and industrial towns. It was Pavlovsky's hypothesis that the potential for specific diseases could be identified from the simple existence of certain environmental conditions, so that planners could design housing, protective clothing, or work scheduling to shield the human population from exposure to the zoonosis. In the United States, there has been limited work on the biogeography of zoonoses. Instead, the silent-zone concept is being extended to nonbiotic environmental hazards, such as toxic land.

As noted earlier, the concept of landscape epidemiology was developed by the Soviet geographer Pavlovsky in his 1966 book. Landscape and connections with Russian culture were a central concern for Russian geographers, not unlike the French geographers of the time. Pavlovsky was one of the leading parasitologists of the former Soviet Union; he was director of the Zoological Institute of the Academy of Sciences and president of the Geographical and Entomological Societies of the Soviet Union. Because of the importance of the landscape epidemiology approach he developed, the Institute of Medical Geography became a separate entity in the Soviet Academy of Sciences. He developed the doctrine of natural nidality, or the natural focus of disease. In a natural nidus, infection is maintained among wild animals and arthropod vectors. These zoonoses, which Pavlovsky determined and established through fieldwork, occur in particular kinds of terrain. There is a biogeography to the life cycles of the arthropods and various other animals involved. Their food sources, soil, climate, slope, exposure, and other ecological parameters determine local distribution and possible occurrence of the disease cycle. By knowing the conditions necessary for specific diseases, scientists can use the landscape to identify disease hazards. Landscape modification can create, or be used to prevent, the establishment of disease cycles.

Landscape Epidemiology and Human Ecology of Malaria

Malaria is by far the most serious vectored disease in the world. It is one of the three diseases targeted by the WHO with the Global Fund. In the 1950s it was possible to dream about eradicating malaria because it has no reservoir. In the 1960s eradication or control over large areas of Asia and South America was in fact achieved. In the 1990s, however, malaria reclaimed land and again grew to infect more than half a billion people and threaten many more (see Figure 4.4). Malaria kills 1–2 million people a year (over 1 million in tropical Africa), totally incapacitates tens of millions, and causes hundreds of millions to struggle in exhaustion with chronically severe anemia. About 40% of the world's people, in more than 100 countries, live in regions where malaria is endemic today. The health improvement in the 1960s in countries in which malaria was almost eradicated was eye-opening. Infant mortality rates plummeted. Fertility rates increased as miscarriages and cases of infertility that were due to malaria disappeared.

The terrible pressure malaria has exerted on the human population over the ages is demonstrated in the population vertex of the triangle of human ecology by the presence of high-mortality genetic diseases, most famously sickle cell anemia. Yet the case mortality rate today for malaria is low compared to that of most serious diseases. Malaria is generally, depending on type, a chronic, debilitating disease. Despite decades in which scientific armament has been marshaled against malaria, it is resurgent in many countries where previously it was virtually eradicated. Cases in Sri Lanka, for example, went from 1 million a year in the 1950s to less than 10, but then rose to 38,500 in 1982 (4% caused by *Plasmodium falciparum*—see below), to 676,000 in 1987 (27% *P. falciparum*), and to 1 million a year in the 1990s. It is still estimated that about 80% of all clinical cases and 90% of parasite carriers live in tropical Africa. More than 1.7 billion people live in once-endemic areas in which transmission had been greatly reduced but is now reinstated, and in which the ecology is unstable and the situation is deteriorating. Today malaria kills one child every 30 seconds. The mosquito vectors are resistant to all the major insecticides; the agent is resistant to all the major drugs; and the ancient scourge is upon us again.

The disease cycle consists of direct transmission of the agent between humans by various species of *Anopheles* mosquitoes. There is no animal reservoir, although monkeys, birds, and other animals have their own forms of malaria. The agent is a protozoan, of a genus called *Plasmodium,* occurring in four species. *P. vivax* causes benign tertian malaria, historically prevalent in Europe. It has a liver-dormant stage (hypnozoites) that allows it to survive even long winters and to reactivate in time to infect the spring mosquitoes. Its case mortality is low enough for it to have been nicknamed "benign tertian." It does not occur in West Africa, where people genetically lack the required antigens in their blood cells and seem immune to it. *P. ovale,* the least common malaria, occurs only in West Africa—perhaps as a niche replacement. *P. malariae* is so old and well adapted as a human parasite that science has yet to discover how it hides and relapses in the blood. The most dangerous form of malaria is caused by *P. falciparum.* It has more than a 10% case mortality rate and often has unusual symptoms of disorientation, coma, shock, and renal failure. Apparently it has most recently transferred from birds. It spreads more slowly because it has not yet fully adapted to *Anopheles,* which are not bird feeders. It lacks the ability to go dormant in the human liver and so must reproduce in enormous numbers—which so block, explode, and shatter the circulatory system that it can kill in hours—in order to infect mosquitoes and stay in hypercirculation and continual reinfection.

All forms of malaria have complicated life cycles within the body, although the cycles differ in detail. Protozoa have much more genetic information, and are thus much more complex, adaptable, and cagy, than viruses and bacteria. The sexual forms, gametocytes, emerge from hiding inside the red blood cells within the gut of the *Anopheles* mosquito after she takes a blood meal. They mate in her gut wall, and reproduce large numbers of sporozoites, which migrate to the salivary glands. These get inoculated into people when, after depositing her eggs, she takes another blood meal. They swim to the liver, within which they grow and mature over a week or two into the blood forms, merozoites, which enter red blood cells and commence gobbling hemoglobin and multiplying by factors of 10. They consume the red blood cells and then break out of the empty husks into the bloodstream in a synchronized manner (hence the alternation of fever and chills). The merozoites inside the cells produce the male and female gametocytes, which do not circulate in the blood (and so are not killed by drugs), but which also do not reproduce until they are taken up by a mosquito to complete the cycle, in about 2 weeks. The complexity of this cycle is such that the hematozoites dormant in the liver were not discovered until 1980, after a century of intense microscopic observation. Each stage has different surface proteins, different antigens, different nutritional needs, and different behaviors. The multiple, changing symptoms, forms of hiding, and details of human immuno-logical response are still of such baffling complexity and mystery that malaria has defied decades of intense effort to develop a vaccine.

The cultural ecology of malaria is also one of the most studied. May (1958) described a village in the highlands of Vietnam that had low-grade, endemic malaria. The disease was a drain on the population's energy, but it was not the worst of these highland villagers' problems. The people lived in houses on stilts and tied their water buffalo and cattle under their houses. They cooked in the houses. These customs were quite different from those of the lowland Vietnamese, who moved into the highland area one year. The lowlanders built their houses on the ground and kept their animals in proper barns. They also had cooking sheds sepa-rate from the houses. They were rapidly driven out of the highlands by epidemic malaria. The local species of *Anopheles* that transmitted the agent was not a strong flier. The height of a house on stilts deterred her. She preferred to bite animals and seldom went past the tethered animals under the house. She, like all *Anopheles,* was active at night and especially in the evening, when the house was full of smoke. The cultural practices of the upland people thus constituted a rather successful protec-tive buffer. Changing nothing in the system except the cultural behavior of the people was enough to render the land uninhabitable.

The experience of the United States illustrates some of the complex forms of cultural interaction with the biotic disease system. Originally malaria was an Old World disease. The British brought *P. vivax* to North America, and the slaves they imported brought *P. falciparum* from Africa. The early, glowing reports of how healthy the colonies were faded as malaria spread. The nuclear settlements of the plantations were especially good foci of disease transmission. The coastal plantations became so deadly that the owners spent their summers in resorts (the foundation of several beach and mountain resort towns) and cities such as Savan-nah, Georgia. The greater resistance of Africans to the disease became a common explanation of why the institution of slavery was "that which is necessary."

People did not know the cause of the disease, but they were keen observers. They knew that as the forest around Savannah was removed and land was con-verted into rice fields, malaria increased. They thought that the cause was a miasma (*mal aria* means "bad air"), which the trees had protected them from. In what was probably the first environmental legislation for public health, Savannah demanded that all land within a mile of the city be used only for dry agriculture. It compen-sated farmers out of tax revenues for the loss they incurred by not being able to grow rice or indigo. Malaria thus decreased in Savannah for a few decades, until the disturbances of the Civil War.

Malaria was *the* American disease of the late 19th century. The area surround-ing the confluence of the Ohio and Mississippi was so deadly that many would travel there only in winter. The census of 1890 undertook a survey of causes of death. Although this census was not up to today's statistical standards of reporting, the rates it recorded for malaria—over 7,000 deaths per 100,000 people across the South, and more than 1,000 in such states as Michigan, Illinois, and California—are impressive (Figure 2.6). An average case mortality rate of 5% means that almost the entire population was infected over large areas. Yet by 1930 malaria had disappeared from the North and West, and caused fewer than 25 deaths per 100,000 people in the South, except for a few counties. What had happened?

The United States has one important malaria vector, *Anopheles quadrimaculatus*. It breeds in water where it can anchor its eggs on vegetation that intersects the surface in open sunlight, and where the currents are not too strong. When the forest was removed across the country, sunlight was let into marshes and into the poorly drained glaciated land of northern states. Developers of rice plantations planted lots of intersecting vegetation, and much occurred naturally elsewhere. Slow travel by rafts and barges turned canals and rivers into linear transmission channels. Susceptible migrants picked up the infection along the waterways, and travelers who were infected spread it to new places. Houses were of poor quality, and their glassless windows and slatted walls offered little obstruction to mosquitoes. People commonly had poor nutrition and had few animals to divert the mosquitoes.

At the turn of the 20th century, however, the poorly drained land of the glacial till country was deliberately drained for agriculture. Houses were more often constructed of brick or other good material and had glass windows and screens

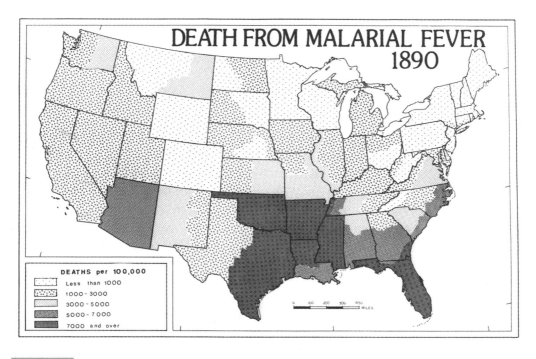

FIGURE 2.6 Malarial deaths in the United States, 1890. Adapted from U.S. Census Bureau (1890, map 17).

(mainly to keep flies out). Transportation shifted from the rivers and canals to the railroads; settlements as well as travelers shifted to the dry upland. The numbers of animals increased. The nutrition of the people improved. Quinine, an antimalarial drug, became available at a more reasonable price when the tariff against its importation (from Java, mainly) was finally dropped. The slower breeding of the mosquito in the cold climate of the North led to the disappearance of Northern malaria. In the South the incidence was greatly reduced as the population became more settled (and therefore less mobile), but some counties near sinkholes or the oxbow lakes of the Mississippi remained highly malarious.

ERADICATION CAMPAIGNS

In the late 19th century, Laveran discovered the agent of malaria, and Ross proved its transmission by *Anopheles* mosquitoes. In the United States, a public health official named Carter (1919) was amazed at how little use had been made of this knowledge to prevent malaria. The disease was so common in the South as to be taken for granted. He had to demonstrate to state and other officials the tremendous economic costs of lost workdays and debilitated farmers. The new Rockefeller Foundation became interested. It supported the training and staffing of entomologists and other experts, as well as the creation of public health departments at the county level. Carter himself determined the flight range of the U.S. vector (1 mile), its water preferences, and the effects of water impoundment for mills. Mills were commonly closed on Sunday, and on Monday the elevated mill ponds were drawn down. Carter determined that this was good because it stranded the flotage that anchored larvae.

Carter's findings became relevant when the Tennessee Valley Authority (TVA) proposed to construct dams that would impound large bodies of water in the malarious South. The economic value of developing cheap electricity was obvious, but what would the cost to the public health be if malaria again became epidemic? The Army Corps of Engineers worked closely with health officials. All people near the proposed reservoirs were checked for malaria and treated, and all dwellings within a mile of each reservoir were screened. All vegetation was cleared within 10 feet above maximum and below minimum water levels to prevent intersecting vegetation. The water level had to be fluctuated, in spite of navigational needs, at fixed times of the year in order to strand flotage. It all worked, and malaria almost disappeared from the TVA area, instead of becoming epidemic.

Similar measures were adopted elsewhere, and levels of malaria were reduced. When threatened by a resurgence of malaria brought by soldiers returning after World War II from fighting in malarious countries, the United States mobilized for eradication of the disease. The military office that had tried to eradicate malaria around military bases where soldiers were training was located in Georgia. It was transformed to head the mobilization against malaria. The weapon was the new insecticide, DDT, which could be sprayed on houses and would for months afterward kill the mosquito vector when it rested on the walls after taking its blood meal. In a massive effort, workers sprayed millions of houses across the South and detected and treated thousands of cases. Malaria was eradicated in the United States. The institution that had coordinated the attack became the Centers for Disease Control (CDC), located in Atlanta, Georgia (a city in the previously malarial South), rather than Washington, DC.

When soldiers returning decades later from the war in Vietnam brought in relapsing malaria cases by the thousands, public health concerns were again raised. The plasmodia, however, found the environment so modified that they could no longer get established in the United States. Few people now sleep where mosquitoes can fly into their bedrooms and bite them; detection is swift, and treatment is intensive. The control of malaria in the United States has been so complete that many people forget why reservoir landscapes have exposed banks and fluctuating water levels. Uses for recreation, livestock, and fish cultivation are coming into conflict with water management laws that were enacted in every state in the South for control of malaria. Public health workers have grown complacent and government officials ignorant. Yet today, because of population mobility, more than 1,000 cases of malaria a year occur. Some cases have been fatal when physicians failed to recognize the disease. There have, after a gap of 30 years, been a few cases of indigenous transmission in Texas, California, Michigan, Florida, New Jersey, and even New York City. Homeless people sleeping without shelter are again exposed.

Beginning in the late 1950s, the WHO attempted to eradicate malaria worldwide. The effort achieved notable successes, especially in Southeast Asia and Central America. By the 1970s, however, health officials talked about control instead of eradication, and in the 1980s they lost control. In most places the eradication campaign had made little progress at all, especially in tropical Africa, where *P. falciparum* predominates and prevalence remains at the high traditional levels that induced a lethal genetic adaptation. Many reasons for the failure of eradication have been offered by authorities. In some cases maintenance and surveillance failed when, for budgetary reasons, the special malaria teams were combined with multipurpose health programs. The international availability and costs of insecticides have changed. Some (e.g., Farid, 1980) believe that when the reduction in malaria led to faster population growth, the capitalist countries cut back their financial support; some even suggest that support was intentionally withdrawn from the eradication effort because of a desire to keep Third World countries poor, sick, and dependent. Whatever faults there were in administration and in financing, however, the original hopes of eradication were probably unrealistic because they failed to appreciate the complexities of tropical ecosystems. Tropical ecology is very complex, and eradication was modeled on the simple U.S. system described above.

In temperate countries, such as the United States, there are few vector species. The United States had only one. Sheets of open, fresh water with intersecting vegetation were the breeding areas. Spraying walls of houses with a residual insecticide interrupted transmission because the vector bit at night in the houses and then rested on the wall. Most *Anopheles* species are normally the vectors of endemic, chronic malaria at low levels. Some species such as *A. maculatus,* however, breed in streams exposed to sunlight and readily bites humans. It rapidly increases to very large numbers when given the simple, open habitat it prefers. Thus, when the forest is cut down for land development schemes or plantations and the forest streams are exposed to sunlight (see Plate 6), *A. maculatus* rapidly increases in numbers; as a vector of epidemic malaria, it is one of the most dangerous in the world.

The objective of the eradication campaigns was to interrupt transmission, using some of the known intervention points. Residual insecticides sprayed on houses were used to kill the adults after they had taken a blood meal. Humans were examined, and often chloroquine was given as a general prophylaxis to the entire population, to kill the plasmodia in their blood. It was recognized that the mosquitoes could not be suppressed forever. The hope was that when they came back, the agent of the disease would no longer exist in that place. This strategy had worked in temperate lands, but the complex ecology of the tropics was frustrating. Although some vectors bit people in their bedrooms at night, rested on the wall, and died, for example, others bit people only outdoors or flew out the window after biting people and never touched the insecticide. Broader attempts to interrupt breeding did not work. Oiling water bodies or stocking them with tilapia fish, which eat mosquito larvae, did not affect mosquitoes breeding in tree holes or hoof prints. People cutting firewood in the mangrove swamps continued to be bitten by *A. sundaicus*. For many such reasons, when the wall-resting, common vector became resistant to the insecticides and rebounded, some malaria was still present in the community, and the transmission chain resumed. Often the population had lost any partial immunity, however, and epidemics, higher infant mortality, and intensified health problems resulted.

Landscape Epidemiology and Human Ecology of Tick-Borne Diseases

Ticks vector an array of viral and rickettsial diseases, from Russian spring–summer encephalitis to boutonneuse fever in India and South Africa (see Table 2.2). Each has its own characteristic landscape ecology. The landscape epidemiology of central European encephalitis, for example, was studied by Wellmer and Jusatz (1981). That encephalitis is caused by a togavirus vectored mainly by a common tick, *Ixodes ricinus*. Wellmer and Jusatz delimited the temperature, humidity, precipitation cycles, vegetation, animal associations, and other environmental conditions needed for its maintenance. They found that lower mountain elevations with dense and diversified vegetation are required by the disease system, which is limited by such factors as an annual isotherm of 46°F (8°C), homogeneous vegetation such as occurs in pine forest biomes, and light soils with humus for sufficient ground moisture. They determined that the foci are limited to a few square kilometers each, and that a density of only 1 virus-infected tick per 1,000 ticks was required to maintain a natural nidus. The German geographers called this a *standortraum* (multifactor location space). They mapped and analyzed the geoecological conditions for the repeated infection of specific regions by infected ticks and the recurrent infections of people who entered the natural nidus. American studies have not been so specific, although there are two excellent candidate diseases to research: Rocky Mountain spotted fever (RMSF) and Lyme disease. Both of these tick-borne serious diseases are increasing in prevalence in the United States, yet their distribution pattern is disjunct. Lyme disease occurs, but mysteriously has not become prevalent in the heartland of RMSF. What makes them ecologically distinct? Is it only a matter of time until they merge?

Tick vectors pose a four-stage ecological puzzle. RMSF is generally vectored by the "dog tick," *Dermacentor variabilis,* and Lyme disease by the "deer tick," *Ixodes scapularis* (now considered to include *I. dammini* in the South). There are important differences between the two disease systems, but they share basic characteristics. Three separate blood meals are required for the life cycle. The adult female lays up to 10,000 eggs, which hatch into six-legged larvae vernacularly called "seed ticks." These must feed for a week on an animal. These animals are usually small and close to the ground; mice are favorites, as well as birds and reptiles. Then the larvae metamorphize into eight-legged nymphs. The nymphs climb grass and bushes and "quest" vigorously, waving their legs at vibration, heat, or carbon dioxide, to latch onto an animal host and get a second blood meal. After feeding, they drop off and change into adult ticks. According to climate and species, larvae and nymphs as well as adults may crawl under leaves or other cover to spend the winter. In spring they emerge hungry for another meal and, as they feed, infect new rodents before new tick larvae hatch. Adult ticks need the third blood meal to reproduce. They may also feed on rodents, rabbits, opossums, or even birds (useful for dissemination), but they tend to quest higher in vegetation and to be more discriminating in what they latch on to. They prefer raccoons, foxes, deer, or other large mammals such as dogs, which can disperse them over a wider territory. Each of three blood meals provides an opportunity for the tick to become infected and in turn infect its next host. When a large animal is infected, it can become an amplifying host, able with high viremia to infect many arthropods that feed on it.

Here RMSF and Lyme disease diverge, however, in an important way. The agent of Lyme is a spirochete, a bacteria, *Borrelia burgdorferi.* Each larva hatches out "innocent" and must begin the cycle of infection again with a blood meal from an infected host. If the host it feeds on is not infected, then the nymph stage is not infective. If the second blood host is not infected, then the adult tick seeking its third meal is not infective. The enzootic disease is maintained by 25–50% Lyme infection rates of nymphs. In great contrast, the rickettsial agent of RMSF, *Rickettsia rickettsii,* can pass transovarially in ticks to the new generation. Larvae are born infected and able to infect a host. Even infection rates lower than 10% among ticks seem adequate to maintain the disease system because the tick itself through its generations is a reservoir.

RMSF, sometimes called tick-borne typhus, was first identified in Montana and Idaho (hence its name), but it is most common today in the Piedmont region of the U.S. Southeast. It has been isolated from the chipmunk, meadow vole, pine vole, white-footed mouse, cotton rat, cottontail rabbit, opossum, and snowshoe hare, and serological (antibody) evidence for infection has been found in many other mammals (Burgdorfer, 1980). Better publicity has enabled RMSF to be diagnosed earlier and treated with antibiotics, sometimes even before diagnosis. Although this complicates statistical reporting, it has brought the case mortality down considerably.

The onset of the disease involves fever, chills, aches, nausea, and a rash that spreads from the wrists and ankles. Many of these symptoms are common to other diseases and are easily misdiagnosed without antibody identification. Many, probably most, cases are so mild they are asymptomatic. Of those who do develop clinical symptoms and are not promptly treated, however, 6–15% may die.

RMSF has been increasing in the Southeast, but it is hard to determine how much of the increase is due to greater incidence and how much to better reporting. The increase and focus of the disease is attributable to the rapid extension of suburbs into the wooded, open-field succession habitat of the dog tick. The Piedmont is one of the most rapidly growing regions within the southeastern Sunbelt. The dispersed settlements of new suburban houses, along with parks, jogging paths, and family dog sojourns, are *in* the natural nidus of RMSF. When human residential land use and recreation are extended into the woods, infected ticks can be brought home.

The first U.S. cases of Lyme disease were identified in 1962 in Lyme, Connecticut. It is not certain whether Lyme disease was relocated from Europe, where the spirochete has been well established, or had an indigenous variant here. It causes initial symptoms of the generic headache/bodyache/fatigue variety. If not treated, serious symptoms of crippling arthritis and heart abnormalities or neurological complications can develop months or years later. Lyme disease already has spread to the Pacific Coast and the Gulf Coast. It has established three main foci: New England and the surrounding area, the upper Midwest, and the Pacific Northwest (Herrington, 1995). In the East, the reafforestation that has occurred since midcentury has brought the white-tailed deer and other forest animals back to levels not seen in two centuries. Deer, raccoon, and other animals favored by the deer tick (which is somewhat misnamed, as there are ticks that spend their entire life on deer, whereas *I. scapularis* needs three separate hosts) are those most quickly domiciliating to suburban residential areas. The forest fringe–suburban grass intersection can support an intense nidus, with more than half the adult deer ticks infected. Diffusion has occurred along deer paths in river corridors and electric pylon corridors into suburban areas. The infection has, in ticks, probably ridden birds to jump over large areas and spread across the country. The West Coast nidus is not so dangerous for people. The main tick involved there, *I. neotomae,* is the most infected but feeds chiefly on woodland rats, not people. A second tick—*I. pacificus,* which feeds on many different hosts, including rats and people—is needed to connect the nidus to humans. Because of its diverse diet, as it were, the prevalence of infection in this linking tick is low, however, and so is the risk to people.

Why has Lyme disease not become highly prevalent in tick heaven, the Piedmont where RMSF thrives? The ecology of transmission is in the details. The southern deer tick, *I. dammini,* in its larval stage also commonly eats lizards and snakes, and they do not carry the bacterial infection. Therefore, while a nymph may occasionally bite an infected host, perhaps a migrant, and become infected, and Lyme transmission does occur now, nevertheless the nymphs do not reach the population-infection level necessary to maintain a natural nidus of the disease. Since the dog tick is born infected with the RMSF rickettsia, it does.

The cultural interventions and buffers for the two systems are similar. Both diseases can be treated successfully with antibiotics if caught early. Theoretically, reservoir hosts can be destroyed; even if people wanted to exterminate the local deer, however, taking on the white-footed mouse would seem impossible. Almost uniquely among vectors, ticks can serve as reservoirs that survive winters and even pass infections generationally. Residential efforts to interrupt disease transmission usually focus on the tick. Because ticks get dehydrated on their questing roosts in the summer sun, by late afternoon they usually have climbed down to the cooler and humid ground. Therefore, walking into the nidus at the end of the day is protective. The different stages of the tick can be killed with insecticide, which can be sprayed in the yard, or put on the dog in collars. Around the house, keeping the lawn mowed short and brush out of the dog's fenced yard helps keep the ticks away. For a human, body buffers while walking into the nidus include long white pants (for easy viewing) tucked, even taped, into socks and boots; long sleeves; and repellent, perhaps. Because it takes hours of feeding for the tick to pass the infection, the best single body buffer is careful examination of children and dogs, as well as oneself, a couple of times a day. For the pinhead-size deer tick nymph, this is, of course, easier said than done. In a study in Maryland, the best landscape predictor of Lyme disease risk has been identified in a remote sensing/GIS study (Jackson, Hilborn, & Thomas, 2006) as the proportion of land cover that is forest–grass edge. Landscape planning guidelines have then suggested that "casual peridomestic contact with tick and host habitat" could be reduced by clustered forest and herbaceous cover, as opposed to current high forest–herbaceous interspersion. This landscape change would minimize Lyme disease risk in low-density residential areas.

The landscape epidemiology of both diseases rests on the tick's habitat, abundance of small mammals, winter shelter, dehydrating sun, bird predators, and such. RMSF in the past has seemed to involve the intrusion of people into the natural nidus of the disease (May's silent zone). People and their dogs walked into the old field habitats and hunted on the forest's edge. Lyme disease, and today RMSF, can be said to result from the expansion of the nidus to include people, as deer have moved into suburbia. (This distinction, if it ever were true, is no longer clear, as houses are pushed into the forest's edge; displaced deer *are* then in suburbia.)

There are both great opportunities and a great need for geographers to delimit the natural nidi of tick-borne diseases in North America by using GIS-integrated remote sensing and human data. In a study about predicting the risk of Lyme disease in Wisconsin, for example, habitat suitability for *I. scapularis* was determined by using maps of soil orders (determined years ago) to get at such characteristics as soil acidity, coniferous–deciduous forest cover, excessive moisture content, fungi adverse to ticks, abundance of winter leaf litter, and the like, to estimate densities of ticks in specific habitat (Guerra et al., 2002). Statistically significant factors were then used in a GIS to generate a risk map predicting the presence of the tick vector. The extensive area of suitable habitat in western Wisconsin corresponded to areas of increased incidence of human Lyme disease. It explained the disease's axis of rapid expansion. Locating human settlement would further have identified the risk of expansion into new areas. Pavlovsky would have been appreciative.

QUICK REVIEW 2

✓ The specific ecologies of vectors (such as preferred breeding sites, preferred biting times, etc.) are significant in determining where and when and who experiences disease.

✓ Landscape epidemiology integrates details about humans, vectors and environments to understand disease patterns.

✓ Silent zones of disease are places where disease circulates but humans have not entered the transmission cycle.

✓ Understanding the natural nidus of a disease and the factors necessary for disease transmission can also highlight places for public health efforts to intervene in infection or cultural buffers that could be developed.

CONCLUSION

The customs, beliefs, and socioeconomic structures that characterize each global culture realm and local ethnic group create the environmental conditions and exposure patterns that result in geographic distribution of health and disease. Genetic factors, to be sure, often underlie susceptibility and resistance, but the geographic distribution of genes is also a result of adaptation to environment, population mobility, and cultural selection. Influences of the physical environment and of insect/other animal communities and habitats are crucial, but they are buffered and even formed by human agency. Entirely natural landscapes scarcely exist, but the natural environment still plays a large role in human disease, particularly in zoonotic emerging infectious diseases (see Chapter 7).

Health geography explains the distribution of health and disease, and can identify efficient ways to intervene and distribute trained personnel and technology. Every disease and quality of health has its ecology, its geographic regionalization, and its patterns of diffusion and change. We discuss how these patterns change in time and are affected by technological and economic development in Chapter 6. This has already been a long and complex chapter, presenting concepts and ideas as well as a lot of information. It should serve as a basis and continuing reference for the considerations of health and disease in the following chapters.

REVIEW QUESTIONS

1. What are the biological agents that cause disease? List them in order of their size. Which medicines or treatments are effective against which agents? Which agents can be transmitted as fomites or survive in the environment outside of a host cell (note that there is differentiation in this ability even within classes of agents).

2. For each transmission chain shown in Figure 2.1, brainstorm diseases that are transmitted in that way. How do the different types of transmission offer different opportunities for public health interventions (such as vaccines, quarantines, etc.)?

3. Choose an infectious disease not discussed in the chapter and diagram the triangle of human ecology. What population, behavior, and habitat factors matter for this disease? What buffers to the disease, if any, have humans developed? What enhances our risk?

4. Choose a noninfectious disease not discussed in the chapter and diagram the triangle of human ecology. What population, behavior, and habitat factors matter for this disease? What buffers to the disease, if any, have humans developed? What enhances our risk?

5. Why does cholera continue to circulate endemically in Bangladesh? Cholera was introduced to Haiti by Nepalese soldiers acting on behalf of the UN, and has since circulated in that country as well. What similarities are there between Bangladesh and Haiti?

6. May's silent zone of disease describes a transmission ecology where humans are not present. What newly emergent diseases had a silent zone of disease that was only recently intruded upon by humans, producing novel disease in human populations? What factors brought humans into these silent zones? Thinking historically, what diseases previously cycled unknown to human populations?

7. Given what you have read about malaria in the United States in previous decades, is there reasonable concern about malaria once again becoming endemic as global climate change alters temperature and humidity patterns? What has changed about our behavior and habitat since malaria was endemic?

8. Visit the Centers for Disease Control (CDC) Lyme Disease page and examine their prevalence maps from 2001 onward (*www.cdc.gov/lyme/stats/index.html*). How did Lyme disease spread in the United States? What ecological conditions are conducive to the transmission of Lyme disease? Do you think the ecology of Lyme disease differs between the Northeast and the Midwest? Where do you think Lyme disease will next spread in the United States, if at all?

REFERENCES

Amato, J. A. (2000). *Dust.* Berkeley & Los Angeles: University of California Press.

Audy, J. R. (1971). Measurement and diagnosis of health. In P. Shepard & D. McKinley (Eds.), *Environmental: Essays on the planet as a home* (pp. 140–162). Boston: Houghton Mifflin.

Carrel, M., Emch, M., Streatfield, P. K., & Yunus, M. (2009). Spatio-temporal clustering of cholera: The impact of flood control in Matlab, Bangladesh, 1983–2003. *Health and Place, 15*(3), 741–752.

Carrel, M., Voss, P., Streatfield, P. K., Yunus, M., & Emch, M. (2010). Protection from annual flooding is correlated with increased cholera prevalence in Bangladesh: A zero-inflated regression analysis. *Environmental Health, 9*(13).

Carter, H. R. (1919). The malaria problem of the South. *Public Health Reports, 34,* 1927–1935.

Dubos, R. (1965). *Man adapting.* New Haven, CT: Yale University Press.

Emch, M. (1999). Diarrheal disease risk in Matlab, Bangladesh. *Social Science and Medicine, 49,* 519–530.

Emch, M., Feldacker, C., Islam, M. S., & Ali, M. (2008). Seasonality of cholera from 1974 to 2005: A review of global patterns. *International Journal of Health Geographics, 7*(31), 1–33.

Emch, M., Feldacker, C., Yunus, M., Streatfield, P. K., Thiem, V. D., Canh, D. G., et al.

(2008). Local environmental drivers of cholera in Bangladesh and Vietnam. *American Journal of Tropical Medicine and Hygiene, 78*(5), 823–832.

Evans, R. G., Barer, M. L., & Marmor, T. R. (1994). *Why are some people healthy and others not?: The determinants of health of populations.* New York: Aldine de Gruyter.

Farid, M. A. (1980). Round table: The malaria programme—from euphoria to anarchy. *World Health Forum, 1,* 8–33.

Guerra, M., Walker, E., Jones, C., Paskewitz, S., Cortinas, M. R., Stancil, A., et al. (2002). Predicting the risk of Lyme disease: Habitat suitability for *Ixodes scapularis* in the north central United States. *Emerging Infectious Diseases, 8*(3), 289–298.

Herrington, J. E. (1995). An update on Lyme disease. *Health and Environmental Digest, 9,* 29–32.

Hunter, J. M. (1973). Geophagy in Africa and in the United States. *Geographical Review, 63,* 170–195.

Hunter, J. M. (1980). Strategies for the control of river blindness. In M. S. Meade (Ed.), *Conceptual and methodological issues in medical geography* (pp. 38–76). Chapel Hill: University of North Carolina, Department of Geography.

Islam, M. S., Drasar, B. S., & Bradley, D. J. (1990). Survival of toxigenic *Vibrio cholerae* O1 with a common duckweed, *Lemna minor,* in artificial aquatic ecosystems. *Transactions of the Royal Society of Tropical Medicine and Hygiene, 84,* 422–424.

Jackson, L. E., Hilborn, E. D., & Thomas, J. C. (2006). Toward landscape design guidelines for reducing Lyme disease risk. *International Journal of Epidemiology, 35,* 315–322.

Jusatz, H. J. (Ed.). (1968–1980). *Medizinische landerkunde* (Vol. 16; Geomedical Monograph Series). Berlin, Germany: Springer-Verlag.

Koelle, K., & Pascual, M. (2004). Disentangling extrinsic from intrinsic factors in disease dynamics: A nonlinear time series approach with an application to cholera. *American Naturalist, 163,* 901–913.

Marmot, M. G. (1986). Social inequalities in mortality: The social environment. In R. G. Wilkinson (Ed.), *Class and health: Research and longitudinal data* (pp. 21–33). London: Tavistock.

Marmot, M. G., Kogevinas, M., & Elston, M. A. (1987). Social/economic status and disease. *Annual Review of Public Health, 8,* 111–135.

May, J. M. (1958). *The ecology of human disease.* New York: MD Publications.

McKeown, T. (1988). *The origins of human disease.* Oxford, UK: Blackwell.

Meade, M. S. (1977). Medical geography as human ecology: The dimension of population movement. *Geographical Review, 67,* 379–393.

Pavlovsky, E. N. (1966). *The natural nidality of transmissible disease* (N. D. Levine, Ed. & Trans.). Urbana: University of Illinois Press.

Rodenwaldt, E., & Jusatz, H. J. (Eds.). (1952–1961). *Welt-Seuchen atlas.* Hamburg, Germany: Falk.

Simoons, F. J. (1969). Primary adult lactose intake and the milking habit: A problem in biological and cultural interrelations. *American Journal of Digestive Diseases, 14,* 819–836.

Simoons, F. J. (1970). Primary adult lactose intolerance and the milking habit: Part 2. A cultural historical hypothesis. *American Journal of Digestive Diseases, 15,* 695–710.

Ward, A. (2007). *The cultural ecology of asthma in Orange County, North Carolina.* Unpublished master's thesis, University of North Carolina at Chapel Hill.

Wellmer, H., & Jusatz, H. J. (1981). Geoecological analysis of the spread of tick-borne encephalitis in Central Europe. *Social Science and Medicine, 5D,* 159–162.

FURTHER READING ●

Abrahams, P. W., & Parsons, J. A. (1996). Geophagy in the tropics: A literature review. *Geographical Journal, 162*, 63–72.

Cockburn, A. (1967). *Infectious diseases: Their evolution and eradication.* Springfield, IL: Charles C Thomas.

Colwell, R. R., & Huq, A. (1994). Environmental reservoir of *Vibrio cholerae*—the causative agent of cholera. *Annals of the New York Academy of Sciences, 740*, 44–54.

Crosby, A. (1972). *The Columbian exchange: Biological consequences of 1492.* Westport, CT: Greenwood Press.

Diamond, J. (1997). *Guns, germs, and steel.* New York: Norton.

Dubos, R. (1987a). *Mirage of health: Utopias, progress, and biological change.* New Brunswick, NJ: Rutgers University Press. (Original work published 1959)

Dubos, R. (1987b). *The white plague: Tuberculosis, man, and society.* New Brunswick, NJ: Rutgers University Press.

Ewald, P. W. (2000). *Plague time: The new germ theory of disease.* New York: Free Press.

Farmer, P. (1997). Social scientists and the "new tuberculosis." *Social Science and Medicine, 44*(3), 347–358.

Farmer, P. (1999). *Infections and inequalities: The modern plagues.* Berkeley & Los Angeles: University of California Press.

Howe, G. M. (1972). *Man, environment, and disease in Britain.* New York: Barnes & Noble.

Learmonth, A. (1988). *Disease ecology.* Oxford, UK: Blackwell.

Logan, M. H., & Hunt, E. E., Jr. (Eds.). (1978). *Health and the human condition: Perspectives on medical anthropology.* North Scituate, MA: Duxbury Press.

Matossian, M. K. (1989). *Poisons of the past: Molds, epidemics, and history.* New Haven, CT: Yale University Press.

May, J. M. (1958). *Studies in disease ecology.* New York: Hafner.

McNeill, W. H. (1976). *Plagues and peoples.* New York: Doubleday.

Meade, M. S. (1976). Land development and human health in west Malaysia. *Annals of the Association of American Geographers, 66*, 428–439.

Prothero, R. M. (1961). Population movements and problems of malaria eradication in Africa. *Bulletin of the World Health Organization, 24*, 405–425.

Schofield, C. J., Briceno-Leon, R., Kolstrup, N., Webb, D. J. T., & White, G. B. (1990). The role of house design in limiting vector-borne disease. In S. Cairncross, J. E. Hardoy, & D. Satterthwaite (Eds.), *The poor die young* (pp. 189–212). London: Earthscan.

CHAPTER 3

Expanding Disease Ecology

Politics, Economics, and Gender

Political ecology examines the intersection of political, economic, social, and environmental systems that shape health across spatial and temporal scales. It explores the political and economic realities surrounding decisions to transform the natural environment by examining the links between actors that occur across multiple scales (King, 2010). Having a theoretical framework for how social and economic power create and structure the processes of risk and the expressed conditions of health and disease is important. Geography plays a role because many political and economic processes operate at a national or international scale, while the health and environmental ramifications are felt at a subnational, village, and individual scale. Understanding the complex interrelationship between these forces requires a unique geographic understanding of the problem.

The purpose of this chapter is to expand the theory of disease ecology laid out in Chapter 2 to explicitly include upstream political and economic forces that change these ecologies and create patterns of unequal mortality at multiple spatial scales. The chapter begins with a discussion of how political forces and economic interests helped shape the Ebola epidemic in West Africa. Although the majority of political ecologies have been written about infectious and communicable diseases, such as Ebola, chronic diseases have a political ecology too. Much of the remainder of this chapter looks at how age, gender, race, poverty, and class, interacting at different spatial scales and influenced by political/economic forces, give pattern to the outcome of individual disease ecologies. We provide a quick examination of epidemiological study design and methodology in order to broaden the reader's understanding of how we can study such complex interrelated processes that produce health outcomes. With reasoning clarified, the ecologies of cancers and

HIV/AIDS are discussed, and several studies of these diseases with different spatial methodologies are examined. In this chapter, we wish to stress that all diseases have an ecology, and political and economic forces are only one (very important) aspect of this ecology. As such, we present the whole ecology of the disease, making sure to point out the political and economic ties that affect the spread and distribution of the disease. We do not, however, neglect the many factors discussed in Chapter 2 in our treatment of HIV and cancer. A discussion of the *precautionary principle* and its relevance to putting environmental factors back into studies of disease ecology ends this chapter.

POLITICAL ECOLOGY

When Ebola was first reported in West Africa in March 2014, no one could have predicted how massive the outbreak would be. Small outbreaks of the disease had occurred in the region since the mid-1970s, but no single outbreak had resulted in more than 500 human cases. By April 2016, the widespread epidemic had been controlled, though a few minor flare-ups were still being reported across the region. All told, nearly 29,000 human cases were reported in Guinea, Sierra Leone, and Liberia. Ebola has a complicated disease ecology, but the magnitude of the 2014 outbreak was deeply dependent on larger political and economic forces that set the conditions for both the initial introduction of the disease into the human population and the inability of world health authorities to quickly bring the outbreak under control.

The latest emergence of the virus has been linked to forest encroachment in isolated Guinean communities, which placed people in contact with several species of fruit bats thought to be the carriers of the disease (see Chapter 7 for more on this topic). But this encroachment is driven by land use shifts connected to globalizing food production in the region, in particular deforestation to produce soybean and maize crops and, more recently, the commodification of palm oil (Bausch & Schwarz, 2014; Wallace et al., 2014). This expansion of the agricultural frontier and shift in food production has been largely supported by the state and financed by the European Investment Bank and private international agribusiness companies (one major player is based in Nevada in the United States). Throughout the process, many traditional farmers lost their land and forests have been turned from shared resources to privately owned commodities. In the midst of these changes, enter Ebola. Some researchers have suggested that the fruit bats are attracted to oil palm plantations for food and shelter (Shafie, Mhod Sah, Abdul Latip, Azman, & Khairudden, 2011). In addition, villagers hunt and butcher bush meat (meat taken from wild animals), perhaps more so now due to the loss of traditional farmland and increasing poverty, which places them at greater risk for contracting Ebola directly from bats. Once it had jumped into people, Ebola then traveled out of the forests as individuals circulated from rural villages to small towns to large cities, and swept into the urban hierarchy (see Chapter 6). Whatever the mechanisms, the unprecedented spread of the disease was largely due to inadequate rural health facilities, poor village sanitation, and the slow response of the WHO, which had limited economic resources to mount a large-scale response. Investigations by several major news outlets also revealed that political forces clouded the WHO's willingness to declare an international health emergency; the health agency was worried about provoking conflict in the affected region and the economic impacts of such a declaration, and were concerned about interfering with the Muslim pilgrimage to Mecca. It wasn't until Ebola threatened countries like the United States that the magnitude of the problem became widely known and the international community finally took action. But the response was also slowed by small-scale social and cultural structures. Guineans were frightened by and distrusted the health workers, a legacy of colonial medicine and negative experiences with inept bureaucracies. Many tried to hide sick family members and refused to stop traditional mourning and burial practices.

Ebola is an excellent case for demonstrating the role of social process in shaping human health. Health is structured by political and economic systems that influence the transmission of the disease and the ability of health care agencies to effectively respond. Many geographers have examined the tensions between **structure** and **human agency**. In this framework, structures are the social, political, and economic contexts that constrain (or enable) an individual's ability to act on his or her own choices. The ability to act in any given environment is referred to as "human agency." There are many forms of structures that constrain and enable behaviors, and these operate at many different geographic scales. At the highest level, global organizations such as the WHO, UNICEF, the World Bank, and the International Monetary Fund influence political and economic decisions made by the political leaders of countries. Even well-meaning nongovernmental organizations (NGOs) often dictate how the resources they provide must be used within a country, and their priorities often do not address the most pressing health problems. Countries govern citizens and make political and economic decisions that directly affect land rights, the crops individuals can grow, whether children can go to school, or how people receive health care. Finally, social and cultural norms also constrain human behaviors and choices. All of these forms of structure affect health. All of these played out in the 2014 Ebola outbreak.

The Ebola outbreak is also an excellent example of how political and economic structures influence human interactions with the environment. In the discussion above, individuals are no longer at the center of the outbreak. Rather we see how larger political decisions, which were supported by a variety of economic interests both national and international, led to changes in land use and movement of people to support new economic endeavors on the frontier. We have expanded disease ecology to include the political economy, which results in a political ecology of Ebola. Brookfield (2005) explains, in his review of Robbins's (2004) introduction to political ecology, how political ecology developed separately from cultural ecology because of the "inability of apolitical forms of explanation to answer critical questions concerning why landscapes are degraded, constructed, or conserved" (p. 286). The development of ideas in political ecology connected with peasant studies, feminist development studies, and others, but not the field-generated cultural ecology studies in anthropology and geography described in the previous chapter.

Neo-Marxist influences predominated in the early writings of political ecology, Brookfield notes, but a pluralism of thought has since developed and class analysis need not be its central issue. Political ecology is itself a large subject about many aspects of geography. Its relevance to health is more clearly presented by some studies focusing on such topics as household ecologies of disease transmission and others focusing on the differential distribution of risk of disease (Inhorn & Brown, 1997).

Mayer identified the political ecology of disease as an important new focus for health geography, arguing that political dimensions of health and disease, while not absent from the field, were not a major focus (Mayer, 1996). Political dimensions were part of the field early on in the work of Hughes and Hunter (1970) on negative health impacts of development projects, Turshen's (1977, 1984) works on the relationships between health and colonial history and unequal power relations in Tanzania, Rosenberg's (1988) work on political dimensions of health care delivery in Canada, and more recently Kalipeni and Oppong's (1998) work on political ecological dimensions of refugee crises in Africa. The more than 60-year tradition of breadth and integration continues today with a growing focus on the political ecology of health (King & Crews, 2013; King, 2010, 2015) that will ultimately make the subfield of health geography stronger. Geography and anthropology again come together in the study of these ecologies, but anthropologist/physician Paul Farmer (1992, 1997, 1999, 2003) has shaken the public health establishment with his political ecologies. As Farmer points out, globalization of economic and political power is increasing gaps in income, knowledge, and health everywhere, at every scale. In China, astoundingly, more than half the population has been lifted out of grinding poverty. Those who remain rural and poor, however, are more economically, socially, and politically marginalized; are less well nourished; are exposed to more dangerous chemicals; and have less access to good health care than before. Those who are now living and working in the industrial cities are exposed to some of the world's worst air pollution from smokestacks and exponentially growing numbers of automobiles, as well as socially promoted behaviors like smoking tobacco. The health consequences of all this, such as asthma and cancer, are already increasing.

Political ecology is concerned with understanding these higher-level or "upstream" political causes of health. However, the process of ecological change stemming from higher-level political causation is not limited to the effects of capitalism or globalization. The policies of Soviet communism created industrially poisoned forests and lakes across Eastern Europe. Centuries earlier, feudal agriculture responded to population growth with continental deforestation, erosion, filthy crowded cities, malnutrition, and other conditions that facilitated bubonic plague to remove a third of the population. Millennia earlier the domestication of plants and animals and the permanent settlements this allowed created the ecological conditions that created most of our contagious diseases. While the field of political ecology may be only a few decades old, the phenomena it explores have been with humans since the beginning. The "downstream" questions of cultural ecology are about what, exactly: What changed ecologically, from what behaviors, altering what feedback and regulation systems, intensifying what, exposing people to what local alterations that affect their lives and to which they must adapt? What information is needed? Where can interventions be made? How can those conditions that produce disease be diminished, and those that improve the population's state of health be promoted? All social scientists concerned with health agree that poverty is the major, primordial "cause of premature death," and that alleviating it "requires that society takes transcendent [policy] actions" (McMichael, 2001, p. 335). As McMichael (2001, pp. 265–267) further explains, two dimensions of complexity about why "this population, at this time" has a health problem involve (1) "deciding the level and type of analysis of disease causation," balancing the upstream and downstream causes; and (2) "envisioning the evolution of health risks over time, over the course of lifetimes."

THE POVERTY SYNDROME

Most diseases occur consistently more often among the poor than among the affluent. There seems to be a **sociopathological complex**, made up in large part of stress, lifestyle, poor diet, bad housing (with old paint and pipes), and polluted air. In urban areas across wealthy nations there are similar patterns of physical and mental disorders. Some diseases, such as tuberculosis, are traditionally associated with the poor. Others, such as breast cancer, have more often afflicted the affluent. And some, such as lung cancer, have recently decreased at high socioeconomic status (SES) and increased at lower SES.

Many of the associations of health problems with poverty occur throughout the world. While noting the major improvements in health in the latter part of the 20th century, the World Health Organization's (WHO's) Commission on Social Determinants of Health (WHO, 2007) concluded that country averages conceal persistent and significant differences according to "social status" (defined as consisting of education, gender, and occupational hierarchy). As Harpham, Lusty, and Vaughan (1988) modeled them, a wide range of the diseases of the poor have universal associations with causal factors, characterized as direct, environmental, or psychosocial. Their direct factor included unemployment, low income, limited education, inadequate diet, and prostitution. Their environmental factor was composed of water quality, sanitation, overcrowding, poor housing, rubbish accumulation, lack of garden land, traffic, industrial hazards, pollution, and accidents. Their psychosocial factor included stress, alienation, insecurity, depression, smoking, alcoholism, drugs, and abandoned children. Harpham et al. found these characteristic associations occurring from the slums of Bombay (Mumbai) to those of London.

There is a higher-order causality to these associations, a political economy, which in geography includes study of the relations of societal process with spatial form. Usually in the United States such studies address the inner city; rural poverty is rarely addressed. Dear and Wolch (1987) have described how "landscapes of despair" were formed when mentally ill people and others were "deinstitutionalized" in the late 1970s and released into unprepared and largely unfunded communities for service and support. At the turn of the last century, human services policy was spatially expressed in the location of public asylums in major cities. As the cities grew into industrial centers, the growth of downtown areas and suburbanization of people working there led to a clear identity of central business districts and their fringe zones as areas that housed noninstitutionalized, service-dependent populations (e.g., blocks of single-room-occupancy hotels in which elderly, poor, alcoholic, transient, and other populations congregated). The general philosophy of treatment and care of such people had become a "specifically urban phenomenon" across North America, a social construction of "the service-dependent ghetto." In the United States in the early 1980s, community building, housing subsidy, and welfare funds were substantially reduced. This was part of a larger conservative political movement in the country that supported a reduction in most government services. People who were deinstitutionalized and did not return home to supportive families, churches, or other social groups gravitated toward "inner-city landscapes where increasing demands for assistance were met by diminished capacity to supply both shelter and services" (Dear & Wolch, 1987, p. 199). Over 20 years later, new issues for mental health are demanding attention: the impact of sprawl, the emergence of an independent scatter-site housing model for poor urban households, and in general the impact of place on community integration and trust (Yanos, 2007).

Homelessness, defined as the lack of basic shelter, is a condition affected by economic, social, and political forces that get localized in specific places. People commonly become homeless because they lose a job and thereby the ability to pay rent. They may also flee physical abuse; lose a supportive relative; become incapacitated by substance abuse; or become susceptible for many other reasons. Some people are homeless for only short periods; some cyclically find and lose jobs/apartments/families; some become chronically homeless. Wolch and Dear (1993) describe the following dualism that results in poverty, with overwhelming health risks around the world. On the one hand, there is rising economic marginality (of persons or groups) that results from economic restructuring, welfare state restructuring, and demographic change. On the other hand, there is a decrease in affordable shelter because of national housing policy, urban housing markets, and loss of affordable units. The U.S. financial crisis and housing foreclosures that started in the fall of 2008 overwhelmed shelters (both human and pet) even before bitter cold set in.

Housing itself is probably the easiest part of the sociopathological complex to understand. The poor are often overcrowded. Older residences have frequently been subdivided into apartments for numerous families and individuals. Larger families, whether large because of deliberate fertility decisions or because of lack of access to contraception, mean that small children add to the high-room-density measure frequently associated with influenza, bronchitis, and tuberculosis. Old pipes may have lead solder and occasionally are themselves composed of lead; they tend to be made of metals whose cadmium and other trace elements are easily eroded. Although lead is no longer allowed in house paint in the United States, older buildings still have lead paint dust (for more on lead, see Chapter 11).

Usually considered a northern inner-city problem, in the rapidly growing South lead may be more of a rural risk (Hanchette, 1998). Figure 3.1 shows Hanchette's spatial modeling of the risk of lead for children. In her research she identified the

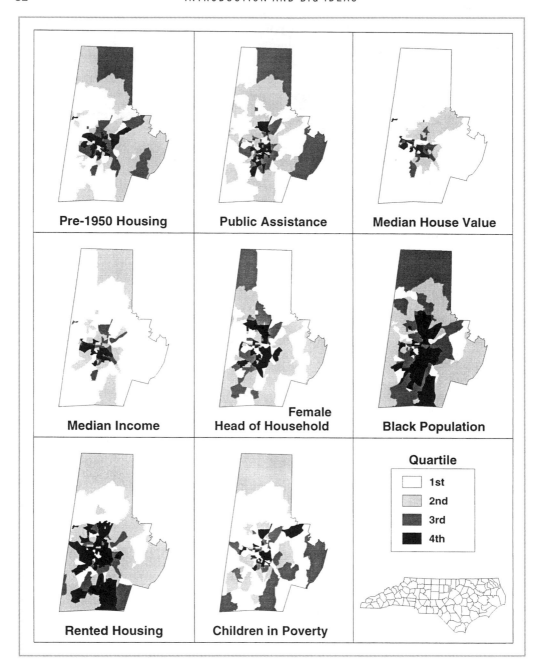

FIGURE 3.1 Lead poisoning risk and risk factors for Durham County block groups, classified by quartiles, and the predicted lead poisoning risk with overlay of lead poisoning incidents. From Hanchette (1998). Used by permission.

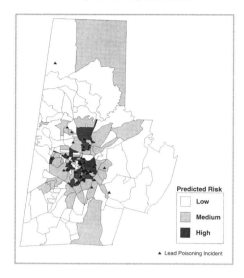

FIGURE 3.1 *(continued)*

census variables that had been repeatedly statistically associated with high blood lead levels in schoolchildren. They were somewhat redundant: low income, rental housing, having a single mother, being black, neighborhood percentage of children in poverty, low median household value, public assistance (welfare), and housing older than 1950. In what is known (from physical geography) as **suitability mapping**, she mapped each of the variables in quartiles and used a GIS to overlay them, so that each census tract could be classified as low, medium, or high risk according to the number of highest-quartile risk factors in it. The known incidents of high blood lead were then overlaid on the map of risk factors. All but a few incidents occurred in the highest-risk blocks; those few occurred in children who were going to school outside of their neighborhoods.

In mapping Durham, Raleigh, and other North Carolina cities, however, Hanchette came to realize how little pre-1950 housing (the main "paint chip" risk nationally) was present in North Carolina. Most of the in-migration, and the great increase in population, has taken place in the last 30 years in the cities. When she broadened her scope, Hanchette realized that the old housing with lead risk was in small towns, especially "down East" on the coastal plain. When the state looked there, North Carolina found high blood lead levels in black children living in the old, often historic houses of the small towns on the coast, where slavery had been strongest as an institution. Pursuing the political ecology, Hanchette (2008) found that in the 1960s as schools were integrated, whites fled to new suburbs and blacks moved into the old historic housing. Poverty prevented the painting, renovation, and other needed upkeep, so that black children came to be exposed to paint chips and lead.

Exposed asbestos is a more recently recognized hazard that is still being removed from old buildings as each is renovated. Central heating systems are frequently obsolete or in poor repair and, besides failing to counteract drafts, contribute to mold, dust, and other allergens. A diet high in starchy foods, sugar, and saturated fat seems in international comparisons to contribute to a greater tendency toward diabetes and coronary heart disease. Obesity, a major risk factor for both, is even more frequent among black females than among other groups composing the U.S. population (see the North Carolina mortality rates in Table 3.1) and more frequent among poor whites than among rich ones. In the past decade, however, it has been increasing in all population groups, including the college-educated. The role of diet is closely tied to lifestyle factors. When people in the United States started jogging, biking, and aerobics for health reasons, these cultural practices, like most others, diffused down the socioeconomic scale from people with higher levels of access to information to those more tied to traditional ways. In Figure 3.2, look at the disproportionately higher contribution of cardiovascular disease to mortality among poor rural southern whites, poor inner-city urban blacks, and poor rural southern blacks. A new definition of a healthy diet seemed to appear among the educated just when dietary behavioral patterns once associated with the well-off became widely practiced among the poorly educated. There are also socioeconomic impediments to lifestyle change. Health spas, gyms, and safe jogging paths are not readily available to the poor. Night shifts, double jobs, and lack of child care impede the uptake of these health behaviors. Crime and fear of it can lead to deprivation of social life, especially for the elderly (see the discussion of heat waves in Chapter 12). Stress also results from anxiety over jobs, fear of inscrutable institutional policies, insecurity over Social Security or Aid to Dependent Children payments and medical care/insurance coverage, and hassles with landlords.

TABLE 3.1. North Carolina's Leading Causes of Death					
Cause[a]	Deaths	Rank	Whites' AADR	Rank	Blacks' AADR
Heart disease	75,969	1	240.1	1	295.9
Cancer	63,709	2	194.8	2	245.7
Female breast[b]		1	148.9	2	130.6
Prostate[b]		2	135.4	1	216.8
Lung[b]		3	71.2	3	65.2
Colon/rectal[b]		4	48.1	4	52.6
Stroke	21,921	3	68.3	3	96.5
Chronic lung disease	14,402	4	50.2	6	31.8
Pneumonia/influenza	7,481	5	25.2	7	25.2
Diabetes	8,510	6	21.5	4	55.6
Motor vehicle injury	6,404	7	19.2	8	20.7
Kidney disease	5,204	8	13.4	5	34.9

Note. AADR, age-adjusted death rates.
[a]Data from North Carolina Department of Health and Human Service (2004, Table 5), except as noted.
[b]Data from North Carolina Department of Health and Human Service (2004, Table 3).

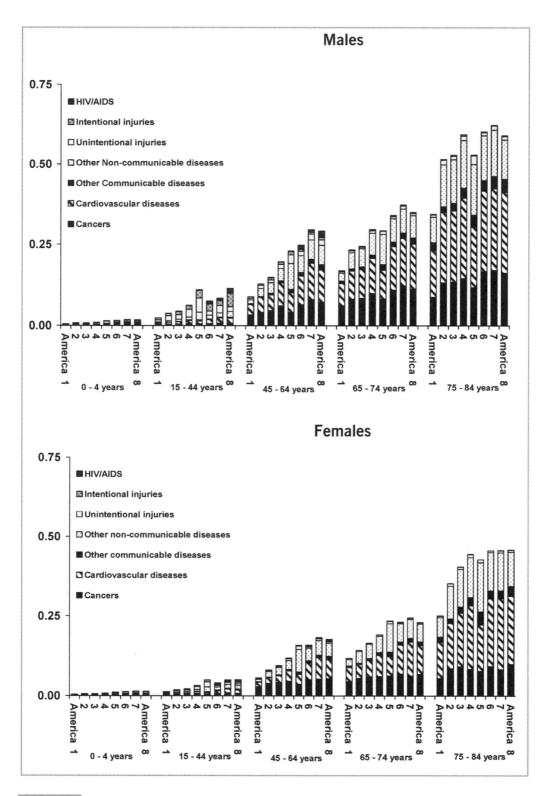

FIGURE 3.2 U.S. probability of dying by age, sex, and disease group. From Murray et al. (2006). Reprinted by courtesy of *PloS Medicine*.

One way of viewing the whole sociopathological complex is in terms of the accumulation of insults to health (Chapter 2). Some people with high levels of health can cope with the insults and emerge stronger and more creative for the struggle. Others, their health depressed by psychological and social insults, are unable to rally from infectious, chemical, or further mental insults. They become part of the small percentage of the population that accounts for a large percentage of illness and repeated health care needs.

RACE IN THE STUDY OF HEALTH RISKS

As a genetic classification system, race has limited relevance to disease etiology and today is not found to be biologically accurate (see Chapter 2). However, a few genetically caused diseases, such as sickle cell anemia or Huntington's chorea, do seem to be linked with specific racial groups. In some other diseases, such as "essential hypertension" in blacks, a special susceptibility of one group or another seems to be involved. In still others, such as prostate cancer, genetic susceptibility may be part of the etiological puzzle. Too often, however, automatic separation into racial categories masks a superficial socioeconomic or class-based, analysis that could be more etiologically significant.

African Americans are a heterogeneous population, and they constitute a minority (less than 13%) of the U.S. population. This means that although a higher *proportion* of blacks live in poverty, receive welfare, are single mothers, are unemployed teenagers, are high school dropouts, and so forth, substantially larger *numbers* of people who are in these categories are white. Millions of African Americans are college-educated professionals leading the lifestyles of the affluent. However, if even college-educated black women have lower-weight babies, is there a role of genetics acting separately from the social process of lower access to adequate prenatal care? Race is indubitably important as a social construction, a category with ascribed characteristics and meaning. If race is taken as a social construction that affects the behavior both of people within the group and of those outside toward it, is there another kind of cause? The work of Evans, Barer, and Marmor (1994) and those who have followed them, for example, strongly indicates that it is not absolute deprivation but relative status that affects health.

In the United States a higher proportion of minority group members are poor, but the majority of poor people are white. The main groups of Americans living in poverty are single mothers, the mentally and physically disabled, the aged, and children. There is a much smaller group of those afflicted with substance abuse. There has been a long-term trend of deterioration in the relative health circumstances of the poor versus the rich, the uneducated versus the educated, and blacks versus whites, expressed as increasing socioeconomic inequalities in mortality (Figure 3.2). Demographic projections make it obvious that there will be future increases in the proportions of old persons, disabled people, minority group members (not just blacks), and probably single mothers (not just never married, but divorced, separated, or widowed) living in poverty, as well as in the proportion of children doing so. The structure and pattern of statistical relations are complicated by their strong spatial variation, with concentrations differing among states, counties within states, and neighborhoods within cities and counties. Scales of analysis addressing "population" without place can easily miss significant associations or opportunities for intervention.

The well-recognized inverse relationship of SES and mortality has many underlying factors: risk from environment or behavior; lack of preventative public health and other health services, which would include early detection, aggressive treatment, social support, and access to care overall. Besides genetic susceptibilities and socioeconomic associations, there are cultural differences in such things as perception of symptoms, compliance with medical regime, and family networking. Not only patients but also their health care providers can be influenced by belief systems and perceptions in ways that affect early detection, aggressive treatment, or simply recognition of problems. Separating the racial, socioeconomic, cultural, and institutional/policy components of the mortality differential is not simple. Nevertheless, as Ezzati and colleagues (2008) found, an increase in mortality inequality across U.S. counties from 1983 to 1999 resulted from an increase in mortality among the worst-off segment of the population.

The differences in this area are not all socioeconomic or racial. As Figure 3.2 shows, whites in the rural northland and in Appalachia/the Mississippi Delta had the same race and comparable incomes, but divergent mortality rates. An earlier study (Geronimus, Bound, Waidmann, Hillemeier, & Burns, 1996) featured in the second edition of this text used standardized mortality ratios to compare the poorest and nonpoor areas and towns (e.g., Harlem vs. Queens, Watts vs. Los Angeles, central vs. northwest Detroit, and "black belt" Alabama vs. northern Alabama for blacks, and the Lower East Side of New York City vs. Queens, Detroit vs. Sterling Heights, northern vs. southwest Alabama, and Appalachia vs. western Kentucky for whites) from 1989 to 1991 and found shocking dimensions of inequality: The probability of a 15-year-old black male in Harlem surviving (in 1990) to age 65 was only 37%, whereas that of a white female in Sterling Heights, near Detroit, was 93%. For both sexes the death rate in each poverty area was excessive. But, as the study authors noted, the "black belt" in Alabama had the highest rate of poverty but the lowest excess mortality; Harlem had the lowest rate of poverty but the highest excess mortality. The social, economic, cultural, racial, and regional geography of mortality in the United States is *very* complex.

GENDER AND SEX: WOMEN'S HEALTH

Male mortality is higher for all racial and ethnic groups and at every age than female mortality. Human ecology recognizes that the universal human sex ratio at birth (104–105 males for every 100 females) probably has evolved to counter the higher youthful male mortality and so to approach an equal sex ratio by the reproductive years. In earlier times, this was countered later in the life course by the high female mortality in pregnancy and childbirth. With the low fertility rates described (in Chapter 4) for the end of the demographic transition, women have been largely relieved of that stress and now have longer life expectancies than men, so much so that the sex ratios (males per 100 females) in nursing homes can approach 30 for populations over 80. The higher mortality rates of males have also, through numerous studies, been associated with greater resistance to seeing a doctor and admitting sickness or weakness. Women in turn have been described as having higher morbidity but less mortality because they do readily seek medical attention. In fact, in one public health survey after another, women rate their health worse than men from adolescence to late middle age, and they do visit doctors and even hospitals more often; yet at every age, women have lower mortality than men.

Women's health has become a focus of interest because in the past it was buried in the totality of health and disease, so that little was learned about it. Sometimes *men's* health was considered synonymous with *human* health. Clinical trials of the efficacy of pharmaceutical drugs were conducted only on men but assumed to be valid for all. Sometimes studies of causation, prevention, and treatment of a major disease, such as heart disease, were focused on men because they had higher rates of that disease and the economy suffered greater loss as a result. Although such studies were cost-effective and the results were clearer and stronger precisely because the variability and confounding posed by including women's differences were avoided, nevertheless the results were considered applicable to the general population across all differences. Women are both the same as and different from men. Differences in health or treatment of disease are not simply due to money, power, and social construction of interpretation (gender), but also due to sex (biology). Examples involving women's health have been used throughout this text. This section focuses on how the concepts and approaches of this text are all relevant to women through the integrative framework of the triangle of health ecology (Figure 2.2), with its three vertices of (and interactions among) population, habitat, and behavior.

Population

Sex, genetics, age, immunological status, and nutritional status constitute the basic vertex of population. The genetics of being female result in a few sex-specific diseases, especially those of the reproductive organs. Female hormones are genetically given, of course, as well as the elaborately orchestrated capacity for pregnancy and lactation. Several genetic conditions are sex-linked at the chromosome level. For example, genes that cause a serious condition (hemophilia) and those that cause a mild condition (color blindness) both occur predominantly in males because their Y chromosome contains no counter gene to repress such recessive characteristics when they are carried on the longer X chromosome. The full expression of female, however, is gradually revealed through the life cycle.

Historically, as noted above, women died at high rates from the rigors and cumulative risks of pregnancy and childbirth. The recent reduction of fertility has done a great deal to spare their lives. When, as in India and China, young girls die at a higher rate than boys, it is usually taken as prima facie evidence of severe discrimination in nourishment, nurturance, and care—discrimination that on occasion even results in female infanticide.

Between menarche and menopause, women are biologically vulnerable to a range of insults associated with their reproductive role. These include such things as infection of the kidneys/urinary tract, scarring, and infertility resulting from sexually transmitted disease; nutritional shortages of iron, vitamin A, or calcium; and so on (Koblinsky, Timyan, & Gay, 1993). Besides blood lost to menstruation, malaria, and intestinal parasites, nutritional deficiencies are made worse by pregnancy, and greatly exacerbated by multiple ones. More than half the pregnant women in tropical Africa and Asia are anemic (which, in addition to the usual energy and health complications, can result in oxygen crisis for a mother and child at birth). Advanced vitamin A deficiency is usually monitored as xerophthalmia, but in women depleted by recurrent pregnancy, lower levels of deficiency may pass unnoticed while they damage the immune system. Even in developed countries, despite good nutrition and prenatal care, most embryos do not survive to become fetuses. Nevertheless, in the developing world, crude and septic induced abortions are a significant cause of maternal mortality for women without means of birth control.

The high rates of maternal mortality evident in Table 3.2 illustrate the heavy burden that such mortality still exacts. As described in more detail in Chapter 4, the summary measures of **disability-adjusted life years** (DALYs) is used by the WHO to best estimate and portray the burden of disease. DALYs for a disease or health condition are calculated as the sum of the years of life lost (YLL) due to premature mortality in the population and the years lost due to disability (YLD) for people living with a chronic health condition or its consequences (e.g., a decrease in the quality of life or limit to functioning). DALYs place more emphasis on some illnesses, particularly those that impact people earlier in life or during their prime working years, compared to those that impact the elderly. In this case, DALYs are also reminders of the large number of children rendered motherless by the lack of birth control. The number of women who die in childbed is presented as a rate per 100,000 births: for Nigeria, the estimate is 560 per 100,000 births, for Japan it is just six. The WHO often reports standardized DALY rates, which are a weighted average of the age-specific DALY rates per 1,000 persons, where the weights are the proportions of persons in each age group of the WHO standard population. The table portrays the 10 largest countries in the world, in various stages of the demographic transition.

Cyclic exposure to estrogen seems to affect cholesterol metabolism and act protectively against heart disease. On the other hand, cumulative estrogen exposure is one of the clearest insults increasing the risk of breast cancer. Hormonal fluctuation can affect mental health, especially postpartum depression, and can be a risk factor for suicide. Female sex hormones are also thought to be the cause, at least partially, of many autoimmune diseases.

In the developed world, two-fifths of a woman's life is lived after menopause. Fertility rates of fewer than two children per couple also mean that most of a woman's life is not engaged in reproduction. Medicine and health research in general have been so preoccupied with women's reproductive processes, reproductive organs, and infections and diseases that little has been learned about their health in other respects. Even clinical trials of drugs were carried out only on men, as noted earlier; one reason for this was the expressed fear that a fetus might be harmed (some would say that this fear was a construction of the value of a woman's life, that childbearing is the worth of a woman). It has recently been learned that women's hearts, bones/calcium metabolism, and patterns of fat deposition are different from men's, and that so are the effects of relevant drugs. Several large longitudinal studies, such as the Nurses' Health Study and the Women's Health Initiative, were undertaken in the late 20th century to research the effects on women of estrogen supplementation, calcium supplementation, various exercise regimens and diets, and common drugs such as aspirin. The Women's Health Initiative study was stopped several years early because of clear evidence that the estrogen supplementation then being recommended by doctors not only for menopausal symptoms, but in order to reduce heart attacks (as presumably the women's own estrogen previously did, since men had higher rates), actually increased the chance of stroke.

There are notable interactions of population with habitat and behavior. Customs about the seclusion of women and definitions of modesty affect diffusion of contagion, exposure to vectored disease, and even biometeorological influences. Rickets in Saudi Arabia, for example, is purely a cultural artifact of purdah, wherein women do not receive enough sunlight to produce vitamin D because they are cloistered in their homes. (On the other side of the world, more open advertisement about new products for menstruation created and spread a new toxic shock syndrome almost overnight.) Cultural beliefs also determine at what age marriage is suitable, and such things as whether the degree of inbreeding to cousins

TABLE 3.2. Impact Measures of Maternal Mortality		
Country	Age standardized DALY per 1,000 population[a]	Maternal deaths per 100,000 births[b]
Bangladesh	388	170
Brazil	316	69
China	248	32
India	479	190
Indonesia	360	190
Japan	157	6
Nigeria	848	560
Pakistan	505	170
Russian Federation	399	24
USA	228	28

Note. DALY, disability-adjusted life years.
[a]Data from WHO Global Health Observatory data repository.
[b]Data from Population Reference Bureau 2015 World Population Data Sheet.

or outcrossing to other ethnic groups is tolerated. Values about desired family size, use of contraception, provision of medical services, accessibility of education to girls, job opportunities, and status/financial contribution from outside the home for women have all changed enormously, as expressed at the 1994 International Population and Development Conference in Cairo and the follow-up International Women's Conference in Beijing.

Increases in income, trade, and agricultural production, as well as the technology of food preservation, storage, shipment, and marketing, have changed nutrition enormously. One consequence in the developed world of continually abundant, high-protein, high-fat food has been the lowering of the age of menarche. It used to be 16 or more rather universally, but now in places it occurs at 11 and even younger (extremes possibly due to pseudoestrogens in food and water today), with consequences for reproductive behavior and attendant risks even as the age of marriage has been culturally raised to more than 20 in much of the world. Probably the most notorious cultural control on nutrition is the widespread custom of the husband's eating first and then the children. When at last the woman eats the leftovers, the most nutritious foods are gone, and adequate calories may not remain. In the developing world, when a mother's milk is changed to a bottled substitute, a baby's antibodies to disease and exposure to diarrheal agents also may change.

Habitat

The natural, built, and social components of human habitat affect women's health in ways already described for people in general, and yet somewhat differently than they do men's. Of course, this is the least studied and least understood aspect of women's health. There is different exposure to insults because of different roles (gender roles based upon sex differences). In many countries, women are responsible for washing clothing and so are exposed to schistosomiasis when men are not. Mobility, jobs, and relationships are among the behavioral components that structure the exposure to habitat conditions.

Radon and other radiation occur naturally in many environments but may be a greater hazard for fetuses. Women suffer more from hay fever and chronic sinusitis caused by natural allergens. The chlorinated hydrocarbons and polychlorinated biphenyls (PCBs) distributed in the environment have an affinity for storage in fatty tissue, which puts women's bodily deposits, especially in breast tissue, especially at risk (Calabrese, 1985).

The greatest risk factor of the built environment for women is simply that they generally still spend much time at home, despite the increase in working women. The exposure to infectious agents of children occurs predominantly at home. Whatever the insults from cleaning agents, formaldehyde, or household allergens, they especially affect women. To the extent that women have different types of factory or office jobs than men do, the insults of the workplace environment would also differ, but this topic has been little studied. The built environment includes the presence or absence of such facilities as health clinics and family planning centers, day care facilities, schools, nursing homes, and hospitals.

The most obvious habitat differences for women are social. Activity spaces, mobility patterns, and isolation versus social contact mediate exposures to such things as air pollution, cigarette smoke, motor vehicle accidents, contagious disease, fresh air, and social support, to name a few.

Behavior

The most complex and yet clearest interactions for women's health fall into the behavioral vertex of beliefs, social organization, and technology. Indeed, most of the approaches to health promotion informed by social theory focus on this vertex alone (Doyal, 1995; Koblinsky, Timyan, & Gay, 1993). A woman's multiple social roles as child nurturer, caregiver, homemaker, wage earner, career professional, community organizer, lifelong learner lie here. Should the woman deal with the infant's toilet needs (and get worms), tend the child's fever (and catch it), deal emotionally with the elder's deterioration and needs, and wash the cadaver (and get cholera)? Has the woman any control over whether to have five children or two or none? Must she cope with full-time job demands, whether law clerking or gutting chickens on a conveyor belt, as well as with domestic chores, parenting, and a connubial relationship? What are these relationships and duties, and how are they determined, and by whom anyway?

Cultures have strong belief systems about everything from what foods are taboo for pregnant women to how thin or plump women should be; from what caretaker role is allowed women to what home remedies mothers should use; from what exercise is appropriate for women to how much domestic violence must be tolerated. Should menstruating women be shut away in separate buildings? Should postpartum women be "roasted" daily in overheated rooms, as is done in East and Southeast Asia, to reduce the womb? Should female genitals be mutilated so that husbands can know their spouses will be pure and uninterested in sex? Should women burn themselves alive on their husbands' funeral pyres, or are widows allowed to remarry, if only to be able to care for their children? Is it appropriate for women to expose themselves to the sun? Is it appropriate for them to smoke cigarettes? The behaviors that result from such cultural beliefs have fundamental implications for health. The beliefs can sometimes, perhaps unexpectedly, be strong cultural buffers against harm.

Social organization that affects women's health can be as personal as customs of polygamy or monogamy sanctioned by religious systems, or as bureaucratic as welfare aid to homeless women with dependent children. Broad social structures and institutions affect women's health: child care services; programs of higher education and of job training; and programs for health care, vaccination, life enrichment for senior centers, and even nursing home pet visitation. Legal structures work laws, minimum wage, safety regulations, and antidiscrimination/affirmative action laws also channel health effects. There is a lifetime of social science research to be done in political economy about such relations.

The complexities of interactions even within just this behavioral vertex are inscrutable when the context of major concerns, such as mental health, is addressed.

Consider simply the dimension of substance abuse. Women have the same range of alcohol metabolism that men do, although their generally lower body weight gives them generally less capacity for absorption. Under the influence of alcohol, women are especially susceptible to rape and to physical abuse.

Technology can involve the full workplace panoply of computers and chemicals, or it can involve the labor-saving gadgets and chemical innovations of homes in developed countries. Medical technology, however, is of special importance. The removal of birth from homes to hospitals and the development of intensive care for premature infants have greatly reduced mortality and transformed the birthing experience. The development of infant formulas to substitute for mothers' milk has removed the mortal crisis of inadequate lactation; has freed infant care from the sole proprietorship of a mother; and has transformed the experience of bonding, as well as that of work. Unfortunately, such formulas, aggressively promoted for profit, have also resulted in infant death from diarrhea and malnutrition where the human resources of adequate money, potable water, fuel, and cleanliness did not comport with safe usage.

Heart disease continues to be a greater cause of both mortality and morbidity among men than women, but after menopause it becomes the major cause of death and the most common life-threatening morbidity for women, too. Numerous other differences between the health status of men and women continue. What component of the differences in arthritis is due to different experience of occupation and what to hormones or diet? Is men's greater hearing impairment due to occupational hazard, or to cultural behavior regarding music, guns, or power motors? What could be the cultural ecology of more hay fever, sinusitis, and bronchitis among women? Or are these differences all due to women's willingness to see a doctor about such problems, while men choose not to?

QUICK REVIEW 1

✓ All diseases have an ecology, and political and economic forces are a very important aspect of this ecology.

✓ There are many social, political, and economic contexts that constrain or enable behaviors related to health. These structures operate at many different geographic scales.

✓ Many political and economic processes operate at a state, national, or international scale and trickle down to affect the health of individuals in cities, villages, and households.

✓ Poverty unequivocally affects health. But poverty is caused by higher-order political and economic forces, working in conjunction with social, cultural, and environmental contexts. This also produces a spatial distribution of poverty-related health outcomes.

✓ Though race has limited relevance to disease etiology, it is highly correlated with poverty and other social, cultural, political, and historical processes that affect health and mortality.

✓ Women experience differences in treatment and treatment options due to power relations formed through money, social roles, cultural norms and restrictions, and social constructions around gender.

CAUSAL REASONING AND EPIDEMIOLOGICAL DESIGN

It is difficult to study the differences among ethnic groups and between races. It is even harder to identify causal exposures to the environment, social, political, and economic relationships. In particular, higher-level, upstream factors, such as policies and economic conditions, are difficult to integrate into causal studies of health and disease (this topic is discussed in more detail in Chapter 9). Age standardization of rates, and often sex or race standardization as well, are usually the first steps for comparison and pattern discernment among differences in risk and exposure. Etiological analysis, however, often involves analyzing developments over long periods or among people with diverse behavior and habitats. Determining the cause of an acute infectious disease can often be helped by interviews and by the small numbers and specific nature of the cases involved. Determining the cause of a disease with a long latency period can be even more difficult, especially given the mobility of population and changing exposure over time.

Four of the logical canons developed by John Stuart Mill (1856) underlie most causal reasoning in health studies.

1. *Difference.* When all conditions among the study populations are alike except for one, that one condition is implicated as either causal or preventive of the disease. This is the classic logic of laboratory experimental studies. Rats with the same inheritance are kept under identical conditions, except for exposure to the chemical that is being studied. The best application of such an experimental design to study of human populations can be found in the clinical trial, in which one group of people is given a new drug and another group is given a placebo, such as a valueless sugar pill. In this way psychological attitude, as well as the healing or exposure risks of time, are kept the same for both groups except for effects of the drug being studied.

2. *Agreement.* When all circumstances are different except for the variable being studied, that variable is implicated as causal.

3. *Concomitant variation.* When a factor varies systematically with the frequency of the disease, that factor is implicated as being causal. When more or less of the variable is associated with more or less of the disease, it is varying systematically.

4. *Residue.* When the effect of the known causal factor is removed in order to isolate and measure the variation remaining, successful explanation of the remainder supports the factor's causality. This is the method that geographers use when they map the residuals of a regression or the factor scores of a factor analysis, in order to see whether the pattern of unexplained variation remaining elicits any further hypotheses.

When repeated studies support these logical canons, causation is gradually established. Studies of cigarette smoking, for example, have found that when groups matched for age, sex, ethnicity, income, education, occupation, personality, and activity differ only in smoking cigarettes or not, there is a great difference in the incidence of cancer between them (i.e., the difference canon is supported). If people are studied who live in totally different cultures and environments—if they are Muslims and Buddhists and Christians; if they are poor and rich; if they are literate and illiterate; if they do and do not eat meat; if they live in cold, dry places and hot, wet places, or in urban places and rural places—the smokers always have a higher incidence of lung cancer than do the nonsmokers (i.e., the agreement canon is supported). When populations of smokers are subdivided according to how many cigarettes they smoke, at every increment there is associated greater risk of getting cancer (i.e., the canon of concomitant variation is supported). When the effect of cigarette smoking is statistically removed, it is possible to identify patterns of exposure to air pollution (i.e., the residue canon is supported). Thus, although statistics cannot prove anything, the accumulated and consistent evidence along different logical paths makes any doubt that cigarette smoking causes lung cancer unreasonable.

It is difficult to be sure, however, that all other factors are the same. Many early animal studies of the dietary effects of water hardness, for example, had to be redone when scientists learned how to measure trace elements. The water had not only differed in calcium carbonate, as intended and measured, but also in the trace amounts of molybdenum, cadmium, selenium, and other elements. The amounts of trace elements had not been held constant and plausibly were causally related to the outcome. In addition, many factors are not so easily measured. Precisely measuring trace elements in a lab setting is one thing, but a major limitation of many epidemiological studies is that variables including policy, culture, and SES are difficult to express quantitatively. These factors undoubtedly affect health as well. We address this topic in more detail in Chapter 9 when we examine the social determinants of health.

A **confounding variable** is one that varies in a systematic way with the hypothesized causal relationship being studied. Although *A* seems to cause *B*, in fact another variable, *C*, is affecting both *A* and *B*; the relationship between *A* and *B* is therefore spurious. Sometimes one is aware of a confusing interaction. For example, soft water has been associated with higher risk of stroke. In the United States, soft water occurs in the Southeast on the coastal plain, which of course is a little above sea level. When researchers study stroke in soft-water regions and hard-water regions, they also find strong associations with altitude and with the range of temperature changes. Which is truly causal? Or do these characteristics (variables) affect each other and merely happen to vary in the same way as stroke, perhaps because of still another variable that has yet to be identified? One of the most difficult tasks in social science research on health is identifying which variables need to be controlled so that they will not confound the relationship being studied. This, of course, is the role of theory, but at early levels of understanding of disease etiology, theory may be inadequate.

Confounding can be controlled both by analysis and by research design. The most confounding variable of all, as discussed earlier, is age. Whether one studies the prevalence of antibodies to a virus, life stresses, activity pattern, cholesterol deposition, or public health knowledge, one needs to know the age of those involved. We have seen how analytical technique can control for age by standardizing data. The devilish thing about confounding factors, however, is that the researcher may not know they exist. One of the major purposes of epidemiological research strategy is to control for confounding factors even when they are not clearly specified.

The ideal epidemiological evidence is to find that the different disease frequencies in two populations are dependent on a difference in a certain factor, and that furthermore, within each of the populations, this factor is more common among those with the disease than those without it. Two broad strategies address these questions: **cohort studies** and **case-control studies**, both diagrammed in Figure 3.3. In a cohort study, a population is studied. It is divided into those with and without a particular exposure, and the frequency of the disease outcome is noted. When this study is started before the exposure and people are followed forward in time, it is called a **prospective study**. When the disease outcome has already happened and the history is reconstructed through interviews and records, it is a **retrospective study**. A case-control study starts with people who have the disease or who died of it, and compares their behavior and exposures with those of the rest of the population. Although all the nondiseased population may be used, usually it is

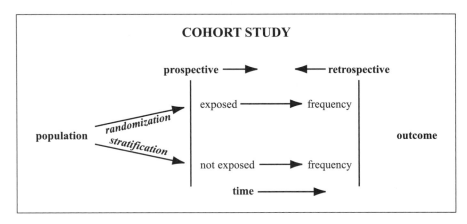

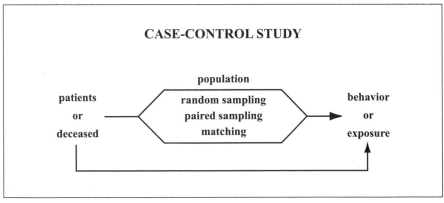

FIGURE 3.3 Epidemiological design.

sampled. The whole population may be sampled randomly, or cases may be paired in a systematic way with people from the general population. One may interview the neighbor next door to the right, for example, or the next patient admitted to the hospital for a nonrelated reason. Alternatively, one may pair the sample deliberately by matching the patient on a series of criteria (age, sex, ethnicity, income, education, etc.).

Each of these research strategies has advantages and disadvantages. A prospective cohort study has the great value of allowing direct estimation of the risk associated with causal factors. As one follows the population through time, for example, one can measure how many times they ate a certain food or drank a certain source of water and how long they spent in a certain occupation. There are no spurious relationships added by the means of collecting data, such as a bias introduced when people who are suffering are motivated to remember events more carefully. The great disadvantages of such a study, however, are the costs in time, skilled personnel, and money. Following a population is laborious, time-consuming, and expensive, especially if the disease is not too common and large numbers of people must be involved in order to get enough data for a definitive analysis.

A case-control study is relatively quick and inexpensive, and these characteristics make it easily repeatable and able to include large numbers of people. This approach can introduce certain biases into data collection, however, and researchers are less likely to find things they did not set out to find. Questions are formed by researchers, and people recall their specific experiences in reaction to the questions.

In both these research strategies, randomization and stratification are used to convert confounding variables into control variables. When a population is randomly assigned to two groups or individuals are randomly chosen for comparison, the effects of the confounding variable are converted into residual variation equally distributed between groups, so there is no need for analytic control of it. For example, consider a test of a new vaccine. Some children are naturally resistant to the disease, and some may have had it already. Some children are in more contact with other children, some have better diets, and some have more psychological stress. All these things may affect whether or not a child gets the disease. A researcher wants to compare the proportion of children given the vaccine who contract the disease with the proportion of unvaccinated children (given a placebo) who contract the disease, in order to assess the vaccine's effectiveness. By assigning children randomly, the researcher converts the difference in contracting the disease that is due to relative natural resistance or to differences in exposure into unexplained variation common to both groups equally, and so it is removed as a confounding factor. Matching cases or stratifying the population with regard to certain characteristics has the same effect of making sure that those characteristics do not introduce confounding variance. Stratifying or matching by ethnicity or education, for example, eliminates variance due not only to those characteristics but to every variable associated with them, such as diet, home environment, and neighborhood influences. These epidemiological methods of analysis have been developed and standardized over decades.

Cancer: Genetics, Political Authority, and the Environment

Cancer is a family of diseases. There is no one cause or one cure for cancer; there are only multiple effects of multiple treatments and outcomes. Our knowledge and classification of cancer are analogous to our knowledge and classification of infectious disease in the 19th century before germ theory. Then cholera was not differentiated from yellow fever or typhoid from diphtheria. There were only fever, "flux," vomiting, rash, flatulence, cramps, and so forth. The best doctors of the day could write of malaria "becoming typhous in its course." It was only after germ theory gave us a microbe for each disease that the separate entities were classified, and causation and treatment were defined appropriately. The long latency of these diseases has bedeviled etiological research because of the many changes over time and space that occur. Geographers, whose study usually encompasses areas large enough to have significant population mobility, need to appreciate these time–space processes. Cancers are caused by **carcinogens**, substances or agents that cause cancer in living tissue. Carcinogens can be viral such as human papillomavirus (HPV) which causes an estimated 70% of all cervical cancer cases. But carcinogens can also come from the environment, either as naturally occurring substances or as by-products of manufacturing and mining, asbestos, arsenic, chromium, formaldehyde, and trichloroethylene, to name a few. Since people move throughout their lifetimes, especially in wealthy countries, it is nearly impossible to monitor all the substances a person has been exposed to. This is what makes cancer so difficult to study.

Figure 3.4 illustrates a model of cancer that explains why it takes cancer so long to develop. When a carcinogen enters the body, it is usually detoxified by enzymes, broken down, and excreted. Occasionally these enzymes and related processes can activate the carcinogen and enable it to enter a cell. If it binds to anything in the cell except the DNA, it can only affect that individual cell. If it binds to the DNA, many repair mechanisms attack it. The bonds between the carcinogen and the DNA are broken, and the carcinogen is broken up, transported, and excreted. Only if the DNA is replicated before the carcinogen is removed can the carcinogen affect the new cell information, creating a cancer gene. The evidence clearly indicates that one altered gene is usually not enough to cause cancer. The process must be repeated, seemingly against all odds, by another carcinogen in a cell that already has the altered genetic material. The odds are changed if substances known as promoters (sometimes metal elements, fractions of hydrocarbon chains, fatty acids, or substances such as saccharin) encourage the altered cells to proliferate faster than normal cells. This process creates an increasing number of targets for another carcinogen to enter. The rare chance that such a contact will occur, among all the normal cells of the body, helps to explain why decades may elapse before expression of the disease. It also explains why cancer is most common in tissues with normally high rates of proliferation, such as skin or the lining of the intestines and uterus, and rare in nonproliferating tissues such as nerves. Proliferation increases the number of cells containing the altered genetic information. Every time a carcinogen enters the body, the whole chain of chance events must repeat itself.

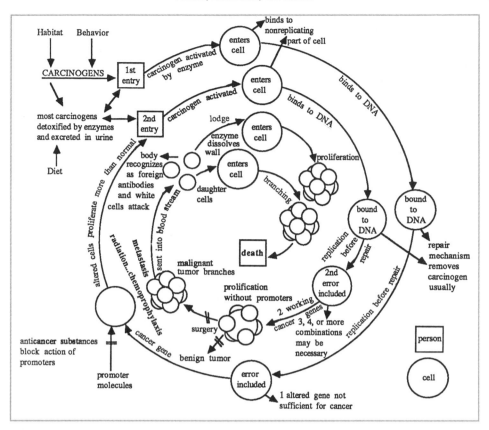

FIGURE 3.4 Cancer causation. Repeated alteration of genetic material is usually required to cause cancer, and the generally low probability that the necessary alteration will be repeated is affected by the presence of promoter substances and the dosage and duration of carcinogens, as well as the presence of protective nutrients for repair and buffering mechanisms.

In 1999, scientists succeeded in turning normal human cells into cancerous ones by altering four genes. They turned on an **oncogene** to make the cell multiply quickly, shut off a tumor-suppressor gene that limited replication, disabled a gene that caused abnormal cells to self-destruct, and turned on a gene to rebuild the telomere caps on chromosome ends that normally shorten and age with each division. It took these four changes in one cell to create a cancer. An oncogene is a special gene that has the potential to cause cancer.

As we continue to follow Figure 3.4, we see that a benign, noninvasive tumor can be removed by surgery. If the tumor sends out branches and envelops other tissue, it is cancer. The malignant tumor can send "daughter" cells into the bloodstream, an event known as **metastasis**. The daughter cells drift with the bloodstream, in danger of being recognized as foreign and attacked by the body's antibodies and white cells, until they lodge. They produce an enzyme that dissolves the blood vessel wall, which allows them to enter normal tissue and begin proliferation and invasive branching in the new site. Eventually vital tissues and processes are involved, and death results.

Cancer's need for repeated defiance of the odds in an orderly but improbable sequence of events explains the long latency period. The need for separate carcinogens, promoters, and overcoming of at least four defense mechanisms of the body explains the often baffling etiology and outcome. The adequacy of vitamins (and other antioxidants) in the diet is important for the body's ability to detoxify a carcinogen, especially after it has bonded within a cell. Another important aspect of cancer is the body's own immune system. While carcinogens or genes may lead to cancer, the body's ability to identify cells with altered genetic (cancerous) information and destroy them is a vital key to why some people get cancer and why some don't. All the interactions and balances of this complex process change over time.

The Population Base

Some kinds of cancer have long been suspected of being genetically linked. There is evidence that some cancers affect men more than women when there has been equal exposure. Others are common in some population groups (e.g., nasopharyngeal cancer among the southern Chinese) and rare in others. Some cancers are more common in a family than in the general population. It is for these **familial** cancers that genetic markers are most likely to be found. The specific genes involved in familial disease may not be involved at all in the **sporadics**, or cancer cases that occur without an apparent hereditary role.

Cancers develop when cells' DNA is somehow altered by a carcinogenic substance or agent. Cells must divide and replicate if tissue is to grow and maintain health. Each replication involves the exact reproduction of their DNA. Accidents happen, however. The four bases of the cells' DNA may spell out G, C, A, T, for example. In this example, G matches with C, and A with T, to form a three-dimensional molecular structure that locks with a specific enzyme. If during replication somehow G matches with T, an error of mismatch has occurred. Alternatively, radiation may knock a proton off, changing G and causing an error of excision. Genetic excision or damage may eliminate the ability to receive and regulate a chemical, as when a damaged cell without ability to respond by negating a cell growth signal turns cancerous. In either case, proteins patrolling the DNA have to recognize the damage and call in repairs. Figure 3.4 implies several mechanisms of genetic involvement.

The importance of damage recognition and enzyme combinations in DNA repair and carcinogen detoxification is genetically given. Differences may make people more or less susceptible. A variant of the p450 gene enzyme that is poor at metabolizing a certain carcinogen, for example, may increase a person's susceptibility to the toxic effects of cigarette smoke. Men may be more susceptible to many mutagenic agents because they bind more carcinogen to their chromosomes. A **mutagen** is any agent (a chemical, radiation, UV light) that can change the genetic material of DNA and increases the frequency or extent of mutations above the normal background level. This has been most studied with regard to lung cancer and the carcinogens and promoters in cigarette smoke. Men have been found to have greater stimulation of DNA repair synthesis, resulting from the increasing amount of damage. Damage and repair activities increase with age for both sexes, but are higher at all ages for men. Not all mutagens cause cancer, but the ones that do are also considered to be carcinogens.

The best single example of the processes above lies with *Science*'s 1993 "molecule of the year," p53, which is resident on chromosome 17. When it was first found that p53 was involved in 60% of cancers, and even 80% of colon cancers, it was implicated as a carcinogen binder. Advancing research, however, demonstrated that p53's role is to halt the replication of any cell with DNA damage, thus stopping abnormal growth processes. It is a tumor killer. Damage to p53 itself, however, can sabotage this critical control and defense system. It is now the gene scientists turn off when they are studying the process of carcinogenesis discussed previously.

The most notorious gene in the class of tumor suppressor genes is BRCA1 (a BRCA2 has also been recently discovered). It makes a protein that binds to DNA and is directly involved in repairing DNA damage. It interacts with many other genes for purposes that are as yet uncertain, but include development of the embryo. More than 600 mutations of this gene have been identified. The mutated gene makes proteins too short to fix the damage to DNA of other genes, as shown in Figure 3.4. Damaged BRCA1 causes greatly increased risk of early-onset breast cancer (breast cancer is bimodal in incidence; most occur in women over 60, but damaged BRCA is associated with breast cancer in 30-year-olds). It also increases the risks of ovarian, fallopian tube, and prostate cancer.

One of the most active areas of cancer research concerns *oncogenes* and the breaking of chromosomes. Molecular biologists have identified more than 20 of these mysterious genes, which normally are inactive. They are found in such diverse organisms as flies, fish, mammals, and yeast. Thus, although their functions are unknown, they are presumed to be vital because they have been so carefully conserved throughout evolution. Oncogenes seem to be found consistently in tumorous cells, often in multiple copies. Sometimes they are involved in turning on growth genes, and possibly in chromosome breaking. It seems clear that chromosomes have a propensity to break in certain places. Genetic control of these weak places may be involved in susceptibility to cancer.

The Behavioral Base

Cultural risk factors for cancer involve a variety of customs and practices, as well as economic goods and occupations. Given that solar radiation and genetic susceptibility by population group have not changed, the propensity of modern youth to tan themselves aggravates risk of skin cancer, compared to earlier practices of wearing long sleeves and broad hats and otherwise modestly covering the body. Tobacco smoking, or chewing, is the most obvious endangering behavior and leads to lung cancer, and cancers of the face, mouth, and throat. The history of tobacco in the United States is long and sordid. The Tobacco Master Settlement Agreement (MSA) of 1998 found that the tobacco industry had known for decades the carcinogenic effects of smoking but failed to make this known to the public. This negligence was deemed fraudulent for many reasons, but in particular Medicaid and Medicare (public insurance programs) were footing the bill for smoking-induced illness among the poor and elderly. The MSA was enacted 40 years after the first study was published on the link between smoking and lung cancer! Hundreds of lawsuits were filed during this time, most of which were quashed by tobacco company lawyers. In fact, only two people ever won their case and both were eventually reversed on appeal. The tobacco companies were simply too large, had seemingly inexhaustible resources at their disposal, lobbied state and federal policymakers, and, as a result, had significant influence over government regulations.

Diet can both help and harm. A diet poor in fiber means that undigested food passes too slowly through the intestines; it is thought that this allows toxic bacterial products to become concentrated and increase risk for colon cancer. In many settings, poverty is intimately linked to dietary choices; people who are poor typically eat diets low in fiber, high in fat and calories, and lower in antioxidants that prevent cancer. The hydrocarbons and nitrosamines of smoked fish (and presumably other smoked products) and charred meat seem to be risk factors for stomach cancer. Diet may also be preventive: substances in green, yellow, or purple vegetables serve as antioxidants to protect cell walls. The vitamin E in fish skin or the antioxidants in green tea may do the same. The human diet contains a variety of natural carcinogens, promoters, antimutagens, and anticarcinogens. Science has barely begun to address the relative risk or protection of the enormous variety of dietary habits around the world.

The myriad factors that place people at risk of contact with or exposure to carcinogens are deeply dependent on a variety of societal forces shaped by political and economic interests. In the book *Poisoned Profits,* Phillip and Alice Shabecoff (2010) question why the U.S. government has failed to regulate industries that produce and release toxic substances. They document "the toxic assault" on American children by pesticides, car pollution, waste-site runoff, industrial plant emissions, and chemicals found in consumer goods, which they, and other scientists, claim have led to the steep increases in asthma, autism, and cancer. Since its inception in the 1970s, the environmental regulatory system in the United States has been chronically underfunded and slowly divested of its ability to police industries that produce toxic chemicals. Chemicals are now "innocent until proven guilty" and there are very few limits on the production and disposal of some of the most concerning toxins for human health. As a result, enforcement of what few regulations exist is weak and often slow because contamination (and potential health problems) must be proved. Furthermore, the chemical industry pulled in approximately $120 billion dollars (after taxes) in 2010, an amount that has doubled since 10 years prior. The sheer size and economic importance of the chemical industry translates to significant political influence, and Shabecoff and Shabecoff document the "revolving door" between this industry and the government, where lobbyists influence and even write legislation and regulations. They also suggest that corporate lawyers are engaged to discredit scientists or suppress research that shows links between chemicals and poor health. For example, in 2009 a study published in

Environmental Health, a very well regarded and influential journal, revealed detectable levels of mercury in nine of 20 samples of commercial high-fructose corn syrup (HFCS) (Dufault et al., 2009). Dr. David Wallinga, a coauthor of the study, wrote that "mercury is toxic in all its forms. Given how much high-fructose corn syrup is consumed by children, it could be a significant additional source of mercury never before considered. We are calling for immediate changes by industry and the [U.S. Food and Drug Administration] to help stop this avoidable mercury contamination of the food supply" (Institute for Agriculture and Trade Policy [IATP], 2009, p. 1). The Corn Refiners Association, a large and influential organization in the United States, immediately set to work disputing and discrediting the report, stating that the "study appears to be based on outdated information of dubious significance" (Study Finds High-Fructose Corn Syrup Contains Mercury, 2009) because the industry had switched to mercury-free versions of certain chemicals used to make HFCS. But evidence suggests that many processing plants have not switched and continue to use mercury. The U.S. government has yet to regulate this industry.

The Habitat Base

The behavioral base in economic structure, occupation, industrial processes, and regulatory control has dominated risk assessment but it is in the habitat, especially the built environment, that carcinogens lurk. Workers exposed to asbestos, to the manufacture or use of industrial chemicals, to agricultural sprays, or to the other multiform dusts and radiations connected with earning a livelihood in an industrialized society have long served as the guinea pigs for most of our threshold and effect data. The larger-scale behavior (discussed above) of locating certain enterprises (e.g., oil refineries, nuclear power plants, or toxic waste incinerators) in certain populous areas and not others has rarely been evaluated from such an experimental perspective. An increasing concern for issues of environmental racism, highlighted by geographers and other social scientists, has demanded such research by the U.S. Environmental Protection Agency (EPA) (for a further discussion of environmental justice, see Chapter 11).

Mostly because of the high incidence of breast cancer in the Northeast, including Long Island, New Jersey, and Cape Cod, major studies (both large-scale and local) have been carried out in recent years. As the map of breast cancer in Plate 7 shows, the industrial shores of the Great Lakes (Chicago, Detroit, Buffalo) and the Northeast coast have concentrations of the most significantly high rates in the country. Ethnically, these areas include many people of eastern and southern European descent. But they also include concentrations of many chemicals, both present and past. The Silent Spring Institute on Cape Cod has organized at a grassroots level and worked to develop GIS and other analytical systems and environmental monitoring and analysis to include studies of local settlement in research on local environments. One of these projects was an exhaustive analytical study of existing PubMed databases and publications on the connection of chemical pollutants identified as important in toxicological (laboratory, animal) studies and breast cancer (Brody et al., 2007). Many environmental pollutants cause mammary gland tumors in animals and are hormonally active, especially mimicking estrogen. A small number of studies with confirmed evidence were found to support an association between breast cancer, on the one hand, and polycyclic aromatic hydrocarbons and PCBs (the latter, though not allowed since the 1970s, have accumulated in fish to become women's main carcinogen exposure), on the other. The evidence of a connection with dioxins and organic solvents is suggestive but methodologically awkward, including disasters of mass exposure of children (not yet reproducing) in Russia and Italy (following accidents in chemical plants making herbicides). Methodological problems abound, especially centered on inadequacy of exposure assessment, lack of access to highly exposed (usually occupational) populations, and the almost total lack of investigation of chemicals in humans that have been found toxic and carcinogenic in animals. Associations of breast cancer with organochlorine pesticides (aldrin, DDT, dieldran, heptachlor, etc.) have also been suggestive but not conclusive. Researchers urge greater use of toxicology data for control and prevention.

Patterns and Change

There is amazing geographic variation in the occurrence of cancer. For decades it has been recognized that differences in incidence rates, not only between men and women but also between urban and rural populations and among countries, hold etiological clues. The highest female cancer rate by country is far lower than the top 10 male rates. Japan has the highest stomach cancer rate but is not ranked in the top 10 for lung, breast, or rectal cancer. Given Australia's biometeorology, its white population's genetics, and such behavioral patterns as ranching and surfing, one can understand why that country is the highest-ranked in skin cancer—but why should countries as diverse as Uruguay, France, and Singapore be ranked so highly for esophageal cancer? Furthermore, within each country, urban and rural rates can differ as greatly as international rates.

The extraordinary data richness at the microscale in China was made clear in an atlas that portrayed the most dramatic spatial patterns of concentration (Editorial Committee, 1979). Not only was nasopharyngeal cancer concentrated among Cantonese on the Guangdong plateau, but there were extremely high foci of esophageal cancer west of the Taihang Shan in Shanxi and of liver cancer around the mouth of the Yangzi and the southeastern coast. At the national level, these cancers, which are almost unknown in the United States, have been the major causes of cancer death until this last decade. With economic liberalization in the late 1970s to the 1980s, however, cigarette smoking reached prodigious levels, and the familiar consequence of lung cancer has built to a wave that is breaking now greatly compounded by exposure to extraordinary levels of air pollution in the major cities.

The patterns of cancer incidence have changed over time. Lilienfeld and Lilienfeld (1980) have outlined several possible reasons. The changes may be **artifactual**; that is, they may result from errors due to changes in the recognition, classification, or reporting of the disease, or from errors in enumerating the population.

Real changes may result because the age structure of the population has changed, because people survive diseases that were once incurable and thereby live long enough for cancer to become manifest, or because genetic or environmental factors have changed. (As epidemiologists, the Lilienfelds include behavioral changes under environmental factors.)

Greenberg (1983) found a spatial convergence of cancer mortality in the United States between 1950 and 1975. There was a strong parallel trend for most types of cancer: rates diverged by sex and race subgroups of the population, but converged geographically. He contended that this spatial convergence of cancer mortality (i.e., development of a homogeneous pattern) in the United States and other industrialized countries has been caused by changes in the geography of risk factors associated with the diffusion of urban culture. These risk factors include air and water pollution, cigarette smoking, alcohol consumption, diet style, occupation, SES, stress, and medical practices. Ezzati et al. (2008) identified lung cancer as being responsible for the "reversal of fortune" as female life expectancy at the county level was decreased for more than a decade after 1983.

Mapping and Measuring Risk

The appropriate scale for comparative mapping of cancer has been especially problematic. Rare cancers, such as brain tumors, have a low range of rates compared to common cancers, such as colon cancer. The distribution of the African American population varies greatly across the United States, leaving regions of counties with almost no population denominator when cancer deaths are broken down by age and gender. One solution, used by the first National Cancer Institute county-level maps (Mason, McKay, Hoover, Blot, & Fraumeni, 1975) as well as a more recent cancer atlas by the U.S. EPA (1987), is to map white rates at the county level and black rates at the scale of state economic areas (which aggregate contiguous counties into clusters, creating an intermediate scale between the county and state levels). The *Atlas of United States Mortality* (from which our Plates 8a and 8b are taken) addressed the small-number problem by using health service areas as the scale of aggregation for all its disease maps (National Center for Health Statistics, 1997). As explained in other chapters, the rates are age-adjusted and mapped as mortality ratios compared to U.S. national rates, and the distribution of death rates by health service areas is presented in a graph. Hatching indicates where the data are sparse, mostly because of small populations at risk, to facilitate an honest interpretation.

Very often risk is measured in ways that take small numbers and emphasize their importance, so that people lose sight of the fact that very small numbers were involved in a study or event. One result is that the public gets scared, and then numbed and contemptuous, as chemicals and especially foods pop up as risks of cancer and then disappear (or earlier findings are reversed by larger studies). Yet, given that large-scale experiments can't be done with people and cancer, such small indicator studies are often the best possible way to connect with the animal toxicology results. Those reading any of this literature should understand how two measures work in particular: the relative risk and the odds ratio.

- **Risk** is the number of times (frequency) an event occurs relative to the total number in the study population, for example, one in five.

- **Relative risk** is the risk of an event in those with a risk factor divided by the risk of an event in those without the risk factor.

- **Odds ratio** is the number of times an event occurs relative to the number of times it doesn't occur; that is, the ratio of event to no event in exposed divided by unexposed individuals.

Consider a population of 1,000, of whom there are 400 smokers and 50 cases of cancer. Forty cases of cancer occur to smokers, and 10 cases of cancer occur to nonsmokers. We can put these numbers into a **contingency table**, which is a two-way table useful for examining relationships between categorical variables. A contingency table can show us the frequency of a health outcome among one or more risk groups; in this case, cancer among smokers versus nonsmokers.

	Cancer	No cancer	Total
Smoker	40	360	400
Nonsmoker	10	590	600
Total	50	950	1,000

From this table, we can calculate the risk, relative risk, and the odds ratio of cancer among smokers versus nonsmokers.

$$\text{Risk} = \frac{\text{Total Cancer Cases}}{\text{Total Population}} = \frac{50}{1000} = 0.05 \text{ or } 50 \text{ per } 1,000 \text{ people}$$

$$\text{Relative Risk} = \frac{\dfrac{\text{Cancer Cases in Smokers}}{\text{Total Number of Smokers}}}{\dfrac{\text{Cancer Cases in Nonsmokers}}{\text{Total Number of Nonsmokers}}} = \frac{\dfrac{40}{400}}{\dfrac{10}{600}} = \frac{0.1}{0.017} = 5.9$$

$$\text{Odds Ratio} = \frac{\dfrac{\text{Cancer Cases in Smokers}}{\text{Smokers with No Cancer}}}{\dfrac{\text{Cancer Cases in Nonsmokers}}{\text{Nonsmokers with No Cancer}}} = \frac{\dfrac{40}{360}}{\dfrac{10}{590}} = \frac{0.111}{0.0169} = 6.6$$

Some Studies of Cancer

Geographers have contributed a significant amount to the study of cancer, mostly through the identification of spatial patterns and the analysis of these patterns to identify risk factors. There are virtually no examples of geographers applying a political ecology framework to cancer research. As discussed above, the research questions and methodology may involve determining regions and investigating common or different factors by using *analogue area analysis* (trying to hold possible associations constant, except for the study variable) and applying the causative reasoning described above. For example, several regions can be identified in which stomach cancer has a high incidence. Iceland, Japan, and northern Minnesota/ Wisconsin are three such regions. Copper seems to be unusually concentrated in the soils of one of these; is it also high in the others? People in Iceland eat smoked and salted fish and are of Scandinavian descent. Do people in Japan eat salted and smoked fish, and do people of Scandinavian descent in Minnesota and Wisconsin continue ethnic dietary practices? Similarly, some people in eastern Africa have a high incidence of esophageal cancer in a culture area in which home-distilled alcohol is made. People outside that culture area have lower cancer rates; perhaps the alcohol or method of its production is involved. The region around the Caspian Sea in Iran, however, has some of the world's highest rates of esophageal cancer, and people there do not consume home-distilled alcohol. Research suggests that the Iranian risk area is characterized by nutrient-deficient saline soil, halophytic vegetation, and poor crops. Alcohol could easily absorb elements from the pottery and cans in which it is distilled. Could trace element deficiency or toxicity be common to both regions? Geographers in Japan, Europe, and North America have analyzed the pattern of cancer occurrence in relation to environmental, socioeconomic, and demographic factors by means of multivariate statistical procedures. There are quite a few of such studies, but results have been inconclusive about the associations found.

Studies that are based on general populations, that do not select disease categories for research carefully and deliberately, or that target fortuitous places run into many data and methodological problems. They so often suffer from the ecological fallacy (see Chapter 1) that "ecological" (i.e., multivariate in statistical use) studies as a whole have earned a reputation for scientific weakness. Such multifactor, population-based studies are, however, a potentially powerful means for choosing places for case-control and other detailed studies. An important research question regarding disease distribution concerns the degree and scale of spatial clustering. These issues are explored in the discussion of neighborhood analysis in Chapter 9. Map patterns alone cannot eliminate chance in clustering, but observed patterns can be compared to randomized ones by using stochastic approaches, chi-square tests, and several spatial statistical methods.

In the discussion of time–space geography, exposure, and population mobility in Chapter 4, the analytic benefit of large, well-defined streams of migration that can be identified and followed is illustrated by the studies of breast cancer and stomach cancer among generations of Japanese in Hawaii. Environmental factors, genetics, lifestyles, and resistant customs such as favorite foods were separated by movement and generation. As noted there, however, such research is rarely possible. The study of microscale mobility and exposure to different environments can also be used to identify possible etiological factors and to test new hypotheses. Armstrong (1976, 1978) studied nasopharyngeal cancer among the Chinese in Malaysia. Using a case-control method (discussed above), he compared the exposure patterns of people with the disease and those of controls (people without the disease) to what he called "self-specific environments." He measured how much time was spent daily in agricultural areas, in squatter housing or middle-class housing, in shop workplaces, in factories, in shopping on the street, in various means of transportation, in public places, and so forth; he also measured how much time was spent in smoky places, in crowded places, around chemicals, and under conditions of bad air pollution. He examined such indexes of traditional or modern/assimilated cultural patterns as having an altar in the house and eating meals in a formal manner, as well as details of what people ate and how they prepared it. He found that among genetically susceptible Chinese, important stimuli were associated with industrial or trade occupations, smoky workplaces, poorer housing, traditional lifestyle, a diet with less variety of foods, and childhood consumption of salted fish.

There are two major classes of skin cancer. Basal or squamous cell carcinomas are fairly common and relatively innocuous because they can be successfully treated without hospitalization. Malignant melanoma, a rare form of skin cancer, is often fatal. Glick (1979) had data from four special surveys of skin cancer. He used regression techniques to estimate the effects of age and of ultraviolet radiation (UV) levels and found that nonmelanoma skin cancer has a strong age effect, as one would expect if the carcinogenesis process requires repeated exposures and alterations of the cell (as illustrated in Figure 3.4). Glick reasoned that because nonmelanoma skin cancer involves more stages of exposure, it should have a steeper spatial gradient of mortality. Indeed, he found that the gradient in mortality rate for a transect (line) through Minnesota-Iowa-Missouri-Arkansas-Louisiana was very steep and constant for nonmelanoma skin cancer, going from about 1.0 to 2.6 per 100,000; mortality from melanoma ranged from 1.1 in Minnesota to 1.8 in Louisiana, with a less constant gradient.

Figure 3.5 presents a spatial correlogram for two types of cancer that illustrates the spatial pattern of point and area exposure to risk. Across the X-axis are the spatial lags in association. For example, at lag 1 the average autocorrelation of each

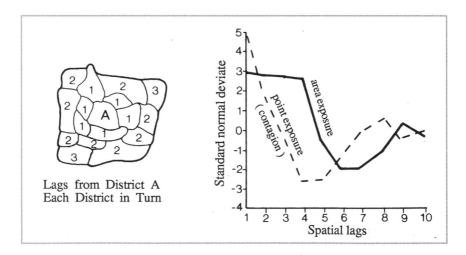

FIGURE 3.5 Spatial correlograms. Point and areal (regional) sources of exposure are reflected in different levels of correlation between spatial units when the units are lagged incrementally.

county in the state with adjacent counties is expressed, whereas at lag 5 the auto-correlation is with five counties away. The autocorrelation is expressed as a normal (*Z*) score on the *Y*-axis. The farther away a score is from zero, either positively or negatively, the stronger the association among counties. The line that shows point exposure has a steep distance decay in autocorrelation from high positive (similar to the adjacent county) to high negative (different from the adjacent county). It represents a spatial pattern of isolated foci of incidence, such as might result from a contagious disease occurring in large cities but not yet diffused across the county line, or from behavior limited to ethnic enclaves. Stomach cancer in Pennsylvania shows such a pattern, whereas bladder cancer is characterized by a nearly horizontal correlogram for lags 1–3 (Glick, 1980). Such a pattern, as illustrated, portrays a large region of similar rates and suggests an environmental carcinogen found over large areas (e.g., as water catchment basins), rather than a diffusion process.

Prostate cancer is one of the most prevalent forms of cancer in the United States, with especially high rates among African Americans. It is of unknown etiology, although strongly related to age. Hanchette (1988; Hanchette & Schwartz, 1992) used regression and trend surface analysis to draw out the association of ultraviolet light, prostate cancer, and a possible role of vitamin D and its associated enzymes. The strong north–south gradient of this disease is also correlated with colon cancer. Schwartz and Hanchette (2006) repeated this trend surface analysis for a 45-year period, cutting the map at 40° latitude because vitamin D synthesis does not occur in winter months at higher latitudes (Figure 3.6). They thus confirmed and extended their earlier results and found strong support for the hypothesis that vitamin D insufficiency increases risk for prostate cancer. Logically (again, see Mill's canons above), this is supported by the greatly higher rates for African American men.

HIV AND AIDS: GENDER, MOBILITY, AND POLITICAL ECOLOGY

Acquired immune deficiency syndrome (AIDS) was first identified in 1981 when increasing numbers of young homosexual men were diagnosed with unusual opportunistic infections and Kaposi's sarcoma (an extremely rare form of cancer). AIDS is caused by the human immunodeficiency virus (HIV), which damages the human immune system and interferes with the body's ability to fight disease. HIV is a sexually transmitted infection, but can also be spread by contact with infected blood or through breastfeeding. Billions of dollars have been invested in developing an HIV vaccine, but at the time this book was written there was no successful candidate. There is no cure for HIV/AIDS but there are medications that can slow the progression of the disease. In fact, in many countries where antiretroviral therapy (ART) is available, HIV has become more of a chronic condition and people are living indefinitely with the disease. It is widely recognized that prevention is the only way to slow the AIDS epidemic. Use of condoms, male circumcision, and ART have all slowed the AIDS epidemic over the past decade. Beyond these Western "medicalized" interventions, culturally targeted education campaigns in some sub-Saharan countries have shown remarkable success at changing or modifying behaviors that lead to HIV. For example, in the book *Tinderbox* (2012), authors Tim Halpern, an

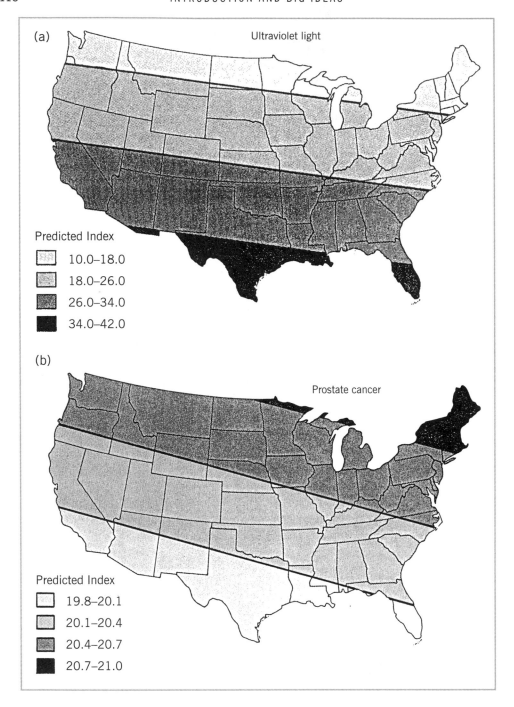

FIGURE 3.6 Maps of trend surface analysis of ultraviolet light as protective of prostate cancer. (a) Linear trend surface by county observations of ultraviolet light measured by epidemiological index. (b) Linear trend surface by county observations of prostate cancer mortality for white males, aggregated from 1970 to 1979. From Hanchette (1988). Used by permission.

epidemiologist/anthropologist, and Craig Timberg, a reporter, examine how Ugandan health officials slowed the HIV epidemic in their country through a public awareness campaign called "Zero Grazing." They encouraged people to be faithful to their spouses and not seek sexual partners beyond their polygamous marriages (so called "sexual grazing"). According to the authors, part of the reason it worked is that the government didn't try to completely change the traditions and culture of the polygamous Ugandan people. Over the course of the campaign, researchers found, casual sexual relationships in Uganda declined by roughly 60% and when behaviors related to casual sex with multiple partners decreased, so did the HIV rate. This is an example of how important the behavioral base of the disease ecology triangle is to HIV. HIV deaths decreased from 1.7 million (3.2%) deaths in 2000 to 1.5 million (2.7%) deaths in 2012 (WHO, 2014). But globally, HIV is still one of the leading causes of death, and ranks number two in low-income countries. HIV continues to decimate populations in Haiti, South Asia, and sub-Saharan Africa.

We now know that AIDS is caused by two distinct lentiviruses: HIV-1 and HIV-2. The origins of HIV were largely unknown, though researchers suggested for decades that HIV had emerged from simian (monkey) populations in West Africa probably through the practice of hunting for bush meat. Advances in genome sequencing allowed scientists to closely examine HIV's genetic material and compare it to lentiviruses found in simian populations (Sharp & Hahn, 2011). The resulting phylogenetic trees clearly show how "related" HIV is to other human and simian lentiviruses. This evidence confirms that HIV-1 is most closely genetically related to a simian immunodeficiency virus (SIV) found in chimpanzees (*Pan troglodytes troglodytes)* and gorillas (*Gorilla gorilla),* both of which live in West Africa in modern day Guinea, Gabon, Congo, and the Democratic Republic of Congo (DRC). HIV-2 appears to be most closely related to the sooty mangabey (*Cercocebus atys)* that live in Cote d'Ivoire, Liberia, and Sierra Leone.

Once HIV emerged from the forest, the geography of the AIDS pandemic was driven by land use change, changing mobility patterns, and urbanization. These larger changes were driven by political and economic interests in sub-Saharan Africa and the dynamics of gender and poverty. The earliest known case of HIV-1 in a human was found in a blood sample collected in 1959 from a man in Kinshasa, Democratic Republic of Congo (then Zaire). Cases of the disease have been identified in Europe (England and Norway) as early as 1960 and were discovered in seamen and their families who had traveled extensively to African ports. Amazingly, the tissues of these early deaths were frozen and recently tested so scientists can confirm these were very early cases of HIV. From West Africa, the disease spread east and south along train and trucking routes, eventually covering the whole of sub-Saharan Africa, and then found its way to the United States, Europe, and later to Asia.

The Population Base

The population dynamics of HIV depend on where you are in the world. In the United States, HIV was first discovered among young homosexual/bisexual men and they continue to be the most severely affected group. But even in the United States, approximately 28% of HIV infections are due to heterosexual contact. In Africa, where the disease has been circulating for decades, most transmissions occur through heterosexual sex among younger populations. In 2010, people ages 15 to 24 accounted for 42% of all new HIV infections (UNICEF, 2011). This figure does not include young people who were born with the virus. There are also deep gender inequalities. Globally, women constitute more than half the people living with HIV, and it is the leading cause of death for women ages 15–49. Young adolescent girls are more likely to have older sexual partners and partners who use injecting drugs, thus increasing their potential exposure to HIV. Studies from South Africa show that young women typically only have one sexual partner at a time. Young men, however, are more likely to have multiple partners, which increases the risk of transmission among all the young women they partner with.

Women appear to be biologically more susceptible to HIV infection, though the reasons why are somewhat unclear. Hormone levels appear to play some role. Women are more susceptible to HIV during pregnancy when progesterone levels are higher (Gray et al., 2005); and some evidence suggests that women taking oral contraceptives, which alter women's hormonal structure, have an increased risk for HIV (Heffron et al., 2012; Wang, Kreiss, & Reilly, 1999). Younger women are at higher risk than older women, perhaps because of changes to hormones as women age. Studies have shown that bacterial vaginosis, the most common reproductive tract infection in women worldwide, increases women's risk of having and contracting HIV (Sewankambo et al., 1997; Myer et al., 2005). These biological risk factors, as well as social and behavioral factors discussed below, mean that women bear the burden of new HIV infections.

The Behavior Base

The history of the HIV epidemic has been dominated by cultural ecologies that increase the risk of HIV transmission, complicate the diagnosis and treatment of the disease, and facilitate the spread of HIV around the world. Poverty, cultural values, and gender norms play a significant role in the HIV epidemic. Widespread gender inequalities throughout much of the world contribute significantly to young people's vulnerability to HIV infection and to the spread of the disease. These gender norms and power inequalities are expressed in sexual relationships wherein women have little ability to negotiate safer sex or refuse unwanted advances. At the same time, norms and expectations of masculine sexual behavior increase men's and boys' risk of acquiring HIV and other sexually transmitted infections (Barker & Ricardo, 2005). For adolescent boys, the passage into manhood is frequently associated with sexual experience and knowledge. Multiple sexual partners are evidence of sexual prowess. Conversely, norms related to femininity leave women with little control over their bodies, in extreme circumstances leading to forced or coercive sex, and deny them access to sex education (Heise & Elias, 1995; Varga, 2003). Women are even more vulnerable because they have few economic opportunities and limited autonomy.

Age-mixing and cross-generational (age difference of 10 years or more) sexual relationships are common, and perhaps the norm in sub-Saharan Africa, and increase girls' vulnerability to HIV infection (Luke & Kurz, 2002). These types of relationships are highly dependent on cultural norms around masculinity, whereby men are considered strong and virile if they have sex with younger women. Girls who engage in such a relationship are also more likely to engage in unsafe sexual behaviors. They are less likely to use condoms, discuss the use of condoms, or discuss HIV and other sexually transmitted infections (STIs) with their partner. Demographic and Health Survey (DHS) data from Swaziland, Uganda, Tanzania, and Zimbabwe consistently find a significant association between a young woman's HIV status and the number and age of her partners. For example, studies in Zimbabwe (Gregson et al., 2002) and Kenya and Zambia (Glynn et al., 2001) found that young women with older partners are significantly more likely to be HIV-positive with a 4% increase in risk with each year of age difference.

Open communication about sexual matters is not particularly common, even among men and women in a stable or marital relationship. It is difficult for women to discuss infidelity or condom use and it is difficult for them to *insist* on condom use particularly within marriage (Lambert & Wood, 2005; Maharaj & Cleland, 2004). Despite the dominant global discourse around how marriage will protect against HIV infection (because couples will only have one partner), HIV rates among married women are significantly higher than among single women. In northern Thailand, a study found that 76% of HIV-infected women had no casual sex partners and had acquired the infection from their husbands (Xu et al., 2000).

There is also a strong correlation with sexual- and gender-based violence; a woman's risk of contracting HIV and suffering sexual violence is all too common in many African cultures. A study in South Africa showed that intimate partner violence and high levels of male control in a woman's current relationship were associated with HIV seropositivity (Dunkle et al., 2004). Violence and the threat of violence hinders a women's ability to protect herself from HIV infection and to make smart decisions regarding sexual health. The fear of violence also prevents many women from asking their partners to use condoms, obtaining information on HIV prevention, and accessing testing and treatment.

Poverty and low education are also highly linked to HIV. At the global level, there is a positive correlation between HIV/AIDS rates and income inequality, the Human Poverty Index, and gross domestic product (all country-level measures of poverty). Country-level analysis supports this finding. Poverty is associated with lack of education, and low literacy rates often translate into incomprehensible messages about HIV risk and prevention methods for the most vulnerable populations. For disadvantaged communities, the cost of prevention may be prohibitively high; many poor people are unable to afford condoms or ARTs. Poverty also constrains people's life choices such that engaging in "high-risk behaviors" is often one of the few modes of economic support. Collins and Rau (2000) suggest that poverty-driven labor migration and commercial sex work are activities likely to increase HIV infection. Longfield (2004) documents how young women in Kenya pursue (older) sexual partners who can provide them with money and gifts. Many of these women have legitimate financial needs. Further, perceptions of risk (either current or future) are affected by current economic concerns. Poor prospects for the future often translate into skewed and distorted perspectives of current risk of HIV infection.

While cultural and social norms are responsible for much of the spread of HIV within populations, larger diffusion of the disease has its roots in migration and urbanization. How did a disease that was isolated in small rural forest communities become a global pandemic? Most scientists believe that the city of Kinshasa is the key. A group of researchers at Oxford examined data from colonial archives and determined that Kinshasa was one of the most highly connected of all central Africa cities due to extensive railway links. Over 1 million people a year traveled through the city. HIV first spread to the major population centers in the Democratic Republic of Congo (DRC) and from there spread east and south to dominate sub-Saharan Africa (Faria et al., 2014). The political economy plays a key role in expansion of HIV. Colonial governments, in particular Belgium set on resource extraction in the DRC, built massive railway structures to link major cities and to transport rubber, wood, copper, gold, diamonds, and other raw minerals to major port cities. The Belgian government took much of the Congo basin by force and many Congolese were pressed into forced labor to harvest rubber, to mine, and to build infrastructure. Transporting people to and from work sites resulted in unprecedented population movement, likely stimulating the emergence of HIV from the forested interior. The DRC was not the only country undergoing development at the hands of colonial powers. Infrastructure was being developed across much of sub-Saharan Africa to feed the political and economic aspirations of the British, French, and Dutch colonial powers. Train networks, roads, and shipping routes began to connect African ports to each other and to Europe and the Americas, beginning an expansive era of globalization of food production, resource extraction, and (eventually) manufacturing. People were increasingly mobile, moving to and from new urban centers. Once these African urban centers were connected to the larger global economy, HIV began to spread around the globe.

The process of urbanization and the changes to population structure and migration that are part of this process is also largely to blame for the AIDS pandemic. Rural to urban movement (as discussed in Chapter 4 and Chapter 10) is likely responsible for the introduction of HIV to major urban centers. This movement is driven by perceived economic opportunities in cities; young men in particular often migrate to find work. Increasingly, as neoliberal economic policies have come to dominate African developmental discourse, transnational corporations (or even state governments themselves) have set up shop in urban areas to take advantage of cheap labor and tax-free zones. The privatization of farmland has decreased family farming and economic opportunities in rural areas, forcing people to migrate for work. HIV rates are significantly higher in urban areas than in rural settings. According to The Joint United Nations Programme on HIV/AIDS (UNAIDS), in sub-Saharan Africa HIV prevalence among people ages 15–49 years in the region is, on average, twice as high in urban areas as in rural areas (UNAIDS, 2014). In Ethiopia, HIV prevalence in urban areas was nearly sevenfold higher in 2011!

The Habitat Base

The story of HIV would not be complete without a discussion of the ecological changes that drove its emergence. Early in the epidemic, these ecological changes were largely driven by political economic forces bent on resource extraction. This is not to say that the interactions between human and simian populations that led to the emergence of HIV had not occurred for years. Hunting bush meat had always been a viable source of food, and many scientists believe HIV was introduced into small rural areas at multiple points as far back as the late 1800s (Faria et al., 2014). But the more intense encroachment into forested land came with the identification and extraction of rubber in the Congo basin and the creation of mines in formerly unsettled lands. This economic strategy likely increased human–simian contact both because more people were circulating on the land and the poverty that arose under Belgian colonial rule caused more people to hunt for food and economic gain. Deforestation continued as people began to clear-cut land for crops, further increasing border contact between animals and humans. Plate 9 shows two remote sensing images from Google Earth that show deforestation between 1978 and 2005 in an area on the Rwandan/DRC border.

The built environment has also played a key role in the HIV epidemic. Without the construction of railways and roads, HIV would not have spread so dramatically throughout sub-Saharan Africa. And with major African ports connected to Europe, the Americas, and major Asian cities via both shipping and airplane travel, it was easy for the HIV virus to spread globally. Early in the epidemic, long-distance truckers were identified with much higher rates of HIV and other STIs and the trans-African highway was identified as a major reason for diffusion of the disease (Gould, 1993). Transactional sex along the trucking routes only fueled the epidemic. Conover (1993) describes his journeys along the trucking routes in East Africa where truck drivers routinely stop at town and engage prostitutes for the night. Sometimes they visit the same prostitute each time but often they do not.

The socioeconomic context of urban centers created hotspots for HIV infection. Within urban areas massive social changes created complex cultural ecologies that continue to drive the HIV epidemic. The urban population in the developing world is disproportionately male. Young men migrating for work leave their families behind and contract the services of female sex workers, creating complicated sexual networks that spread the disease within the urban setting. The liberalization of sexual attitudes and behaviors, spurred by political and social connections to the global north, changed attitudes and ideas about sex and other risk-taking behaviors. These socio-behavioral changes, coupled with poverty, inadequate health systems, and lack of education, have created urban HIV hotspots. Although ART is widely available in developed countries like the United States, this is not the case in much of the developing world. Standard ART consists of the combination of at least three different antiretroviral drugs that work in tandem to suppress the HIV virus. Weak public health systems, especially in large urban slums and distant rural areas, make provision of ART difficult. Medications must be transported and stored, regular distribution of the drugs must be monitored, and there must be enough doctors and nurses to serve the needs of the population. Much of sub-Saharan Africa has chronic health worker shortages; in Kenya, there are 1.1 nurses per 1,000 population, but in Sierra Leone, Tanzania, and Liberia this number is closer to 0.35 per 1,000 (Kinfu, Dal Poz, Mercer, & Evans, 2009). Migrant populations are ephemeral, so that managing the health of these populations over time is difficult. Traditionally, HIV treatment has been far too expensive for the majority of infected people in resource-poor countries. This, too, is influenced by the larger political economy. The major drug companies that developed ART charged high prices for a full regime of pills: US $10,000–$15,000 per person per year. The public health system in low-resource countries could not possibly provide ART for all the people in need! Five years after ART was introduced in the United States and Europe, only 2% of people in developing countries were receiving the drugs (Stover & Schwartlander, 2002). In 2001, an Indian drug manufacturer defied international laws related to international property rights (called TRIPS) and provided a generic version of ART (Waning, Diedrichsen, & Moon, 2010). This created competition in the international marketplace and caused the price to plummet. Additional pressure by activist organizations, outcry from the international community, and negotiations with pharmaceutical companies have led to a system of "tiered pricing" whereby low-income countries can purchase drugs at vastly reduced prices.

The Geography of HIV

Spatial studies of HIV/AIDS can roughly be divided into two categories: those that study the origin and diffusion of the disease, and those that have tried to identify risk factors, including spatial risk factors like access to care and the spatial clustering of at-risk populations. The goal of many of these studies is to more effectively target interventions that either prevent the spread of the disease or provide services to those most in need. Early in the epidemic, Shannon, Pyle, and Bashshur (1991) examined the geographic variation in HIV to trace the expansion of the virus and the socio-behavioral reasons for transmission in different parts of the world. They identified central Africa as the index location for HIV (well before any phylogenetic analyses were done!) by mapping HIV infection rates reported by the WHO and comparing these rates to settlement patterns and trade routes. Peter Gould (1993), another notable geographer, followed with a more sophisticated analysis of the AIDS crisis in *The Slow Plague*. What is remarkable about this work is Gould's focus on the complexities of spatial diffusion. He asserts that diffusion is scale-dependent and that different processes drive diffusion at different geographic scales: transportation networks (within Africa and from Africa to Europe/the United States), political–economic ties (between Thailand and the United States), and social ties (within cities in the United States). He weaves in a deep understanding of disease ecology, highlighting how social, economic, and behavioral factors (many of which we discussed above) have fueled the diffusion of the disease in different settings. Gould makes the case that this spatially explicit analysis of the AIDS pandemic is imperative for the effective design and implementation of public health responses.

There are several good examples of how the political ecology framework has been applied to HIV. Kalipeni (2000) uses a vulnerability framework to examine HIV/AIDS in sub-Saharan Africa. He suggests that differential access to resources is at the core of the epidemic and that inequalities are a result of power relations that occur across multiple scales—local, regional, and global—and act on individual health. In particular, the structural adjustment programs of the World Bank and International Monetary Fund (IMF) forced many countries to cut back on public spending for health and education. This decreased accessibility of health services and increased the vulnerability of the population to HIV/AIDS, especially among marginalized or disenfranchised groups. The subsequent increase in the sever- ity of the HIV/AIDS epidemic is a direct result of this economic policy. Kalipeni and Oppong (1998) examined the health among refugee populations in central Africa. They provided a thorough review of the political and economic circum- stances around conflict and refugee movements in the region and then explored how resulting ecologic circumstances (e.g., refugee camps) affected health services delivery. Though they discuss refugee health in general, they specifically argue that refugee settlements create environments where HIV rates are quite high (many populations fled from highly infected regions), poverty leads to transactional sex for food and money, and health services are severely limited including access to condoms and treatment for STDs.

King (2010) examined the HIV/AIDS epidemic in South Africa before and after apartheid. During the apartheid era the forcible relocation of a majority of the African population from traditional lands that were agriculturally productive cre- ated poor, high-density regions with limited access to food. Many young men and women migrated for work, shaping HIV transmission patterns even today. Further, the slow government response to the HIV/AIDS epidemic was "tempered by a colo- nial and apartheid history that used public health as a justification for racial seg- regation" (King, 2010, p. 47). The Western scientific discoveries about HIV/AIDS and messages regarding prevention and transmission were distrusted by the South African government because they were seen as a neocolonial racist discourse con- structed to sell expensive pharmaceuticals (ART). King also examined the smaller- scale cultural norms and traditions that have facilitated the spread of HIV. For example, mortality has been high in South Africa, and social systems and extended kin networks have broken down as parents die, leaving children to be cared for by friends or relatives. In direct relation to the environment, natural resources such as wood, fruits, and herbs are being utilized to offset the decrease in food production when an adult dies.

What is common to all these political ecology studies is an acknowledgment that the "ecology" piece should be very broadly conceived. Political–economic forces act directly on the natural environment but also create spaces such as refu- gee camps, villages with no health care, and regions with poor economic opportu- nities. Cultural ecologies and social and economic contexts are unique. Thus, the whole environment can adversely affect health.

More than a few studies use statistical methods to model spatial patterns of HIV/AIDS. This includes identifying the risk factors that contribute to apparent spatial clustering of the disease. Kalipeni and Zulu (2008) used a GIS to construct a spatiotemporal timeline of HIV/AIDS prevalence rates in sub-Saharan Africa from 1986 to 2010. Maps that interpolate HIV/AIDS rates (e.g., create smooth surfaces from point data) clearly show the eastern and then southern spread of HIV over 25 years. Finally, they used their geographic timeline to forecast rates to 2010, and suggested that the HIV/AIDS epidemic for many countries has reached saturation and has even begun to decline. Tanser, Bärnighausen, Cooke, and Newell (2009) found significant spatial variation in HIV prevalence (6–36%) in KwaZulu-Natal, South Africa. Statistical analyses revealed greatest risk in urban and high-density settlements and found localized spatial clusters of HIV that were contained within geographically defined settlements. They suggested that their results could be used to target communities at greatest risk for HIV.

Feldacker, Ennett, and Speizer (2011) used HIV data from Malawi and estimates of community socioeconomic and accessibility data to determine whether larger community context affects risk behaviors and HIV prevalence and how these relationships vary by gender. They found that for women, community income inequality was associated with an increased risk of HIV. For both men and women, access to a public health clinic and greater connectivity of their community were also related to HIV risk. Thus, HIV risk is related to larger economic forces that result in inequalities in social relationships (particularly for women) and access to care. The authors suggested that integrated microfinance and HIV prevention programs may directly benefit communities in Malawi. Since HIV/AIDS is so highly related to sexual networks, researchers have examined the spatial distribution of sexual networks in order to better understand how to target prevention programs. Zenilman, Ellish, Fresia, and Glass (1999) used GIS to map 286 sexual partnerships and examined the geographic distance between partners. They found two geographically distinct core areas of HIV, and partners in those core areas lived remarkably close to one another.

THE PRECAUTIONARY PRINCIPLE AND SOME POLITICAL ECOLOGY OF RESEARCH

As one reads through hundreds of articles and reports on cancer, heart disease, and HIV, it is remarkable how little is known, studied, or said about prevention. The few exceptions include the recent promotion of diets rich in fruit and vegetables (perhaps organically grown), and finally, after more than half a century of opposition and inaction, serious promotion of smoking cessation. Even now, though, although the tobacco-growing state of North Carolina allows its university in Chapel Hill to ban smoking within a hundred yards of all its buildings (definitely including its hospitals), the legislature steadfastly refuses to raise taxes on cigarettes despite critical budget problems. And despite the transformative research usefulness of GIS in managing and analyzing spatial data sets (discussed in Chapter 5), comparatively little funding is available for analyzing environmental exposures (geographic environment, not epidemiological behavior), developing surveillance of harmful chemicals, or monitoring population risk or public health interventions. Why has so little been done? What drives funding? Why do so much research funding and research promise today lie with genetics and drugs for treatment? The answers to these questions lie in complex ties between economic incentives and political interests, and the influence large corporations (with large profits) have over government policy, research funding decisions, and regulation.

The **precautionary principle** states that when there is a clear indication of harm (to human life and health), that indication should be a trigger for action even if the scientific evidence of proof is not complete, and especially if delay may cause irreparable damage. The idea has been argued for decades, but it received a new life when the principle was endorsed in 1987 by European ministers of the environment who were concerned about deterioration of the North Sea. More recently, it has been important in the arguments over causation of and response to global warming. In the area of health, Sandra Steingraber (1998) advanced the principle powerfully to address the proliferation of known carcinogens in our environment. She described the experience of cancer, putting a human face on the victims and even on the cell lines of research. She described the nature of the data (the small-number problem, the difficulties of cluster identification, the lack of a register, and the mobility of the population). But most of all, she described the chemical soup in which we live, exposed to chemicals in food and drinking water, by showers, and in clothing, house cleaners, paints, agricultural spraying, and industrial food wrapping. She searched for the research on connections of the environmental context: agricultural chemicals in the water, industrial gases and particles belching into the air, uncountable products made of plastics, and countless other industrial, residential, and personal chemical uses (see Table 11.1) to which the population is exposed. By her best tally in 1997, 40 possible carcinogens appear in drinking water around the country, 60 are released by industry into ambient air, and 66 are routinely sprayed on food crops as pesticides. She cited in full the "Wingspread Statement" from a 1997 conference (of scientists, lawyers, farmers, physicians, planners, environmental activists, philosophers, and others) in which she participated. An extensive excerpt from that statement follows (Steingraber, 1998, p. 284):

> The release and use of toxic substances, the exploitation of resources, and physical alterations of the environment have had substantial unintended consequences affecting human health and the environment. Some of these concerns are high rates of learning deficiencies, asthma, cancer, birth defects and species extinctions, along with the global climate change, stratospheric ozone depletion and global worldwide contamination with toxic substances and nuclear material.
>
> We believe existing environmental regulations and other decisions, particularly those based on risk assessment, have failed to protect adequately human health and the environment—the larger system of which we humans are a part.
>
> [We believe that there is] evidence of damage to humans and that worldwide new environmental principles are needed; recognizing [that] human activities may involve hazards, people must proceed more carefully than [they have done in] recent history: corporations, governmental entities, organizations, communities, scientists, [and] other individuals must adopt a precautionary approach to all human endeavors.

Therefore it is necessary to implement the precautionary principle. When an activity raises threats of harm to human health or the environment, precautionary measures should be taken even if some cause and effect relationships are not fully established scientifically. In this context, the proponent of an activity rather than the public, should bear the burden of proof.

The process of applying the precautionary principle must be open, informed and democratic and must include potentially affected parties. It must also involve an examination of the full range of alternatives, including no action.

Funding of Studies: Genetics and Environmental Risk

Many millions have died in the last few decades from preventable cancer. In recent years several shocking analyses and exposés have been published about the suppression of meaningful research on prevention. Two of the most powerful and respectable are described here. Devra Davis, who is now director of the Center for Environmental Oncology at the University of Pittsburgh Cancer Institute and a professor of epidemiology, has looked back over a career of 40 years in epidemiology, research, and government regulation and has reviewed rare unpublished research reports going back to the 1930s. In her book *The Secret History of the War on Cancer* (2007), she reports on how much was known about the carcinogenic importance of benzene in petroleum products, benzidine (which causes bladder cancer) in dyes, chromium, asbestos, and of course tobacco. (The U.S. government even supported research on "safe cigarettes," known [by scientist consensus] to be impossible.) With deep data, and interviews (many from 50 years ago), and detailed scientific explanations, Davis explains how the 1971 launch of the government's War on Cancer "focused on defeating an enemy by detecting, treating, and curing the disease—not preventing it." She goes on to describe how this campaign ignored and/or deliberately suppressed "proof that how and where we live and work affects the chances we may get cancers was basically ignored" (p. 15); how alliances (either naïve or pecuniary) of producers of products, makers of medicines, sellers of medical treatments, and academic researchers ignored how much was known about social and environmental causes of cancer from the sophisticated research done before World War II in their peer-reviewed funding decisions; and how industrial leadership acted "decade by decade" to downplay research on prevention, instead preventing environmental causes from gaining circulation in public information or benefiting the general public.

David Michaels (2008) has recently published a book titled *Doubt Is Their Product*. He is an epidemiologist and director of the Project on Scientific Knowledge and Public Policy at the George Washington University; during the Clinton administration he served as assistant secretary of energy for environment, safety, and health, with responsibility for protecting the health and safety of workers, neighboring communities, and the environment surrounding nuclear weapons facilities. Winner of the 2006 American Association for the Advancement of Science's Scientific Freedom and Responsibility Award, he is no muckraker. Yet he further exposes the failure of legislation, policy development, and agency regulation to prevent cancer and other serious diseases because of the warping and compromising effects of money on judgment and supporting research. He especially takes on the peer review process and the increasing involvement of industry in financing even academic research. Review of one's research by professional peers is critical to all academic research, whether social, physical, or medical. It is how science maintains standards, builds on previous knowledge, and gives vital feedback. Peer review affects research at two stages: the evaluation of research proposals resulting in determining which research to fund, and the publication of research results. Noting how corporations funding university projects in the 1990s blocked publication of research detrimental to their companies until leading journals refused to publish unless the contract left the (academic) research free of commercial interest, Michaels notes how current regulatory agencies (the EPA, the Occupational Safety and Health Administration, etc.) still have no mechanisms to identify conflicts of interest resulting from sponsor control. He makes 10 recommendations, three of which are that there should be full disclosure of any and all sponsors' involvement in scientific studies; that manufacturers should test chemicals for safety before exposing workers and the public to them; and that secrecy should be ended as it affects the health of the public, with chief executives being held responsible for submitting "accurate and complete" reports on what is known about the toxicity of their products.

Davis and Michaels agree that the responsibility for the safety of new products released to the environment (and thus the responsibility for testing these) should be put on the producers of the products. They also agree that those setting priorities for research funding have concentrated on treating the effects of pollution (new medicines, medical specialties, and equipment and tests) and perhaps refining or reducing the pollution (better smokestack scrubbers), but have largely ignored prevention of risk, which would threaten or preclude the production and release of the dangerous chemicals themselves.

Genetics has become the dominant theme of research, funding, and popular explanations in the mass media as well. This is increasingly disturbing to many kinds of health scientists. What British authors Hall (2004) and Hedgecoe (2001, 2002) call the "geneticisation" of heart disease, schizophrenia, and other conditions seems an "unstoppable process." That is, genetic explanations are being expanded to include the role of behavior and environment in the expressions of genes, including those that result in complex conditions. However, such explanations marginalize individual, social, and even environmental influences, which are reduced to mere "triggers" of the underlying genetic process or predisposition. (Or is this merely the old outrage of social scientists that anything about human behavior could be genetically caused?)

Promising new medicines are being developed for regulating Parkinson's disease, starving cancer cells, and treating asthma. Some lives are being saved, no doubt about it. But what was it that threatened those lives in the first place? The academic researchers gain reputations; the pharmaceutical companies and the manufacturers of medical research tools profit; the hospitals fill their beds and coffers; and the manufacturers of the dangerous chemicals themselves, of whatever kind, continue with their growth, profits, and dividends. So many are benefited except the millions of better-treated human beings, who perhaps did not need to be afflicted at all. Some social scientists fear a new eugenics movement, and others fear the loss of health insurance by millions, as new "susceptibility genes" are disclosed. The underfunded, under-researched, unanswered basic question of disease ecology still burns: What is it that people with these genes are susceptible to? And, by the way, why did their society expose them to it?

QUICK REVIEW 2

✓ Causal relationships between health and population, environment, or social/behavioral relationships are very difficult to establish. The field of epidemiology seeks to design studies that allow researchers to establish causality over time.

✓ Epidemiologists study whether there are different disease frequencies in two or more populations and the factor(s) that are more frequent in the population with the higher rates of disease.

✓ While some cancers are genetically linked, many are related to carcinogens we are exposed to in our environment. These exposures are often shaped by political and economic interests, such as the tobacco and petrochemical industries.

✓ Geographers have long studied the spatial distribution of cancers around the world. They use tools such as the relative risk and odds ratio to create maps of risk.

✓ While environmental changes related to forest encroachment and urbanization played a large part in the inception and expansion of the HIV epidemic, HIV transmission is driven by social, cultural, and behavioral structures that increased the risk of transmission and complicated the diagnosis and treatment of the disease.

✓ The precautionary principle places the burden of proving that environmental toxicants do not affect health on the companies that are producing and using these toxicants. It also calls for government policy and regulation of these industries that is free from political–economic influence.

✓ Evidence suggests that decisions about what research to fund, and what research is published in peer-reviewed journals, is influenced by political and economic interests.

CONCLUSION

This chapter has expanded on the concept of disease ecology by explicitly including the political, economic, and gender forces that not only directly impact disease, but also the research on, regulation of, and prioritization of disease. As the demographic, nutrition, and mobility transitions continue, and as the world's population continues to urbanize and age, the etiology, prevention, and treatment of degenerative diseases are increasingly dominating world health concerns. But communicable diseases continue to be a problem, driven by poverty, income, and gender inequalities, and influenced by global, regional, and national economic policies.

This chapter has also examined some of the methodological and epistemological issues for causative reasoning, and some designs for modeling and analyzing multifactorial relationships, which may include upstream political and economic factors. The spatial patterns, causal relationships, and time and scale parameters of these diseases are complex matters. If hypotheses are to have value, the geographic researcher needs to understand the disease processes as founded in the population's biology and its interaction with habitat and behavior. The complexities of spatial pattern and disease etiology demand competence in statistics and demography as well as mapping/GIS. On this foundation rests the ability to project population health needs, as distinct from economic demand, and to plan for the delivery of appropriate health services.

REVIEW QUESTIONS

1. Political ecologies of disease are often multiscalar. Define this term and describe why is it important in studying the upstream factors (including political and economic forces) that affect disease ecologies.

2. How might international policy, including monetary support (e.g., from WHO, UNAID, UNESCO), hinder long-term disease eradication and control efforts in developing nations? Use a specific disease as an example to make your case.

3. Revisit the example of the disease ecology of asthma presented in Chapter 2. How do upstream political and economic forces, including structural barriers related to poverty and race, affect asthma's disease ecology?

4. What is the relationship between structure and agency? Are chronic conditions such as heart disease and obesity more influenced by structure or agency? Why?

5. Drawing on the example of HIV in this chapter, describe the role that politics and economic interests play in deforestation, especially in developing countries, and the emergence of disease.

6. South African president Thabo Mbeki is famous for saying that "HIV is not the sole cause of AIDS." Rather, he suggests that AIDS is a complex syndrome that involves poor nutrition, poverty, and other upstream factors. Do you think that the HIV/AIDS epidemic is the outcome of poverty in sub-Saharan Africa? Why or why not?

7. Describe how political and economic forces might drive the prioritization of health research in the United States. Use a specific disease to make your case.

8. "Race" is a classification that is highly dependent on social and political conventions. Does it have any relevance for discussing disease ecologies? Why or why not?

REFERENCES

Armstrong, R. W. (1976). The geography of specific environments of patients and non-patients in cancer studies, with a Malaysian example. *Economic Geography, 52,* 161–170.

Armstrong, R. W. (1978). Self-specific environments associated with nasopharyngeal carcinoma in Selangor, Malaysia. *Social Science and Medicine, 12D,* 149–156.

Barker, G., & Ricardo, C. (2005). *Young men and the construction of masculinity in sub-Saharan Africa: Implications for HIV/AIDS, conflict, and violence.* Washington, DC: World Bank.

Bausch, D. G., & Schwarz, L. (2014). Outbreak of Ebola virus disease in Guinea: Where ecology meets economy. *PLoS Neglected Tropical Diseases, 8*(7), e3056.

Brody, J. G., Moysich, K. B., Humblet, O., Attfield, K. R., Beehler, G. P., & Rudel, R. A. (2007). Environmental pollutants and breast cancer: Epidemiological studies. *Cancer, 109*(Suppl. 12), 2667–2712.

Brookfield, H. (2005). Political ecology: A critical introduction. *Geographical Review, 95*(2), 296–300.

Calabrese, E. J. (1985). *Sex differences in response to toxic substances.* New York: Wiley.

Collins, J., & Rau, B. (2000). *AIDS in the context of development* (UNRISD Programme on Social Policy and Development) (Paper No. 4), UNRISD/UNAIDS. Geneva, Switzerland.

Conover, T. (1993). Trucking through the AIDS belt. *New Yorker, 16,* 56–75.

Davis, D. (2007). *The secret history of the war on cancer.* New York: Basic Books.

Dear, M. J., & Wolch, J. (1987). *Landscapes of despair: From deinstitutionalization to homelessness.* Princeton, NJ: Princeton University Press.

Doyal, L. (1995). *What makes women sick?: Gender and the political economy of health.* New Brunswick, NJ: Rutgers University Press.

Dufault, R., LeBlanc, B., Schnoll, R., Cornett, C., Schweitzer, L., Wallinga, D., et al. (2009). Mercury from chlor-alkali plants: Measured concentrations in food product sugar. *Environmental Health, 8*(1), 2.

Dunkle, K. L., Jewkes, R. K., Brown, H. C., Gray, G. E., McIntryre, J. A., & Harlow, S. D. (2004). Gender-based violence, relationship power, and risk of HIV infection in women attending antenatal clinics in South Africa. *The Lancet, 363*(9419), 1415–1421.

Editorial Committee for the Atlas of Cancer Mortality in the People's Republic of China. (1979). *Atlas of cancer mortality in the People's Republic of China.* Shanghai: China Map Press.

Faria, N. R., Rambaut, A., Suchard, M. A., Baele, G., Bedford, T., Ward, M. J., et al. (2014). The early spread and epidemic ignition of HIV-1 in human populations. *Science, 346*(6205), 56–61.

Farmer, P. (1992). *AIDS and accusation: Haiti and the geography of blame.* Berkeley & Los Angeles: University of California Press.

Farmer, P. (1997). Social scientists and the "new tuberculosis." *Social Science and Medicine, 44*(3), 347–358.

Farmer, P. (1999). *Infections and inequalities: The modern plagues.* Berkeley & Los Angeles: University of California Press.

Farmer, P. (2003). *Pathologies of power: Health, human rights, and the new war on the poor.* Berkeley & Los Angeles: University of California Press.

Feldacker, C., Ennett, S. T., & Speizer, I. (2011). It's not just who you are but where you live: An exploration of community influences on individual HIV status in rural Malawi. *Social Science and Medicine, 72*(5), 717–725.

Geronimus, A. T., Bound, J., Waidmann, T. A., Hillemeier, M. M., & Burns, P. B. (1996). Excess mortality among blacks and whites in the United States. *New England Journal of Medicine, 335,* 1552–1558.

Glick, B. J. (1979). Distance relationships in theoretical models of carcinogenesis. *Social Science and Medicine, 13D,* 253–256.

Glick, B. J. (1980). The geographic analysis of cancer occurrence: Past progress and future directions. In M. S. Meade (Ed.), *Conceptual and methodological issues in medical geography* (pp. 170–193). Chapel Hill: University of North Carolina, Department of Geography.

Gould, P. (1993). *The slow plague: A geography of the AIDS pandemic.* New York: Blackwell.

Gray, R. H., Li, X., Kigozi, G., Serwadda, D., Brahmbhatt, H., Wabwire-Mangen, F., et al. (2005). Increased risk of incident HIV during pregnancy in Rakai, Uganda: A prospective study. *The Lancet, 366*(9492), 1182–1188.

Greenberg, M. R. (1983). *Urbanization and cancer mortality: The United States experience, 1950–1975.* New York: Oxford University Press.

Hall, E. (2004). Spaces and networks of genetic knowledge making: The "geneticisation" of heart disease. *Health and Place, 10,* 311–318.

Hanchette, C. (1988). *Geographical patterns of prostate cancer mortality: An investigation of the relationship between prostate cancer and ultraviolet light.* Unpublished master's thesis, University of North Carolina.

Hanchette, C. L. (1998). *The geographic modeling of lead poisoning risk in North Carolina.* Unpublished doctoral dissertation, University of North Carolina.

Hanchette, C. L. (2008). The political ecology of lead poisoning in eastern North Carolina. *Health and Place, 14*(2), 209–216.

Hanchette, C. L., & Schwartz, G. G. (1992). Geographic patterns of prostate cancer mortality. *Cancer, 70,* 2861–2869.

Harpham, T., Lusty, T., & Vaughan, P. (Eds.). (1988). *In the shadow of the city: Community health and the urban poor.* New York: Oxford University Press.

Hedgecoe, A. (2001). Schizophrenia and the narrative of enlightened geneticisation. *Social Studies of Science, 31*(6), 875–911.

Hedgecoe, A. (2002). Reinventing diabetes: Classification, division and the geneticisation of diseases. *New Genetics and Society, 21*(1), 7–27.

Heffron, R., Donnell, D., Rees, H., Celum, C., Mugo, N., Were, E., et al. (2012). Use of hormonal contraceptives and risk of HIV-1 transmission: A prospective cohort study. *The Lancet: Infectious Diseases, 12*(1), 19–26.

Heise, L., & Elias, C. (1995). Transforming AIDS prevention to meet women's needs: A focus on developing countries. *Social Science and Medicine, 40*(7), 931–943.

Inhorn, M. C., & Brown, P. J. (Eds.). (1997). *The anthropology of infectious disease: International health perspectives.* Amsterdam, The Netherlands: Gordon & Breach.

Institute for Agriculture and Trade Policy (IATP). (2009). Much high fructose corn syrup contaminated with mercury, new study finds. Retrieved from *www.iatp.org/documents/much-high-fructose-corn-syrup-contaminated-with-mercury-new-study-finds.*

Kalipeni, E. (2000). Health and disease in southern Africa: A comparative and vulnerability perspective. *Social Science and Medicine, 50,* 965–983.

Kalipeni, E., & Oppong, J. (1998). The refugee crisis in Africa and implications for health and disease: A political ecology approach. *Social Science and Medicine, 46*(12), 1637–1653.

Kalipeni, E., & Zulu, L. (2008). Using GIS to model and forecast HIV/AIDS rates in Africa, 1986–2010. *The Professional Geographer, 60*(1), 33–53.

Kinfu, Y., Dal Poz, M. R., Mercer, H., & Evans, D. B. (2009). The health worker shortage in Africa: Are enough physicians and nurses being trained? *Bulletin of the World Health Organization, 87*(3), 225–230.

King, B. (2010). Political ecologies of health. *Progress in Human Geography, 34*(1), 38–55.

King, B. (2015). Political ecology and human health. In T. Perreault, G. Bridge, & J. McCarthy (Eds.), *The Routledge handbook of political ecology* (pp. 343–353). New York: Routledge.

King, B., & Crews, K. (Eds.). (2013). *Ecologies and politics of health.* London: Routledge.

Koblinsky, M., Timyan, J., & Gay, J. (Eds.). (1993). *The health of women in global perspective.* Boulder, CO: Westview Press.

Lambert, H., & Wood, K. (2005). A comparative analysis of communication about sex, health and sexual health in India and South Africa: Implications for HIV prevention. *Culture, Health and Sexuality, 7*(6), 527–541.

Lilienfeld, A. M., & Lilienfeld, D. E. (1980). *Foundations of epidemiology* (2nd ed.). New York: Oxford University Press.

Luke, N., & Kurz, K. (2002). *Cross-generational and transactional sexual relations in sub-Saharan Africa.* Washington, DC: International Center for Research on Women (ICRW).

Maharaj, P., & Cleland, J. (2004). Condom use within marital and cohabiting partnerships in KwaZulu-Natal, South Africa. *Studies in Family Planning, 35*(2), 116–124.

Mason, T. J., McKay, F. W., Hoover, R., Blot, W. J., & Fraumeni, J. F., Jr. (1975). *Atlas of cancer mortality for U.S. counties: 1950–1969.* Washington, DC: U.S. Department of Health, Education and Welfare, National Cancer Institute, Epidemiology Branch.

Mayer, J. D. (1996). The political ecology of disease as a new focus for medical geography. *Progress in Human Geography, 20,* 441–456.

McMichael, A. J. (2001). *Human frontiers, environments and disease: Past patterns, uncertain futures.* Cambridge, UK: Cambridge University Press.

Michaels, D. (2008). *Doubt is their product: How industry's assault on science threatens your health.* New York: Oxford University Press.

Mill, J. S. (1856). *A system of logic.* London: Parker, Son & Bowin.

Myer, L., Denny, L., Telerant, R., de Souza, M., Wright, T. C., & Kuhn, L. (2005). Bacterial vaginosis and susceptibility to HIV infection in South African women: A nested case–control study. *Journal of Infectious Diseases, 192*(8), 1372–1380.

National Center for Health Statistics. (1997). *Atlas of United States mortality.* Washington, DC: U.S. Government Printing Office.

National Center for Health Statistics. (1998). Table 8: Deaths and death rates for the 10 leading causes of death in specified age groups, by race and sex: United States, 1996. *National Vital Statistics Reports, 47,* 26–36.

Robbins, P. (2004). *Political ecology: A critical introduction.* Malden, MA: Blackwell.

Rosenberg, M. W. (1988). Linking the geographical, the medical and the political in analyzing health care delivery systems. *Social Science and Medicine, 26,* 179–186.

Schwartz, G., & Hanchette, C. L. (2006). UV, latitude, and spatial trends in prostate cancer mortality: All sunlight is not the same (United States). *Cancer Causes and Control, 17*(8), 1091–1101.

Sewankambo, N., Gray, R. H., Wawer, M. J., Paxton, L., McNairn, D., Wabwire-Mangen, F., et al. (1997). HIV-1 infection associated with abnormal vaginal flora morphology and bacterial vaginosis. *The Lancet, 350*(9077), 546–550.

Shabecoff, P., & Shabecoff, A. (2008). *Poisoned profits: The toxic assault on our children.* New York: Random House.

Shafie, N. J., Mohd Sah, S. A., Abdul Latip, N. S., Azman, N. M., & Khairuddin, N. L. (2011). Diversity pattern of bats at two contrasting habitat types along Kerian River, Perak, Malaysia. *Tropical Life Sciences Research, 22*(2), 13–22.

Sharp, P. M., & Hahn, B. H. (2011). Origins of HIV and the AIDS pandemic. *Cold Spring Harbor Perspectives in Medicine, 1*(1), a006841.

Steingraber, S. (1998). *Living downstream.* New York: Vintage Books.

Stover, J., & Schwartlander, B. (2002). *Coverage of selected health services for HIV/AIDS prevention and care in less developed countries in 2001.* Geneva, Switzerland: World Health Organization.

Study Finds High-Fructose Corn Syrup Contains Mercury. (2009, January 28). *The*

Washington Post. Retrieved from *www.washingtonpost.com/wp-dyn/content/article/2009/01/26/AR2009012601831.html.*

Tanser, F., Bärnighausen, T., Cooke, G. S., & Newell, M. L. (2009). Localized spatial clustering of HIV infections in a widely disseminated rural South African epidemic. *International Journal of Epidemiology,* dyp148.

Timberg, C., & Halperin, D. (2012). *Tinderbox: How the West sparked the AIDS epidemic over the past decade.* New York: Penguin Books.

Turshen, M. (1977). The political ecology of disease. *Review of Radical Economics, 9,* 45–60.

Turshen, M. (1984). *The political ecology of disease in Tanzania.* New Brunswick, NJ: Rutgers University Press.

UNAIDS. (2014). *The cities report.* Geneva, Switzerland: UNAIDS.

Varga, C. A. (2003). How gender roles influence sexual and reproductive health among South African adolescents. *Studies in Family Planning, 34*(3), 160–172.

Wallace, R. G., Gilbert, M., Wallace, R., Pittiglio, C., Mattioli, R., & Kock, R. (2014). Did Ebola emerge in West Africa by a policy-driven phase change in agroecology? *Environment and Planning A, 46*(11), 2533–2542.

Wang, C. C., Kreiss, J. K., & Reilly, M. (1999). Risk of HIV infection in oral contraceptive pill users: A meta-analysis. *Journal of Acquired Immune Deficiency Syndromes, 21*(1), 51–58.

Waning, B., Diedrichsen, E., & Moon, S. (2010). A lifeline to treatment: The role of Indian generic manufacturers in supplying antiretroviral medicines to developing countries. *Journal of the International AIDS Society, 13*(1), 35.

Wolch, J., & Dear, M. (1993). *Malign neglect: Homelessness in an American city.* San Francisco: Jossey-Bass.

World Health Organization (WHO). (2007). Health inequality, inequity, and social determinants of health. *Population and Development Review, 33*(4), 839–843.

World Health Organization (WHO). (2014). Top 10 Leading Causes of Death, Fact sheet No. 310. Available at *www.who.int/mediacentre/factsheets/fs310/en.*

Xu, F., Kilmarx, P. H., Supawitkul, S., Yanpaisarn, S., Limpakarnjanarat, K., Manopaiboon, C., et al. (2000). HIV-1 seroprevalence, risk factors, and preventive behaviors among women in northern Thailand. *Journal of Acquired Immune Deficiency Syndromes, 25*(4), 353–359.

Yanos, P. T. (2007). Beyond "landscapes of despair": The need for new research on the urban environment, sprawl, and the community integration of persons with severe mental illness. *Health and Place, 13,* 672–676.

Zenilman, J. M., Ellish, N., Fresia, A., & Glass, G. (1999). The geography of sexual partnerships in Baltimore: Applications of core theory dynamics using a geographic information system. *Sexually Transmitted Diseases, 26*(2), 75–81.

FURTHER READING ·····································

Anderson, R. (1984). Temporal trends of cancer mortality in eastern New England compared to the nation, 1950–1975. *Social Science and Medicine, 19,* 749–757.

Andrews, H. F. (1985). The ecology of risk and the geography of intervention: From research to practice for the health and well-being of urban children. *Annals of the Association of American Geographers, 75,* 370–382.

Armstrong, R. W. (1971). Medical geography and its geologic substrate. *Geological Society of America Memoir, 123,* 211–219.

Carson, R. (2002). *Silent spring.* New York: Houghton Mifflin Harcourt.

Casper, M. L., Barnett, E., Halverson, J. A., Elmes, G. A., Braham, V. E., Majeed, Z. A., et al. (2000). *Women and heart disease: An atlas of racial and ethnic disparities in mortality* (2nd ed.). Morgantown: West Virginia University, Office for Social Environment and Health Research.

Cleek, R. K. (1979). Cancer and the environment: The effect of scale. *Social Science and Medicine, 13D,* 241–247.

Cooper, R., Cutler, J., Desvigne-Nickens, P., Fortmann, S. P., Friedman, L., Havlik, R., et al. (2000). Trends and disparities in coronary heart disease, stroke, and other cardiovascular diseases in the United States: Findings of the National Conference on Cardiovascular Disease Prevention. *Circulation, 102,* 3137–3147.

Farmer, P., Kleinman, A., Kim, J., & Basilico, M. (2013). *Reimagining global health: An introduction.* Berkeley & Los Angeles: University of California Press.

Gardner, M. (1976). Soft water and heart disease. In J. Leniham & W. W. Fletcher (Eds.), *Health and the environment* (pp. 116–135). New York: Academic Press.

Gardner, M. J., Winter, P. D., & Acheson, E. D. (1982). Variations in cancer mortality areas in England and Wales: Relation with environmental factors and search for cause. *British Medical Journal, 284,* 284–287.

Howe, G. M., Burgess, L., & Gatenby, P. (1977). Cardiovascular disease. In G. M. Howe (Ed.), *A world geography of human diseases* (pp. 431–476). New York: Academic Press.

Krieger, N., Chen, J. T., Waterman, P. D., Rehkopf, D. H., & Subramanian, S. V. (2005). Painting a truer picture of US socioeconomic and racial/ethnic health inequalities: The Public Health Disparities Geocoding Project. *American Journal of Public Health, 95,* 312–323.

Lewis, N. (1998). Intellectual intersection: Gender and health in the Pacific. *Social Science and Medicine, 46,* 641–659.

Lux, W. E., & Kurtzke, J. F. (1987). Is Parkinson's disease acquired?: Evidence from a geographic comparison with multiple sclerosis. *Neurology, 37,* 467–471.

McLafferty, S., & Tempalski, B. (1995). Restructuring and women's reproductive health: Implications for low birthweight in New York City. *Geoforum, 26,* 309–323.

Murray, C. J. L., Kulkarni, S. C., Michaud, C., Tomijima, N., Bulzacchelli, M. T., Landiorio, T. J., et al. (2006). Eight Americas: Investigating mortality disparities across races, counties, and racecounties in the United States. *PLoS Medicine, 3*(9), e260.

Pyle, G. F. (1971). *Heart disease, cancer, and stroke in Chicago* (Research Paper No. 134). Chicago: University of Chicago, Department of Geography.

Pyle, G. F. (1979). *Applied medical geography.* New York: Wiley.

Shigematsu, I. (1981). *National atlas of major disease mortality in Japan.* Tokyo: Japan Health Promotion Foundation.

UNICEF and UNAIDS. (2011). *Opportunity in crisis: Preventing HIV from early adolescence to young adulthood.* New York: UNICEF.

Waterhouse, J., Correa, P., Muir, C., & Powell, J. (Eds.). (1976). *Cancer incidence in five continents.* Lyon, France: International Agency for Research on Cancer.

Weinstein, M. S. (1980). *Health in the city: Environmental and behavioral influences.* New York: Pergamon Press.

CHAPTER 4

Transitions and Development

The human population of earth—now more than 7 billion—has tripled since World War II, in just over half a century. It is likely to increase by half again before natural increase ceases late in the 21st century. After increasing for centuries, the world population growth rate peaked at a little over 2% a year in the late 1960s and then began to slow down. Although the present growth rate of 1.04% (2015) will continue to decrease, the base human population will get larger, adding more than 800 million in each of the next three decades (United Nations Population Division, 2013). Its age structure will change during this time, as fewer children are born and a higher proportion becomes older. By itself, the change in age structure will alter causes and patterns of death. But population growth does not only change population composition; it leads to population redistribution. New population settlements and the explosive growth of old ones through migration are occurring already, impacting global ecology. The majority of earth's human population became urban in 2007 (see Chapter 10). This is surely one of the most profound alterations ever and has far-reaching implications for human health.

The lives of people, their families, their communities, and their societies are being transformed everywhere. Human consumption is growing exponentially and, especially through energy use from fossil fuels, changing the climate and creating pollution (see Chapters 10, 11, and 12). "Globalization" not only of trade and employment, but of food, information, travel, communication, ideas, and more, is making new connections faster than can be realized and tearing old ones asunder. There is no local disease condition, from sick Asian birds to drug-resistant tuberculosis in fetid squatter slums, that couldn't be in your neighborhood next week. And countries in the developing world can have both a starvation problem and an obesity problem at the same time!

The purpose of this chapter is to explain some of the complexity and dynamics of the population–habitat–behavior interactions that constitute disease systems, and the consequent need for careful evaluation of health consequences in developmental and other changes. The concepts developed in Chapters 2 and 3 on cultural and political ecology are applied here to show how social and economic change affect disease occurrence. First the population transitions of demography, epidemiology, mobility, and nutrition are explained and the conditions of each stage in these transitions are discussed. The impact of mortality and fertility change upon age structure and disease causation, and the consequent methodological need to control the confounding of such change causes, is discussed. Then these transitions are used as the frame to examine the disease ecology of new settlements, new exposures of population to risk, and new health conditions created by agricultural development. Finally, some of the implications of globalization for health status and emerging diseases in the future are considered.

ECOLOGIES OF POPULATION CHANGE: MULTIPLE TRANSITIONS

Healthy living things and systems of life adapt and change. Human population change over the past two centuries has been transformative for life on earth, the major "driving force," even as it has itself been transformed. The next half-century seems certain to continue the transformation. The changes have been systematically described over recent decades as demographic transitions between preindustrial and postindustrial dynamic states of being. What we understand of these changes was originally modeled on the European experience, but there have been enough variations on this theme now to approach the universal. Here is a simple summary: High mortality rates (especially for infants and children) decreased greatly, causing rapid population growth until birth rates also decreased greatly. The fall in mortality was caused by/was associated with a shift in cause of death from infection to noncommunicable, degenerative causes. As first mortality and then fertility fell, the age structure of the population shifted from very young to older; nutritional status and its relationship to health and susceptibility changed; and population movement (or mobility) changed, and with it exposure to risks and opportunities for adaptability.

These population changes are described and explained in this section in terms of several interacting transitions: demographic, mortality (or "epidemiological" [Omran, 1977]), mobility (Zelinsky, 1971), and nutrition (Popkin, 2003). This chapter focuses on the first three of these transitions, leaving the nutrition transition to further discussion in Chapter 8. The falling fertility is fundamental to changes in growth, mobility, and age structure, but this chapter treats the causes of fertility decline only as context. There is a large body of social science research on why human birth rates and fertility behavior have fallen globally and in each population, and there are many questions about the political economy of those changes that can be asked—but that is the subject of another book (about population geography). Health geography is about health, disease, and life.

Change in Population on Earth

The world's population of 7 billion is today not only growing at a lower rate each year, but since 2000 it has increased by fewer people each year than the year before. Depending on the source of population projection data, scientists expect the world's population to reach a maximum of 8–12 billion. Then, if it continues to follow the pattern set by Europe, population may begin to decrease as fertility rates fall below mortality rates. These projections, of course, are based on many assumptions that may not be valid decades into the future. The demographic impact of HIV was not foreseen, and newly emerging and reemerging diseases such as Ebola or influenza may drastically change the future as we can foresee it. Nevertheless, several processes have been started and demographic structures have been changed that transform the state of health. The age structure of the population has been transformed; and, as noted above, the majority of the human race has become urban, with profound implications.

The Demographic Transition

Mortality and fertility have both been high, fluctuating around each other, for most of human existence. The progression from this state of affairs to the present conditions of low mortality and fertility in the most economically developed countries is known as the **demographic transition**. It is a transition from a demographic regime characterized by high birth and death rates to one of low birth and death rates, which has fundamentally altered the age structure of posttransition societies. Countries in less developed regions of the world have experienced the demographic transition differently than countries in more economically developed regions. But the general trends we have observed—a decline in mortality and the subsequent decline in fertility—appear to be near universal.

The Swedish demographic experience as shown in Figure 4.1 illustrates the four stages of the classic demographic transition model. In the first stage (premodern), both mortality and fertility are high and overall population growth is very low. Stage 2 (urbanizing/industrializing) sees a decline in mortality rates while fertility continues at its historically high level, leading to a rapid increase in population growth. During this early stage of the transition, **natural increase** (the difference between births and deaths) accelerates. Then, in stage 3 of the transition (mature industrial), under conditions associated with urbanization and economic demands for a more educated work force, the birth rate falls. In the classic European model, the change to a "small-family norm" happens partly because children, who add to the wealth of a family in a rural agrarian society, drain a family's wealth in an urban manufacturing society. There have been exceptions and differences among regions and countries in the speed of this idea's diffusion. For example, in some parts of rural agrarian 18th-century France, fertility actually declined sooner than in urban industrial England, so this relationship wasn't simple even in Europe.

Recently, Reher (2004) and others have argued that families were always relatively small before the demographic transition, due largely to high rates of infant

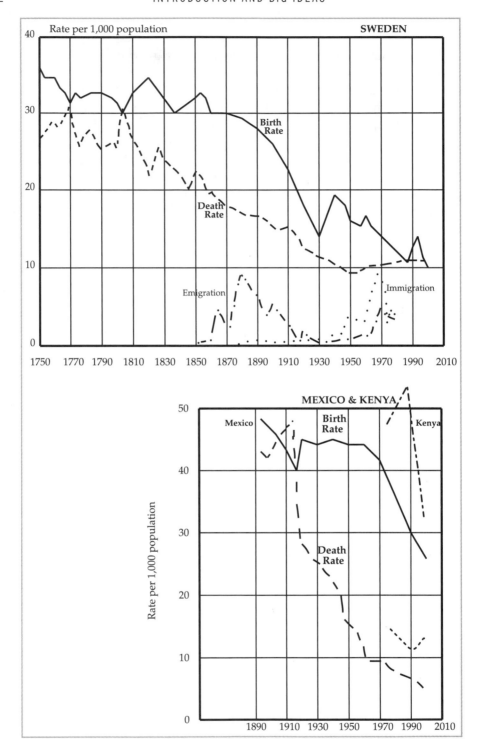

FIGURE 4.1 The demographic transition. Different periods of transition are reflected in the experiences of Sweden, Mexico, and Kenya. Original figure concept and data from Population Reference Bureau.

and child mortality. The decrease in fertility seen in stage 3 of the transition is therefore a response to an increase in the number of surviving children and fertility control was an attempt to maintain family size, not to decrease it. More attention has been paid recently to cultural attitudes about childbearing, to the status and the education (especially literacy) of women, and to diffusion of the very idea of birth control. Cultural values and societal norms about such critical behaviors as age at marriage, divorce and remarriage, women's working for wages, importance of sons, and/or costs of dowries matter a lot. Social–organizational changes from support by extended families to provision of health, family planning, educational, and child care services transform the social environment. The development process corresponds to an increase in demand for human capital and many women chose to actively engage in the labor force rather than stay home with larger families. The transition is completed (stage 4, postindustrial) when birth rates fall to meet low death rates. Countries in this stage often have **replacement level fertility**, where women on average are having only enough children to replace themselves and their partner in the population (~2.1 children per woman). In some European and Asian countries, fertility rates have actually dropped below death rates. The **total fertility rate** (the average number of children a woman would have if she went through her life at today's age-specific birth rates) has fallen as low as 1.3 children in Japan, Germany, Spain, and Italy. The Japanese, German, Spanish, and Italian age structure that results from these low birth and death rates now has a higher proportion of people over 65 than under 15. These changes and those in the following paragraph are illustrated in Table 4.1.

Lesthaeghe (2010) and others have suggested that this "second demographic transition" will bring about declining population sizes and rapid population aging, which has significant implications for health, changing the demands on health systems in these countries.

The circumstances of the demographic transition have been different in several important ways for countries in the developing world. In particular, the gap between the beginning of the mortality and fertility declines is much larger in many developing countries. Reher (2004) suggests that for many economically developed Western countries, the gap between the mortality and fertility declines was much shorter (5–10 years) than it was for less developed countries with later transitions (30–40 years). This has had profound implications for population growth rates in less developed countries—some of which have reached more than 4% a year—and has fundamentally altered the age structure of many countries, creating very youthful populations. As shown in Figure 4.1, in 1988 Kenya reached a total fertility rate of 8.2 children and an astounding 52% of population under age 15—demands on the economy, social institutions, and local environments threatened to increase death rates again. Indeed, in the case of AIDS in Kenya and other countries in sub-Saharan Africa, they have. In most of Latin America, the decrease in infant mortality seemed to plateau for a decade. Social scientists hypothesized that the impact of modern medical technologies (vaccination, assisted delivery and care of newborns, vector control, nutritional supplementation, and other practices from outside the traditional system) had done all that these technologies could. Issues of

TABLE 4.1. Transitions

Country	2014 pop. (mil.)	IMR /1,000 births	% HIV+	% deaths Diarrhea <5	% deaths Malaria <5	% YLL Comm.	% YLL Noncomm.	Age-standardized mortality/100,000 CVD	Age-standardized mortality/100,000 Cancer	Age-standardized mortality/100,000 Injuries	% pop. age <15	% pop. age >65+	% urban	TFR	% <5 underweight	% women 15+ obese	$GNI_{ppp}
Sub-Saharan Africa																	
Nigeria	181.8	60	3.7	9.6	20.6	73	16	267	107	146	43	3	50	5.5	31	11	5,680
South Africa	55.0	34	15.9	7.2	0	62	28	298	109	104	30	6	62	2.6	9	27	12,700
Kenya	44.3	39	6.2	9.9	4.4	72	18	205	140	101	41	3	24	3.9	16	7	2,890
Malawi	17.2	53	10.0	7.3	14.4	76	17	337	104	98	44	3	16	5.0	14	5	780
Niger	18.8	60	0.8	11.8	18.8	77	15	318	56	98	52	4	22	7.6	38	4	950
South Central Asia																	
India	1314	42	0.3	10.5	0.5	42	44	306	72	116	29	5	32	2.3	44	5	5,760
Iran	78.5	25	0.2	3.5	0	18	60	350	97	75	24	5	71	1.8	5	26	16,080
Bangladesh	160.4	38	<0.01	6.4	0.4	45	43	166	88	64	33	5	23	2.3	32	4	3,340
Pakistan	199.0	69	0.1	10.5	0.1	55	31	274	88	97	36	4	38	3.8	32	5	5,100
Afghanistan	32.2	74	<0.01	13.2	0	58	23	512	124	169	45	2	25	4.9	33	3	1,980
Middle East/North Africa																	
Egypt	89.1	22	<0.01	5	0	20	72	445	120	34	31	4	43	3.5	7	29	11,020
Morocco	34.1	26	0.2	5.4	0	30	59	331	96	47	25	6	60	2.5	3	22	7,180
Jordan	8.1	17	<0.01	3.5	0	25	59	326	112	54	37	3	83	3.5	3	31	11,910
Saudi Arabia	31.5	16	<0.01	1.7	0	18	66	338	63	41	30	3	81	2.9	5	35	53,760
Yemen	26.7	43	0.2	9.2	1.0	59	28	377	72	84	41	3	34	4.4	36	17	3,820
Southeast Asia/Pacific																	
Indonesia	255.7	31	0.3	5.5	1.8	36	55	371	111	49	29	5	54	2.6	20	6	10,250

Thailand	65.1	11	0.2	2.8	0.4	22	62	184	103	73	18	11	49	1.6	9	9	13,950
Vietnam	91.7	16	0.5	11.9	0	25	60	193	112	59	24	7	33	2.4	12	4	5,350
Fiji	.9	15	0.1	4.4	0	19	69	372	98	64	29	5	51	3.1	5	36	8,030
Papua New Guinea	7.7	47	0.7	8.8	11.1	58	31	149	143	100	39	3	13	4.3	28	28	2,510

East Asia

China	1,372	12	<0.01	3.9	0	11	77	300	143	50	17	10	55	1.7	3	7	13,130
Japan	126.9	2.1	<0.01	1.9	0	10	77	82	104	41	13	26	93	1.4	3	3	37,920
Mongolia	3.0	21	<0.01	7.5	0	20	65	587	194	69	27	4	68	3.1	2	17	11,230

Americas

United States	321.2	6.0	0.6	1.9	0	8	80	136	121	44	19	15	81	1.9	0.5	34	55,860
Canada	35.8	4.8	0.3	0.4	0	7	83	89	119	31	16	16	80	1.6	—	20	43,400
Mexico	127.0	13	0.2	2.7	0	16	64	148	72	63	28	7	79	2.3	3	28	16,710
Brazil	204.5	19	0.3	1.9	0.1	17	62	214	118	80	24	7	86	1.8	2	20	15,900
Cuba	11.1	4.2	0.2	1.2	0	7	82	185	113	45	17	13	75	1.7	3	25	18,710
Haiti	10.9	42	1.8	12	0.5	57	31	384	103	89	35	4	59	3.2	12	12	1,750

Europe

Russia	144.3	9.3	1.1	0.8	0	10	75	531	147	103	16	13	74	1.8	—	24	24,710
Sweden	9.8	2.2	0.2	0.3	0	5	87	132	110	26	17	20	84	1.9	—	21	46,710
Germany	81.1	3.3	0.1	0.4	0	5	89	143	112	23	13	21	73	1.4	1	20	46,840
United Kingdom	65.1	3.9	0.3	0.0	0	7	86	112	130	22	18	17	80	1.9	—	28	38,370
Italy	62.5	2.9	0.4	0.1	0	5	89	106	116	20	14	22	68	1.4	—	21	34,710

Note. Data from Population Reference Bureau (2015) for population, infant mortality rate (IMR—per 1,000 born); % HIV positive; % population age under 15 and age 65+, % urban, total fertility rate (TFR), gross national income purchasing power parity (GNI_{ppp}). Data are for 2015 except HIV which is for 2011 and 2013. Data from the World Health Organization Global Health Observatory database (*http://apps.who.int/gho/data/?theme=home*) for % of deaths under age 5 years due to diarrhea or malaria; % of years of life lost (YLL) to communicable and to noncommunicable diseases; age-standardized mortality rates per 100,000 for cardiovascular disease (CVD), cancer, and injuries; % children under age 5 underweight and % women age 15 or over obese. Data are for 2015, except diarrhea and malaria for 2013, and % obese from 2014, and % <5 underweight from the NLiS dataset, varying years. – means no data available.

135

internal distribution—of food, housing quality, crowding, literacy, and knowledge—had become critical. Then the fall of mortality and fertility resumed in Latin America. Fertility decline began in sub-Saharan Africa, as Kenya's fertility fell rapidly and population giant Nigeria's total fertility rate fell to 5.9.

The transition has varied in its pace and circumstances. Some countries have been "accelerated," lowering their fertility before economic development and urbanization. Indonesia, for example—the world's largest Muslim country—has lowered its total fertility rate to 2.6 children, even though its world parity gross national income per person (GNI_{ppp}) is only $3,460 and it is even now only 42% urban, having been 20% urban when it began lowering its fertility. More remarkably, Thailand has lowered its fertility below replacement level (1.7 children), even though it is still an overwhelmingly rural, low-income country and was much more so when it began to lower its fertility. Other countries have experienced a "delayed" transition, whereby mortality declined dramatically but fertility failed to follow suit. In many of these countries, use of medical interventions, such as vaccines and antibiotics and maternal and child health initiatives, decreased child mortality. Without the subsequent decrease in fertility, population growth rates soared.

So each country in its own time and way—with cultural behavior and environmental pressures unique to each, with starts and plateaus and leaps—has been progressing through the relentless demographic transition (as studying Table 4.1 reveals). With the important exception of the impact of lowered fertility on maternal mortality, fertility is further considered in this text only as it affects the age structure of populations.

The Mobility Transition

Population movement in both time and space exists at various scales. The concept of **migration**, as used until recently, implies the intention of a permanent move. It is rarely measured unless it crosses an administrative, political border. But migration may be international in scale, or it may involve moving to another building after marriage. **Circulation** is movement that returns to the original point. This may involve minutes to go buy milk, days on a vacation, weeks for a pilgrimage, or years of absence for long-term study or labor. Mobility affects exposure, introduction, and other aspects of disease risk, and has long been studied by population geographers.

Zelinsky (1971) suggested a mobility transition that adds the dynamic relationship of development to mobility. In a premodern society, with high birth rates and death rates, there is little residential mobility except for that following marriage (see Table 4.2). Circulation is limited to travel for customary religious observances, agricultural needs, local commerce, and warfare. As the demographic transition begins and death rates fall, a massive movement to three general destinations is generated by population growth: people migrate (1) to new agricultural frontiers, (2) from the countryside to cities, and (3) to foreign countries. Both social- and labor-related circulation increase. Later in the transition, fertility falls, movement to cities and frontiers slackens, and the emigration virtually ceases; however,

TABLE 4.2. Stages in Transitions				
Stage	**Demographic**	**Mortality**	**Mobility**	**Nutrition**
Historic Stable	Birthrates and death rates fluctuating at >35 per 1,000; population increase imperceptible; >45% population aged <15; <5% aged 65+.	Life expectancy <35 years; infant mortality >180 per 1,000 births; death results mainly from infectious diseases.	Individual migration almost unknown except for marriage; circulation local for basic agricultural, commercial, and religious needs.	Seasonal malnutrition or stable crop deficiency; periodic famine, catastrophic.
Early in Transition	Death rates fall; birth rates remain high; population grows rapidly by natural increase.	Infectious diseases are controlled by sanitation, vaccination, suppression of vectors, and medical treatment; young population.	Migration from rural areas to cities, agricultural frontiers, and labor internationally begins and grows; circulation increases socially.	Growing shortages and vulnerability; heavy labor.
Late in Transition	Birthrates fall as death rates stabilize; population growth slows and population ages.	Degenerative diseases become major causes of death; life expectancy extends to 65.	Migration to cities and agricultural frontier slackens; emigration descreases; circulation increases in structural complexity and intensity.	Income growth increases meat and sugar consumption; great inequality in deficiency and abundance.
Future Stable	Birthrates and death rates fluctuating at <20/1,000; population increase imperceptible or population declining; <20% aged <15, >15% aged 65+.	Life expectancy >70 years; infant mortality <10 per 1,000 births; death results mainly from degenerative diseases.	Multipurpose and vigorous circulation migration mainly between cities and for retirement; immigration of labor.	High consumption of abundant fat and sweet calories; low labor; malnutrition: obesity prevalent.

circulation still increases in intensity and in structural complexity. Later still, when fertility and mortality have stabilized at low levels, migration is primarily between cities, although there is significant immigration of unskilled workers. Circulation is vigorous and accelerating; it includes pleasure-oriented trips as well as moves for social and economic purposes. Zelinsky presciently suggested that in the future the technology of communications and delivery systems may decrease migration and some forms of circulation.

The Nutrition Transition

As conceptualized by Popkin (2003), the **nutrition transition** involves a shift in the behavior of people in lower- and middle-income countries from traditional diets with their associated malnutrition and shortage to convergence on the so-called Western diet of overabundance and its associated malnutrition. As income per capita increases, a dietary preference (likely to be population-based—i.e., deep in our human heritage) for increased animal protein and sweet foods leads to intake of more animal fat and higher energy density from sugar and alcohol. Sweeteners shift (technologically enabled and politically motivated) from mostly disaccharides like sucrose to monosaccharides like fructose and glucose—in other words, from honey and maple syrup to high-fructose corn syrup. At the same time, the hard physical labors of traditional life are replaced with more sedentary occupations and entertainments. Leading chronic diseases such as diabetes, heart disease, and stroke are related to both nutrition and physical activity. Chapter 8 elaborates on the nutrition transition.

Four Stages in the Three Transitions

A child born in the United States today is more likely to die from cancer, heart disease, or stroke than from measles, mumps, or rubella. But for most of human history, infectious disease, famine, and chronic malnutrition have been the leading causes of death. **Epidemiological transition** theory, first proposed by Omran (1971), examines the mortality component of the demographic transition and describes a shift from a cause-of-death pattern dominated by infectious diseases with very high mortality, especially at younger ages, to a pattern dominated by chronic diseases and injuries with lower mortality, mostly peaking at older ages. This change is responsible for the tremendous increase in life expectancy and is intimately linked to economic development and the technological, socioeconomic, and environmental changes that occur as part of the development process. But the transition in mortality is also influenced by transitions in nutrition and mobility and ultimately leads to the overall changes in mortality, fertility, and population growth discussed above. The following description of the four stages in the demographic, mobility, and nutrition transitions interrelates them and their significance for the ecology of health and disease.

Before the Change: Pretransition Conditions

For thousands of years—at least since the Agricultural Revolution that domesticated plants and animals laid the basis for population growth in permanent settlements and the creation of cities, which in turn led to the creation of most of our contagious diseases—birth and death have fluctuated around each other at what seem to us today incredibly high levels. This can be called the *age of pandemics and famine*. Overall death rates of more than 40 per 1,000 population reflected infant mortality rates of over 200 infants dying in their first year for every 1,000 babies born alive, and child mortality of 50% before age 6. Life expectancy at birth was less than 35 years. The children died mostly from acute infectious diseases—especially those that caused repeated and prolonged dehydration from diarrhea, or virulent contagious fevers (such as smallpox, whooping cough, diphtheria, typhoid, or pneumonia). People were made more susceptible to dying from these infections because of at least seasonal malnutrition, and sometimes starvation from crop failure and famine.

Matching this mortality, people had seven or more children (high total fertility rate). Women were married at puberty, and complications of pregnancy and childbirth were major causes of death. Population growth was very slow, mostly on the order of 1% a century! The high birth rates, of over 40 per 1,000, combined with the mortality of children, meant that the age structure was very young. Almost half the population was under age 15. People usually died within 5 miles of where they were born, having migrated only at marriage according to mother- or father-localizing custom. There was little circulation except for marketing, and perhaps religious observance.

Sixty years ago, these conditions still prevailed in much of sub-Saharan Africa and occurred widely in Asia and Latin America as well. More than 20 countries still had infant mortality rates over 200 per 1,000 births. Colombia, Egypt, and India were among those countries with infant mortality rates over 150; in Italy the rate was 74 and in France still 56 (United Nations, 1950). Most of the world's population was rural, and over 40% of population were children under age 15. Today there are fewer than 25 countries in the world with infant mortality over 100. As shown by the rates in Table 4.1, the region of sub-Saharan Africa is still earliest (i.e., closest to the beginning) in the transitions. The death rates in most countries there did not begin to decline until after independence. Throughout the table, notice the relations to age structure (>40% of population under age 15); the high percentage of deaths due to communicable disease, especially diarrhea; and the low gross national income purchasing power parity (GNI$_{ppp}$). The variation within as well as between regions can be seen in this table as the relationships of income, urbanization, fertility, age structure, cause of mortality, and finally proportions of underweight small children and of overweight adults, shift. This complex table can be referred to throughout the following stages.

Early Transition Stage

In the *age of receding pandemics and famines,* deaths from acute infectious diseases decreased in Europe, largely as a result of improved standards of living and public health. This era spanned from the 1750s to the 1920s and saw dramatic changes in our understanding of the causes of disease, germ theory, sanitation, and the role of public health services. In the late 18th/early 19th centuries, England especially experienced the horrible conditions of the early Industrial Revolution. Cities were death traps that took 5–10 years of life expectancy from those who moved into them. They were crowded, lacked adequate housing, and had poor or nonexistent sanitation systems (see Chapter 10). Orphans abounded, with the worst conditions of child labor and poorhouses imaginable. These were the conditions that Thomas Malthus, Charles Dickens, and Karl Marx wrote about.

Over the course of the 19th century and the first half of the 20th, conditions of the cultural ecology of disease changed dramatically. Increased prosperity brought improved housing, increases in agricultural production and in imported food supplies, better clothing, improved education, knowledge of germs, and better government. It brought private philanthropy, the "garden city" movement, public sanitation, chlorination of water, pasteurization of milk, electricity and refrigeration, sterilization and other leaps in medical treatment, and many other changes that improved health conditions. Vaccination for smallpox became routine; in the 20th century other vaccines followed for rabies, typhoid, diphtheria, tetanus, and other deadly diseases. The previous luxury of soap became universal, and the habits of personal hygiene were transformed. Rene Dubos (1965) famously noted that even before sulfa drugs and (after World War II) antibiotics were used, infectious diseases such as leprosy and scarlet fever had all but disappeared and many others, like tuberculosis, had greatly decreased. With public understanding of germs, new vaccines, antibiotics, X-ray and scanning machinery, and new biomedical treatments, deaths and serious illness from infectious disease almost disappeared (see the entries for Europe in Table 4.1). During this stage of the demographic transition, European populations grew as much as 1% a year. As population pressure in rural lands increased, people migrated to the newly industrial cities and to other countries.

The decline of mortality in most of the developing world proceeded in an accelerated manner. By the late 20th century "outside" or international intervention could help with vaccination programs, treatment for previously fatal childhood diseases, pre- and postnatal care, and safe hospital delivery. Mosquito control reduced and in some places even eliminated malaria and yellow fever. Smallpox became the first, and still the only, infectious disease eradicated from earth. Refugee and emergency food programs prevented starvation. For a while in Latin America, as noted earlier, the decline in infant mortality seemed to plateau. Social scientists concerned with population growth thought mortality might not come down further until there were internal structural changes in crowding, social class distribution of wealth, household distribution of food, and changes in power by class and gender. In some regions mosquitoes and microbes developed resistance to control measures: the effort to eradicate or at least control malaria failed, and the disease again increased in parts of South Asia and Central America from where it had been cleared. Mortality decline resumed, however, and life expectancy at birth in almost all countries outside Africa is now in the high 60s or 70s.

During this maximum-rate-of-growth stage (when mortality has fallen but births are still high), the mobility transition posits three kinds of migration destinations to relieve the growing rural population pressure, as noted earlier. First, people push to new agricultural frontiers as farmers who want to farm convert forest and marshland to new cropland. Second, they start migrating to cities in search of jobs and a means of support. Even as the world adds its last 2–3 billion people to developing countries, these and currently rural people will be added to cities. Today's 53% urban is composed of more-developed countries at 77% and less-developed countries at 48% (Population Reference Bureau, 2015). By 2050 the world will approach 66% urban, and the low- and lower-middle-income countries are projected to be 48% and 57% urban, respectively (United Nations, 2014). Some health consequences of this are considered in Chapter 6 and Chapter 10. Third, people living to grow up move internationally in search of jobs—a process that is already transforming nations.

Late-Transition Stage

The late stage of the demographic transition is entered when birth rates begin to fall. Once this starts, fertility continues to decline until the now low birth rates are met, and the transition is completed. During this stage the population stops growing by natural increase and, without migration, it can even decrease. This has happened to counties, provinces even, when accompanied by out-migration to cities and other countries, as described in regard to the mobility transition. Fertility rates may have fallen even to replacement levels, and the base of the age-structure pyramid narrowed, but those born 20 years ago are now looking for jobs. In some countries, like the Philippines, the remittances sent home from those working internationally can even become the major source of foreign exchange. Migration out of rural areas and into cities compounds the lowered fertility to change age structure drastically. The broad base of youth disappears from the age structure pyramids. For a couple of decades, a country can enter the "demographic window" of development when the proportion of child dependents (under age 15 or 16) is greatly reduced and they are healthy; the proportion of elderly dependents (over age 65) is still tiny, and most of the population is composed of healthy, working-age adults. This variation in age structure that follows the transition can be seen in the country pyramids of Figure 4.2.

As the population urbanizes and ages, it also increases its income and consumption. As discussed in later chapters, people become exposed to the air and water pollution so far associated with heavy industrialization. Populations also seem to increase their cigarette smoking. Their diets increase in animal protein, fresh fruit, sweetened breakfast cereals, soft drinks, alcohol, and fast foods. The main causes of disease become noncommunicable and degenerative, such as cardiovascular disease and cancer (Table 4.1).

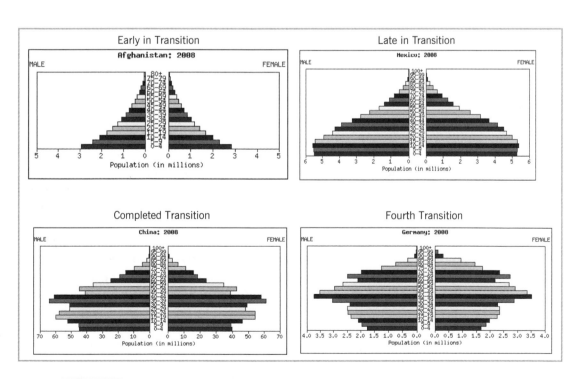

FIGURE 4.2 Population pyramids of age structure. Data from U.S. Census Bureau (2008).

After the Transitions Are Completed

There have been two major surprises in countries past "the end" of the demographic and other transitions. First, fertility so low that population decreases by attrition can characterize regions and countries (Western Europe, Russia, and Japan) that have completed the demographic transition. This was the big demographic surprise of the last two decades. These lowest-low fertility countries have become increasingly concerned with providing health care for an ageing population. Health systems will need to transform to provide for the unique health needs of the elderly, rather than focusing so intensely on maternal and child health. The United States has largely escaped this dynamic (see Figure 4.2) because of the immigration of so many millions, most of whom are working adults having on average 2.8 children instead of 1.8. The rural South was the last area to undergo the transition and still has some of the highest infant mortality rates. There has also been cultural resistance to lowering fertility below two children: American fertility levels are highest among the Mormons of Utah, for example, and in the rural mountain West (and Alaska) generally. A new cultural phenomenon observed in the United States is that having more than two children is a sign of wealth, that high fertility indicates an economic status that can afford to send many children to expensive universities.

The second biggest demographic—and major health—surprise was how fast a major, industrialized, late-demographic-transition country could deteriorate in health. The fragility of the interactions that supported good health and their vulnerability to deterioration in economic and social organization was shocking. When the Soviet Union collapsed in the early 1990s, the turmoil in economy and society was quickly reflected in soaring mortality. Within a year, Russia had epidemics of diphtheria and whooping cough even in Moscow, which affected mostly small children at startlingly high rates. Death from cardiovascular disease has soared (Table 4.1). It is ascribed largely to the impact of alcohol, as is a huge increase in infant deaths from fetal alcohol syndrome. The increase in infectious disease was variously attributed to poor quality control for vaccines and failure of delivery systems; deterioration of nutrition, housing, and heat; failure to pay medical personnel; and failure to produce and supply medicine or blood products. Suicide became epidemic, especially among those with their pensions and their life's work destroyed. In more recent years, HIV infection and drug addiction have become social plagues. Related to this is the reversal of life expectancy gains in many midtransition African countries, largely due to the HIV/AIDS epidemic. In 1985 life expectancy in Zimbabwe peaked at 61 years before falling to 42 years in 2002. As the AIDS pandemic ripped through sub-Saharan Africa, young economically productive men and women experienced the highest rates of infection and mother-to-child transmission of the disease was common. Mortality in these age groups profoundly altered the demographic structure of many countries, and had significant negative economic impacts.

At completion of the demographic transition, rural-to-urban migration is complete, and migration occurs largely between cities. Circulation is constant and intense. The causes of death overwhelmingly become degenerative diseases: heart disease, stroke, cancer, kidney failure, diabetes. Motor vehicle accidents also become a major cause of death, as well as adding a great burden due to **years of life lost** (YLL).[1]

In the last few years, fertility has increased a couple of tenths of a percent in a few northern European countries, perhaps responding to programs of parental leave and child care intended to lessen the burden on parents and interference with professional lives of having children. All low-fertility countries, with at least the legal exception of Japan, have become targets of international labor migrants. Many of these migrants intend to return home, and so are technically circulating— meaning that the movement of infections or ideas between source and destination goes both ways. The main demographic and health development, however, is one of disease ecology. Old diseases are becoming resistant to medicines and insecticides and are even adapting to new behaviors. Led by HIV, new diseases are emerging. The next stage of the demographic transition may yet become the *age of returning pandemics*.

QUICK REVIEW 1

✓ Population change is driven by a series of interacting transitions: the demographic transition, the epidemiological transition, the mobility transition, and the nutrition transition.

✓ Less developed countries have experienced the demographic transition differently than more developed countries, but the decline in mortality and the subsequent decline in fertility appear to happen everywhere.

✓ As countries become more economically developed, the mobility of the population increases. People migrate to urban areas, and even begin to travel abroad.

✓ In the early stages of the demographic transition, when mortality rates are high, life expectancy is short and most people die of conditions related to malnutrition and from infectious disease.

✓ Near the end of the demographic transition, mortality from infectious disease declines and people live longer lives. Chronic diseases, such as heart disease or cancer, become the primary cause of death.

MAJOR IMPACTS OF POPULATION CHANGE

Perhaps the most profound health impact of the progression through the demographic transition is the change in the age structure, or composition, of the population. There are living people who in their own lifetimes have gone from a community population that was 45% under 15 years old to one that is 15% children, and now is also 15% over 65. This transformation is a change for the lives and health

---

[1]The YLL measures the burden of disease by counting not only the number of deaths, but the years (and earning power) lost from those lives. Thus the YLL of AIDS or motor vehicle accidents is more than that of stroke or colon cancer.

of the individuals, especially women, as well as for the structures, norms, supports, and dependencies of society. When we explain the demographic transition, we like to use the "cousin rule" to illustrate the change in experience. In a society in which everyone had six children, including your parents and their siblings, you would have 60 cousins. Many would not live to grow up, but they would be born. In a society in which the majority of parents have two children, you have 4 cousins—and, because of mobility, they live in other states! The social relations that help people survive sickness, raise children, support the elderly, or recover from flood are fundamentally altered.

Who knew that such change was humanly possible? A woman who used to get married by 18 would probably, if fortunate enough to remain healthy and alive, have been pregnant or nursing for most of 20 years and sporadically for another 10. She might have had more than 10 children born alive (although half would have died before age 5), but there would have been several other stillbirths and miscarriages. She would have been suffering from malnutrition and severe iron deficiency even in good times, and the burden on her body would have made her susceptible to infectious diseases. Anemic from intestinal worms, malaria, and other infections, she would have faced obstructed births, hemorrhage, and sepsis from pregnancy and childbirth (and perhaps desperate, unsafe abortion to save her own life). Even today, a woman in Niger, early in the demographic transition, has a one-in-seven lifetime risk of dying from reproduction (World Health Organization [WHO], 2005). When fertility rates fall from a total fertility rate of seven children to five to three, many more women live to grow older.

Age Structure and Its Implications

As people live to grow old and reproduction falls below replacement, the age structure shifts from predominantly youthful to predominantly working age, and then the population ages. The causes of death as a population goes through the mortality part of the demographic transition shift from acute infectious diseases that kill children, to low death rates of healthy adults, to degenerative and other noncommunicable diseases of older age. Just looking at the pyramids in Figure 4.2 can tell you where deaths from diarrhea, from stomach cancer, from violence, or from heart attacks are occurring. To know the total fertility rate of a country is to have a good idea of its major causes of death. Notice also in the pyramids of those latest in the transition the growing problem of who will provide medical care and support in regions and countries with large proportions of elderly people. In the U.S. states of North and South Dakota, for example, some registered nurses have daily drives of more than 200 miles to visit their patients. Chinese civilization has held an ideal of "four generations under one roof." Under the mortality rates that used to prevail, it was usually a dream. The Chinese "one child" policy, however, led to a society that is almost 30% over 65. If that child is a girl, she will provide care for her own child, her parents, her husband's parents, and surviving grandparents.

Look again at the changes in age structure that are happening as the world goes through the demographic transition. Public health needs to grapple with these changes of structure and the changes in health conditions and need for services that they portend. They are often approached in epidemiology and sociology as stages of the life course.

Stages of the Life Course

INFANCY

The greatest risk of dying before age 60 occurs in the first year of life, and within that in the first month after birth (neonatal) or the first month both before and after birth (perinatal). The chief components of this risk in developed countries are respiratory conditions and disorders related to short gestation, low birthweight, and congenital abnormalities. Sudden infant death syndrome, accidents, and pneumonia complete the top six causes of death. Infections pose the greatest risk early in the demographic transition, as discussed earlier.

CHILDHOOD

When fatal infectious diseases have been largely eliminated by public health systems, vaccination, and antibiotics, the major cause of death in childhood is accidents, and even these occur rarely. Leukemia does occur, and genetically caused diseases and congenital abnormalities still take a toll, especially in heart disease. In the United States, deaths from childhood pneumonia and meningitis are quite rare. There are more murders and suicides in the population than children dying from pneumonia.

YOUNG ADULTHOOD

Violence in the United States—accidents, homicides, and suicides—accounts for more than three-quarters of deaths among those ages 15–24. The overall death rate of barely 1 in 1,000 conceals great racial and gender differences in violent death (e.g., white male accidents are 2.7 times those of white female accidents; black male homicide is 8.4 times black female homicide).

MATURITY

During the years between ages 25 and 64, death rates increase ninefold; degenerative diseases rise to dominance; and gender, racial, and ethnic differences increase. Although chronic obstructive pulmonary disease (COPD: mostly emphysema) is relatively more important for whites, liver disease for Hispanics, and AIDS for blacks, the top 10 causes remain the same for all groups—except that suicide continues to be more important for whites and murder for blacks and Hispanics.

SENIORITY

As mortality increases steeply between ages 65 and 74, heart disease, cancer, and stroke come to account for more than 75% of all deaths. Kidney disease, liver cirrhosis, and diabetes as well as emphysema become major causes of death, and functional disability increases. Although three-quarters of seniors have no trouble with such personal care activities as walking, dressing, or bathing, the need for home help becomes more common. Blacks continue to have death rates half again as high as those of whites, and male deaths continue to exceed female deaths.

OLD AGE

In old age, the causes of death in seniority escalate, and pneumonia becomes a common threat. Morbidity increases greatly. More than half the population suffers from arthritis, and a third from serious impairment of vision and/or of hearing. Mental illness also increases steeply. In this group there are high rates of institutionalization, lower proportions of minorities, and a uniquely low sex ratio. The population over 85 is the most rapidly growing group in the population structure of developed countries.

Confounding and the Need for Age Standardization

As conditions of poverty and education, behavioral norms, and environmental exposures change in the future, these stage-of-life associations will be altered in both developed and developing countries. The ability to "decompose" risks of mortality from specific disease by sex, ethnicity, region, education, occupation, and other groupings has made it clear that there are strong *cohort effects* (i.e., effects due to the shared experiences of being born at the same time and progressing through life's experiences together) for mortality and morbidity. One has only to look at the age structure of China, the tobacco being smoked by young adults, and the air pollution being breathed by children to foresee the coming epidemics of lung cancer and heart disease in that country. Through both biology and experience, age structure pervades the experience of health and disease.

A basic four-part age division for four states is presented in Table 4.3. These states have different incidences of heart attacks, accidents, ear infections, pregnancy, mental illness, AIDS, and every other condition and health service need. Consider how the need for heart specialists or obstetricians must differ between Utah and Florida, or how you might try to determine whether North Carolina's historically high exposure to tobacco has affected lung and heart illnesses, when the age structure indicating cumulative exposure is also so different. How then can mortality be compared between countries when they have such different age structures, as shown in the population pyramids of Figure 4.2? Are the millennial goals of the United Nations being more successfully implemented in one country or region than in another? How can "before" and "after" conditions be compared when age structure has changed too?

The etiology or treatment success of almost anything cannot be studied or addressed through any comparison of places without controlling for those differences in age structure. The ubiquitous presence of age as a confounding influence (age is related to nearly everything, from immune status to cumulative carcinogenic burden to behaviors involving exposure, activity, and knowledge) is routinely handled by **standardization**. In Table 4.3, the population age structure of the United States—the total of the parts, in this case—stands as a basis for comparison.

TABLE 4.3. Age Structure of Several States and the Nation as Standard, 2014					
State	**Florida**	**Utah**	**California**	**North Carolina**	**United States**
Population (millions)	18.8	2.8	37.3	9.5	308.7
% 65+	17.3	9.0	11.4	13.0	13.0
% 18–64	61.4	59.5	63.6	63.1	62.9
% 5–17	15.5	22.0	18.2	17.3	17.5
% <5	5.7	9.5	6.8	6.6	6.5

Note. Data from U.S. Census Bureau (2014 population estimates, American FactFinder).

Florida's population of 18 million is 16.8% age 65 or older; this means that 3.0 million people are in this age group. The older populations of Utah are 220 thousand, of California 3.8 million, and of North Carolina 1.1 million. Of course there are going to be more deaths from any cause in that age group in California, since it is not only our most populous state but has the largest number of older people. Dividing the number of deaths from heart disease that occurs in Utah, California, or North Carolina, by the population of each state, respectively, would give crude death rates that are intrinsically not comparable. The "age-specific" death rates (in this example not year by year, but for the category of those age 65 or older) of the four states can be applied to the total U.S. population by age, yielding a hypothetical number of deaths that would occur if the United States had each death rate in turn. When these hypothetical deaths are divided by the standard U.S. population, then a directly standardized rate is created that can be used to compare the mortality experience of the states because they are all being compared on the deaths that would result *if* each state had the age structure of the standard U.S. population.

ENVIRONMENTAL EXPOSURES, THE MOBILITY TRANSITION, AND TIME–SPACE GEOGRAPHY

During the middle of the demographic transition, when rural populations are growing rapidly by natural increase, the fragmented farmland and the traditional social structures and political systems face not only population pressure, depletion, and erosion, but today also new global markets and the new financial structures and entourages of nongovernmental organizations (NGOs). The alterations in ecology that result from mobility to Zelinsky's three destinations are considered at length soon. First the scales, importance, and measurement of population movement in time and space need to be considered.

Analyzing Mobility in Space and Time

As noted earlier in this chapter, circulation is population movement that returns to its place of origin. Such movements occur in a continuum of time and space. People may move from room to room in a house in minutes; they may go out to their garden, walk the dog, visit a neighbor, or go to a playground in hours. They also journey to work, go shopping for food, and pursue other activities (e.g., trips to church, sports, schools, dentists, and friends and relatives). Young adults may go away for college or job training, may work locally, or may move to another state for employment. College students today often study abroad, and people of all kinds travel internationally for tourism, business, research, and other work. (In the first decade of the 21st century, for example, hundreds of thousands of Americans circulated through Iraq or Afghanistan.) Circulation for college takes years, and commonly students do not in fact return to their home towns, and sometimes not even to their home states; in other words, long-term circulation can become migration (permanent). Leaving one's country for wage labor has become a common practice in the developing world. In the community of "illegal immigrants" in the United States, it has become impossible to estimate how many people leave Mexico, for example, with the intention of living permanently in the United States, and how many would circulate home to family and friends if the border were easy to cross. The old definition of migration—crossing political boundaries (or there would be no knowing that it had occurred) and moving with the intention of permanence—is becoming not only difficult to identify but, when circulation extends over years, irrelevant.

Traditionally, early in the mobility transition, agricultural people seldom migrated except for marriage. Sometimes "forced migration" occurred due to natural disasters, or war actions, government labor conscription for building roads or walls, or the like. A lucky few Muslims might earn the esteemed title *hadji* for making the long, arduous pilgrimage to Mecca and back home—a journey that might take over a year. People moved around their home compounds, their village establishments and markets; they circulated in larger areas of regional markets and of religious and business needs.

Such patterns still exist in a few parts of the Third World, but are rapidly being supplanted. As roads and bridges get built, people's activity areas increase. As they extend the area for work in plantations, cities, new agricultural lands, or international education or work, they also intensify the circulation with visits home and visits by village friends and neighbors to them. Villages become connected to the agricultural frontier, as well as to the big city (and thus to the world urban system). Each of these movements in time and space brings exposures to different insults and raises the potential of introducing old disease agents to new places or being exposed to new ones. As illustrated in Figure 4.3, people may move out of an area where a disease is endemic and introduce it to their destination area; if the agent is very contagious or the transit time between origin and destination is long (as in a refugee camp), the transit route may disseminate the disease. Alternatively, people may leave an area in which the disease agent does not occur, only to be susceptible to it when they reach their destination. Or, finally, people may leave a safe area and move to another one, but be susceptible to the hazards—infection, malnutrition—in camps and stops along the way. The various points in time and space of exposure of susceptible individuals to infections identify different locations and different prevention or containment strategies for public health interventions.

Prothero (1961, 1965) was the first geographer to realize the importance of population mobility for disease diffusion and control. He conceptualized a typology of time–space mobility that proved useful in the control of malaria in East

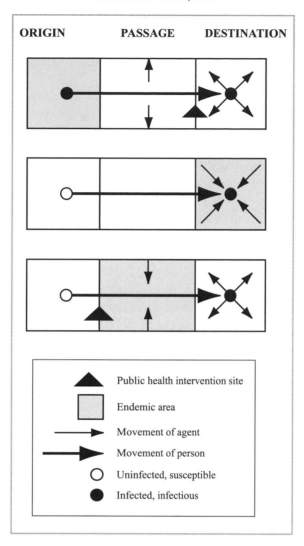

FIGURE 4.3 Mobility and disease agent transmission.

Africa. He mapped animal-herding territories and routes, pilgrimage routes, and other forms of culture-specific population movement that were vital to this effort. Prothero also mapped for the WHO **labor-sheds** within which people moved, even crossing international borders, to work in mines or plantations—areas within which migrants would be continually circulating and introducing or being exposed to the malaria plasmodia. Efforts to control and eradicate malaria internationally, he stressed, had to be coordinated to concentrate on the labor-shed, or one country's eradication efforts might well be thwarted by another's. Gould and Prothero (1975) developed a typology of mobility in which they classified on one axis increasing time (from hours to days, weeks, months, years, and so on into permanence) and on the other scale of movement (from local grazing to international travel). Into this time–space classification, they put purposes (herding, visiting, marketing, school-ing, working, etc.). By their nature, circulation movements have reverse flows about equal to their forward flows and so have greater potential for disease circulation than does unidirectional migration. The relevance of this old typology of mobility for understanding the movement of malaria has recently been rediscovered (Martens & Hall, 2000).

More recently, the massive refugee movements stimulated by violent conflict in regions such as the Middle East have created new migrations of people across the globe. Some of this movement is unidirectional. As of the writing of this book, the UN High Commission on Refugees (UNHCR) estimated that the crisis that began in 2014 in the Syrian Arab Republic led to nearly 4.9 million Syrian **refugees** and 6.6 million **internally displaced people** by the end of 2015 (UNHCR, 2016). These numbers are expected to rise over the next several years. Many fleeing Syrians headed to the bordering countries of Lebanon, Jordan, and Turkey. At the same time, countries in the European Union, Canada, and the United States were strongly encouraged to open their doors and welcome large numbers of refugees. Movement among refugee populations varies. Those that move to countries far away from their homeland often move unidirectionally; that is, they resettle permanently in the country that has agreed to host them. But the vast majority of refugee movement is circular. Refugees intend to return to their homeland once the political crisis is over. Many of these live in refugee settlements in countries that border their homeland. Conflict also creates large populations of internally displaced persons (IDPs), who are forced to leave their home but do not cross an international border. These moves are also circular in nature with IDPs returning to their homes once conflict has ended.

The effects of these politically and economically driven changes in mobility on patterns of disease diffusion are important subjects for future study, as we will see in the discussion of emerging diseases. Understanding the types of mobility and anticipating their transformations provide a basis for conceptualizing disease diffusion systems at any scale.

Measuring Mobility and Exposure

One can visualize different kinds of hazards varying in intensity from absent to intolerable as they are spread out on the surface of the earth. Each can be mapped and then each mapped "surface" can be overlaid on the others. At various scales, from backyard up to national and continental, such overlaid distributions of hazards could be used to define hazard regions. Mapping how people move around and through such a multiple-hazard surface and how long they spend at any point or in any hazardous area/region (i.e., the duration of that exposure) can be used, conceptually at least, to model risk and predict future disease outcomes. For example, people may breathe different air pollution at home, on the highway, and at work; may have benzene exposure at the gas station; and/or may be exposed to chemicals in the office, chemicals on the golf course, or gasoline chemicals in drinking water at home. The individual movements will need to be aggregated in all their variation, and the hazards measured across multiple scales of occurrence, to attempt to approach any population predictions. Population and economic geographers have developed several ways to summarize and analyze such population movement. These are usually classified as **time–space geography**. Future health geographers need to conceptualize and develop this critical area better.

Time–space geographical analyses for epidemiology have been most fully developed and explained by Schaerstrom (1996). They build on the ideas of Torsten Hagerstrand, who first conceptualized such movements and developed "lifelines" in space to analyze them (Hagerstrand, 1957, 1975, 1982). Working with Swedish data, he located the moves made by each person in a town over his or her lifetime among the possible residence sites of that town. A vertical line represented the time spent in that site; a horizontal line indicated distance moved to another site. In these lifelines, people left home to work, got married, moved to new houses, grew old, left houses; these residences were then occupied by the lifelines of others replacing them in those locations. What a wonderful way to conceptualize exposure, to chemicals perhaps, over a lifetime. It has seemed impossible to do this for more than a few people, but now new modules of analysis on GIS have opened up the possibilities. (See the discussion of Bian's work in Chapter 5.)

Population movement in space, mobility, is thus a continuum from the arrangement of furniture in a room to massive flows of people between continents. Perhaps the best way to understand the nature and measure of exposure is to begin at the microscale. Within a district, valley, village, yard, house, or room, areal differences in health hazards can be identified. The child standing in the kitchen next to a pot of boiling water is exposed to a different hazard than the one across the room tending the cat litter; the man driving his car on a busy street in smog is exposed to different hazards than the farmer driving his tractor under direct midday sunshine through the spore-laden dust. Among the key evidence of gasoline as a source of lead hazard, leading to its removal, were the concentrations of lead in dust on apartment windowsills adjacent to roads and intersections. They were many times higher than those of dust on windowsills farthest away, clearly showing distance decay from a source. People in buildings adjacent to landfills, incinerators, or simply industrial smokestacks breathe different air than those living in the same city under suburban trees—but these trees are the ones that have birds infected with encephalitis! In a lifeline approach, the vertical line representing time in place is one of continuous exposure (e.g., to allergens, air pollution, or ultraviolet radiation); the horizontal line across space represents contact with, and perhaps transfer of, infectious agents. Having a goal of understanding who is exposed when and how long to whatever risk is being considered, one still needs to be able to measure the exposure.

Larger areas of some repeated or maintained exposures may be identified by constructing elliptical cells. The axes of these cells are simply measures of the average population movements and individual deviations, such that a cell contains two standard deviations of all trips. They can be calculated to summarize areas within which people circulate for work, school, shopping, and other social activities. These are the areas in which people are actually exposed, rather than just the point locations of their residences. Each area can then be considered for the hazards it contains, or, more commonly, for the medical services and facilities accessible to people moving within the area. Since population mobility is very much involved in

diffusion of disease, several ways of modeling and finding patterns in such movement are developed in Chapter 6 on disease diffusion.

At the opposite end of the time–space scale continuum, massive movements of
population between countries, states, or provinces can occasionally be analyzed over
lifetimes to separate etiological factors of environment, behavior, and genetics. The
argument is that a sudden change in disease incidence upon migration would be
due to some environmental factor; failure to change at all would suggest genetic
disposition; and change over decades or generations would suggest behavioral roles
and exposures. Migration to Hawaii from Japan set the classic epidemiological stage:
stomach cancer was high in Japan but low in the United States, and breast cancer
was the reverse. Stomach cancer fell slightly in the migrants, fell more in their children, and by the next generation had approached American rates. The low Japanese
breast cancer rates at first did not seem to change at all, suggesting less genetic susceptibility. Third-generation Americans of Japanese descent in Hawaii, however, did
not have the low rates of their genetic homeland, but intermediary rates between
Japanese women and European American women. In an updated study of women in
Hawaii and California (596 cases and 966 controls), there was a sixfold gradient in
breast cancer risk by migration patterns (Ziegler et al., 1993). Asian American women
born in the United States had a 60% higher risk than those born in Asia; among those
born in the United States, the risk was determined by the number of grandparents
born in the United States. Of course, habitat in Asia was changing too. Among Asian
Americans born in Asia, the risk was 30% higher if the prior community was urban,
and 80% higher if they had lived in the United States over 10 years. Such evidence
strongly suggests the importance of American lifestyles on health outcomes.

Migration streams have also been used to analyze the remarkable latitudinal
pattern of risk for multiple sclerosis and a few other conditions, and possible etiological factors for esophageal cancer, as we shall see in Chapter 8. The existence of
large (because such diseases are relatively uncommon, large numbers of people are
needed to study them), coherent migrant flows with known origins as well as destinations is a rare phenomenon. Mobility data available in the United States are from
few sources, designed to protect privacy, and offered at improper scales for such
analyses. The decennial census asks respondents where they lived (same house,
county, state?) 5 years previously; changing residence to another state requires paying taxes to license a car in that state, and lists of totals can be accessed; the much-
researched 1:100,000 sample of Social Security address changes offers nothing that
can be connected.

QUICK REVIEW 2

✓ The age structure of the population has changed dramatically as a result of the
demographic transition. A much larger portion of the population is over the age of 65.

✓ The increase in the number of elderly in a population has a significant impact on the health
system. Long-term chronic conditions, as opposed to infectious disease, require more
resources.

> ✓ Health scientists use age standardization in order to examine mortality rates among populations with very different age structures.
>
> ✓ Geographers have developed unique techniques for tracking human mobility in space and time. These techniques have made examining exposure to social and physical environments more precise.

DISEASE ECOLOGIES OF THE AGRICULTURAL FRONTIER

As population grows rapidly in the middle of the demographic transition, and as farmland becomes fragmented by inheritance and/or acquired by larger landowners, farmers in the developing world have often set out for new farmland. Not every country has rain forest or other land that can be converted to intensive agriculture, of course, but surprisingly many do. Farmers in India and China have cleared land higher up the mountains, into more arid land, into colder land, and into marshy land. People have come to live densely in land at the margins—land now flooded in extreme monsoons and cyclones, land newly infected with resurgent malaria, land under sliding mountains, land blowing away in the wind. Hundreds of thousands have recently been thus forced from their homes and their new land in India. China is struggling to rebuild and again resettle people displaced by earthquake, landslides, and flood from such lands. In Bangladesh, devastation seems to have become an annual event (and the seas have yet to rise). The Vietnamese have converted most of the now largely deforested central highlands to coffee or vegetable farming; tribal hill peoples have been displaced everywhere, along with their own particular nutritional ecologies. The Philippines encouraged the resettlement of millions of people from the degraded and impoverished Catholic Visayas (central islands) to Mindanao, to replace its forests with oil palm and marginalize its small Moro population. Indonesia's national endeavor of transmigration moved millions of farmers from Java and Madura to the outer islands of Sumatra, Borneo, Sulawesi, Halmahera, and finally their province of Irian Jaya on New Guinea. Millions of hectares of forest were converted to oil palm in particular, and to rubber, coffee, sugar, and even rice. When Suharto fell and the Indonesian central government lost control, local people in many places tried to throw out the intruders who had received the benefits of development and taken their land. But many thousands of those settlers, days away from any medical care, had already died from malaria, diarrhea, and acute infections compounded by malnutrition. In those six countries alone live more than 3 billion people (45% of earth's total). Their population's growth and push to the agricultural frontier have changed the global environment profoundly.

In Latin America farmers are pushing into the Peten Reserve in Guatemala and the forest sanctuaries of Belize; from all sides they convert the forests of the Amazon to agriculture. Soybean expansion in Brazil has received much of the blame for deforestation, but farmers moving in along new logging roads and highways, slashing and burning forest for the cultivation of a few years, have been at it for decades. Today large numbers of farmers with little experience of wet tropical soils have moved down from the altiplano, where ancestrally they farmed potatoes and other crops at 12,000 feet. They have created one of the world's new hyperendemic malaria hotspots. In sub-Saharan Africa, which is at an earlier stage in the demographic transition and just beginning to lower its fertility, population is growing regionally at 2.5% and will more than double before approaching stabilization. In East Africa the value of tourism has made possible the preservation of wildlife habitat and their ancient ecologies, but farmers still struggle to clear new land and grow crops though it is in the paths of migrating elephants. Roads for logging and mining have opened up some of the Congo forest, followed closely by a great expansion of hunting for bush meat. Butchering of primates was almost certainly the source of HIV transfer to the human population. The conversion and farming expansion are more famous for the episodic eruption of Ebola and Marburg viruses from their disturbances. In Gabon and the adjacent Congo basin remains a vast wild land yet unpenetrated by farmers. From disturbance of the African land and its ancient primate ecologies, surprises are yet likely to emerge (see Chapter 7).

This basic geography is needed to connect what are usually reported and studied as single episodes, unique developments, or unexpected consequences. The following connections and comparison of land development in the agricultural frontiers of Malaysia and Ecuador illustrate both the commonality of the political ecology of ecological disturbances and exposed migrants at the frontier, and the importance of detail (species-specific, cultural, and microscale).

Population Resettlement: Malaysia, 1970s

Some agricultural conversion by migrants is spontaneous, carried out by individuals or families without much organization or legal authority; some is highly planned, organized, and capitalized. One of the best examples of the latter is the program developed by the Federal Land Development Authority (FELDA) in Malaysia. In the 1970s and 1980s it resettled hundreds of thousands of farmers who needed land on "land schemes" converting rain forest land to grow rubber trees and oil palms. Although not without political and economic problems, these land development schemes became models for other countries. They played an important role in making Malaysia internationally dominant in palm oil production, and for a generation they created decent livings for poor farmers who otherwise might have gone to the urban squatter slums. FELDA schemes have been much studied by political scientists and economists for their costs and benefits and for their organizational successes and failures in increasing human capital. Interested in the ecological alterations being wrought worldwide by land development programs, Meade (1976, 1977, 1978) studied the FELDA schemes' impact on health over a 2-year period through engaging in participant observation; conducting long interviews in Malay and English; conducting a census of the settlement; maintaining a register of 60 households interviewed weekly in both the scheme and a comparative existing rubber farming village 10 miles away; and sampling children's stools, mosquito larvae, and field rats and their mites.

People who had applied for land were admitted from all over peninsular Malaysia. They spoke different dialects and, in those schemes in which Chinese and Indian Malaysians were admitted along with Malays, they had different religions, diets, and other cultural practices. Most were far from their home villages for the first time. For the first few years, while the tree crops matured, they faced raw land that had to be cultivated, fertilized, drained, and planted. Some settlers could not speak to their neighbors and in any case were not used to mixing with strangers. It took 2–3 years for communities to form. Occasionally some lonely settlers, unable to get to town or visit home villages on unpaved roads, left the scheme. Women, following the custom of giving birth to their first child in their mothers' houses, were especially stressed. Malay women did not traditionally hoe plantation fields or work on roads, but they did these and other jobs on the land schemes. Sometimes older children had to stay home from school to take care of small children while the mothers worked, leaving the children to eat plain cold rice for lunch. The men were required to report to work early in the morning even in poor weather—a kind of regimentation previously unknown to many of them. There were no local ways to earn money when there was no labor required on their holdings.

Settlers' houses (see Plate 6) at first were too small for large families, but later houses were enlarged and smaller families were selected. The corrugated iron roofs were long-lasting and vermin-free, but because the whole area had been clear-felled and was exposed to tropical sun, the houses frequently heated to over 110°F (43°C) during the afternoon. Each house had land around it for gardening. It takes years for fruit trees to mature, however, and it was difficult for government agents to get the Malay settlers to grow vegetables (which are usually grown by Chinese farmers in Malaysia in the highlands). There were no fresh fruit or vegetables, or meat since men weren't at leisure to hunt, unless vendors made their way in or male settlers out to Saturday market (both on motorcycles). FELDA had a store with dry and canned food, but even that was open only during hours when male settlers were supposed to be in the field. FELDA also provided latrines (Plate 6), which many settlers had not previously had. Almost all the adults did use the latrines, but because toddlers are not diapered, some house compounds were fecally contaminated. A few children's play groups maintained helminthic infections until school age, and the school became a focus for infecting the children who had grown up in more sanitary homes. Every few houses shared a standpipe that brought chlorinated water. Although the system often broke down because of leakage and siltation problems at the source, the enteric infections and fevers in one such scheme, when studied in detail, were clearly reduced compared to those in the nearby traditional village.

The vectored diseases were the worst affected, and the worst among these was malaria. Many of the schemes constructed in rain forest areas repeated the experience of rubber plantations early in the 20th century. The ecosystem was simplified, and epidemic conditions were created (see Plate 6).

Trees were cut down; commercial hardwood logs were removed, and the remainder were burned. This exposed the streams to sunshine and created breeding habitat for *Anopheles maculatus*. There are an abundance of *Anopheles* mosquito species in Malaysia inhabiting different ecological niches. *A. umbrosus*, for example, shared the forest with other *Anopheles* species. It liked the deep shade of the rain forest. Although abundant, it preferred to bite other animals rather than people. It nevertheless occasionally transmitted malaria between people. Only *A. maculatus* liked the direct sunlight on water for breeding that was created by clear-felling. It quickly bred in huge numbers, the only species with only people to eat. When rubber trees were first planted in clear-felled forest along the coast of the Strait of Malacca, Tamil laborers contracted from India were carried out in the thousands.

On the FELDA schemes, houses were sprayed with insecticide twice a year as part of Malaysia's rather successful campaign against malaria, but the initial success of chloroquine and insecticides led to neglect of traditional concern for drainage, shading, oiling, and other techniques that had originally been developed in Malaysia for those rubber plantations. The mosquitoes across Southeast Asia were becoming resistant to the insecticides. The clinics serving the remote areas of low-density population where new land schemes were established were not enlarged before the settlers came; their capacity had been adequate for the presettler, mostly Semai aboriginal population. Settlers sick with malaria were therefore driven in Land Rovers for hours over rutted and flooded land to clinics with no capacity to treat them, were given chloroquine pills, and were sent home. The resulting underdosage, combined with new strains introduced from older settlements, led to increased prevalence of chloroquine-resistant malaria and its unusual expressions. In at least one scheme, unfamiliar cerebral malaria led to a perception of spirit possession and consequent abandonment of the land. As the schemes matured and crop trees grew to shade the water, the epidemic receded.

In the less capitalized settlements of Indonesian transmigration, malaria, diarrhea, respiratory illness, and malnutrition plagued early decades of developments. The responsible ministry learned to do better over time. Construction of houses, roads, and economic supports; health service provision; food marketing; and labor opportunities in the interim before tree crop production matured were improved. Health education, family planning, and bed netting were all improved in established areas, although the insertion of land development and transmigration of ethnic groups into new lands of other islands with different ethnicities and religions caused increased instability. Despite all that was learned about the effects of agricultural conversion on disease ecology in the forests of Southeast Asia, little seems to have translated across the seas by government health or forest ministries, or by academic researchers. Policy issues and disease ecologies from conversion of forests in Latin America or Africa are repeating themselves.

The FELDA scheme stopped recruiting new settlers in 1990. Government funding directly to FELDA was cut under the assumption that the FELDA could generate their own income and support continued development through a variety of businesses. FELDA has since diversified from its original business of land development, to other economic ventures—some of them entirely unrelated to land resettlement—and the program has been criticized for shifting toward a model more like a "plantation company aiming for commercial efficiency than a state organization with social development objectives" (Sutton & Buang, 1995, p. 125). This has changed the way land in the schemes is being used, and may again shift the disease ecologies in the region. In addition some of the FELDAs located near the development corridors of the country (e.g., near Kuala Lumpur) are being redeveloped into residential and industrial satellites. The land prices in these schemes have skyrocketed and some families have become instant millionaires from the sale of their land (Ahmad, 2015). This has been driven by the larger process of urbanization and population growth.

Population Resettlement: Amazon, 2000s

Most Latin American countries are quite late in the demographic transition now, although a few haven't progressed as far as the others. Total fertility rates for the region as a whole are down to 2.5 children per family. Fully 76% of the regional population is urban. Yet, because of the relatively recent decline of fertility, more than 30% of the population is still under 15. This is shown in the 2% net out-migration to Europe and the United States. It is also shown in the changing push to the agricultural frontier. As Brazil, the world's fifth most populous country, has become an urban industrial giant, gold mining, commercial logging, and capitalized agribusiness have become greater threats than are poor farmers from the northeastern region slashing poor plots. Farmers still move into the Peten in Guatemala, but more are turning to cities and international work. In the western Amazon, however, agricultural clearance is increasing as Andean farmers seek new land. The Peruvian Amazon in the late 1990s became one of the world's hot spots for hyperendemic malaria.

South America offers unique challenges to frontier migrants, as well as reprising others from tropical Asia. Before the Columbian exchange brought malaria, yellow fever, river blindness, schistosomiasis, and so many contagious diseases, leishmaniasis and especially Chagas disease "protected" some tropical lowlands from settlement. Chagas disease in particular continues to kill several times more people than malaria in the Western Hemisphere. The wild animal reservoir of this natural nidal disease includes more than 80 species, from sloth to dog. Its vector, a reduviid or "kissing bug," has species variants that allow it to be transmitted from the lower Andes to the grasslands of the Grand Chaco and the central Mexican plateau. Its primary nidus is in the wet tropical lowlands. Chagas disease is also called American trypanosomiasis because its disease agent is a trypanosome, a kind of protozoan like the agent of African trypanosomiasis (sleeping sickness). After causing initial symptoms for a few days, the American trypanosome reproduces in muscle and circulates in the blood for years, even more than a decade, before people die from enormously enlarged hearts and dysfunctional colons. Metacolon has been found in many pre-Columbian mummies, indicating that the person had been unable to defecate for more than a month.

The reduviid bug, which transmits Chagas disease, resembles in its habitat, behavior, and size a cone-nosed cockroach. This bug sucks blood and prefers the lips of sleeping people (hence its moniker, "kissing bug"). After its meal the bug defecates, and people then rub the trypanosome-laden bug feces into the wound or mucous membranes. The reduviid bug is one of the disease vectors most clearly affected by substitution of iron roofs for thatch ones. A memorable image from a reading decades ago (a lost reference) is of a "rain of feces upon the mucous membranes of people sleeping under the thatch roofs." Like cockroaches, reduviid bugs live in piles of firewood, old leaves, cracks, and other such dark and dirty places (e.g., behind refrigerators). They have accompanied migrants to the urban slums and spread Chagas disease there. The trypanosome has even accompanied migrants to the United States and contaminated the U.S. blood supply. When blood is donated to the Red Cross now, a series of screening questions is asked in an attempt to eliminate possible exposure to Chagas. Remember that the latent period of this disease, during which hosts are unaware of growing infection and infectivity of their blood, is more than 10 years.

The best protection against Chagas disease is good housing in which the bugs find no habitat. Usually when settlers leave their home villages and family houses and strike off to the frontier today, they move to poor housing (see Plate 10), but at least seldom to thatch roofs. Houses in these settlements do offer entrance to mosquitoes, though—*Anopheles* at night and *Ae. aegypti,* the vector of yellow and dengue fevers, by day. The mosquitoes rest under furniture and clothing hanging from nails on the wall, and breed in abundant "container" water around homes. Water-borne disease is common. Latrines are rather haphazard, and wells are often nearby. The main improvement between the Ecuadorian frontier homes of Plate 10 and the Malaysian ones of Plate 6 from 30 years earlier is the presence of electricity; the worst comparative note is the absence of safe drinking water, which the Malaysian government did provide.

The satellite image in Plate 10 shows two rivers, two small urban areas marked with the orange circles, and two of the settlements studied by Polo (2009) and pictured. The roads and agricultural clearings (including a large planting of oil palm in the blue circle) can be seen to be cut from the depths of the Amazon rain forest in fields and pieces, not clear-felled and logged over 20 miles at a time. As in Malaysia, rodents were disturbed and dispersed. Local field species must have been changed, along with their viruses and the likelihood of human contact with their urine aerosols, which can cause Bolivian (Machupo), Venezuelan (Guanarito), and Argentinian (Junin) hemorrhagic fevers. Monkeys and other mammal reservoirs (like bats); many species and whole ecologies of birds; and uncountable species of mosquitoes, bugs, acarids, and amphibians have been exterminated, replaced, displaced, promoted, and transported in this ecologically simplified habitat. With all this disruption, the potential for human disease and its epidemic eruption has been fundamentally altered.

The cultural ecology of malaria offers dramatic contrasts and similarities to those of agricultural conversion and population resettlement in Southeast Asia. Let's start with the fact that the large clear-felled conversions for agribusiness in Brazil have often changed dense rain forest into monocropped soybean fields or cattle grazing, but have resulted in relatively little malaria. For more than a century, there has been massive deforestation for cattle ranching, mining, and agricultural conversion. Malaria has been associated with this change in land use and land cover, but the powerful repeated links of epidemics with forest clearing have not been clear, as they were in the Malaysian experience.

Efforts to use satellite images of water flooding and malaria potential in Belize found that *Anopheles pseudopunctipennis* existed at densities too low for sampling change. The main vector of epidemic malaria in South America is *Anopheles darlingi*. Singer and Caldas de Castro (2006) explain how the linkage of deforestation, conversion, and malaria must be analyzed at several temporal and spatial scales, starting with the specific ecology of the vector and how it is affected by transformation of the ecosystem. These researchers characterize the larval habitat of *A. darlingi* "as zones of partial shade, proximal to a forest fringe or river edge, in relatively deep and clear water of high pH (relative to the more acidic Amazon streams), and with temperature greater than 25°C" (p. 1). Contrast this with the Malaysian vector, *A. maculatus,* which breeds in shallow water in direct sunshine and has intersecting vegetation to anchor eggs. The large area, fast clearance, and conversion of the corporate sector development in the Amazon has created habitats largely inimical to the larvae of the most efficient vector. In contrast, it is the government-sponsored and spontaneous colonization of the forest that has resulted in epidemic malaria. Compare the settlement habitats in Plates 6 and 10: notice the trees in the settlement and along the riparian background.

Singer and Caldas de Castro (2006) pick up and extend the relevance of Sawyer's (1988) stages of *frontier malaria*. At the micro/individual level of *colonistas* moving into the primary forest, the shaded habitat and riverine areas are very suitable for *A. darlingi* larvae, and indeed their population is dense. In the small clearings where the settlers work, the exposure is intensive. *Plasmodium falciparum* is the primary *Plasmodium* species in these conditions, spreading more rapidly than *P. vivax*. Settlers have limited understanding of transmission. Their houses readily admit mosquitoes, and the insides provide poor surfaces for residual spraying of insecticides (mosquitoes leaving). Little medical care or malaria prophylaxis is available. At the community level there are poor organization, few management efforts to minimize malaria risk, little health service availability, and high rates of both in- and out-migration (which circulate infected and susceptible individuals). Some settlers try to escape poor land and malaria by development of new unplanned areas, but this process only promotes more transmission. After 6–8 years the population settles, and roads and impervious surfaces create habitat inhospitable to *A. darlingi* larvae. People move away from riparian areas. Clinics and health care increase. Population immunity grows, nutrition improves, and the epidemic subsides into endemic malaria as background.

This scenario of frontier malaria in transition also fits the experience of Southeast Asia in both time and space scales as well as human processes. There is one exception, posing a paradox that is important for conservation. In Malaysia conservation of forest habitat, which promotes biodiversity, thereby promotes multiple species as vectors of malaria—each with fewer members, less contact, lower intensity, and lower disease transmission. When it is pointed out that in the Amazon "altering the landscape likely plays an even larger role than people migrating into the jungle," and reported that *A. darlingi* bites people 300 times more after areas are cleared than before, what is being described is the frontier malaria of colonization (Vittor et al., 2006). The statement "conservation policy is one and the same with public health policy. It's probable that protected conservation areas may ultimately be an important tool in our disease prevention strategies," describes tropical Asia more than the Amazon and raises the paradox. There, it seems, the corporate large-scale total clearing and replacing are what most protect the workers and most threaten biodiversity; the individual settlers on their still-forested clearings are most at risk.

These examples are representative of how, time and again, developmental changes result in environmental changes that impact human health. Further exploration of these ecologies in the remainder of this chapter focuses on two aspects: other rural impacts, mostly as expressed in African ecologies and especially from impoundment of water; and globalization of movements, including people, food, and lifestyles.

OTHER DEVELOPMENT IMPACTS ON RURAL ECOLOGIES

Many of the health problems of both new settlers and old villagers are simply those of protracted poverty. Farmers' land becomes fragmented, leached of nutrients, eroded by wind and rain; their livestock are depleted. They become landless laborers, who must move to the frontier or the city or internationally to sell their labor. Little health care is available. The water is not safe to drink. The billion people living on less than US$1 a day cannot afford to eat balanced diets. Those who become involved in capital-intensive agribusiness may have money to buy food grown elsewhere; to get vaccinations and prophylactic drugs; and to have screened housing, deeper wells, less dangerous work, and access to primary health care when they do get sick. Their actions can induce different adverse disease ecologies, however. Increased cultivation of cotton and other crops for export usually leads to a greatly increased exposure to insecticides, the use of which produces disease vectors highly resistant to it. The contaminated environment can initiate the late-transition-stage diseases of cancer and genetic alteration in fetuses. The stresses resulting from enormous change itself, and from ever-increasing economic and social inequity, have (as in India) caused suicide to become a major cause of death.

As has been mentioned earlier, sub-Saharan Africa is the world region at the earliest stage in the demographic transition (see Table 5.1). Shocking data easily accessible from the Global Health Observatory (GHO) database on the WHO website (www.who.int/gho/en) confirm this. Infant mortality is almost 1 in 15; maternal mortality is high (45% of the world's); and life expectancy for the region as a whole is 57 years. Multiple infections with different types of helminths and protozoa remain common. Figures 4.4, 4.5, and 4.6 show the global prevalences of HIV, malaria, and tuberculosis, and it is clear that sub-Saharan Africa has more than its share of all these diseases. Southern Africa is hardest hit by all of them. With only 12% of the world's population, more than 40% of the world's diarrheal mortality occurs in this region. Much of the rural African population continues to have close contact with vectored disease. Yellow fever still kills "only" 20,000–30,000 annually.

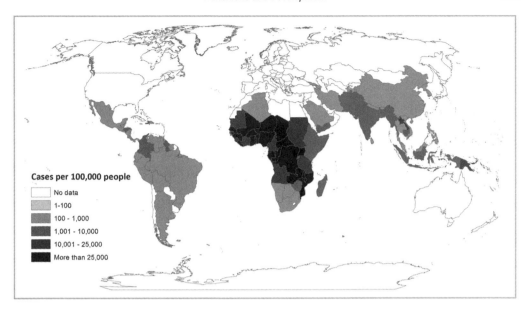

FIGURE 4.4 Sub-Saharan Africa has the highest malaria prevalence.

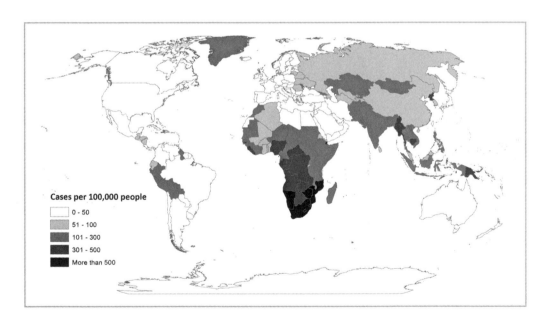

FIGURE 4.5 Sub-Saharan Africa also has the highest tuberculosis prevalence. Southern Africa is especially hard hit. Asia also has quite high tuberculosis prevalence levels.

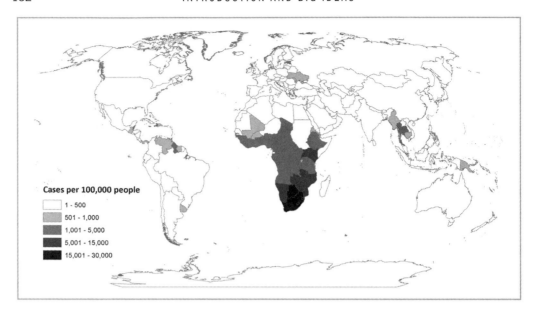

Cases per 100,000 people

1 - 500
501 - 1,000
1,001 - 5,000
5,001 - 15,000
15,001 - 30,000

FIGURE 4.6 HIV/AIDS is a global epidemic; however, sub-Saharan Africa has the highest HIV prevalence. Southern Africa is especially hard hit.

The three diseases that can render good agricultural land totally uninhabitable—malaria, river blindness, and sleeping sickness—evolved in the African realm. Indeed, 90% of the world's malaria mortality, and now 75% of its HIV mortality, occur in this region; only 6% of its cancer does, however.

In recent years, major progress has been made through the United Nations's millennial goals; WHO initiatives and global funds targeting malaria, HIV, and tuberculosis; the development of GIS for integration and surveillance; and major funding from the Gates Foundation for research on new drugs and treatments for the neglected tropical diseases. Onchocerciasis in West Africa has been the target of one of the most successful eradication efforts in world history; surveillance is now maintained by vigorous community participation and powerful GIS integration, so that in 2001 more than 40 million people, living in 117,000 hard-to-reach villages, regularly received ivermectin (first developed in the United States to combat the filaria responsible for canine heartworm) for prophylaxis. The GHO database (see above) also shows a 75% drop in measles deaths between 2000 and 2013 owing largely to the fact that 84% of the world's children receive the measles vaccine; leprosy prevalence has dropped by 90% and has been nearly eliminated in all but 14 countries (but 215,557 people were newly infected in 2013).

The health geography of developmental changes in Africa has been both the first, thanks to the work of John Hunter (see Chapter 1 for a description of his work), and the least investigated of any region. The initial comprehensive review by Hughes and Hunter (1970) of the literature on development and its impact on health in Africa sensitized a generation of scholars and experts. Hughes and Hunter described, for example, how the evacuation of tribes from their agricultural and grazing land was carried out by colonial governments because of outbreaks of rinderpest or smallpox or in order to stop tribal warfare. Such evacuations allowed brush and tsetse to claim the abandoned territory, so that returning herdsmen or farmers years later found the land denied to them by trypanosomiasis. They also described how roads had become linear transmission sites for disease. Tsetse flies (which are attracted to moving vehicles), blackflies, mosquitoes, and the disease agents they vector became concentrated where roads crossed streams and people stopped to eat, wash, and urinate. New water reservoirs and roads meant new exposures and transmission to new areas. Hughes and Hunter also first described some of the nutritional and social–psychological consequences of development, as well as the difficulty of intervening in the complex disease systems and multiple levels of infection. They emphasized most of all, however, the need for a comprehensive, ecologically informed approach to developmental projects designed to break the synergism between poverty and disease. Since then, examples of the inadvertent consequences of developmental changes in Africa and everywhere else have multiplied. Yet their political ecology and the health promotion efforts of current massive, vigorous educational campaigns have been little studied. Guinea worm, river blindness, and sleeping sickness have been almost eradicated, for example, even as blood pressure, diabetes, and violent death soar. It is not only the rapidly urbanizing population that is so neglected by research, but the rural conditions of modernizing agriculture also.

Agricultural development in Africa, for example, affects the cultural–ecological process of animal herding. There are increased penning, greater control of breeding, and higher-quality feed. The nature of barns and enclosures often changes. Animals do less scavenging. They are harvested or culled more regularly, so that the age structure of the herd becomes younger, and the proportion of immune animals to susceptible ones changes. The socioeconomic role of contact with animals may change. It may come to involve more men than boys and may become an occupational specialty rather than a source of general population exposure. Replacement of draft animals by machines may deprive mosquitoes of their source of blood meals and displace them to less favored meals from people, causing new or intensified disease transmission. As developmental changes affect the quality and resistance of animal and human hosts, increase the intensity of animal herding, displace animals, change occupational connections and gender roles, modify water and food pathways, and alter habitat for arthropod vectors, the significance and distribution of various diseases transmitted from animals to humans will also be affected. These interactions, interventions, and promotions remain in great need of research.

Irrigation and Water Impoundment

Nothing except the use of fire can compare with the ecological consequences of the human ability to control water. There are today tens of thousands of large dams and water impoundments; millions of tube wells pumping water up from the depths; and hundreds of millions of hectares of land under irrigation. Mountainsides have been terraced for wet rice cultivation; swamps have been drained over millions of square miles. The particular health hazards resulting from impoundment of water and irrigation of crops have long been recognized, although not often evaluated by engineers and economists. Much of the present appreciation of their importance stems from the construction of the Aswan High Dam in Egypt and the spread of schistosomiasis.

Basin irrigation was an ancient practice in upper Egypt. The basins (*tanks*, in British English) constructed to store the floodwaters of the Nile were eventually drained to supply a continuing flow of water to irrigated fields. Subsequently they were themselves planted to crops, and eventually dried up completely before again being flooded by the rising river. At the height of Ethiopia's rainy season, the flooding Nile brought vital silt and water to the lower river valley in Egypt. It also washed out and damaged the snails and their eggs that had survived the dry season, thus serving to control the snail population. With the coming of the Aswan High Dam and perennial irrigation, the snail population expanded greatly. Because the nutrient-rich irrigation water flowed only slowly through the agricultural fields, was cleared of silt by reservoir sedimentation, and often supported aquatic vegetation, it was an excellent snail habitat. Schistosomiasis, which had been endemic at low levels in the region from the time of the pharaohs, exploded to become a universal scourge. The same situation has been repeated monotonously, from the Gezira irrigation scheme in the Sudan to Lake Volta in Ghana. Africa is a dry continent. Irrigation is absolutely crucial to countries' hopes of feeding their growing populations, but the price is high. In Kenya, in a district on the shores of Lake Victoria, the rate of schistosomiasis infection for children entering elementary school went from virtually nothing to 100% in less than 10 years when an irrigation system was developed.

In Brazil, too, expanding irrigation means expanding the territory of schistosomiasis. In Southeast Asia there are only a few foci, including a small but critical one in the Mekong River. As some of the plans for damming and controlling that great river materialize, now that there is peace and many dams on tributaries have been completed, one of the feared consequences is the spread of schistosomiasis throughout the Mekong River basin. Much research concern for the spread and control of schistosomiasis (*S. japonica* in the Yangzi lake basins) has also followed the completion of the Three Gorges Dam in China.

Water impoundment and irrigation have affected vectored diseases ranging from California encephalitis to lymphatic filariasis in Ghana, from clonorchiasis (liver fluke) in Thailand to the release of heavy metals and poisons from sediments as old dams are removed in the United States today. Throughout the ecologically more diverse tropical world, however, impoundment continues to increase malaria hazards.

Water impoundment schemes/projects/structures have undoubtedly contributed to economic development in many Third World nations. In Bangladesh, for example, government and NGO-funded water containment structures have protected rural populations from annual floods caused by monsoon rains and typhoons. Earthen embankments circling households and fields decrease deaths due to drowning and allow farmers to double- or triple-crop in the growing season, as well as to plant genetically modified high-yield varieties of rice. At the same time, however, such structures may act to increase the risks of infection from bacteria and viruses, causing diarrheal diseases like cholera or dysentery. By disrupting the annual cycling of water through the system, embankments prevent the flushing and replenishing of the ponds and streams that supply households' drinking and cooking water. In some cases, this increases the densities of microorganisms to which residents are exposed.

GLOBALIZATION OF MOVEMENTS

One of the fundamental activities of human behavior in disease ecology is to move people and elements of the habitat from place to place (see Chapter 2). In the midst of "globalization" of economic activities and communication of every kind, this movement counts as a new thing if only because of the speed with which it is happening. This aspect of human agency is more important than ever. We have seen it in this chapter in the relocation of disease agents of schistosomiasis and malaria through impoundment of water and irrigation. To this should be added the transformation of the world's food supply.

Consider first, as one example of new development changes, what has to be called "the industrialization of food animal production." Like the urbanization of the human population, the herding of animals into enormous, densely packed populations reprises the ecological consequences of the first domestications. Cattle in feedlots at the giant slaughterhouses and meat-processing plants can be penned by the hundreds of thousands at a single facility. Chicken raising has been specialized into brooding; egg production; chicken (broiler) raising; and assembly-line slaughter, cleaning, parts production, packaging, and marketing. Fowl being raised for food are penned so densely that their beaks commonly must be cut. Following the business success with chickens and turkeys, in recent years swine production (according to the so-called North Carolina model) has followed suit. Piglets are now nursed through bars on trapped sows, raised densely packed by tens of thousands into a facility, and fed by conveyor belts; their wastes are flushed into septic "lagoons" and eventually sprayed on the land. Food contamination from such industrialized production and marketing poses new challenges for inspection and for risks of pandemic. Food thus produced is shipped not only across the United States but internationally, as from Thailand to Japan.

Even if we put aside for present purposes ethical issues about the humane treatment of animals, serious public health consequences need consideration. The waste discharge and spills thus produced have attracted a lot of attention through putrefaction of streams, river, and offshore fish kills, and recently attacks by *Pfiesteria* (a genus of dinoflagellate protozoa) on fish and production of neurological toxins. Animals cannot be herded so densely without massive use of antibiotics, which are passing newly resistant strains of microbes on to people. The salmonella and other intestinal fauna are increasingly contaminating carcasses and parts in the mass, speeded-up, assembly-line processing. The combination of birds and pigs close together at such densities raises several red flags. The influenza virus is thought to "shift" when avian flu and other strains get combined in pigs, producing pandemics. The emergence of bird flu in chickens in Hong Kong in 1998 caused a panic and the slaughter of all the chickens; similar situations have continued to occur elsewhere. Avian encephalitis viruses, such as Japanese encephalitis (and probably Eastern equine encephalitis), can be amplified by the high viremia in pigs into major epidemics. Animals are not raised on food grown on the same farm any more, but by agribusiness production that spans continents. When cattle in Britain began to get bovine spongiform encephalopathy (BSE; also known as "mad cow disease"), it was traced through epidemiological methods to feed that included sheep brains and body parts (although cows are famous as vegetarians). This British feed had already affected cows in Italy and other European countries in trade. Although the string of protein molecules that seems to be involved (prions) defies any definition of life, it seems to reproduce the same brain destruction in chains of animals that consume it. Epidemiologists have made an analogy with *kuru,* previously found among cannibals in Papua New Guinea. The Western public's greatest alarm, however, is about a rare human brain disease, Creutzfeldt–Jakob, which may be part of the same food chain as BSE. More common, if little understood, health challenges from animal feed include synthetic estrogens, growth hormones, and similar elements of high-tech mass production. In such ways, new international technologies of agricultural production, processing, and marketing are creating a new disease ecology.

The globalization of food production and marketing thus brings not only the benefits of abundant food available in any season, but risks of worldwide exposure to risky local conditions, whether these stem from industrial-model chicken and pig raising or the intensive use of pesticides on fruits and vegetables. It has been said that you may eat in the most sanitary gourmet restaurant of the 21st century, but your food was grown in the 16th. The dubious sanitary conditions and unsafe water sources of the developing world are daily encountered on the vegetables of the French table (from Mali) or on the grapes of the American table (from Chile). Food is distributed from centralized sources as never before. Twenty years ago there was concern that a sick milk cow could spread contamination through a dairy and its equipment to a supermarket outlet system, or that contaminated machinery in a meat-, candy-, vegetable-, or dairy-processing system could poison vending machines, school cafeterias, and kitchen counters nationally through the system of marketing. Now the threat is international. National food inspection systems themselves are largely antiquated bureaucracies. Contamination of beef with the intestinal bacteria of cattle, of water with runoff from feedlots, of apple juice with untested insecticides, and of Chinese children's milk with an industrial chemical for plastics that tests "protein" are not random events. For more on the nutrition transition and health and environmental impacts of food production and consumption, see Chapter 8.

We need briefly to return to the third destination of the mobility transition—international wage labor—and to ever-growing circulation. Globalization has many important effects, whether on wage structures and employment, information flow, belief systems, or scale of the pollution syndrome. One of the most profound for the ecology of disease has been the escalation of international migration and circulation. The numbers are "staggering," as the cliché goes. There are over 250 million people legally working outside their home countries. There are over 250 million people a year who move between the tropics and developed latitudes for business, tourism, study abroad, and so on. Nigerians are trying to reach work in Germany, Tongans in New Zealand, Samoans in California, and Pakistanis in the Persian Gulf. There are countries in which more than a fifth of the labor force consists of foreign workers (Switzerland, Côte d'Ivoire); in Saudi Arabia and other Gulf states, more than half of it does. There are other countries, like the Philippines, in which more than half the population over large areas can be overseas working—and actually making remittances the largest source of foreign exchange. Look at how the age structures of countries (Figure 4.2) fit together to connect countries with young adults needing work to countries without young adults to do needed work. People go from villages to the agricultural frontier and back; they go to cities and back; they circulate internationally for tourism, religious obligations, or education. Most of this they do at the speed of jet aircraft, arriving before symptoms can develop. Quarantine has become almost useless (as illustrated again by the emergence of the H1N1 influenza from Mexico in spring 2009). In addition to this, the United Nations High Commission on Refugees (UNHRC) reported that worldwide there were more refugees, asylum seekers, and internally displaced people than there had been since World War II. Some 51.2 million people have been forcibly displaced, up from 45.2 million just two years earlier (UNHCR, 2014). This was largely driven by a war in Syria and conflicts in South Sudan and the Central African Republic, as discussed earlier in the chapter. These 51 million people (astounding, unprecedented numbers) often live under desperate conditions. They are a virtual nation of the most vulnerable people struggling to live in health conditions that render our interconnected and mobile world susceptible to their pathologies of body and mind.

QUICK REVIEW 3

✓ Rapid population growth has led many countries to develop land that is not ideal for farming. This encroachment into the "agricultural frontier" has caused the emergence of new diseases, and has shifted the disease ecologies in these regions.

✓ In both Malaysia and the Amazonian Brazil, migration to the rural frontier, supported and actively encouraged by the government, significantly increased risk for vectored diseases such as malaria.

✓ The globalization of food production and marketing has environmental consequences. Large-scale farming and industrial livestock operations fundamentally alter environments and create harmful waste and alter disease ecologies and lead to the emergence of new diseases.

✓ Globalization also increases population movement across the globe so that diseases can more easily spread.

CONCLUSION: EMERGING DISEASES IN YOUR FUTURE

Human population growth and change, driven by economic development, cultural shifts, and technology, are the driving forces in changing the earth's climate, biological diversity, water availability and quality, agricultural production systems, food marketing, settlement patterns, urban forms, movements of disease exposure and introduction, creation of disease systems, and health status. This chapter has tried to explain the multiple transitions the world's population is undergoing and some of the consequences for human health that have resulted.

The end of the demographic transition may well be ushering in a stage of renewed parasitic and infectious disease. Infectious agents are newly discovered to be involved in everything from ulcers to hardening of the arteries. Instead of disappearing at the end of the demographic transition, diseases that were once almost eradicated are reemerging. Some of these are **iatrogenic** (from the Greek word for "cure"), or caused by medicine and doctors. The most notorious of these is the multiple-drug-resistant new tuberculosis, but resistant strains of staphylococcus (MRSA) threaten to make hospitals places to get sick (see Chapter 7). The widespread use of, and demand for, antibiotic treatments even when useless (as against viruses), and the general overprescription of drugs, have helped create drug resistance. Inadequate medical follow-up and patient compliance have also contributed to this resistance. New technologies of transferring human organs and tissues have created new means for agents to pass; future use of pig and other animal parts for humans (**xenografts**) may create **xeno-zoonoses**. Use of immunosuppressant drugs to prevent tissue rejection or fight the rapidly growing cells of cancer has allowed a new set of infectious agents to become unsuppressed. Although these health hazards develop from the behavioral dimension of the triangle of human ecology, they have not been examined in this chapter.

The simplifications of ecosystems produced by agricultural conversion and, even more, by urban conversion when connected with the speed and multiple dimensions of population movement, have established new global epidemic propensities. Some disease agents are as yet purely zoonotic. However, as farmers in the frontiers contact the viruses of the rodents, bats, and birds they have disturbed, they bring them to home villages connected to cities where the farmers' siblings live amid densities of human pathogens, market animals, and water/sanitation/housing conditions that couldn't be better designed to promote the jump by disease agents to the human species—and then into global connections. Children in the most developed places are again dying from intestinal fauna acquired in the meat and water of their countries.

Certainly the emergence of diseases can be approached as a political ecology (see Chapter 3). Governance, capitalism, and the struggles for power by groups at every scale are inherent. What those forces do, however, is shape the cultural ecology. Whatever the power and whoever holds it, the needs of the local disease vectors, the adaptability of enteric microorganisms, the need for more food production and for employment in cities, and the changes in age structure and distribution of susceptible individuals will create new disease systems. This chapter has presented how population change is resulting in changes in the creation and maintenance of disease systems and their emergent possibilities. Health geographers need to become involved in developing prospective health geographies of health and disease.

REVIEW QUESTIONS

1. Why is there dramatic population growth during the early transition stage of the demographic transition?

2. Discuss four major factors that have contributed to the emergence of chronic diseases as the leading cause of death. Choose a chronic disease (e.g., cancer, heart disease, stroke) and discuss how these factors have specifically impacted that disease.

3. What are the most prevalent diseases in a preindustrial society? How are they related to settlement patterns? To mobility?

4. How have postindustrial mobility patterns affected the types of diseases people in wealthy countries are worried about?

5. Indonesia is a lower income country (LIC) where the per-capita income is increasing rapidly due to strong manufacturing and agricultural sectors. As Indonesia becomes wealthier, what mortality, demographic, and mobility changes do you foresee and why?

6. Which two populations experience the most profound changes in mortality during the epidemiological transition? Why?

7. Over the next 50 years, what will be the major causes of mortality in high-income countries? Compare this to the likely future causes of mortality in low-income countries.

8. What are some potential reasons that fertility is high in the early phase of the demographic transition? What are the drivers for the drop in fertility as the demographic transition progresses?

9. What is the second demographic transition and how will it affect health systems in developed countries?

10. What are the reasons that a country may experience a "delayed" epidemiological transition?

REFERENCES

Ahmad, T. A. T. (2015). Dimensions of food and livelihood security of agricultural trade: The case of Malaysia. In P. Shome & P. Sharma (Eds.), *Emerging economies* (pp. 113–132). India: Springer.

Dubos, R. (1965). *Man adapting.* New Haven, CT: Yale University Press.

Gould, W. T. S., & Prothero, R. M. (1975). Space and time in African population mobility. In L. A. Kosinski & R. M. Prothero (Eds.), *People on the move* (pp. 39–49). London: Methuen.

Hagerstrand, T. (1957). *Migration and area in Sweden* (Lund Studies in Geography). Lund, Sweden: Royal University of Lund, Department of Geography.

Hagerstrand, T. (1975). Space, time and human conditions. In L. Karlquist, L. Lundquist, & F. Snickars (Eds.), *Dynamic allocation of urban space* (pp. 3–12). Farnborough, UK: Saxon House.

Hagerstrand, T. (1982). Diorama, path and project. *Tijdschrift voor Economische en Sociale Geograpfie, 73*(6), 323–339.

Hughes, C. C., & Hunter, J. M. (1970). Disease and development in Africa. *Social Science and Medicine, 3,* 443–493.

Hunter, J. M. (1993). Elephantiasis: A disease of development in northeast Ghana. *Social Science and Medicine, 35,* 627–649.

Keiser, J., Caldas de Castro, M., Maltese, M. F., Bos, R., Tanner, M., Singer, B. H., et al. (2005). Effect of irrigation and large dams on the burden of malaria on a global and regional scale. *American Journal of Tropical Medicine and Hygiene, 72*(4), 392–406.

Lesthaeghe, R. (2010). The unfolding story of the second demographic transition. *Population and Development Review, 36*(2), 211–251.

Martens, P., & Hall, L. (2000). Malaria on the move: Human population movement and malaria transmission. *Emerging Infectious Diseases, 6*(2), 103–109.

Meade, M. S. (1976). Land development and human health in west Malaysia. *Annals of the Association of American Geographers, 66,* 428–439.

Meade, M. S. (1977). Medical geography as human ecology: The dimension of population movement. *Geographic Review, 67,* 379–393.

Meade, M. S. (1978). Community health and changing hazards in a voluntary agricultural resettlement. *Social Science and Medicine, 12,* 95–102.

Omran, A. R. (1971). The epidemiologic transition: A theory of the epidemiology of population change. *Milbank Memorial Fund Quarterly,* 509–538.

Polo, P. E. (2009). *Nature–culture interactions: Understanding the prevalence of malaria in the northern Ecuadorian Amazon.* Unpublished master's thesis, University of North Carolina at Chapel Hill.

Popkin, B. M. (2003). The nutrition transition in the developing world. *Development Policy Review, 21*(5–6), 581–597.

Population Reference Bureau. (2015). DataFinder: U.S. and World Data [data file]. Retrieved from *www.prb.org/datafinder.aspx.*

Prothero, R. M. (1961). Population movements and problems of malaria eradication in Africa. *Bulletin of the World Health Organization, 24,* 405–425.

Prothero, R. M. (1965). *Migrants and malaria.* London: Longmans.

Reher, D. S. (2004). The demographic transition revisited as a global process. *Population Space and Place, 10*(1), 19–41.

Sawyer, D. R. (1988). *Frontier malaria in the Amazon region of Brazil: Types of malaria situations and some implications for control.* Brasília: PAHO/WHO/TDR.

Schaerstrom, A. (1996). *Pathogenic paths?: A time geographical approach in medical geography.* Lund, Sweden: Lund University Press.

Singer, B., & Caldas de Castro, M. (2006). Enhancement and suppression of malaria in the Amazon. *American Journal of Tropical Medicine and Hygiene, 74*(1), 1–2.

Sutton, K., & Buang, A. (1995). A new role for Malaysia's FELDA: From land settlement agency to plantation company. *Geography, 80*(2), 125–137.

United Nations. (1950). *Demographic yearbook 1949–50.* New York: Author.

United Nations High Commission on Refugees (UNHCR). (2014). *UNHCR global report 2014.* New York: United Nations.

United Nations High Commission on Refugees (UNHCR). (2016). *Global Trends: Forced Displacement in 2015.* New York: United Nations.

United Nations Population Division (UNPD). (2013). *World population prospects: The 2012 revision.* New York: United Nations.

Vittor, A. Y., Gilman, R. H., Tielsch, J., Glass, G., Shields, T., Sánchez Lozano, W., et al.

(2006). The effect of deforestation on the human-biting rate of *Anopheles darlingi,* the primary vector of falciparum malaria in the Peruvian Amazon. *American Journal of Tropical Medicine and Hygiene, 74,* 3–11.

Wolfe, B. L., & Behrman, J. R. (1983). Is income overrated in determining adequate nutrition? *Economic Development and Cultural Change, 31,* 525–549.

World Health Organization (WHO). (2005). *World health report 2005: Making every mother and child count.* Geneva, Switzerland: Author.

Zelinsky, W. (1971). The hypothesis of the mobility transition. *Geographic Review, 61,* 219–249.

Ziegler, R. G., Hoover, R. N., Pike, M. C., Hildesheim, A., Nomura, A. M. Y., West, D. W., et al. (1993). Migration patterns and breast cancer risk in Asian-American women. *Journal of the National Cancer Institute, 85,* 1819–1827.

📗 FURTHER READING •

Akhtar, R. (Ed.). (1987). *Health and disease in tropical Africa.* London: Harwood Academic.

Audy, J. R. (1965). Types of human influence on natural foci of disease. In B. Rosicky & K. Heyberger (Eds.), *Theoretical questions of natural foci of disease: Proceedings of a symposium* (pp. 245–253). Prague: Czechoslovak Academy of Science.

Desowitz, R. S. (1981). *New Guinea tapeworms and Jewish grandmothers.* New York: Norton.

Drexler, M. (2002). *Secret agents: The menace of emerging infections.* Washington, DC: National Academy Press.

Farvar, M. T., & Milton, J. (Eds.). (1972). *The careless technology: Ecology and international development.* New York: Stacey.

Gade, D. W. (1979). Inca and colonial settlement, coca cultivation and endemic disease in the tropical forest. *Journal of Historical Geography, 5,* 263–279.

Garrett, L. (1994). *The coming plague: Newly emerging diseases in a world out of balance.* New York: Farrar, Straus & Giroux.

Gopalan, C. (1992). *Nutrition in developmental transition in South-East Asia.* New Delhi: World Health Organization.

Harpham, T. (1994). Urbanization and mental health in developing countries: A research role for social scientists, public health professionals, and social psychiatrists. *Social Science and Medicine, 39,* 233–245.

Howe, G. M. (1972). *Man, environment and disease in Britain.* New York: Barnes & Noble.

Howe, G. M. (Ed.). (1977). *A world geography of human diseases.* New York: Academic Press.

Hubbert, W. T., McCulloch, W. F., & Schnurrenberger, P. R. (Eds.). (1975). *Diseases transmitted from animals to man* (6th ed.). Springfield, IL: Charles C Thomas.

Karlen, A. (1995). *Man and microbes: Diseases and plagues in history and modern times.* New York: Putnam.

McMichael, T. (2001). *Human frontiers, environments and disease.* Cambridge, UK: Cambridge University Press.

Meade, M. S. (1992). Implications of changing demographic structures for rural health services. In W. M. Gesler & T. C. Ricketts (Eds.), *Health in rural North America* (pp. 69–85). New Brunswick, NJ: Rutgers University Press.

Mena, C. F., Barbieri, A., Walsh, S. J., Erlien, C. M., Bilsborrow, R. E., & Lu, F. (2006). Pressure on the Cuyabeno Wildlife Reserve: Development and land use/cover change in the northern Ecuadorian Amazon. *World Development, 34*(10), 1831–1849.

Mena, C. F., Bilsborrow, R. E., & McClain, M. E. (2006). Socioeconomic drivers of deforestation in the northern Ecuadorian Amazon. *Environmental Management, 37*(6), 802–815.

Moeller, D. W. (1997). *Environmental health* (rev. ed.). Cambridge, MA: Harvard University Press.

Murray, C. J. L., & Lopez, A. D. (Eds.). (1996). *1990 global burden of disease: A comprehensive assessment of mortality and disability from diseases, injuries, and risk factors in 1990 and projected to 2020* (Global Burden of Disease and Injury Series, Vol. 1). Cambridge, MA: Harvard University Press.

Mutatkar, R. K. (1995). Public health problems of urbanization. *Social Science and Medicine, 41,* 977–981.

Pan American Health Organization. (1996). New emerging and re-emerging infectious diseases. *Bulletin of the Pan American Health Organization, 30,* 176–181.

Phillips, D. R., & Verhasselt, Y. (1994). *Health and development.* London: Routledge.

Reardon, T., & Berdegue, J. A. (2002). The rapid rise of supermarkets in Latin America: Challenges and opportunities for development. *Development Policy Review, 20*(4), 371–388.

Rodriguez, A. D., Rodriguez, M. H., Hernandez, J. E., Dister, S. W., Beck, L. A., Rejmankora, E., et al. (1996). Landscape surrounding human settlements and *Anopheles albimanus* (Diptera: Culicidae) abundance in southern Chiapas, Mexico. *Journal of Medical Entomology, 33,* 39–48.

Sawyer, D. R. (1992). *Malaria and the environment.* Brasília: Instituto SPN.

Sawyer, D. R., & Sawyer, D. O. (1992). The malaria transition and the role of social science research. In L. C. Chen (Ed.), *Advancing health in developing countries: The role of social research* (pp. 105–122). Westport, CT: Auburn House.

Sharma, V. P. (1996). Re-emergence of malaria in India. *Indian Journal of Medical Research, 103,* 26–45.

Singhanetra, R. A. (1993). Malaria and mobility in Thailand. *Social Science and Medicine, 37,* 1147–1154.

Stanley, N. F., & Alpers, M. P. (Eds.). (1975). *Man-made lakes and human health.* New York: Academic Press.

Takemoto, T., Suzuki, T., Kashiwazaki, H., Mori, S., Hirata, F., Taja, O., et al. (1981). The human impact of colonization and parasite infestation in subtropical lowlands of Bolivia. *Social Science and Medicine, 15D,* 133–139.

Thompson, K. (1969a). Insalubrious California: Perception and reality. *Annals of the Association of American Geographers, 59,* 50–64.

Thompson, K. (1969b). Irrigation as a menace to health in California: A nineteenth century view. *Geographical Review, 59,* 195–214.

Vachon, M. (1993). Onchocerciasis in Chiapas, Mexico. *Geographical Review, 83,* 141–149.

Weil, C. (1981). Health problems associated with agricultural colonization in Latin America. *Social Science and Medicine, 15D,* 449–461.

Williams, B. (1990). Assessing the health impact of urbanization. *World Health Statistics Quarterly, 43,* 145–152.

Wills, C. (1996). *Yellow fever, black goddess: The coevolution of people and plagues.* Reading, MA: Addison-Wesley.

World Health Organization (WHO). (1976). *Water resources development and health: A selected bibliography* (Document No. MPD/76.6). Geneva, Switzerland: Author.

World Health Organization (WHO). (1993). *The urban health crisis: Strategies for health for all in the face of rapid urbanization: Report of the technical discussions.* Geneva, Switzerland: Author.

PART II

MAPS and METHODS

Health geographers use the same methods that others use in the field of geography, including cartography, geographic information systems (GIS), and spatial analysis, but with specific focus and application of these methods to understanding health and disease. In particular, health geographers have paid special attention to the movement of diseases through space and over time in studies of diffusion. The emergence and reemergence of infectious diseases, combined with the increasing availability of data on genetic characteristics of pathogens, such as drug resistance, provides new avenues for health geographic research.

Chapter 5 discusses both the history and the practice of the cartography of health and disease from the 1800s to modern Web-based mapping. It includes an introduction to basic cartographic principles and operations and then focuses on the spatial data management and analysis tools available through a GIS, including visualization and spatial analysis. Related technologies including the global positioning system (GPS) and satellite remote sensing are also discussed. Spatially referenced data add a significant amount of important information to studies of health, but can also be challenging to use without a proper understanding of some fundamental spatial concepts, such as the modifiable areal unit problem and spatial autocorrelation. The chapter concludes with a discussion of several spatial statistical methods commonly used in health geography studies.

Chapter 6 focuses on the methods that explore the spread of diseases and health care over space and through time. The fundamental geographic concepts and literature on waves of innovation and acceptance of social phenomena are introduced and then applied to the study of health outcomes. These geographic concepts are then supplemented with background on epidemiological models of disease transmission. Different types of spatial diffusion processes are delineated and spatial diffusion modeling techniques discussed. Several case studies of disease diffusion are offered, including measles, poliomyelitis, and influenza.

Chapter 7 describes the forces that drive the emergence and reemergence of infectious diseases (e.g., globalization, trade, migration, expansion of humans into animal habitats) and explores these forces in the context of specific diseases such as lymphatic filariasis, HIV, and dengue fever. The convergence of high-resolution spatial data and high-resolution molecular data enables health geographers to understand not only the population–environment drivers of disease emergence and occurrence, but also those specific landscapes that result in the evolution of pathogens, such as the emergence of drug resistance. The field of landscape genetics, and its application to questions from health geography, are introduced and explored through examples of drug-resistant malaria and the evolution of avian influenza viruses.

CHAPTER 5

Maps, GIS, and Spatial Analysis

Geography has a fundamental interest in variations over space and in how things interact across different geographic scales. Maps have been used for hundreds of years to describe spatial variations of health and disease. Koch (2005) provides a meticulously documented history of disease mapping, from the Naples plague in the 1690s to the contemporary HIV/AIDS epidemic in Thailand. As historians are concerned with actions and influences considered over time, analyzed in periods of various lengths, so geographers are concerned with actions over distance and interrelationships considered across space, analyzed at various scales. The capabilities of computers and the capacity for digital transmission of information available today have transformed not so much the questions geographers ask as their ability to address them. Although, like other social scientists, geographers have for decades used statistics, only recently have technological developments allowed the power of the computer to be directed at spatial data management through new software packages. The map has become the tool known as a geographic information system (GIS). This chapter initially offers a brief history of the cartography of disease, followed by an introduction to basic cartographic principles. Next the chapter describes some of the cartographic and spatial data integration capabilities of a GIS and then an overview of spatial analysis within a theoretical context of basic geographic concepts, including scale.

The purpose of this chapter is to increase awareness of the ways distributions can be analyzed spatially across scales, the ways queries and findings can be portrayed and presented, and the potential for error and abuse that lurks in computerized analysis. There is not enough space to review all the ways spatial analysis has been applied to measuring health and disease distributions, and we could not do such a review as well as the tour de force of Cliff and Haggett (1988). Nor do we attempt to explain how to make maps or conceptualize and manage a GIS. Those specialized kinds of knowledge have many sequential courses and texts of their own (e.g., Cromley & McLafferty, 2012; Longley, Goodchild, Maguire, & Rhind, 2010). Some of our readers will learn new questions they can ask and ways that these can be addressed. Whether or not we do research, we all see maps and research results in journals, books, and papers. These are proliferating as the use of GIS spreads in studies of health and disease (see "Further Reading" for a few examples of empirical papers that use GIS and spatial analysis). People dealing with health and disease matters today need to become more spatially sophisticated thinkers, and this chapter is intended to help them do that.

CARTOGRAPHY OF HEALTH AND DISEASE

The use of GIS has its roots in the field of cartography, which is a field highly developed as both an art and a science. Cartography has advanced through the centuries along with mathematics, astronomy, and navigation. It has invoked the curiosity and imagination of countless minds. It has recently been computer-automated and has been revolutionized by satellites, including those that take complex digital pictures of the Earth's surface and those that constitute the **global positioning system** (GPS). People working in health sciences need to become aware of more and different ways they can spatially analyze and portray their data, and of the ways such maps not only can be used but can be abused to lie and distort, or simply to err.

The beginnings of health geography in the United States lie in mapping, specifically in Jacques May's project of mapping major vectored, parasitic, and nutritional diseases at the world scale for the American Geographic Society in New York. As discussed in Chapter 1, between 1950 and 1954 he produced a series of 17 maps that were published with limited commentary as supplements to the *Geographic Review* (see the May entries in "References"). Several maps were later reduced and published in May's book *The Ecology of Human Disease* (1958). These maps were for most geographers the first ones that had ever brought distributions of disease and their ecological relationships with cultural patterns and physical environments to their attention. A few years later, Rodenwaldt and Jusatz (1952–1961) edited the more comprehensive, three-volume *Welt-Seuchen Atlas* (*World Atlas of Epidemic Disease*) in German and English. This atlas included global and regional maps (with a special emphasis on Europe) and a commentary on the diseases. It had a wide impact in public health and international medical and development centers. Disease mapping has come a long way since May created his maps. Disease maps are now readily available on websites from governments and other sources.

In the GIS era, disease maps often display the results of sophisticated spatial–analytical operations performed in a GIS, such as the work by Openshaw and colleagues (Openshaw, Craft, Charlton, & Birch, 1988) that identified clusters of childhood leukemia in Great Britain. Also in Great Britain, Peter Haggett, Andrew Cliff, and their colleagues have written several books and atlases that use spatial–analytical techniques to reveal patterns of infectious disease transmission. These volumes include *Spatial Aspects of Influenza Epidemics* (Cliff, Haggett, & Ord, 1987), *Atlas of Disease Distributions* (Cliff & Haggett, 1988), *London International Atlas of AIDS* (Smallman-Raynor, Cliff, & Haggett, 1992), *Measles: An Historical Geography* (Haggett, Cliff, & Smallman-Raynor, 1993), *Deciphering Global Epidemics* (Haggett, Cliff, & Smallman-Raynor, 1998), *Island Epidemics* (Cliff, Haggett, & Smallman-Raynor, 2000), and *World Atlas of Epidemic Diseases* (Haggett et al., 2004). In the United States, Peter Gould (1993) documented the progression of the AIDS epidemic in the United States during the 1980s in *The Slow Plague: A Geography of the AIDS Pandemic.* These aforementioned cartographers of disease were the pioneers; there have been many since who have used maps and GIS to understand epidemics, and some of their work is listed in the "Further Reading" section of this chapter.

As an abstraction of reality, a map is a distillation of complex interrelationships that the mapmaker chooses as important to represent. That is, all of reality cannot be put on a map. The cartographer must generalize. The teacher or researcher must decide which elements to represent: If roads, what type? If vegetation, how specific? If water courses, what size or permanence? If houses, what type, and in what detail? Although health and disease data are usually collected according to administrative organization, such as counties, a map can portray the patterns of habitat, windborne pollutants, or water sources and flows that render the administrative borders dangerous irrelevancies. Again, a map is a model of what a mapmaker believes crucial for analyzing etiology or communicating important relationships.

Geographers sometimes forget the power of a simple map. Yet much of the early geographic disease studies' impact on our understanding of health stems from the use of maps. To quote an 1852 cholera study:

> Geographical delineation is of the utmost value, and even indispensable; for while the symbols of the masses of statistical data in figures, however clearly they might be arranged in the Systematic Tables, present but a uniform appearance, the same data embodied in a Map, will convey at once, the relative bearing and proportion of the single data together with their position, extent, and distance, and thus, a Map will make visible to the eye the development and nature of any phenomenon in regard to its geographic distribution. (Petermann, 1852, cited in Gilbert, 1958, p. 178)

Modern health cartography (frequently called *medical topography* during the 1800s) began in Europe during the late 18th century, when most of today's academic disciplines were not separately defined. Investigators, who were usually medical practitioners, described a place's topography and climate as they related to health and disease. This, of course, was in the Hippocratic tradition as exemplified by the quotation at the start of this book. Although disease distributions were often described in detail, and the reports sometimes did contain detailed topographic maps, they did not contain disease maps. The yellow fever epidemics of the late 18th/early 19th centuries and the cholera outbreaks of the 19th century generated the first disease maps. *Dot maps* of the distribution of yellow fever victims (which were clustered in filthy port and sailor lodging areas) were used by both contagionists and anticontagionists in their argument over the nature of what caused that dreaded disease. Contagionists considered it a single disease brought by travelers from places already afflicted with yellow fever, whereas anticontagionists thought that it simply emerged from crowded, filthy urban areas. Apparently the first such map was produced by Dr. Valentine Seaman in his anticontagionist treatise on yellow fever in New York City in 1798. His work was continued over the next half century by many other people.

In 1852, Heinrich Berghaus published his *Physikalischer Atlas*. One of its eight sections included a number of medical maps and charts. These were the first medical maps included in an atlas, the first to show the distribution of a variety of epidemic and endemic diseases, and the first published by a major cartographer. The extraordinary quality of these maps represents a singularly important development in health cartography. The most famous 19th-century disease map, used to argue the contagionists' analysis, was John Snow's (1855) **dot map** of cholera around the Broad Street water pump in London in 1854 (Figure 5.1). The clustering of cholera in the vicinity of the well supported Snow's contention that cholera was a water-borne disease, with the pump the local source of infection. He urged that the handle be removed and the pump shut down. It was, and the local incidence of cholera declined quickly.

Figure 5.2 is a more recent example showing two ZIP-code-level morbidity ratio maps, standardized for age and gender, of anxiety-related International Classification of Diseases, ninth revision (ICD-9), diagnoses before and after the September 11, 2001, terrorist attacks in New York City (DiMaggio, Galea, & Emch, 2010). These maps and the associated research project show that distance from the World Trade Center site in the postattack time period was associated with increased risk of anxiety-related diagnoses. In fact, each 2-mile increment in distance closer to the World Trade Center site was associated with a 7% increase in anxiety-related diagnoses in the population and no similar association was found during a similar time period in the year prior to the attacks. The map reader can develop hypotheses about the reasons for the spatial distribution of one of these maps or for changes between the two dates. (ICD-9 is used to code and classify morbidity data from inpatient and outpatient records, physicians' offices, and surveys from the National Center for Health Statistics, e.g., 1997.)

Disease Maps on the Web

GIS, in conjunction with the Internet, has made it possible to display disease maps inexpensively and sometimes interactively to a wide audience. For instance, the World Health Organization (WHO) Communicable Disease Global Atlas (*apps.who.int/globalatlas*) allows users to map infectious diseases at country, regional, and global levels, using data and mapping software available on their server via the Web. Figure 5.3 displays a map of reported global cholera cases by country from 2009 to 2013, created on the WHO Global Atlas website. The system allows users to browse, view, query, and search the contents of the WHO's communicable disease

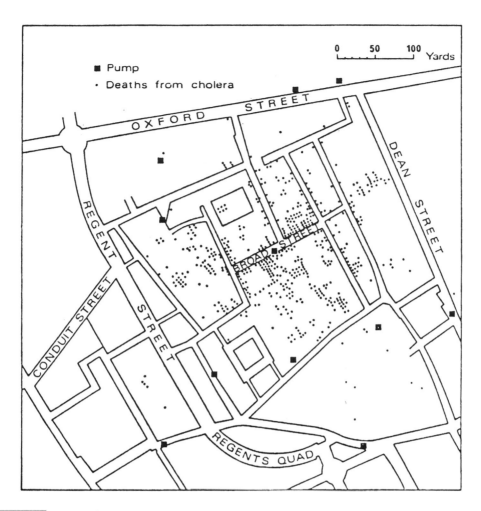

FIGURE 5.1 Snow's map of cholera. The affected well is clearly identified by the concentration of cases in its vicinity. From Howe (1972, p. 178). Copyright 1972 © G. M. Howe. Reprinted by permission.

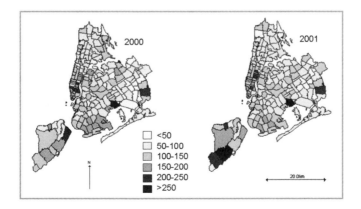

FIGURE 5.2 Fitted standardized morbidity ratios, anxiety-related diagnoses, New York City, September 11–December 30, 2000 and 2001. New York City outpatient Medicaid data. From DiMaggio, Galea, and Emch (2010). Copyright © 2010 Elsevier. Reprinted by permission.

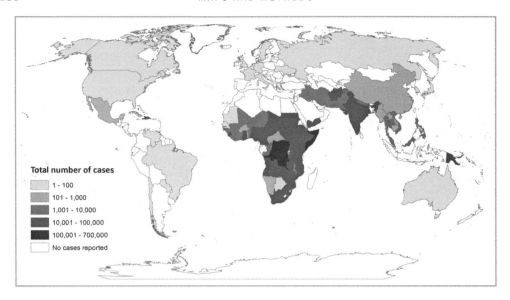

FIGURE 5.3 Global cholera distribution from 2009 to 2013. From the WHO *Communicable Disease Global Atlas* website (*apps.who.int/globalatlas*).

global database, and to output data in reports, charts, and maps. It has a user-friendly mapping interface that can be used to select geographic areas of interest and create maps of disease outbreaks; the locations of health facilities, schools, and roads; and other geographic features. The website also provides access to static maps of infectious diseases. Other data that can be interactively displayed include demographic variables, socioeconomic conditions, and environmental factors; this capability allows the user to explore the broad range of determinants that influence patterns of infectious disease transmission.

Another example of disease-mapping capabilities available on the Web is the USGS Disease Maps site (*diseasemaps.usgs.gov/mapviewer*). This site provides close to real-time statistics on several vector-borne diseases in the United States, including West Nile virus (WNV). It provides a series of county-level maps and dynamic mapping tools that display confirmed WNV cases in humans, birds, mosquitoes, and animals throughout the United States. For example, the grey counties in Figure 5.4 are those that had at least one confirmed human, bird, or animal WNV case in 2015. The online map allows the user to zoom in on individual states displaying the numbers of confirmed cases in a particular county. The USGS Disease Maps site also provides similar information for other vector-borne diseases, including St. Louis encephalitis, Eastern equine encephalitis, Western equine encephalitis, La Crosse encephalitis, and Powassan virus. The newest additions to the site are dengue fever and chikungunya, which have emerged in the southern and western United States in recent years, due primarily to "imported" rather than domestically transmitted cases of disease. This site and others like it are valuable tools to communicate information to the public.

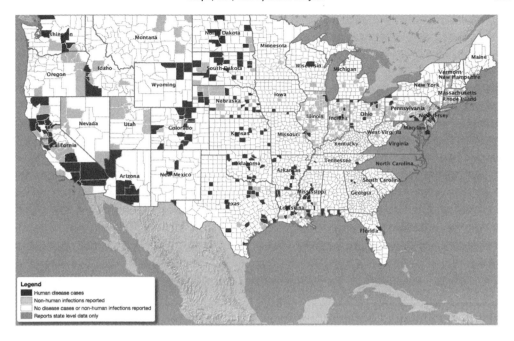

FIGURE 5.4 Human West Nile virus (WNV) cases in 2015. From the USGS Disease Maps website (*diseasemaps.usgs.gov*).

The Health Literacy Data Map site (*healthliteracymap.unc.edu*) allows users to interactively display census block group-level health literacy scores for communities. The site includes choices of display options by national and state quartiles and allows raw data to be downloaded into spreadsheets nationally or by state. Thus this site serves both as a data visualization tool and a data source. Another example of an online interactive disease mapping tool is AIDSVu (*aidsvu.org*) which is a national map of HIV/AIDS by county, state, and ZIP code/census tract/neighborhood in selected cities in the United States by year. Users can interactively visualize spatial distributions of HIV prevalence and incidence by age, race, transmission category (e.g., drug use, sex), and social determinants (e.g., poverty, not having insurance). Some data are suppressed in particular when there are small numbers of cases in certain jurisdictions to protect privacy.

Concepts of Cartography

Although a map can be visually effective and authoritative, it really can be no better than its data. The importance of political or administrative borders for the collection of data, and their frequent irrelevance to the patterns of distributions on the surface of the Earth, have been intensely spotlighted by the availability of digitized satellite imagery of such distributions. The effort to reconcile this difference is at the core of much research in GIS and health geography today. The origins, however, lie in the basic question of scale of analysis and of how different types of information are portrayed.

Scale

We can think most simply of **map scale** as the amount of area being considered, while realizing that the area under consideration is at once a component of a larger area and is itself made up of smaller areas: space is a continuum. The word *scale* itself originates from the Latin word for ladder, which was both a means of ascent and descent and a set of marks used for measuring. A scale is the proportional relation a map bears to the thing it represents.

A scale can be a ratio between the dimensions of a representation and those of the object being represented, as in "1 inch to a mile." Its technical use in geography, through cartography, is sometimes at variance with common usage. This can be a source of confusion. If 1 centimeter on a map represents 100,000 centimeters on the earth's surface, the scale is 1:100,000; if 1 centimeter represents 1,000 centimeters, the scale is 1:1,000. Because the former fraction is a much smaller number than the latter, it is referred to as a "small scale" and the latter as a "large scale." Note, however, that the small-scale map covers a much larger area, and the large-scale map covers a smaller area. Thus a map of the United States is small-scale, and a map of a house's interior is large-scale. To avoid confusion, it is not uncommon today to speak of "macro" and "micro" studies, referring to the area covered in terms more familiar to common usage (large areas or populations being macro, and small populations or areas being micro).

Things vary over space. The amount of that variation that can be captured and studied is necessarily dependent on the proportion being represented (i.e., on the scale of analysis). This causes both the fascination and the frustration—the inherent challenge—of spatial analysis.

The type of projection used to portray the round globe on a flat surface remains important for global mapping of distributions or diffusion. Most health researchers of course will—and can—just use the base maps of the city blocks, counties, or countries that are readily available to them. However, the most convenient map may not be of the best scale, and the easiest route may not yield the best outcome. Base maps—blank maps waiting for data—need to be sought out for the scale most appropriate for aggregation and analysis of the topic at hand.

- A small scale (1:250,000 to 1:1,000,000 or more) is appropriate for continents, realms, and other large regions.
- A medium scale (1:25,000 to 1:250,000) is appropriate for states, state economic areas, health service areas, standard metropolitan statistical areas, and some minor civil divisions (counties, cities); topographic maps and coverage from the U.S. Geological Survey (USGS); and satellite coverage of vegetation, watersheds, and geological formations.
- A large scale (1:200 to 1:25,000) is appropriate for some minor civil divisions (cities, census tracts, etc.); satellite imagery of transportation routes, buildings, agricultural fields, and local vegetational features; USGS topographic coverage; air photography special coverage; U.S. Census Bureau **Topographically Integrated Geographic Encoding and Referencing (TIGER) files;**

and engineering blueprints (of electrical, sewerage, water, and other infra-structure).

Types of Maps and Data

Point *data* represent information that has no spatial dimension, but of course this is a matter of scale. An individual with a case of tuberculosis may be represented by a point on a city map; a city's population can become a point on a smaller-scale national map. These kinds of data are most commonly presented in dot maps (like John Snow's cholera map). The dots can become quite elaborate such as a set of graduated circles indicating population or incidence size (Figure 5.5) or a set of pie graphs with proportions for outcomes (Figure 5.6) or types of people.

Lines are used to show direction of movement (**flow maps**), often with the thickness of the line being scalar (proportional to the amount that is moving, as in

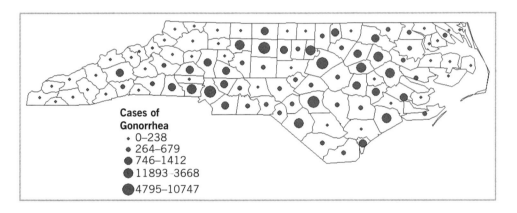

FIGURE 5.5 Graduated circle map of gonorrhea cases by county in North Carolina.

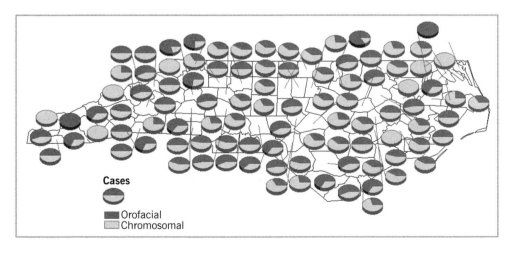

FIGURE 5.6 Pie graph map of birth defect type by county in North Carolina.

the arrow thickness in Figure 6.7). Most commonly, lines are used to connect points and so portray new information. *Isolines* (also known as *isopleths* or *isarithms*) can connect points of equal data value (see Figures 5.7 and 6.10). These are most familiar as lines of equal altitude on topographic sheets or lines of equal atmospheric pressure on weather maps. Isolines are two-dimensional representations of three-dimensional reality because they have both extensions in space and values (i.e., *X, Y,* and *Z* axes). The area that lines separate into different spaces is continuous, and points are discrete. Connecting points with a line through an area containing no data points, values are *interpolated*. For example, a line connecting points of value 10 could be drawn midway between points of value 5 and value 15, as well as through points of value 10 itself. This used to be a laborious as well as a specialized art, but computer programs do the calculations routinely now. Careful inspection and human art are still required, however, as there can be more than one way to configure such interpolated line patterns, and the simplest computer solution is not always the most sensible. Figure 5.7 is an isoline map of blood lead levels determined by laboratory screening tests, allocated to the residential locations of the

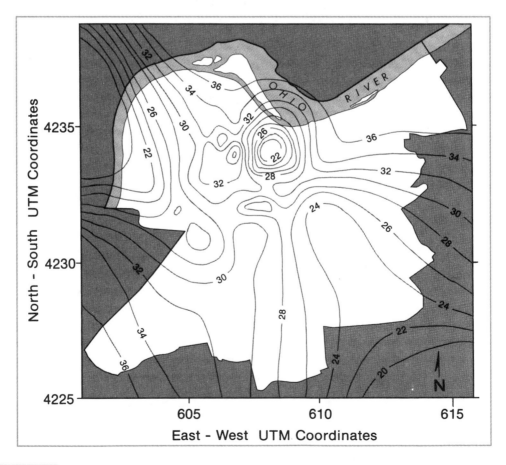

FIGURE 5.7 Lines of equal pediatric blood levels, Louisville, Kentucky, 1979. Values are micrograms per deciliter of blood.

children. The pattern of distribution identified is much richer and more intricate than a mere dot distribution of cases that exceeded a certain limit.

Area information is symbolized on thematic maps by shades, patterns, or colors covering the area and symbolizing the class or category to which the information value belongs. This kind of map is known as a **choropleth map**. Choropleth maps are the familiar maps of data distribution by county, state, or nation, but the areas do not have to be politically delimited. Vegetation regions, soil regions, activity areas, or population circulation regions—any areas that can be discretely defined— may be appropriate units for choroplethic thematic mapping.

Critical Intervals: Mapping Patterns

Choropleth maps are the most commonly used maps for presenting disease or health service information. The question is how to divide this data distribution into intervals (i.e., create classes) to be symbolized with different shading on a choropleth map. Figures 5.8a through 5.8d are examples of choropleth maps of North Carolina infant mortality rates classified in different ways.

A map made well can communicate a lot of information. The information, however, can be distorted if the intervals are not appropriate to the data distribution. First, there can be too many intervals. Students sometimes think there will be more detail in the distribution if they map 10 or 15 categories, but cartographic research has determined that the mind can only differentiate patterns from six or seven categories in black and white, and fewer in color. If too few categories are used, of course, detail and information will be lost. Regardless of how many categories are used, selecting shading, colors, or patterns to grade continuously from one class to another will eliminate the need to consult the key for every observation and will allow whatever pattern there is to appear.

Cynthia Brewer, a cartographer at Pennsylvania State University, developed a set of color schemes for maps that have been adopted by most U.S. government agencies, including the U.S. Census Bureau, National Cancer Institute, and the Bureau of Labor Statistics (available online at *colorbrewer2.org*). These color schemes are based on extensive research on how people see and interpret color on choropleth maps; some schemes even work for color-blind individuals! When a series of maps compares distributions over time periods or different areas, consistent class selection is necessary for comparison; otherwise, clusters and patterns will seem similarly significant, even though actual prevalence varies enormously. The most popular method in health sciences is to map quantiles (quantitative divisions), especially quintiles (20% categories) or quartiles (25% categories). There can be confusion about whether the data range (from the lowest point to the highest) or number of observations (i.e., geographical units) is being so divided and classified. Sometimes the default option of computerized mapping programs does the former (i.e., divides the range of data into five categories), whereas the researcher intends the latter (i.e., to divide the number of observations into five groups).

Figure 5.8a illustrates the most intuitive way of categorizing classes of infant mortality rates: division by natural breaks. When a frequency distribution is plotted with the values on one axis and the number of observations accumulated on the

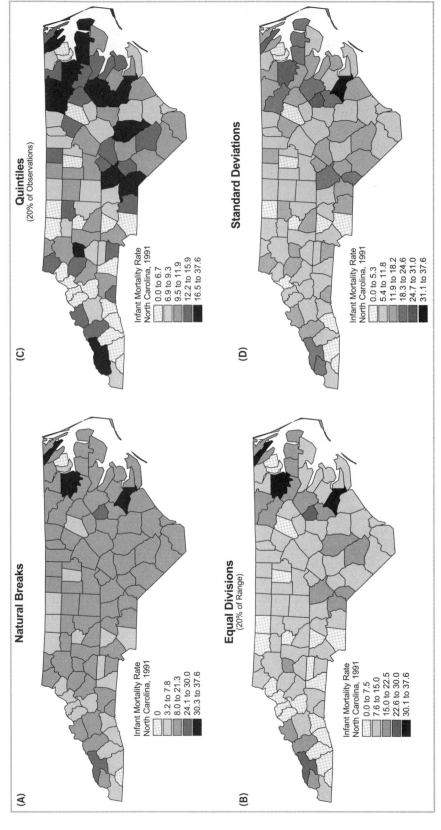

FIGURE 5.8 Choropleth maps of infant mortality in 1991 in North Carolina according to different means of interval determination. (A) Intervals are determined by natural breaks. (B) The range of rates is divided into five equal divisions. (C) The number of observations (100 counties) is divided into five equal groups, or quintiles. (D) Standard deviations from the mean value for infant mortality determine the intervals.

Natural Breaks

(A)

Infant Mortality Rate
North Carolina, 1991

0
3.2 to 7.8
8.0 to 21.3
24.1 to 30.0
30.3 to 37.6

Equal Divisions
(20% of Range)

(B)

Infant Mortality Rate
North Carolina, 1991

0.0 to 7.5
7.6 to 15.0
15.0 to 22.5
22.6 to 30.0
30.1 to 37.6

Quintiles
(20% of Observations)

(C)

Infant Mortality Rate
North Carolina, 1991

0.0 to 6.7
6.9 to 9.3
9.5 to 11.9
12.2 to 15.9
16.5 to 37.6

Standard Deviations

(D)

Infant Mortality Rate
North Carolina, 1991

0.0 to 5.3
5.4 to 11.8
11.9 to 18.2
18.3 to 24.6
24.7 to 31.0
31.1 to 37.6

other, "breaks" usually appear in the curve. That is, many observations accumulate over a range of values, and then the rate of accumulation slows and even levels out until a new range of values is reached. These natural breaks in the data distribution can sometimes be more interesting in an etiological sense than the standardized breaks are; the problem with this method is the difficulty of comparing the patterns on one map with those on another done the same way.

Figure 5.8b shows the classification of infant mortality rates into five equal divisions of the range of the data from lowest to highest values. These equal divisions are a common default setting on commercial mapping programs and are easily interpreted if the data distribution is normal. Otherwise, this method can result in categories that are either empty or have few observations. In this map, the interval from 22.6 to 30.0 contains only one county.

Figure 5.8c is truly a map of quintiles—classes containing 20% of data observations (counties). Figure 5.8d shows the infant mortality as mapped by standard deviations, not by acres of corn or trees or desert. A county-level map of the United States communicates the disease rates of small numbers of people over large areas west of the Mississippi, and crams the rates of millions of people in places on the East Coast into fractions of an inch. Especially when the disease or health service concern is targeted at a subgroup of the population—whether this subgroup is defined by age and life stage, ethnicity, or behavioral activity—it makes better sense to construct a map in which the size of the units is proportional to population instead of land area. Such demographic base maps are types of *cartograms*.

Demographic base maps are used much less often than they should be because they are time-consuming to construct and often difficult for people who are unfamiliar with them to interpret. Figure 5.9 shows the North Carolina infant mortality

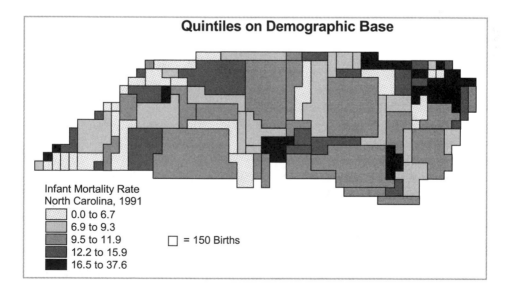

FIGURE 5.9 Demographic base map of infant mortality in North Carolina, 1991. Quintiles of data are mapped on a cartogram of county size proportional to county births.

of Figure 5.8c mapped on a demographic base. Because high rates often apply to small populations that are less accessible or economically developed, a population base often portrays conditions very differently. On a global map of infant mortality mapped on a demographic base of births, for example, it is clear that many more infants are dying in South Asia than in Africa.

QUICK REVIEW 1

✓ A map is a distillation of complex interrelationships that the mapmaker chooses as important to represent. All of reality cannot be put on a map so the cartographer must generalize.

✓ Disease maps have been made for centuries but the modern cartography of disease in the United States started in 1950 with Jacques May's project on mapping major vectored, parasitic, and nutritional diseases at the world scale for the American Geographic Society. Many disease atlases have been published since then by Rodenwaldt, Jusatz, Haggett, Cliff, and Smallman-Raynor, among others.

✓ The most famous historic disease map is an 1855 dot map of cholera deaths around the Broad Street water pump in London, England.

✓ Today disease maps are often provided via Web-based servers with interactive choices for users. Organizations that provide such online mapping tools include the World Health Organization and the U.S. Centers for Disease Control.

✓ Choropleth maps are one of the most common types of maps used to show health and disease distributions; they are thematic maps symbolizing areas by shades, patterns, or colors.

✓ The data on maps must be divided into classes using various methods to create intervals including natural breaks, equal divisions, quintiles, and standard deviations.

GEOGRAPHIC INFORMATION SYSTEMS

A **geographic information system** (GIS) is an information system specifically created for maps. GIS are much more than computerized cartography systems, since they facilitate the manipulation of spatial data in many different ways. They are automated systems for the capture, storage, retrieval, analysis, and display of spatial data. Spatial data are stored in either **vector** (i.e., points, lines, and polygons) or **raster** (i.e., cell-based) data structures, allowing them to be integrated on the basis of common geography. In Figure 5.10a each raster cell has a value that has a corresponding land cover class represented as vector polygons in Figure 5.10b.

Spatial data integration is one of the fundamental reasons for using a GIS. A nonspatial information system stores data via a common identifier, called a *key* in relational database management terminology. An example of a common identifier is a student number by which a university may keep track of each student. Records such as grades are maintained in academic offices, and financial information such as whether a student has paid her or his tuition is stored in another office. Information from the different databases can be linked via the database key (the student number, in this case) for different purposes at different times, such as before graduation. A student who has gotten high enough grades in the right classes and paid his or her tuition bills will be given a diploma; this information requires integration of databases from the different offices before the student graduates.

Now consider trying to link two different databases together that have no common identifier. People live in different communities with varying social and biophysical environments. If we have information about a person's disease status and address, and we want to link this information to the presence and quality of sanitation systems around the person's residence, then we can integrate these two types of information based on common geography. In other words, we can integrate two different maps: one of the locations of homes and the disease status of people living in them, and the other of sewerage and septic systems. Having all this information for many people will allow us to investigate whether there are associations between sanitation practices and disease distributions.

The other main purpose of using a GIS is to facilitate spatial analysis. Simple spatial calculations can be automated, such as distance measures to a source of contamination (e.g., a toxic waste site). However, more complex techniques can also be employed that take into account wind currents from such a plant at different times of day. Disease clusters can be located, and hypotheses about their causes can be formulated; given the right ancillary spatial data layers, these hypotheses can also be tested. Although spatial data integration and analysis are the main reasons for the rapid expansion of GIS technology, during the last two decades spatial visualization has also blossomed such as through the online mapping tools described above and the creation of many commercial 3D visualization tools (e.g., ArcScene, ArcGlobe).

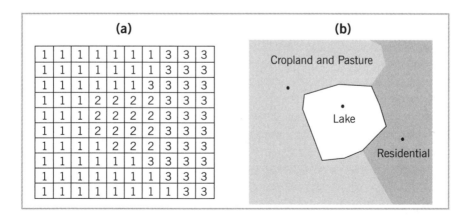

FIGURE 5.10 (a) Raster data structure. (b) Vector data structure. From the USGS (*erg.usgs.gov/isb/pubs/gis_poster*).

Spatial Data Sources

There are many different sources of GIS data for health studies most commonly involving address matching, the global positioning system (GPS), and remote sensing.

Address Matching

The majority of studies use health data that have some geographic location associated with each observation, and they are georeferenced after the fact. For instance, when you are hospitalized or see a doctor, the hospital or doctor's office usually has information about where you live. Studies sometimes have address-level data that permit individuals' houses to be located. The result of an address-matched database of a health outcome would be the household point locations of all people with a particular disease. Data at the individual level are not available, however, except under strict research guidelines governed by university institutional review boards. One of the main sources of geocoded health data in the United States is based on **address matching** using the U.S. Census files called Topologically Integrated Geographic Encoding and Referencing (TIGER) (*www.census.gov/geo/maps-data/data/tiger.html*). They are block-level street files that have their own unique data structure that GIS software can read and convert into various GIS formats (e.g., shapefile, ArcInfo coverage, geodatabase). Figure 5.11 is an example of the GIS structure of one hypothetical city block that is comprised of several streets, represented by lines in a GIS.

Block 103 is bounded by four streets. Each of the corners is treated as a node. The first node—the corner of McGlashan and Shannon Streets—has a precise location in space, which has a unique X, Y coordinate (e.g., latitude, longitude coordinate pair). The node at the corner of McGlashan Street and Hunter Boulevard defines the other end of the line segment, and has its own X, Y coordinate. If one

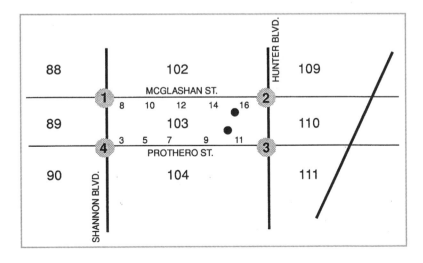

FIGURE 5.11 GIS file structure.

continues "chaining" from node 2 to 3, node 3 to 4, and node 4 back to 1, a polygon that is block 103 will be defined. This specific locational information for the nodes, line segments (or arcs), and polygons, however, also comes associated with certain topological attributes. Table 5.1 shows the so-called topology tables (including the from–to nodes and the left–right polygon tables) that give lines direction. The software "knows" that polygon 102 is across the line segment from polygon 103. If house numbers have been associated with that line segment—say, odd numbers 7 through 15 on the left, and even numbers 8 through 16 on the right—then the home addresses of cases in the cancer registry can be allocated to the separate blocks. Furthermore, if a line segment of Hunter Boulevard from Prothero Street to McGlashan Street is blocked by a reported fallen tree, the GIS can be used to determine the minimum route (distance along line segments) necessary to route an ambulance around the blockage to the hospital. The GPS navigation instruments used in cars routinely calculate minimum distances, and their rerouting functions can easily determine alternate routes in such situations.

Separate files are created of attributes associated with these points, lines, and areas (polygons), and it is the job of the GIS to manage and interrelate all of these files. Thus there are files that give each line segment its context of other line segments and polygons, so that an ambulance can be routed. Whether the polygons are census tracts, portions of marketing areas, or congressional districts, all the attributes a researcher desires and can specify are linked to it. Block 103, for example, is a part of a census tract. Individual household data are not released, but information for the block may include median family income, numbers of single-female-headed households, numbers of white children ages 5–9, room density, age of housing, percentage of households with air conditioning, proportion of the population that has graduated from high school, and numbers speaking Spanish at home. Some information may be confidential at the block level, especially by sex/age subdivisions of the population, which can get very small (and so inappropriate to use for mapping incidence rates); however, because the block is part of a tract, the data for the larger census tract can be attached to it.

There are now several Web-based geocoding services that can be used to geocode a small number of addresses. For those with a large number of addresses to match, options include geocoding the addresses by using GIS software, or paying a commercial vendor to do the geocoding. In developed countries such as the United States, many health and disease data include the addresses of individuals,

TABLE 5.1. Topology of Block 103 of the Town of Search

	Coordinate			Coordinate		Polygon to	
From node	X	Y	To node	X	Y	Right	Left
1	2.0	8.0	2	3.0	8.0	103	102
2	3.0	8.0	3	3.0	7.7	103	110
3	3.0	7.7	4	2.0	7.7	103	104
4	2.0	7.7	1	2.0	8.0	103	89

and through geocoding a locational component can be added by geocoding these addresses.

The GPS

Locations where people live and work can also be identified during fieldwork by using a GPS receiver that measures the X and Y coordinates (e.g., latitude, longitude). The GPS is a satellite-based system made up of a network of 24 satellites placed into orbit by the U.S. government. A GPS receiver takes this information and uses triangulation (more precisely, something similar to triangulation, called trilateration) to calculate the receiver's exact location. A GPS can be used to create a GIS map of points or combinations of points (i.e., lines or polygons). The GPS is therefore a vector-GIS input device. Inexpensive GPS receivers are accurate to approximately 10–15 meters, and more expensive receivers can be accurate to the centimeter. The GPS has revolutionized GIS database creation by making it easy and inexpensive. Figure 5.12a shows an inexpensive handheld GPS that can be used to identify point locations to the nearest 10 meters. More sophisticated and expensive receivers are accurate to the nearest meter, decimeter, or even centimeter. Figure 5.12b shows a much more expensive model that is accurate to the nearest decimeter.

Since all smartphones now have GPS receivers embedded in them, apps have been created to use the receivers to create spatial databases. One called GIS Cloud (*www.giscloud.com*) allows users to create databases on their website and store data on the Cloud.

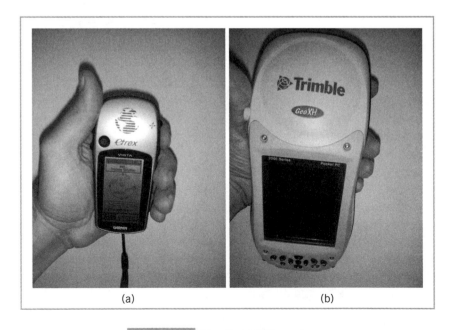

(a)   (b)

FIGURE 5.12 Handheld GPS receivers.

Remote Sensing

Remote sensing is the science of obtaining information about objects or areas from a distance, including by satellites and by airplanes. Satellite images are complex digital photographs taken from satellite sensors orbiting the earth. However, satellite sensors do not take these digital photographs only in the visible part of the spectrum that humans can see; they can also sense electromagnetic wavelengths (see Figure 12.1) that humans cannot detect, such as near-infrared wavelengths. Because different surface features of the earth reflect electromagnetic energy differently, these satellite images can be used to distinguish between them and therefore to map them. Satellite imagery is made up of pixels (rasters), which represent small square areas of the Earth's surface and define the spatial resolution of the satellite. Figure 5.13a is a Landsat ETM+ satellite image of a study area in Bangladesh where there is a large health and demographic surveillance system for health research. The imagery has a spatial resolution of 30 meters, which means that each pixel represents 30 meters on the ground. This image shows information from electromagnetic radiation that has been sensed in the near-infrared and visible parts of the spectrum. Figure 5.13b is a Google Earth image overlaid with the same study area boundary and hospital locations. The satellite imagery that is displayed is Quickbird imagery, which has a 2.44 meter resolution.

Other types of resolutions that define the quality of satellite imagery are the temporal and spectral resolutions. The temporal resolution is how often the satellite images a particular area of the Earth's surface; the spectral resolution is the

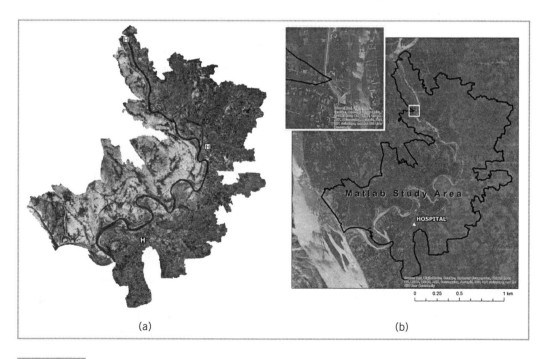

(a)   (b)

FIGURE 5.13 Satellite images of Matlab, Bangladesh, overlaid with hospital and health center locations. (a) Landsat ETM+ image. (b) Quickbird image.

detail of the electromagnetic energy measured, and therefore represents how well the imagery can distinguish between Earth surface features. Satellite imagery is a raster GIS data source.

Digitizing Legacy Data

The last important GIS data source to be discussed in this chapter consists of legacy data that were originally collected as analog data (i.e., on paper maps) and subsequently digitized. Almost the entire history of cartography occurred before the computer age, and disease maps were originally created as paper maps. Many of these paper maps represent important sources of data on spatial distributions of health and disease. Thus many such maps have been converted to GIS files by using one of a couple of available technologies. Digitizing boards are now an obsolete technology, but much of the digitization of paper maps was done by using these boards for the conversion process. Digitizing boards were drafting tables with electronic grids inside them; as the digitizer dragged a mouse along the points, lines, and polygons of the map features, these were electronically captured. The modern technology for digitizing consists of a large-format scanner and subsequent "heads-up digitizing." A large-format scanner is a raster input device through which the user feeds a map and it captures a digital picture, which is displayed on a computer screen. The digitizer then drags the computer mouse over the screen and identifies the points, lines, and polygons that make up the map, creating a vector GIS map.

The Ecological Fallacy and Modifiable Areal Unit Problem

Two fundamental theoretical concepts that are inherently geographic are discussed here to provide a basis for understanding the results and limitations of spatial analyses described in the rest of this chapter and book. The infamous ecological fallacy has bedeviled geographic comparisons for a long time. It crops up when statistics are compared across scales. We cannot use a state-level risk factor to predict what will happen to specific individuals, and we cannot interview friends and neighbors to predict national opinion. Even given bona fide cause and effect, measurable statistical association varies differently at different scales. Geographers also confront it by aggregating areas to different scales and trying to analyze zones (the **modifiable areal unit problem**).

Variance changes with scale. If information is based on individual questionnaires, we are aware of the diversity of response. When these answers are aggregated to county level, central tendency leveling takes place. There will be less variation among 100 counties in a state than there was among the millions of individuals who contributed the information. There is less variance among state economic areas than among counties, and less among states than among state economic areas. These changes in variance have little to do with cause and effect, and a lot to do with the way we measure and generalize our findings.

Cleek (1979) and others have pointed out that correlations are especially affected by changes in aggregation. Data are often aggregated by an independent variable. This inflates correlation coefficients, but usually does not affect regression coefficients. For example, as individuals are aggregated into the county in which they live, socioeconomic measures are generalized. Several adjacent counties are likely to be more similar with regard to such generalized characteristics as median income than are the thousands of individuals who compose them. In Cleek's example, the coefficient of variation (standard deviation/mean) for mortality rates from leukemia was 7.0% when states were compared at the national level and 20.9% when counties were compared within the state of Wisconsin. Similarly, the coefficient of variation of mortality rates for cancer of the nasopharynx is 188.9% for Wisconsin at the county level and 24.4% for the United States at the state level. Colon cancer, however, has higher state-level variation (25.8%) than county-level variation (17.3%). The message, Cleek says, is clear: report regression coefficients, and treat correlation coefficients with care.

What this all means is that patterns of association are different at different scales of analysis, and it is an error to take an association that is true at one scale and infer that it will be true at any other scale. On the other hand, the change in association patterns can itself sometimes be used as an etiological clue to separate the expression of different causal factors. In one case there may be great between-state variation but little within-state, between-county variation; in another there may be great local variation but little state-level variation. These patterns can be used to address different possible causes that relate to individual behavior, broader patterns of water sources or occupations, or still broader regional patterns of atmospheric pollution or population migration.

Choosing Units of Observation

Among the first things a researcher must decide are the units of observation that will be used to define the scale at which a question will be addressed. Individuals may be a source of information, but their information must be merged and aggregated. Socially, people may be grouped into families, play or work groups, ethnic groups, and so on. Spatially, people may be aggregated into households (which occupy a living space), census blocks, census tracts, cities or minor civil divisions, counties, state economic areas (which are merely groupings of counties that create a larger areal aggregation), states, regions, and so on. Some of the theoretical issues involved in analyzing area-level distributions are discussed in Chapter 9. Such units of observation or data collection may be split up or put together to make other units. This may be done either deliberately, as when regions are defined for a specific research purpose, or inadvertently, as when census tract boundaries are altered from one decennial census to the next. Regions can be defined that cut across borders, raising an issue that is sometimes referred to as "modifiable units." Changes in boundaries often alter the level of analysis, make comparisons difficult, and affect the interpretation of results, as illustrated in the following examples.

The first example shows what might happen when one is looking for the cause of a disease (Figure 5.14). Suppose an industrial chemical is being dumped into a river. This causes people who live along the river and drink its water to contract a

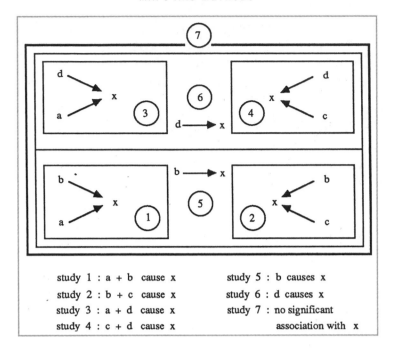

study 1 : a + b cause x   study 5 : b causes x

study 2 : b + c cause x   study 6 : d causes x

study 3 : a + d cause x   study 7 : no significant

study 4 : c + d cause x   association with x

FIGURE 5.14 Disease causation factors at different scales of analysis.

certain disease. This disease can also be contracted by breathing car emissions or by breathing the chemical put into the air by a smelter. Because people who live near the smelter, along the river, or in congested parts of this town tend to have similar socioeconomic and demographic characteristics, ethnicity, income, and race may be associated with this disease. If one were to study this disease of unknown etiology at different scales (i.e., using different geographic units of analysis), different associations would be identified.

Study 1 focuses on a minor civil division where people breathe the air of a smelter (a) and car emissions (b), and so a and b are found to be associated with the disease. Study 2 examines another minor civil division where people drink water from a stream (c) and breathe car emissions (b). Study 3 is carried out in a rather rural minor civil division wherein people work in a smelter and breathe its air (a), and are German in ethnic origin (d). In study 4 the rural minor civil division has ethnic German people (d) who drink the stream water (c). These associations are found at the microscale, but they are puzzling because there is so little consistency. Suppose that study 5 looks at county data, combining areas 1 and 2, and finds that car emissions (b) are important. Study 6 also examines county data, combining areas 3 and 4, and finds that being German (d) is important. Finally, study 7 uses state data and combines all these areas but finds no important associations.

The second example of modifiable units deals with health care delivery. Suppose a health systems agency consists of 10 counties, as shown in Figure 5.15. Table 5.2 shows the number of physicians and total population for each county. Planners in the agency may wish to organize their counties in different ways, and then

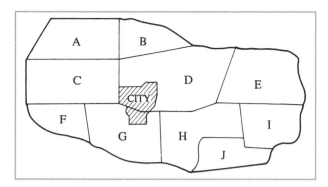

FIGURE 5.15 Ten counties within a health systems agency jurisdiction.

compare groups of counties to determine whether they are being relatively well or poorly served by physicians. Suppose they compare a "northern tier" (counties A–E) with a "southern tier" (counties F–J). The northern tier has an overall physician-to-population ratio of 128.8, and the southern tier one of 171.2. Now suppose that there is an interest in comparing the standard metropolitan statistical area (SMSA) (counties D, F, and G) with the other, more rural counties. The SMSA has a physician-to-population ratio of 187.5, and the other seven counties a ratio of 84.9. Choosing the way of grouping the counties clearly changes the assessment of the resource situation. Urban–rural investigations confront this issue constantly. Do we use the SMSA, urban area, political city, central county, or minor civil division in the analysis? The results can vary significantly. Whatever scale of unit is used for analysis, researchers face this problem: Even census tracts, for example, have spatial variation within them, such that values may be more similar between

TABLE 5.2. Health System Agency County Physician-to-Population Ratios

County	Number of physicians	Population	Physician-to-population ratio (per 100,000)
A	12	19,122	62.8
B	15	25,639	58.5
C	2	4,553	43.9
D	126	62,798	200.6
E	10	16,014	62.4
F	40	29,528	135.5
G	74	35,670	207.5
H	22	9,608	229.0
I	5	5,926	84.4
J	8	6,321	126.6
Totals	314	215,179	145.9

two blocks across the tract border (often a street) than across the census tract unit of observation.

Spatial Autocorrelation

Because units of observation exist in spatial relationships with each other (adjacency, distance, etc.), another issue that needs to be addressed is **spatial autocorrelation**. That is, census tracts or counties when used as units of observation may not be truly independent of each other, as most statistical procedures require (e.g., regression analysis). Adjacent territories usually influence each other, as the property value of one house affects that of those properties adjacent. Time series analysis also has to confront this problem, as the value of something in one year usually affects the value of that thing in the next year, so that using sequential years as units of observation can be statistically invalid because of temporal autocorrelation. Because more scientists worry about time than about space, statistical procedures have been developed to "remove" or adjust for the autocorrelation in time periods, and geographers usually adapt these techniques to their study in space.

Spatial autocorrelation is often discussed among geographers because it can be a serious hindrance to their work, but it can also be used as an analytical tool. Spatial autocorrelation means that observations from places next to each other are influenced by each other, in the same way that the real estate value of one piece of property affects that of the properties around it. If one city block is poor, the adjacent one is probably similar; if one county is affluent and has low unemployment, the adjacent county is probably better off than average as well. One important problem with spatial autocorrelation is that the assumption of independent observations, required for most statistical procedures such as linear regression and correlation, may be wrong.

It is perhaps easiest to explain spatial autocorrelation by mapping a dichotomous variable—one that is either present or absent in each spatial unit of a study area. Consider Figure 5.16, which diagrams three situations in which a certain disease is either present (black) or absent (white) in each of 16 square units. In the first situation, there is positive spatial autocorrelation; black and white units are grouped together. Perhaps the illness is quite contagious in the black part of the study area, but has come up against some type of physical or human barrier in the white area. The second situation illustrates negative autocorrelation; adjacent units are dissimilar. In the third diagram, a random pattern of black and white units indicates no autocorrelation, either positive or negative.

There are two important things to consider in measuring autocorrelation: whether units are adjoining ("have a join"), and what the value of a variable or phenomenon is in each unit. One can say that there is a join if units have a common nonzero boundary (rook's case, to borrow some terminology from chess), a common vertex or point (bishop's case), or either of these (queen's case). In the figure, the unit values are simple presence or absence of a phenomenon. How can one tell, statistically, whether there is autocorrelation in a particular situation? Basically, one counts the number of black–white (BW), black–black (BB), and white–white (WW)

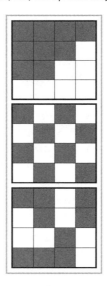

FIGURE 5.16 Spatial autocorrelation. Three types of spatial autocorrelation with a dichotomous variable are illustrated.

joins, and compares these with the number of joins that would be expected if the black and white units were distributed randomly. In the third diagram of Figure 5.16, there are six BB joins, four WW joins, and 14 BW joins (rook's case). If there are significantly more BB or WW joins than expected, there is positive autocorrelation; if there are significantly more BW joins than expected, there is negative autocorrelation; if there are a random number of each type of join then there is no autocorrelation.

Unit values need not only represent absence or presence of a disease; they can also represent high and low disease rates. In addition, more than two nominal data categories can be considered and the analysis taken from the "two-color" to the "k-color" case. Furthermore, definitions of a join can be altered in innovative ways. For example, Haggett (1976), in a study of measles diffusion in England (discussed in Chapter 6), defined joins in seven ways based on different types of paths along which the disease might be diffusing (e.g., along journey-to-work routes). If a certain path type indicated positive spatial autocorrelation, that particular path type could be important in the spread of measles—and, indeed, Haggett identified the importance of different pathways at different stages of the epidemic.

Spatial autocorrelation techniques have been used in a constructive manner to determine links between spatial patterns and causal processes. In particular, the techniques can help identify connections between disease rates and environmental and socioeconomic factors. Disease rates, which are interval data, have been most commonly used. **Moran's *I* statistic**, which is used for interval data, can be calculated to quantify spatial autocorrelation. If units with similar rates (high or low) tend to be next to each other, the *I* statistic will be relatively large (positive autocorrelation); if the opposite is the case, the *I* statistic will be relatively small (negative

autocorrelation). The *I* statistic ranges from –1 to 1 with 0 representing no autocorrelation. It can be tested for a significance as a standard normal deviate (*Z*-score)
after determining the mean and variance of its distribution. Appropriate formulas
can be found in books dealing with this subject.

These issues of scale dependency are complex but critical to spatial analysis.
What is the range of scales at which relationships are alike, and at what scale do
things appear and act differently? That is, at what scale does spatial autocorrelation
break into randomness and scale independence occur? There are various specialized techniques to determine this problem, using such things as semivariograms to
analyze scale relationships, and fractal analysis to describe the nature and degree of
the spatial relationships. The very extent of the spatial associations, whether limited
just to adjacent units or extending over broad regions, can be used to illuminate
etiology. Just as change in variability through scale can be usefully or fallaciously
used, so the phenomenon of spatial autocorrelation can be used.

For instance, Figure 5.17a displays the locations of households in a study area
of rural Bangladesh called Matlab, where all of the authors of this book have conducted extensive health geography studies. In one study the location of each cholera case was mapped, as shown in Figure 5.17b. (Note that the locations of cases in
Figure 5.17b are not real, as there are ethical problems associated with identifying

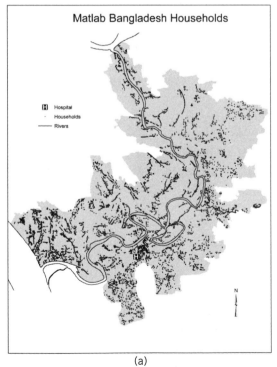
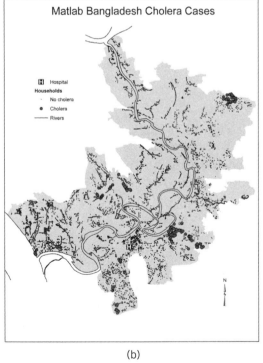

(a)   (b)

FIGURE 5.17 Household-level maps. (a) Household points for which disease data can be mapped.
(b) Cholera cases.

the locations of individual households and thus revealing exact locations of where people live who contracted a disease.)

Visualizing and Summarizing Disease Distributions

Often when disease distributions are reported, they are aggregated up to an area level. In Matlab, Bangladesh, for example, cholera disease distributions can be shown at the village level. Figure 5.18a shows the total number of cases of cholera by village in Matlab in a choropleth map. Figure 5.18b shows the number of cases per population by village.

Disease distributions can also be summarized by using summary statistics such as mean centers. A mean center takes a distribution of cases such as the cholera points shown in Figure 5.17b; the average X coordinate (i.e., the longitude or horizontal projected coordinate) and the average Y coordinate (i.e., latitude or vertical projected coordinate) are then calculated and represent the center of the epidemic. Figure 5.19a depicts the mean center of cholera cases for the distribution shown in Figure 5.17b. The map shows the original distribution of cholera cases as well. We can also measure whether the disease distribution exhibits a directional trend—in other words, whether cases are farther from the mean center in one direction than in another direction. Figure 5.19b shows the standard deviational ellipse, which

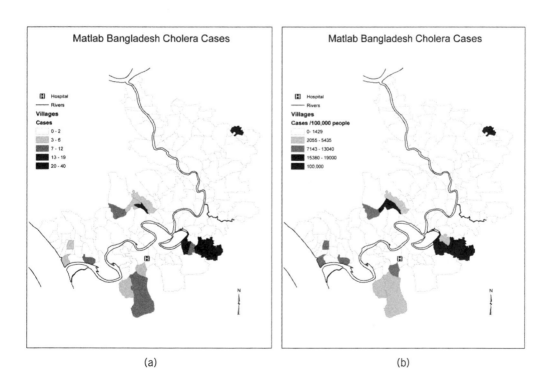

(a)   (b)

FIGURE 5.18 Area-level maps. (a) Cholera cases by village. (b) Cholera cases per population by village.

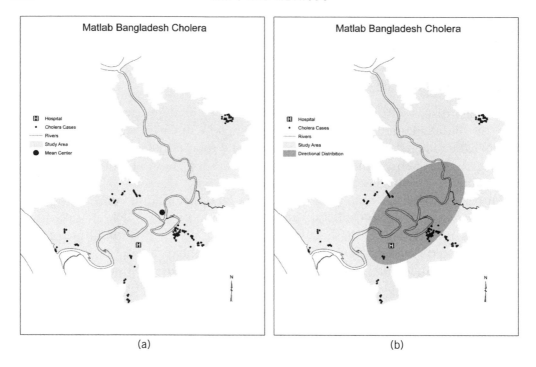

FIGURE 5.19 Spatial summary statistics. (a) Mean center of cholera epidemic in Matlab, Bangladesh. (b) Standard deviational ellipse of cholera.

summarizes the directional distribution of cholera in Matlab within one standard deviation of the mean.

Another tool that can be used to display distributions is raster density calculation. The user starts with a vector point file and sums the points that fall within the search area; that number is then divided by the search area size to get each cell's density value. The search area represents the raster resolution and is the scale at which the map will be produced, and the cell size is the area that each raster cell represents in ground units. Figure 5.20 is a raster GIS map of cholera density in three classes. The map shows the hot spots of cholera within the study area. Notice that the map looks boxy or "pixelated" around the edges. That is because each cell represents 100 meters on the ground. The density calculation was made using a 500-meter search radius; thus the software counted the number of cases of cholera within 500 meters of each cell, and each cell has a value in the GIS database that is the number of cholera cases in the 500 meters around it. Then all of the pixels in the raster GIS were classified into three categories, which were given the subjective names of low, medium, and high.

Spatial interpolation uses existing observations to estimate values at unsampled sites. Spatial interpolation is based on **Tobler's law of geography**, which is that things that are close together in space are more likely to have similar values than things that are far apart. Spatial interpolation is a common tool in most commercial GIS packages. The user starts with a series of points such as those shown

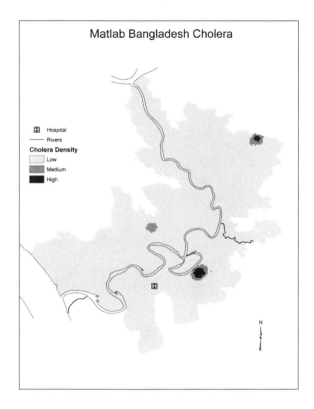

FIGURE 5.20 Cholera density in Matlab, Bangladesh.

in Figure 5.21a, which represents socioeconomic status (SES) values for groups of households. The values in this hypothetical example range from 1 to 9, with 1 being very low SES (poorer) and 9 being very high (richer). Because people who are poor tend to live next to other poor people, we can guess that if there is a cluster of poor people, a household with an unknown SES value will have a value similar to those for the ones around it. Figure 5.21b is an interpolated SES map. This interpolated map was created using the inverse-distance-weighted interpolation method, which determines cell values by using a linearly weighted combination of a set of sample points. The weight is a function of inverse distance, which means that known observation points that are closer have more of an influence on the value at the unknown location. The input is a vector point map of known observations, and the output is a raster surface map of interpolated cell values.

Spatial Analysis

All GIS software can be used not only to create maps to visualize spatial distributions but also to do various types of spatial analyses. There are many different types of spatial analyses that cannot be discussed comprehensively in this text (see "Further Reading" for many diverse examples). Much has been written about spatial

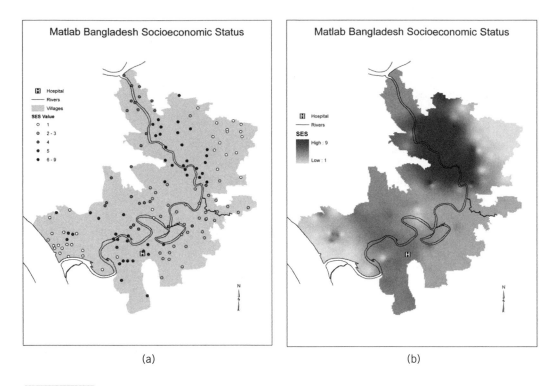

(a)   (b)

FIGURE 5.21 Interpolated SES map. (a) Known SES observation points. (b) Interpolated inverse-distance-weighted map.

analysis with a GIS. A few examples are given below and citations are provided in the "References" and "Further Reading" sections. Four types of spatial-analytical operations are discussed here including buffering, overlay analysis, and map algebra, which is really an entire suite of raster spatial-analytical tools.

Buffers are new polygons that are zones of a specified width around a point, line, or polygon. For example, Figure 5.22 is a map showing areas in Lilongwe, Malawi, that were chosen to conduct a malaria vaccine trial. The circles around the market points represent the areas within 1 kilometer of those market centers. These buffer areas were used to define the area to recruit patients in the trial.

Overlay analysis is a fundamental method in a GIS and simply means that the user is mathematically integrating multiple map layers together, which can be done using both raster and vector operations. One type of overlay analysis is in a vector GIS: two different polygon layers can be combined through a geometric intersection, and only features in the area common to both layers will be preserved. Figure 5.23 shows how two layers are integrated through raster overlay: The spatial intersection of the two map layers results in a composite layer. For example, if the cells in layer 1 represent areas near toxic waste dumps, and the cells in layer 2 represent childhood leukemia cases, then the composite can pinpoint the locations where there are both. There are numerous mathematical overlay methods (e.g., intersection, union).

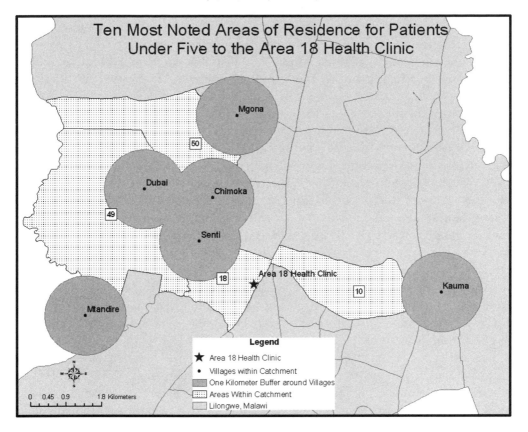

FIGURE 5.22 Patient recruitment around market areas in Lilongwe, Malawi.

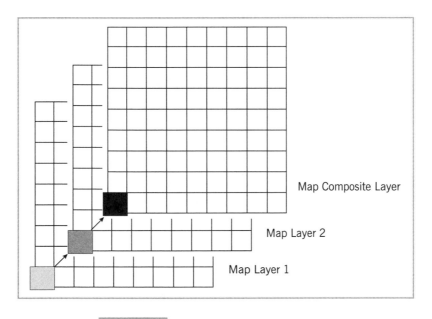

FIGURE 5.23 Overlay analysis in a raster GIS.

An example of how overlay analysis can be used in a study investigating the human–environment ecosystem drivers of an avian influenza virus (AIV) is offered below. Plate 11a is a map of global H5N1 AIV distribution from 2003 through 2015. As described in Chapter 6, the highly pathogenic H5N1 AIV was first identified in 1997 and has spread across Asia, Africa, and Europe since 2003. Scientists and public health officials are trying to determine where in the world this pandemic flu could emerge next. A GIS approach has been devised to answer the "Where?" question about a future strain of pandemic H5N1 AIV by spatially integrating population, environmental, and genetic information (Wan, Chen, Luo, Emch, & Donis, 2007; Wan, 2007; Wan et al., 2008; Carrel et al., 2011, 2012). Figure 5.24 shows the types of layers that might be integrated in a GIS to understand what human–environmental contexts are associated with appearance of the H5N1 virus and in genetic change in the virus. The idea is that certain land use patterns, population characteristics, and climatic patterns might lead to strains of the virus that are dangerous to humans.

Another genetic variant of the influenza A virus, H1N1, of swine origin appeared for the first time in March and April 2009; because humans had never been exposed to the virus before, it quickly became a pandemic. Plate 11b displays the geographic spread of the H1N1 flu after it had spread to 61 countries throughout the world.

Map algebra is another system of spatial data integration, done in the raster data model (DeMers, 2000; Tomlin, 1990). In map algebra each cell in a raster map has a cell value that represents a characteristic of an earth surface feature. For example, in Figure 5.25, map 1 (top left) might represent areas with contaminated groundwater, and map 2 (bottom left) might represent areas with sick people. In

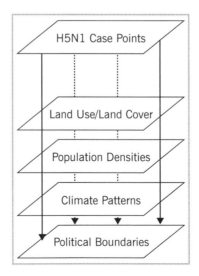

FIGURE 5.24 Human–environment ecosystem factors that can be overlaid with H5N1 virus locations and genetic information.

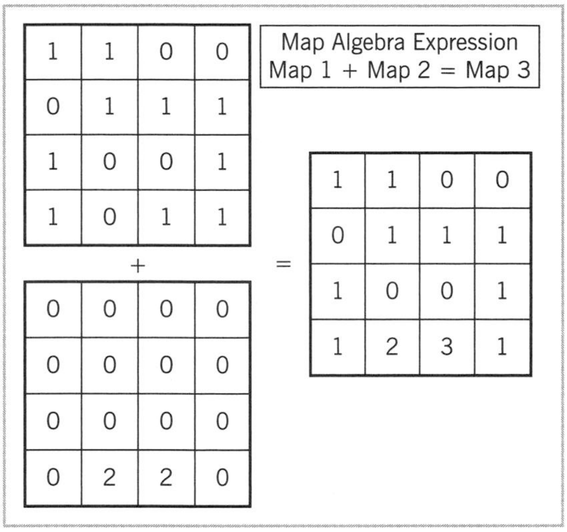

FIGURE 5.25 Map algebra expression.

map 1 a value of 1 represents a contaminated area, and in map 2 a value of 2 represents an area with sick people. Adding the values of the raster cells that correspond to the same areas in maps 1 and 2 would result in the values shown in map 3 (right). In map 3 values of 3 would therefore represent areas with contaminated groundwater and sick people; values of 2 would indicate areas with sick people without contaminated water; values of 1 would indicate areas with contaminated water and no sick people; and values of 0 would represent areas without contaminated water or sick people.

Map algebra is basically math applied to rasters that are geographically referenced. The language facilitates simple arithmetic combinations of rasters such as in Figure 5.25, or complex combinations of many layers. Arithmetic combinations include addition, subtraction, multiplication, division, and others. An arithmetic operator might be used to convert feet to meters by multiplying a raster in which cells represent elevation in feet by 0.3048, resulting in an elevation raster in meters. Other types of map algebra operators include Boolean and specialized. Boolean operators can be used to make true–false statements, resulting in 1 (true) and 0 (false) values in the output raster map. For example, in Figure 5.25 a Boolean statement could be created in the map algebra language that would result in values of 1 for areas with contaminated water and sick people, and the rest of the cells would have values of 0. Specialized operators are those that do something very particular. For instance, if an input raster represents elevation, the slope operator identifies the rate of maximum change in elevation from each cell. Map algebra is a powerful spatial modeling tool, and the analyst's combination of tools in innovative ways defines the spatial-analytical operation. Commercial GIS software packages, including ESRI's ArcGIS, implement combinations of map algebra tools.

SPATIAL STATISTICS

Geographic distributions require specialized statistics such as spatial autocorrelation or mean center calculations described above. Some spatial statistical methods build off of nonspatial statistical methods such as regression analysis, which allows the user to relate one distribution to another. For instance, a regression formula might look at how many calories per day people eat and how their caloric consumption is related to how much they weigh. The regression formula will no doubt show a positive relationship, but it will not be perfect because the amount of exercise that people do modifies the effect of eating. One of the assumptions of regression analysis is independence of observations. What this means is that one observation cannot be dependent on another. But with spatially autocorrelated map distributions, units are spatially dependent by definition. For example, if you were doing a study on how pollution affects health and you had data at the census tract level, neighboring census tract pollution values would be related to one another, and therefore the assumption of independence of observations would be violated. To deal with this problem, specialized spatial regression models have been created to deal with the nuisance of spatial autocorrelation (Anselin & Bera, 1998). These specialized models attempt to control for the autocorrelation (Anselin, 2002, 2005). One freeware software package that can be used to implement such spatial regression models is called GeoDa (see *spatial.uchicago.edu*).

Another type of spatial regression analysis called "geographically weighted regression" (GWR) explores the nature of relationships (Brunsdon, Aitkin, Fotheringham, & Charlton, 1999; Fotheringham, Brunsdon, & Charlton, 2003). For instance, the relationship between educational attainment and poverty might be strong in one part of a country and not in another. In this hypothetical example, it's possible that poverty explains low educational attainment in areas where there isn't a lot of government spending on education. Therefore, since poorer people do not have money, they would not be able to attend college. In other parts of the country, where the government provides free college education to its people, the relationship might not be as strong because even poor people can attend college. GWR can thus be used as a tool to understand how spatial relationships vary in space and to hypothesize why they vary the way they do. One implementation of GWR is available in ArcGIS software.

Spatial clusters of disease are aggregations of disease in space or in space and time. Disease clusters occur when there are more cases than are expected by chance. Identifying disease clusters can be used to identify possible risk factors related to the disease. For instance, as described earlier in this chapter, John Snow identified a cluster of cholera near the Broad Street pump in London in 1854. The cluster was attributed to a contaminated water supply, and the pump handle was taken off. There are many different types of cluster methods for different types of distributions. Global spatial cluster methods detect clustering of a disease, but they do not identify the particular locations of the clusters. Local cluster methods identify the particular locations of clusters. The method of focused clusters determines whether there is a cluster around a particular location, such as the Broad Street pump.

One problem with spatial cluster analysis is that a user does not always have an idea of the spatial scale at which the cluster exists. The spatial scan statistic was developed to allow the user to look for clusters across spatial scales. This statistic detects clusters by gradually scanning an elliptical window across a study area, determining the number of observed and expected cases inside the ellipse around each location (Kulldorff, 1997; Kulldorff & Nagarwalla, 1995). An example is shown in Figure 5.26, which shows spatial clusters of a birth defect called gastroschisis in North Carolina. Notice that two clusters of different sizes were identified (Root, Meyer, & Emch, 2009). A freeware software implementation of the spatial scan statistic is SaTScan, which was developed at the National Cancer Institute (see *www.satscan.org*).

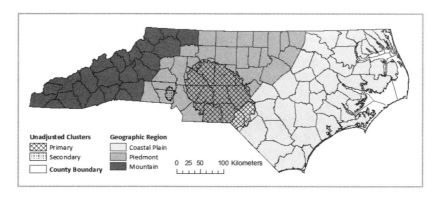

FIGURE 5.26 Spatial clusters of gastroschisis births, identified by using the spatial scan statistic.

CONCLUSION

Geographic methodology has been revolutionized by the information now available and the speed and complexity with which it can be manipulated in GIS. Software has been developed to make GIS available on a point-and-click basis in windows and menus. Uses are spreading rapidly into a variety of disciplines, including epidemiology and other areas of public health. The careful and accurate specification and analysis of disease data are complex undertakings. Scale is an important property that is often overlooked in disease studies. Much of the most important analysis in geography is of patterns and dynamics in space or over distance. Spatial analysis has some special complications of its own, but also some of the most powerful methods of portrayal and communication. When studies do not consider cartographic principles or theoretical issues like scale effects and the ecological fallacy, they are often not well conceptualized.

Computer and software development continues to advance, as does the development of technologies like the GPS for spatial data collection. The present state is one of euphoria over possibilities. With all of these advances, GIS-based health studies have become quite easy to implement, but there are also many ill-conceived projects that produce spurious findings by unthinking users. To put it into a software context, beware the "dangers of defaults." Health researchers need to proceed with due care, with appreciation of the complexities of scale, but the possibilities are revolutionary. The power of a GIS lies in the quality of the data and the innovation of the operator. Just as cartography is both an art and a science, so is the spatial analysis performed in a GIS. An expert in GIS is someone who not only has an understanding of the breadth of spatial-analytical tools available, but also can put them together in innovative ways to answer spatial questions.

REVIEW QUESTIONS

1. John Snow wasn't the first person to produce a disease map, but his map is the most commonly cited historic example. How was Snow's map used to change public health policy in the mid-19th century in London?

2. Develop a plan to create a disease map for a current health problem. How could the disease map be used to answer a question that could lead to a current health policy change?

3. Find an online disease map site on the Internet. List the Web address and describe the online map. Is the map interactive? In other words, can you change what is displayed on the map? What is the purpose of the online mapping site?

4. Scale is the amount of area being considered on a map. Find several examples of static maps (fixed scale) either on the Internet or paper maps. List their scales and whether they are small-, medium-, or large-scale maps. For each, what type of analyses would be most appropriate: local, regional, or national/global level. Are there different types of spatial questions that you can answer with each?

5. Points, lines, and areas represent different kinds of spatial data. Give examples of each type of

spatial object that you might use to represent spatial Earth surface features for a hypothetical health geography study.

6. Data displayed on choropleth maps are classified in different ways. Explain some of the decisions that you make when creating a choropleth map (e.g., number of classes, classification methods). Describe a hypothetical example of a choropleth map that you might create in a health geography project and describe the classification method and number of classes you chose and why.

7. Address matching, the GPS, remote sensing, and digitized legacy maps are the four most common spatial data sources used in health geography studies. At your university library find copies of the following journals: *Health & Place, Social Science and Medicine,* and the *International Journal of Health Geographics.* Look at the titles and abstracts of some of the articles in recent editions and find some that you think could be categorized as health geography studies that use GIS. Read the data and methods sections of the papers that you choose. Describe the spatial data source and methods used, and then build the kind of spatial database that would be used in a GIS analysis.

8. Spatial data are useful for understanding the geographic context of health issues. The ecological fallacy and the modifiable areal unit problem are issues that need to be considered when doing GIS analysis in health studies. Explain each issue and why they can be problematic in health geography studies.

9. The scale of analysis of health geography studies is often defined by the unit of observation. In the health geography papers you found in review question 7, what is the unit of analysis? Sometimes the unit of observation is chosen because the data are only available at that level and not because it is the best unit of observation. Do you think that the unit of analysis is the ideal unit to study the spatial phenomenon being analyzed in the paper? Why or why not?

10. Spatial autocorrelation can be measured using Moran's I statistic. What is it and why is it so fundamentally important to health geography studies?

11. Using spatial analysis in health geography studies has grown dramatically in recent years because of the availability of spatial data and spatial computing tools. In review question 5 you gave examples of types of spatial objects that you might use to represent spatial features for a hypothetical health geography study. Develop the example further by conceptually describing the types of spatial analyses you could do and the questions you could address using the data.

REFERENCES

Anselin, L. (2002). Under the hood: Issues in the specification and interpretation of spatial regression models. *Agricultural Economics, 27*(3), 247–267.

Anselin, L. (2005). *Exploring spatial data with GeoDA: A workbook.* Urbana: Spatial Analysis Laboratory, Department of Geography, University of Illinois, Urbana–Champaign.

Anselin, L., & Bera, A. (1998). Spatial dependence in linear regression models with an introduction to spatial econometrics. In A. Ullah & D. Giles (Eds.), *Handbook of applied economic statistics* (pp. 237–289). New York: Dekker.

Berghaus, H. (1852). *Physikalischer atlas.* Gotha, Germany: J. Perthes.

Brunsdon, C., Aitkin, M., Fotheringham, A., & Charlton, M. (1999). A comparison of random coefficient modeling and geographically weighted regression for spatially nonstationary regression problems. *Geographical and Environmental Modeling, 3*(1), 47–62.

Carrel, M., Emch, M., Nguyen, T., Jobe, R., & Wan, X.-F. (2012). Population–environment drivers of H5N1 avian influenza molecular change in Vietnam. *Health and Place, 18*(5), 1122–1131.

Carrel, M., Wan, X.-F., Nguyen, T., & Emch, M. (2011). Genetic variation of highly pathogenic H5N1 avian influenza viruses in Vietnam shows both species-specific and spatiotemporal associations. *Avian Diseases, 6*(4), e25–e27.

Cleek, R. (1979). Cancer and the environment: The effect of scale. *Social Science and Medicine, 13D,* 241–247.

Cliff, A., & Haggett, P. (1988). *Atlas of disease distributions: Analytical approaches to epidemiological data.* Oxford, UK: Blackwell.

Cliff, A., Haggett, P., & Ord, K. (1987). *Spatial aspects of influenza epidemics.* London: Routledge & Kegan Paul.

Cliff, A. D., Haggett, P., & Smallman-Raynor, M. R. (2000). *Island epidemics.* Oxford, UK: Oxford University Press.

Cliff, A., Haggett, P., & Smallman-Raynor, M. (2004). *World atlas of epidemic diseases.* London: Arnold.

Cromley, E., & McLafferty, S. (2012). *GIS and public health* (2nd ed.). New York: Guilford Press.

DeMers, M. (2000). *GIS modeling in raster.* New York: Wiley.

DiMaggio, C., Galeo, S., & Emch, M. (2010). Spatial proximity and the risk of psychopathology after a terrorist attack. *Psychiatry Research, 176,* 55–61.

Fotheringham, A., Brunsdon, C., & Charlton, M. (2002). *Geographically weighted regression: The analysis of spatially varying relationships.* Chichester, UK: Wiley Europe.

Gilbert, E. (1958). Pioneer maps of health and disease in England. *Geographical Journal, 124,* 172–183.

Gould, P. (1993). *The slow plague: A geography of the AIDS pandemic.* Oxford, UK: Blackwell.

Haggett, P., Cliff, A., & Smallman-Raynor, M. (1993). *Measles: An historical geography of a major human viral disease from global expansion to local retreat, 1840–1990.* Oxford, UK: Blackwell.

Haggett, P., Cliff, A., & Smallman-Raynor, M. (1998). *Deciphering global epidemics: Analytical approaches to the disease records of world cities, 1888–1912.* Cambridge, UK: Cambridge University Press.

Haggett, P., Cliff, A., & Smallman-Raynor, M. (2000). *Island epidemics.* Oxford, UK: Oxford University Press.

Koch, T. (2005). *Cartographies of disease: Maps, mapping, and medicine.* Redlands, CA: ERSI Press.

Kulldorff, M. (1997). A spatial scan statistic. *Communications in Statistics: Theory and Methods, 26,* 481–1496.

Kulldorff, M., & Nagarwalla, N. (1995). Spatial disease clusters: Detection and inference. *Statistics in Medicine, 14*(8), 799–810.

Longley, P., Goodchild, M., Maguire, D., & Rhind, D. (2010). *Geographic information systems and science* (3rd ed.). Chichester, UK: Wiley Europe.

May, J. (1958). *The ecology of human disease.* New York: MD Publications.

National Center for Health Statistics. (1997). *Atlas of United States mortality.* Hyattsville, MD: Author.

Openshaw, S., Craft, A., Charlton, M., & Birch, J. (1988). Investigation of leukemia clusters by use of a geographical analysis machine. *The Lancet, 331*(8580), 272–273.

Root, E., Meyer, R., & Emch, M. (2009). Evidence of localized clustering of gastroschisis births in North Carolina, 1999–2004. *Social Science and Medicine, 68,* 1361–1367.

Smallman-Raynor, M., Cliff, A., & Haggett, P. (1992). *London international atlas of AIDS*. Oxford, UK: Blackwell.

Snow, J. (1855). *On the mode of communication of cholera*. London: John Churchhill.

Tomlin, C. (1990). *Geographic information systems and cartographic modeling*. Englewood Cliffs, NJ: Prentice Hall.

Wan, X.-F., Chen, G., Luo, F., Emch, M., & Donis, R. (2007). A quantitative genotype algorithm reflecting H5N1 avian influenza niches. *Bioinformatics, 23*(18), 2368–2375.

Wan, X.-F., Nguyen, T., Smith, C., Zhao, Z., Carrel, M., Davis, C., Inui, K., et al. (2008). Evolution of highly pathogenic H5N1 avian influenza viruses in Vietnam between 2001 and 2007. *PLoS ONE, 3*(10), 1–12.

Wan, X.-F., Wu, X., Lin, G., Holton, S., Desmone, R., Shyu, C.-R., et al. (2007). Computational identification of reassortments in avian influenza viruses. *Avian Diseases, 51*(s1), 434–439.

FURTHER READING •

Aldstadt, J., Yoon, I.,Tannitisupawong, D., Jarman, R., Thomas, S., Gibbons, R., et al. (2012). Space–time analysis of hospitalized dengue patients in rural Thailand reveals important temporal intervals in the pattern of dengue virus transmission. *Tropical Medicine and International Health, 17*(9), 1076–1085.

Ali, M., Emch, M., Tofail, M., & Baqui, A. (2001). Implications of health care provision on acute lower respiratory infection mortality in Bangladeshi children. *Social Science and Medicine, 52*(2), 267–277.

Ali, M., Emch, M., von Seidlein, L., Yunus, M., Sack, D., Rao, M., et al. (2005). Herd immunity conferred by killed oral cholera vaccines in Bangladesh. *The Lancet, 366*(9479), 44–49.

Ali, M., Emch, M., Yunus, M., Sack, D., Lopez, A., Holmgren, J., et al. (2008). Vaccination of adult women against cholera protects infants and young children in rural Bangladesh. *Pediatric Infectious Disease Journal, 27*(1), 33–37.

Ali, M., Goovaerts, P., Nazia, N., Haq, M., Yunus, M., & Emch, M. (2006). Application of Poisson kriging to the mapping of cholera and dysentery incidence in an endemic area of Bangladesh. *International Journal of Health Geographics, 5*(45), 1–11.

Armstrong, R. (1973). Tracing exposure to specific environments in medical geography. *Geographical Analysis, 5,* 122–132.

Bailey, A., Sargent, J., Goodman, D., Freeman, J., & Brown, M. (1994). Poisoned landscapes: The epidemiology of environmental lead exposure in Massachusetts children 1990–1991. *Social Science and Medicine, 39,* 757–766.

Bailey, T., & Gatrell, A. (1995). *Interactive spatial data analysis*. New York: Wiley.

Barnett, R., Pearce, J., & Howes, P. (2006). "Help, educate, encourage?": Geographic variations in the provision and utilization of diabetes education in New Zealand. *Social Science and Medicine, 63,* 1328–1343.

Barnett, R., Pearce, J., & Moon, G. (2005). Does social inequality matter?: Changing ethnic socio-economic disparities and Maori smoking in New Zealand, 1981–1996. *Social Science and Medicine, 60*(7), 1515–1526.

Bernett, R. (1979). *Spatial time series*. London: Pion.

Bian, L. (2013). Spatial approaches to modeling transmission of communicable diseases—A review. *Transactions in GIS, 17*(1), 1–17.

Bian, L., Huang, Y., Liang, M., Lim, E., Lee, G., Yang, Y., et al. (2012). Modeling individual vulnerability to communicable diseases—A framework and design. *Annals of the Association of American Geographers, 102*(5), 1016–1025.

Blackburn, J., Hadfield, T., Curtis, A., & Hugh-Jones, M. (2014). Spatial and temporal patterns of anthrax in white-tailed deer, Odocoileus virginianus, and hematophagous flies in west Texas during the summertime anthrax risk period. *Annals of the Association of American Geographers, 104*(5), 939–958.

Blackburn, J., McNyset, K., Curtis, A., & Hugh-Jones, M. (2007). Modeling the geographic distribution of Bacillus anthracis, the causative agent of anthrax disease, for the contiguous United States using predictive ecological niche modeling. *American Journal of Tropical Medicine and Hygiene, 77*(6), 1103–1110.

Burrough, P., & McDonnell, R. (1998). *Principles of geographical information systems.* Oxford, UK: Oxford University Press.

Carrel, M., Emch, M., Streatfield, P., & Yunus, M. (2009). Spatiotemporal clustering of cholera: The impact of flood control in Matlab, Bangladesh, 1983–2003. *Health and Place, 15*(3), 771–782.

Carrel, M., Eron, J., Emch, M., & Hurt, C. (2014). Spatial epidemiology of recently acquired HIV infections across rural and urban areas of North Carolina. *PLoS ONE, 9*(2), e88512.

Carrel, M., Escamilla, V., Messina, J., Giebultowicz, S., Winston, J., Yunus, M., et al. (2011). Increased tubewell density decreased diarrheal disease risk in rural Matlab, Bangladesh: A zero-inflated and geographically weighted analysis. *International Journal of Health Geographics, 10*(1), 41.

Clarke, K. (1995). *Analytical and computer cartography* (2nd ed.). Upper Saddle River, NJ: Prentice Hall.

Clarke, K. (1997). *Getting started with GIS.* Upper Saddle River, NJ: Prentice Hall.

Clarke, K., McLafferty, S., & Tempalski, B. (1996). On epidemiology and geographic information systems: A review and discussion of future directions. *Emerging Infectious Diseases, 2,* 85–92.

Cleek, R. (1979). Cancer and the environment: The effect of scale. *Social Science and Medicine, 13D,* 241–247.

Cliff, A., & Ord, J. (1973). *Spatial autocorrelation.* London: Pion.

Comrie, A. (2005). Climate factors influencing coccidioidomycosis seasonality and outbreaks. *Environmental Health Perspectives, 113,* 688–692.

Curtis, A. (1999). Using a spatial filter and a geographic information system to improve rabies surveillance data. *Emerging Infectious Diseases, 5*(5), 603–606.

Curtis, A., Copeland, A., Fagg, J., Congdon, P., Almog, M., & Fitzpatrick, J. (2006). The ecological relationship between deprivation, social isolation and rates of hospital admission for acute psychiatric care: A comparison of London and New York City. *Health and Place, 12,* 19–37.

Curtis, A., Mills, J., & Blackburn, J. (2007). A spatial variant of the basic reproduction number for the New Orleans yellow fever epidemic of 1878. *Professional Geographer, 59*(4), 492–502.

Curtis, S., Southall, H., Congdon, P., & Dodgeon, B. (2004). Area effects on health variation over the life-course: Analysis of the longitudinal study sample in England using new data on area of residence in childhood. *Social Science and Medicine, 58,* 57–74.

Delmelle, E., Zhu, H., Casas, I., & Tang, W. (2014). A Web-based geospatial toolkit for the monitoring of Dengue fever. *Applied Geography, 52,* 144–152.

DeMers, M. (1997). *Fundamentals of geographic information systems.* New York: Wiley.

DiMaggio, C., Galea, S., & Emch, M. (2010). Spatial proximity and the risk of psychopathology after a terrorist attack. *Psychiatry Research, 176*(1), 55–61.

Dunn, C., Bhopal, R., Cockings, S., Waker, D., Rowlingson, B., & Diggle, P. (2007). Advancing insights into methods for studying environment–health relationships: A multidisciplinary approach to understanding Legionnaires' disease. *Health and Place, 13,* 677–690.

Dunn, C., Kingham, S., Rowlingson, B., Bhopal, R., Cockings, S., Foy, C., et al. (2001). Analysing spatially referenced public health data: A comparison of three methodological approaches. *Health and Place, 7*(1), 1–12.

Emch, M. (1999). Diarrheal disease risk in Matlab, Bangladesh. *Social Science and Medicine, 49,* 519–530.

Emch, M. (2000). Relationships between flood control, Kala-azar, and diarrheal disease in Bangladesh. *Environment and Planning A, 32,* 1051–1063.

Emch, M., Ali, M., Park, J., Yunus, M., Sack, D., & Clemens, J. (2006). Relationship between neighborhood-level killed oral cholera vaccine coverage and protective efficacy: Evidence for herd immunity. *International Journal of Epidemiology, 35,* 1044–1050.

Emch, M., Ali, M., & Yunus, M. (2008). Risk areas and neighborhood-level risk factors for Shigella dysenteriae 1 and Shigella flexneri: Implications for vaccine development. *Health and Place, 14,* 96–105.

Emch, M., Ali, M., Yunus, M., Sack, D., Acosta, C., & Clemens, J. (2007). Efficacy calculation in randomized vaccine trials: Global or local measures? *Health and Place, 13,* 238–248.

Emch, M., Feldacker, C., Islam, M., & Ali, M. (2008). Seasonality of cholera from 1974 to 2005: A review of global patterns. *International Journal of Health Geographics, 7*(31), 1–33.

Emch, M., Feldacker, C., Yunus, M., Streatfield, P., Thiem, V., Canh, D., et al. (2008). Local environmental drivers of cholera in Bangladesh and Vietnam. *American Journal of Tropical Medicine and Hygiene, 78*(5), 823–832.

Emch, M., Root, E., Giebultowicz, S., Ali, M., Perez-Heydrich, C., & Yunus, M. (2012). Integration of spatial and social network analysis in disease transmission studies. *Annals of the Association of American Geographers, 102*(5), 1004–1015.

Escamilla, V., Emch, M., Dandalo, L., Martinson, F., Miller, W., & Hoffman, I. (2014). Community level sampling using Google Earth satellite imagery and geographical methods in the absence of a demographic surveillance system. *Bulletin of the World Health Organization, 92,* 690–694.

Escamilla, V., Knappet, P., Yunus, M., Streatfield, P., & Emch, M. (2013). Influence of latrine proximity and type on tubewell water quality and diarrheal disease in Bangladesh. *Annals of the Association of American Geographers, 103*(2), 299–308.

Fagg, J., Curtis, A., Streatfield, P., & Congdon, P. (2006). Psychological distress among adolescents, and its relationship to individual, family and area characteristics in east London. *Social Science and Medicine, 63,* 636–648.

Fotheringham, A., & Rogerson, P. (1994). *Spatial analysis and GIS.* London: Taylor & Francis.

Gesler, W. (1986). The uses of spatial analysis in medical geography: A review. *Social Science and Medicine, 23,* 963–973.

Giggs, J. (1983). Schizophrenia and ecological structure in Nottingham. In N. McGlashan & J. Blunden (Eds.), *Geographical aspects of health* (pp. 197–222). New York: Academic Press.

Grady, S. (2006). Racial disparities in low birth weight and the contribution of residential segregation: A multilevel analysis. *Social Science and Medicine, 63,* 3013–3029.

Grady, S., & Ramirez, I. (2008). Mediating medical risk factors in the racial segregation and low birth weight relationship. *Health and Place, 14*(4), 661–677.

Griffith, D., & Amrhein, C. (1997). *Multivariate statistical analysis for geographers.* Upper Saddle River, NJ: Prentice Hall.

Haggett, P. (1990). *The geographer's art.* Oxford, UK: Blackwell.

Hanchette, C. (1999). GIS and decision-making for public health agencies: Childhood lead poisoning and welfare reform. *Journal of Public Health Managment and Practice, 5*(4), 41–47.

Hanchette, C., & Swartz, G. (1992). Geographic patterns of prostate cancer mortality: Evidence for a protective effect of ultraviolet radiation. *Cancer, 70*(12), 2861–2869.

Hay, S., Randolph, S., & Rogers, D. (2000). *Remote sensing and geographic information systems in epidemiology.* London: Academic Press.

Heywood, I. (1990). Geographic information systems in the social sciences. *Environment and Planning A, 22,* 849–854.

Hohl, P., & Mayo, B. (1997). *ArcView GIS exercise book* (2nd ed.). Santa Fe, NM: On-Word Press.

Howe, G. (1970). *National atlas of disease mortality in the United Kingdom* (2nd ed.). London: Nelson.

Howe, G. (1972). *Man, environment, and disease in Britain.* New York: Barnes & Noble.

Jacquez, G. (1993). Introduction to the special issue on statistics and computing in disease clustering. *Statistics in Medicine, 12,* 1751.

Jerrett, M., Burnett, R., Kanaraglou, P., Eyles, J., Finkelstein, N., & Giovis, C. (2001). A GIS–environmental justice analysis of particulate air pollution in Hamilton, Canada. *Environment and Planning A, 33,* 955–973.

Jerrett, M., Burnett, R., Ma, R., Pope, C., Kewski, D., Newbold, K., et al. (2005). Spatial analysis of air pollution and mortality in Los Angeles. *Epidemiology, 16*(6), 727–736.

Kalkbrenner, A., Daniels, J., Emch, M., Morrissey, J., Poole, C., & Chen, J.-C. (2011). Geographic access to health services and autism diagnosis. *Annals of Epidemiology, 21,* 304–310.

King, P. (1979). Problems of spatial analysis in geographical epidemiology. *Social Science and Medicine, 13D,* 249–252.

Kolivras, K., & Comrie, A. (2003). Modeling valley fever (coccidioidomycosis) incidence based on climate conditions. *International Journal of Biometeorology, 47,* 87–101.

Lam, N., & Quattrochi, D. (1992). On the issues of scale, resolution, and fractal analysis in the mapping sciences. *Professional Geographer, 44,* 88–98.

Marble, D., Calkins, H., & Peuquet, D. (1984). *Basic readings in geographic information systems.* Williamsville, NY: SPAD Systems.

Marshall, R. (1991). A review of methods for the statistical analysis of spatial patterns of disease. *Journal of the Royal Statistical Society, 154,* 421–441.

Matthews, S. (1990). Epidemiology using a GIS: The need for caution. *Computers, Environment and Urban Systems, 14*(3), 213–221.

May, J. (1983). The role of spatial analysis and geographic data in detection of disease causation. *Social Science and Medicine, 17,* 1213–1221.

Mayer, J. (1981). Problems of spatial analysis in geographical epidemiology. *Social Science and Medicine, 13D,* 249–252.

McLafferty, S., & Grady, S. (2004). Prenatal care need and access: A GIS analysis. *Journal of Medical Systems, 28*(3), 321–333.

McLafferty, S., & Grady, S. (2005). Immigration and geographical access to prenatal clinics in Brooklyn, NY: A geographic information systems analysis. *American Journal of Public Health, 95*(4), 638–640.

Meade, M. (1979). Cardiovascular mortality in the southeastern United States: The coastal plain enigma. *Social Science and Medicine, 13D,* 257–265.

Meade, M. (1980). An interactive framework for geochemistry and cardiovascular disease. In M. Meade (Ed.), *Conceptual and methodological issues in medical geography* (pp. 194–221). Chapel Hill: Department of Geography, University of North Carolina.

Meade, M. (1983). Cardiovascular disease in Savannah, Georgia. In N. McGlashan & J. Blunden (Eds.), *Geographical aspects of health* (pp. 175–196). New York: Academic Press.

Messina, J., Mwandagaliwa, K., Taylor, S., Emch, M., & Meshnick, S. (2013). Assessing the spatial, social, and disease-related determinants of anemia in Congolese women. *Health and Place, 24,* 54–64.

Messina, J., Taylor, S., Meshnick, S., Linke, A., Tshefu, A., Atua, B., et al. (2011). Population, behavioral and environmental drivers of malaria prevalence in the Democratic Republic of Congo. *Malaria Journal, 10*(161), 10–1186.

Misch, A. (1994). Assessing environmental health risks. In L. Brown et al. (Eds.), *State of the world* (pp. 117–136). New York: Norton.

Moon, G., & Barnett, J. (2003). Spatial scale and the geography of tobacco smoking in New Zealand: A multilevel perspective. *New Zealand Geographer, 59*(2), 6–15.

Odland, J. (1988). *Scientific geography: Vol. 9. Spatial autocorrelation.* Newbury Park, CA: Sage.

Openshaw, S., Charlton, M., Wymer, C., & Craft, A. (1987). A Mark 1 geographical analysis machine for the automated analysis of point data sets. *International Journal of Geographical Information Systems, 1*(4).

Ord, K., & Getis, A. (1995). Local spatial autocorrelation statistics: Distributional issues and an application. *Geographical Analysis, 24,* 286–306.

Pearce, J., Barnett, J., & Jones, I. (2007). Have urban/rural inequalities in suicide in New Zealand grown during the period 1980–2001? *Social Science and Medicine, 65,* 1807–1819.

Pearce, J., Barnett, J., & Kingham, S. (2006). Slip! slap! slop!: Cutaneous malignant melanoma incidence and social status in New Zealand, 1995–2000. *Health and Place, 12,* 239–252.

Pearce, J., Hiscock, R., Moon, G., & Barnett, J. (2009). The neighbourhood effects of geographical access to tobacco retailers on individual smoking behaviour. *Journal of Epidemiology and Community Health, 63*(1), 69–77.

Perez-Heydrich, C., Braly, J., Giebultowicz, S., Winston, J., Yunus, M., Streatfield, P., et al. (2013). Social and spatial processes associated with childhood diarrheal disease in Matlab, Bangladesh. *Health and Place, 19,* 45–52.

Pope, K., Rejmankova, E., Savage, H., Arredondo-Jimenez, J., Rodriguez, M., & Roberts, D. (1994). Remote sensing of tropical wetlands for malaria control in Chiapas, Mexico. *Ecological Applications, 4,* 81–90.

Ricketts, T., Savitz, L., Gesler, W., & Osborne, D. (1994). *Geographic methods for health services research.* Lanham, MD: University Press of America.

Riva, M., Apparicio, P., Gauvin, L., & Brodeur, J.-M. (2008). Establishing the soundness of administrative spatial units for operationalising the active living potential of residential environments: An exemplar for designing optimal zones. *International Journal of Health Geographics, 7*(1).

Riva, M., Curtis, C., Gauvin, L., & Fagg, J. (2009). Health variation among categories of urban and rural areas: Evidence from a national sample in England. *Social Science and Medicine, 68,* 654–663.

Riva, M., Gauvin, L., & Barnett, T. (2007). Toward the next generation of research into small area effects on health: A synthesis of multilevel investigations published since July 1998. *Journal of Epidemiology and Community Health, 61,* 835–861.

Robinson, V. (1978). Modeling spatial variations in heart disease mortality: Implications of the variable subset selection process. *Social Science and Medicine, 12D,* 165–172.

Rodenwaldt, E., & Jusatz, H. (Eds.). (1952–1961). *Welt-Seuchen Atlas* (3 vols). Hamburg, Germany: Falk.

Root, E., Meyer, R., & Emch, M. (2011). Socioeconomic context and gastroschisis: Exploring associations at various geographic scales. *Social Science and Medicine, 72*(4), 625–633.

Rushton, G. (1988). *Improving public health through geographic information systems: An instructional guide to major systems.* Iowa City: Department of Geography, University of Iowa.

Rushton, G., Armstrong, R., Gittler, J., Greene, B., Pavlik, C., West, M., & Zimmerman, D. (2006). Geocoding in cancer research: A review. *American Journal of Preventive Medicine, 30*(2, Suppl. 1), S16–S24.

Rushton, G., Peleg, I., Banerjee, A., Smith, G., & West, M. (2004). Analyzing geographic patterns of disease incidence: Rates of late stage colorectal cancer in Iowa. *Journal of Medical Systems, 28*(3), 223–236.

Shisemstsu, I. (1981). *National atlas of major disease mortalities in Japan.* Tokyo: Japan Health Promotion Federation.

Vine, M., Degnan, D., & Hanchette, C. (1997). Geographic information systems: Their use in environmental epidemiologic research. *Environmental Health Perspectives, 105,* 598–605.

Vine, M., Stein, L., Weigle, K., Schroader, J., Degnan, D., Tse, C., et al. (2000). Effects on the immune system associated with living near a pesticide dump site. *Environmental Health Perspectives, 108*(12), 1113–1124.

Walter, D. (1992). The analysis of regional patterns in health data: II. The power to detect environmental effects. *American Journal of Epidemiology, 136,* 742–759.

Washino, R., & Wood, B. (1994). Application of remote sensing to arthropod vector surveillance and control. *American Journal of Tropical Medicine and Hygiene, 50*(6), 134–144.

Winston, J., Meyer, R., & Emch, M. (2014). A geographic analysis of compositional and contextual risk factors for hypospadias births. *Birth Defects Research Part A: Clinical and Modecular Teratology, 100*(11), 887–894.

Yoo, E. (2014). Site-specific prediction of West Nile virus mosquito abundance in Greater Toronto Area using generalized linear mixed model. *International Journal of Geographical Information Science, 28*(2), 296–313.

Ziegenfus, R., & Gesler, G. (1984). Geographical patterns of heart disease in the northeastern United States. *Social Science and Medicine, 18,* 63–72.

CHAPTER 6

Disease Diffusion

There are only two ways in which anything—whether it be an idea, a type of mosquito, a food crop, a sophisticated technology for hospital care, or an infectious agent of disease—can come to be found in a particular location. Either it developed there independently, or it somehow moved there from another place. Movement is by far the more usual explanation of the existence of phenomena at a particular location. An understanding of the mechanisms that influence the spread of any phenomenon and its spatial pattern is at the core of the geographic study of diffusion. The term **diffusion** implies a spread or movement outward from a point or beginning place. In public health and medical fields the term *diffusion* is seldom used and instead "disease spread" is used. Geographers have for decades studied diffusion of all manner of things: fish hooks, types of barns, musical instruments, religious beliefs, agricultural seeds, and communications technology, among hundreds of others. Health geographers have studied diffusion of new medical technologies, equipment, treatments, hospices, and the like. Mostly they have focused on the diffusion of infectious diseases. In these studies many of the conceptual and methodological concerns of epidemiology and geography have been combined. Geographers, however, want to know not only how a disease spreads in a population, but how that spread occurs over space; not only how many cases might occur at a future point in time, but where those cases are likely to occur.

Western civilization has long generally recognized the existence and importance of disease diffusion. Disease **quarantine**, which could involve an individual, a household, or even an entire community or country, at least tacitly implies the existence of disease diffusion. Indeed, the term *quarantine,* which means to restrict movement of people, comes from the medieval port city of Raguna (in present-day Italy), when in 1377 it tried to seclude incoming vessels for 40 days in an effort to prevent importation of plague. In the United States, Noah Webster's (1799) *A Brief History of Epidemic and Pestilential Diseases* traced the spread of several late 16th-century epidemics across New England. At about that same time, William Currie (1792, 1811) was also describing the geography of a number of epidemics in the United States. The contagionists investigating the distribution of yellow fever in the early 1800s recognized that the disease was carried from place to place, spread by an infected individual. Although we might find their medical explanations of causation amusing, the quality of their geographic descriptions was, given the severe data limitations, excellent.

Most studies of geographic diffusion have emphasized an explanation of events in a space–time context. The extent and form of human interaction are critical for both innovation and diffusion. Variables such as distance, the locations of settlements, and the distribution of jobs and facilities all influence the level and direction of interaction. Diffusion is patterned by the configuration of the networks that encourage movement and the barriers that discourage it. A major prerequisite of disease diffusion, for example, is the existence of a sufficiently large susceptible population. A sufficiently large immune population (or, more precisely, a sufficiently small susceptible population) can serve as an effective barrier to disease transmission. Knowledge of where a disease has been before and how it moved through population settlements, therefore, can help epidemiologists and medical personnel target vaccination campaigns on transmission channels or establish buffer zones of immune people as barriers.

New infectious diseases and newly virulent or antibiotic-resistant strains of old ones have appeared in recent decades. (The evolution and emergence of new disease agents are discussed in Chapter 7.) Whether they spread only locally (such as Legionnaires' disease) or diffuse as great world pandemics (such as HIV/AIDS), the threat of emerging diseases challenges the development of diffusion prediction. This chapter includes some background theory on diffusion processes and epidemiological concepts and then describes different types of diffusion and the main forms of diffusion modeling. Examples are then offered for HIV/AIDS, cholera, and influenza.

DIFFUSION BACKGROUND

It is often convenient to think of diffusion as waves of innovation and acceptance spreading geographically (Hagerstrand, 1952). These innovation impulses tend to lose their energy with distance from the source of the innovation. If we plot the acceptance of a new idea, or the onset of a disease contagion, for a series of time periods against the distance from the source, we can see how the innovation gradually fades with increased distance (Figure 6.1). During each successive time period, the locus of greatest initial acceptance of contagion is further from the source. During the first several periods, the total volume of acceptance increases; after that, the number of new acceptances decreases with each successive period. The summation of all these time curves across space and through time results in a pair

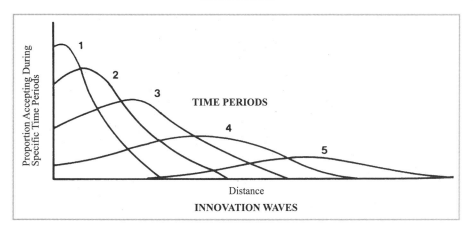

FIGURE 6.1 Innovation waves. From Gould (1969, p. 11). Copyright ©1969 Association of American Geographers. Adapted by permission.

of bell-shaped curves. The geographic bell is centered over the original innovation point and identifies a declining share of the total population's ever accepting the innovation with increasing distance. The second bell graphs the overall pattern of volume of acceptance (or contagion) from a small number of innovators through the great bulk of acceptors to a final few laggards.

Innovators and acceptors infect several new individuals each generation, with time periods being set by the incubation period of the agent. Suppose the underlying population is unevenly distributed. This population can be used to create a set of normalized probabilities, with each area's probability based not only on distance from the innovator but also on its share of the total population. Barriers can be incorporated into the model as well. A permeable barrier (such as immunity levels) might mean that only every fourth attempted passage of the innovation into an area beyond the barrier is successful. Other types of barriers can be similarly incorporated.

EPIDEMIOLOGICAL BACKGROUND

The broad term **communicable disease**, as used by the Centers for Disease Control and Prevention (CDC), includes all diseases due to a specific infectious agent that is transmitted from an infected host to a susceptible host, whether directly or indirectly through an intermediary vector. Epidemiologists are concerned with disease infections moving through a population (one that seems not to exist distributed in space but as a whole occupying a point). Several of the terms from Chapters 1 and 2 apply to this process (see Figure 1.4). A person can be infected but never develop symptoms (subclinical), or can develop symptoms (clinical). The person can be even clinically sick, but not pass the infection to another person (insufficient viremia, a dead end). How likely is a disease to develop in an infected person (pathogenicity of the agent, resistance of the host)? How virulent is the infection that occurs (**case fatality rate**)? How likely is the infection to be passed on (**secondary attack rate**)? The **basic reproduction number** or R_0 is the average number of secondary infections produced when one infection occurs in a virgin population (i.e., one in which everyone is susceptible). How long is the time period between infection and when a person becomes infectious to others (incubation period)? People who can be infected are susceptible. Those who have already had the disease, in reality or through vaccination, and can no longer be infected are immune. Those who are actively able to pass the agent on are infectious. Mathematical epidemiologists build SIR disease transmission models that partition populations into three groups, susceptible (S), infected (I), and removed (R) (Figure 6.2). S's are individuals who have not been infected, but are at risk; I's are those who are infected and are capable of transmitting the disease; and R's are those who can no longer contract the disease because they have recovered with immunity, have been quarantined, or have died.

The R_0 is number of cases one case generates on average over the course of its infectious period in the uninfected population, and this varies by disease. When the R_0 is less than one the infection will burn itself out, and when it is greater than one it will spread. Biological and genetic characteristics of disease agents partially determine what the R_0 is for a particular disease. However, the context in which the disease occurs also influences how the disease spreads. For instance, the first reported Ebola virus outbreak was in Zaire (now the Democratic Republic of Congo[DRC]) in 1976 and there have been a couple of dozen outbreaks since including the 2014–2015 West African epidemic where more than 27,000 cases resulted in over 14,000 deaths. In late 2014, at the same time as the West African Ebola epidemic, there was an unrelated Ebola outbreak in the DRC where there were only 66 cases, which did not spread beyond several villages in that country. Why did the disease spread in West Africa and not in the DRC? In other words, why did the disease diffuse through the population in one place and not another? The reason is complex but at its core involves differential interaction between people, which results in differences in transmission. West Africa is very different than the DRC. It is densely populated with many roads connecting populations to one another compared to the DRC where there are very few roads connecting places to one another. There have been at least eight outbreaks in the DRC since 1976 and they have all quickly burned out.

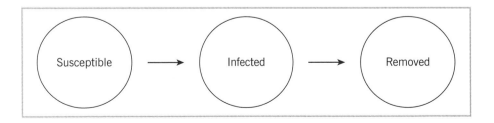

FIGURE 6.2 SIR disease transmission model.

The characterization of infectious disease as either endemic or epidemic is often inadequate because a disease can be both. A disease heretofore nonexistent in an area may suddenly appear, often virulently and in the absence of cultural buffers. More commonly, an endemic disease develops in epidemic proportions, perhaps seasonally. How or even whether it develops depends on several well-known factors. The first of these are the qualities of the microorganism that influence its relationship with a human host. These qualities include the number of organisms needed to initiate symptoms and produce clinical disease; for most agents, subclinical infections also produce immunity. Second are the number of organisms shed by a carrier of the infection, the period of shedding, the method of shedding into the environment, and the survival time of infectious organisms in the particular environment outside a human host. Third is the probability of contact with a susceptible host during the period of infectiousness—a factor influenced by the agent's **virulence**, as well as by the numbers and distribution of susceptible people. A fourth factor is the population's resistance, whether from prior infection, vaccination, or genetic predisposition.

Spread of contagion requires contact in space. Distance and density are important parameters. The likelihood of an infection's spreading from one host to another depends on such things as whether the potential host is susceptible to the infection (age immunity, vaccination, nutritional status, etc.); whether the distance between them (survival of the agent in the environment) is bridgeable; and how likely one or more contacts are (density of population with the opportunities for other secondary infections). Likelihood of contact can be related to social networking as well as "herding" (into schools, prisons, nursing homes, etc.). The cultural process of herding, putting many animals into close proximity, helped create the intensity of transmission that led infectious agents to cross species and infect humans back at the beginning of domestication, densely settled agricultural villages, and the first cities. **Herd immunity** is the proportion of immunity in a specific population needed to prevent the infectious contact from finding a susceptible host, thereby ending the epidemic. "Epidemic" originally meant a disease's spreading in a population, but in recent years has been used (so as to refer also to nontransmissible diseases) to mean a disease's occurring in excess in a population. **Pandemic** means that a disease is spreading among populations across national borders. The World Health Organization (WHO) today technically uses the term *pandemic* to mean sustained spreading within communities in two or more countries (stage 5 in the WHO's rating system) or in two or more WHO regions of the world (stage 6, globally pandemic).

Knowledge of these concepts and parameters enables researchers to model and predict future developments, as well as possible intervention strategies when an epidemic starts. If *this* many people are infected at a certain point in time, the argument goes, and this is the incubation period, the secondary attack rate, and the proportion of the population susceptible, then at the next point in time *that* many people will be infected. Such calculations can be complicated. The proportion of susceptible persons may vary with such characteristics as age, vaccination history (sometimes substituted for by income and education levels), or prevalent health status, for example. Nevertheless, epidemics in a population are often mathematically reduced to a set of differential equations. Mathematical epidemiologists with such knowledge can predict the need for hospital beds, the economic impact of work absenteeism, and such matters for the whole population. Researchers can model the comparative advantage of targeting various population subgroups for a vaccination drive, for example. Far too often in such efforts, however, the question of where is left out.

Much has been learned about the movement of a new infection through a population. An epidemic typically follows an S-shaped curve, or logistic curve (Figure 6.3). At first only a small proportion of the population is infected ("early innovators," in geographic parlance). The numbers of infected people increase exponentially at first, but then growth slows down as the proportion of susceptible individuals decreases and such individuals become increasingly difficult for the agent to contact. The epidemic finally wanes, leaving a small proportion of the population uncontacted. If the population is not of a critical size, the disease then becomes extinct. This explains why Ebola outbreaks have always burned out in isolated villages in the DRC. If a population is large enough to continually produce enough new susceptibles (often children), the disease may become endemic.

Whole populations do not, of course, occupy a point in space. Diseases diffuse over space because population in all its diversity is distributed over space. All such parameters as vaccination coverage, age, income, education, ethnicity, density, and

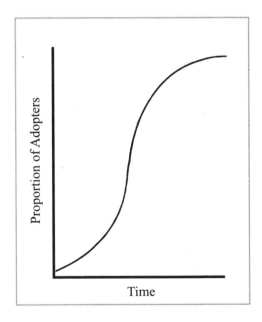

FIGURE 6.3 Logistic curve. Many diffusion studies have identified the characteristic S-shaped curve of volume of adoption through time. From Gould (1969, p. 20). Copyright ©1969 the Association of American Geographers. Adapted by permission.

settlement size vary over space. Day care centers, schools, and medical facilities are located in space. Contact is made in space and must overcome distance to occur. The remainder of this chapter is about geographic components of disease diffusion.

TYPES OF DIFFUSION

The geographic process of diffusion goes on constantly and affects many aspects of our lives. Information about someone's death or illness is usually spread from mouth to mouth, by friend calling friend or neighbor telling neighbor. If a new product becomes available (e.g., a new contraceptive or medical treatment), at first the supply will be limited, and people will need to travel to a large urban area or specialized medical center because the innovation will not be available locally. Perhaps a foreign student has finished medical study in a U.S. university and is returning home to practice and teach the new techniques he or she has learned. These examples are medically related, but they could as well be about political rumor, a new sports franchise, or the ongoing restoration of the American bald eagle to regions in which it was recently extinct. They are examples of relocation, expansion, and hierarchical diffusion.

Relocation diffusion involves the introduction of an innovation to a location that is not part of the network or system of interactions from which it comes. It often involves leaps over great distances and intervening populations. People with knowledge may migrate to a new place. The movement of plague bacilli from burrowing rodents in central Asia to rodents in European cities to burrowing rodents in American, Argentinian, and South African grasslands constituted a whole series of relocation diffusions. When hunters who had depleted the local raccoon population captured raccoons from a Florida population in which rabies circulated and introduced them into the woods of West Virginia, they set off an epizootic of raccoon rabies that has diffused across a region currently extending from Massachusetts to North Carolina, where raccoon rabies is now enzootic.

The terms **expansion diffusion**, contact, and **contagious diffusion** are often lumped together or used interchangeably. Adoption of innovations, which takes place through contact between people, is strongly affected by distance. Geographers speak of the friction of distance and measure distance decay. These terms simply mean that the farther apart people are, the less likely they are to have contact or interact. The strong effect of distance causes the process of contact and expansion diffusion to have a distinctive pattern, which is sometimes likened to an ink spot spreading on paper or to ripples moving out from a stone thrown into water. The major difference among these types of spatial diffusion is that ideas can travel over the Internet, telephones, electric lines, and so forth with almost no friction of distance, whereas a contagious microbe must be transmitted from one individual to another within some physical proximity. Ideas are frequently disseminated to millions of people at a time via television or the Internet. The movement of real people is required for contagion. Although most geographic literature is concerned with expansion diffusion, it is not the only kind of contact diffusion.

The final type of diffusion, **hierarchical diffusion**, is characterized by disease agents' moving by contact among people through urban areas according to the areas' relative size. Some of the most important theoretical constructions in geography have addressed the regular spatial arrangement of settlements (villages, towns, cities) of different sizes, and have described how population size relates to economic functions and the distance decay (and hence market regionalization) of such activities as retailing. The settlements of different size and function are referred to as "central places." These exist in a framework of hierarchical relationships among which goods, ideas, money, and especially people move up and down.

Some goods and services, like getting gas, buying bread, or seeing a dentist, are needed by everyone frequently. These are known as low-order goods and services and need to be located as close to people as possible, or people will go to an intervening opportunity. Low-order goods are widely distributed. But every location with a gasoline pump can't support a skilled mechanic or a supermarket. People have to travel to a central place, a town. Every central place with a shopping center and a garage with a skilled mechanic can't support a piano store. A small proportion of the population buys pianos, and usually only once or twice in a lifetime. The threshold of population for specialized goods and services, which are high-order goods, is high. For example, the number of people it takes to keep a laser eye surgery facility is large, and so they must be drawn from a wider market area. Small central places, through the people living there, provide low-order goods and services frequently for local people. The people of many small towns travel to a larger nearby city to get more variety, greater choice, and specialized goods and services. These cities in turn look to higher-order central places for certain more specialized services, such as financing or information media. In health services, for instance, general practitioners and pediatricians should be widely dispersed, and every county hospital should be able to tend broken legs and heart attacks. But to get a skilled brain surgeon or radiation and chemotherapy for cancer, one has to go to a higher-order center. A burn ward dealing with a very specialized service of high training and cost (one that is fortunately needed by few people very frequently) is a very high order service that usually serves a large area. The area necessary to provide the population needed to support the highest-order goods and services is very large, sometimes regional.

Gerald Rushton at the University of Iowa got the idea of modeling the "market" area over which people traveled to their health services like a watershed for rainfall—a market-shed. People do not necessarily go to the dentist or general practitioner in a small town in their own county, but instead go to the higher-order central place, which offers more variety and skill (e.g., an orthodontist and a pediatrician). When a disease agent diffuses, it moves through such interaction of people up the urban hierarchy to the largest city, and then is said to cascade down the hierarchy to other cities. In Figure 6.4, where was the agent introduced?

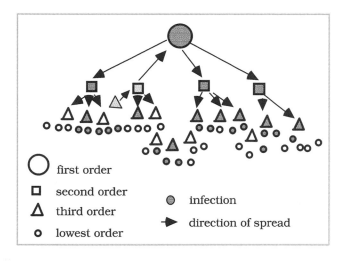

FIGURE 6.4 Infectious disease cascading down the urban hierarchy.

NETWORKS AND BARRIERS

Diffusion occurs over a web, or network, of places, people, and the communication links through which information, people, and goods flow. The diffusion of a given phenomenon may be studied at several different time periods in order to gain knowledge about changes in network structure. Alternatively, research on the simultaneous diffusion of the same or similar phenomena in different networks can illuminate the impact of types of network structures.

In opposition to networks, which pattern and support diffusion, barriers slow and shape the process (see Chapter 7 for discussion of barriers to genetic change of pathogens in the sections on landscape genetics). Barriers have three basic effects. *Absorbing barriers* simply stop an innovation. Populations vaccinated against smallpox were an example of an absorbing barrier. Unable to contact susceptible persons, isolated cases had no chance to spread, and smallpox disappeared. Sparse populations in arid lands can behave like absorbing barriers to disease diffusion when the low population density does not facilitate contact transmission. *Reflecting barriers* channel and intensify the local impact of a diffusion process while blocking its spread to another locale. For example, the presence of lakes or unbridged rivers may cause rabid raccoons to turn back and limit an epizootic to a territory, or the presence of a hostile gang may cause people to shop at a certain place or travel in one direction instead of another. Finally, the barrier may be *permeable,* allowing some diffusion but slowing the process. Most international borders are permeable barriers.

Barriers are often physical: oceans, deserts, mountains, rivers, and so forth. Broad barriers, such as oceans or deserts, were effective barriers to contact diffusion during the period of slow transportation. The disease might run its life course during the journey and never reach the destination. Aboard a ship, the susceptible population might be too small to maintain the contagion through the course of a trip. Rapid transport today has almost eliminated the effectiveness of such barriers.

Cultural barriers are especially important in the diffusion of styles of health care. The Chinese carried their medical system with them as they migrated to Southeast Asia in the early 20th century (relocation diffusion). Comparatively few people indigenous to that area have chosen to adopt the Chinese system, however, perhaps indicating the cultural separation of the population groups. It has spread more, for example, among the Thai, who are Therevada Buddhists, than among Malays, who are Muslims. Programs such as child vaccination and family planning efforts diffuse in a similar way, with varying types of constraints.

QUICK REVIEW 1

✓ Mathematical epidemiologists build SIR disease transmission models that partition populations into three groups, susceptible (S), infected (I), and removed (R). The relative proportions of S, I, and R individuals in different places can influence the spread of disease over space.

✓ In SIR models, the R_0 is number of cases one case generates on average over the course of its infectious period, in the uninfected population, and this varies by disease. When the R_0 is less than 1 the infection will burn itself out and when it is greater than 1 it will spread. Geographic context affects R_0, which helps explain why Ebola epidemics in remote areas of the Democratic Republic of Congo were small and short while in more densely populated West Africa there was a major epidemic.

✓ Diseases diffuse through space in several different forms. Relocation diffusion is the introduction of a disease to a new location that is not part of the network or system of interactions from which it comes. Expansion diffusion occurs when a disease spreads outward from an area. Hierarchical diffusion is characterized by diseases spreading by contact among people through urban areas according to the areas' relative size.

✓ Vaccines reduce the proportion of the population that is susceptible to a disease, thus reducing diffusion to an area through direct protection or herd immunity. Herd immunity is the proportion of immunity in a specific population needed to prevent the infectious contact from finding a susceptible host.

MODELING DISEASE DIFFUSION

The diffusion of infectious disease moves both through a given population and over a particular space. Most studies and efforts to model and predict the course of diffusion start with a time series. An epidemic starts when the incidence of cases exceeds expectation, whether that expectation is no disease at all or an endemic clutter of occasional cases. Many diseases follow a cyclical pattern in which a virus or other agent contacts a cohort of susceptible children, born since the last epidemic, and spreads rapidly within the group. As the proportion of immune members of the group increases, it becomes more difficult for a disease agent to contact a susceptible person, and the rate of spread slows down. The periodicity of measles epidemics, for example, especially in less populated areas, was regulated by the fairly consistent number of children born each year and the substantial minimum susceptible population needed to maintain the infection (Cliff & Haggett, 1983; Cliff, Haggett, & Smallman-Raynor, 1993). Immunization has greatly reduced the number of measles cases today, but any relaxation in the completeness of vaccination could reestablish a susceptible population. Such was the case in the early 1980s when numbers of college students began coming down with measles. The relatively new vaccine they had received as small children was found to confer less than permanent immunity. The college epidemics were stopped by requiring all students to get booster shots immediately. In developed countries, the most pressing concern today is over parents choosing to leave their children unvaccinated for diseases such as measles, mumps, and rubella, increasing the number of susceptible individuals. Given that parents choosing to exempt their children from childhood vaccines tend to cluster spatially, this creates pockets of populations with high densities of susceptible individuals (Carrel & Bitterman, 2015). Figure 6.5 shows the clustering of personal belief exemptions to vaccination in schools in Southern California.

The remainder of this chapter discusses the ways by which the diffusion of contagious disease over space and through time can be visualized and modeled. For the most thorough and graphic explanation and presentation of the subject, readers should consult Cliff and Haggett (1988) or, more mathematically, Thomas (1992). Geographic disease diffusion research is often based on models developed to study other types of innovations and applied outside the area of health concerns. These general models have been extended and expanded to fit the special circumstances of disease diffusion. The studies by Brownlea (1972) on hepatitis; Haggett and others on measles (Cliff et al., 1993; Cliff & Haggett, 1982, 1983, 1988; Cliff, Haggett, & Ord, 1986; Cliff, Haggett, Ord, & Versey, 1981; Haggett, 1976); Stock (1976) on cholera; Pyle on cholera (1969) and influenza (1984, 1986); and Gould (1993) on AIDS are used in the following section to look at the diffusion of disease as a geographic subject. The historic application of these methods of analysis is addressed with poliomyelitis (Trevelyan, Smallman-Raynor, & Cliff, 2005a, 2005b). The more recent method pioneered by Ling Bian of using individual-based models developed out of Hagerstrand's lifelines and stochastic processes is also briefly discussed (Bian, 2003, 2007; Bian & Liebner, 2005). A subsequent section presents the diffusion modeling with some ecology for various influenzas.

Most of the epidemiological research in disease diffusion has involved microscale field investigations, which rely on relatively small numbers. The range of scales used by geographers in diffusion studies offers a different view. A long-range advantage to broad-scale models of disease diffusion is their possible importance for forecasting, even for a specific place. If we can identify those variables (centers, barriers, population distributions, spatial processes) that consistently influence the diffusion of a particular disease, or if the geographic diffusion pattern of a place is repetitive and thus predictable, health officials could focus vaccination programs and health education on blocking or minimizing outbreaks. In this section we look at models of hierarchy, distance decay, stochastic diffusion, gravity models

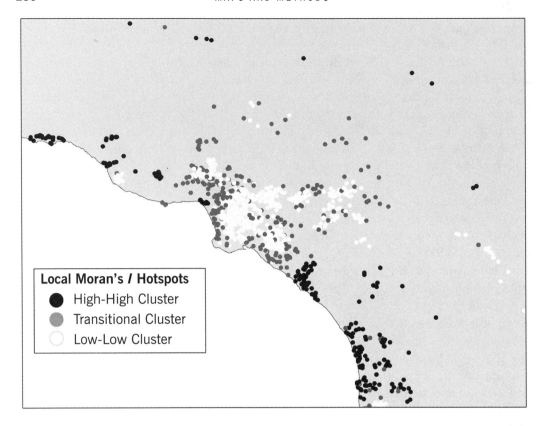

FIGURE 6.5 School locations in southern California (Los Angeles and San Diego metro areas) indicating personal belief exemption (PBE) cluster membership. High-high cluster schools are those where a large number of children in kindergarten remain unvaccinated by choice with other high exemption rate schools as neighbors. Low-low cluster schools have low PBE rates and are surrounded by low PBE schools.

of population potential and spatial transformation, and spatial autocorrelation. We also examine how these approaches have been employed to better understand measles, poliomyelitis, and influenza.

Diffusion over Distance and through Hierarchy

The two major diffusion processes—contact diffusion as expansion diffusion, which involves traversing distance over time, and hierarchical diffusion, which involves the channeling of contact through the population interactions across the urban hierarchy—have been discussed previously. Figure 6.6 illustrates how they are expressed. Try to identify four types of diffusion in this figure before reading further.

Contact diffusion is best expressed as taking place when the disease agent arrives later at places farther away from the source. Traversing distance dominates the movement of the disease along the coast and up the lower river in Figure 6.6.

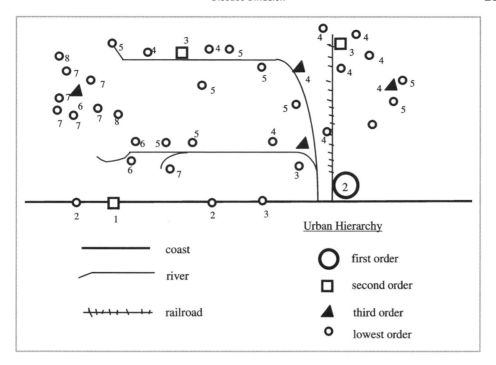

FIGURE 6.6 Distance decay and urban hierarchy. Numbers indicate time periods of a disease's arrival.

The movement along the upper river becomes more complex at the appearance of hierarchical diffusion. The infection moves up the urban hierarchy from the second-order coastal city to the first-order city quickly. All the large cities of the second functional order get the infection immediately after the coastal *primate* (first-order) city does. As the disease cascades down the hierarchy, the size of the urban place becomes more important than the distance to be traversed. Of course, contact diffusion begins to occur locally from each of the infected urban places too. Radial diffusion from a settlement to the nearest settlements is obvious, especially among the cluster of villages in the west. They get the disease so much later than the other places, despite the third-order urban place and despite the proximity to the initial settlement infected, that a strong barrier effect is clearly present. It is easy to visualize rugged interfluvial mountains and perhaps a hill tribe ethnic group. Relocation diffusion, of course, occurs when a ship arrives at the coastal port and introduces the first infection to the coastal settlement system. This composite diagram of diffusion types was inspired by the patterns of cholera diffusion in West Africa that were actually mapped, diagrammed, and analyzed by Stock (1976), whose work is discussed soon. The relationships of distance and hierarchy can be represented on a graph.

Figure 6.6 portrays distance from the coastal primate city on one axis and the period of the disease's arrival on the other. The symbols used to represent places indicate their status in the urban hierarchy. It is clear that hierarchical diffusion has become most important as size of place dominates distance in time of disease arrival. The minor (mathematical) vector, however, suggests the presence of an unintegrated system within the country—perhaps similar to the lower river valley of Figure 6.6—in which distance is still the dominant factor for diffusion.

This perspective on disease diffusion was first developed through the studies of cholera diffusion by Pyle (1969), whose map of cholera spreading in 1832 and 1866 appear in Figures 6.7a and 6.7b. These two conceptually important studies, by Stock and Pyle, are described following a short history.

Cholera is an acute intestinal bacterial disease that is internationally report-able to the WHO (see Chapter 2 for more on the disease ecology of cholera). Its symptoms of severe diarrhea and vomiting result in mortality from severe dehydra-tion and the consequent shock and cardiovascular collapse. Historically, mortality (for untreated cases) can be more than 50%. The bacterial agent, *Vibrio cholerae*, has two biotypes including classical and El Tor. For millennia the classical strain of *V. cholera* has broken out epidemically throughout South Asia. By the early 19th century international trade networks and population mobility had developed to spread cholera to southeast and southwest Asia, where the Hajj (pilgrimage) to Mecca became a center of diffusion. Six 19th-century pandemics of cholera spread from the Ganges to the great ports of India, and then through Suez and on to Europe and North America. Each pandemic was cleared elsewhere and collapsed back to only the Ganges.

Cholera epidemics in the early 20th century seemed effectively limited to Asia. Then a World War II–era epidemic introduced cholera to the eastern islands of the Indonesian archipelago, where it stayed and evolved into the El Tor biotype. El Tor cholera has a lower case mortality (i.e., it is evolving toward commensalism). Infections may involve severe disease or only diarrhea, or even no symptoms at all. The new strain of cholera spread after 1961 to the southern states of Europe and the (former) Soviet Union, entered the United States and Japan (where sanitation contained it), and apparently for the first time spread to West Africa (where it has stayed). In the 1990s the agent was brought from Southeast Asia to Peru, and it then diffused through South America. In 2010 following a larger earthquake and a damaged public health infrastructure, cholera spread to Haiti. It was thought to have resulted from the relocation of Nepalese UN peacekeeping soldiers who were infected with the disease. The diffusion of El Tor to a susceptible world population set the stage for studies of disease diffusion. Smallman-Raynor, Cliff, and Barford (2015) meticulously describe the diffusion of cholera in Haiti from 2010 to 2013, showing diffusion down river after the initial introduction and then from the river basin to other parts of the country. The spread to other parts of a country followed a "gravitational" pattern to other regions proportional to their population size and distance.

Pyle was the first to realize that the diffusion of a disease such as cholera could be used to analyze the spatial structure of a country. That is, the pattern of the diffu-sion could illustrate the type of transportation system, the development of national economic integration, and the evolution of a functioning urban hierarchy of com-mercial and mobility interactions. Even before the development of germ theory and vital statistics, the arrival and presence of cholera in a place were noted in the

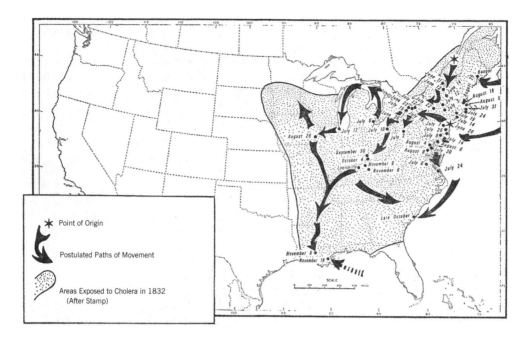

FIGURE 6.7a. Diffusion of the 1832 cholera epidemic in the United States. From Pyle (1969, p. 63). Copyright ©1969 Ohio State University Press. Reprinted by permission.

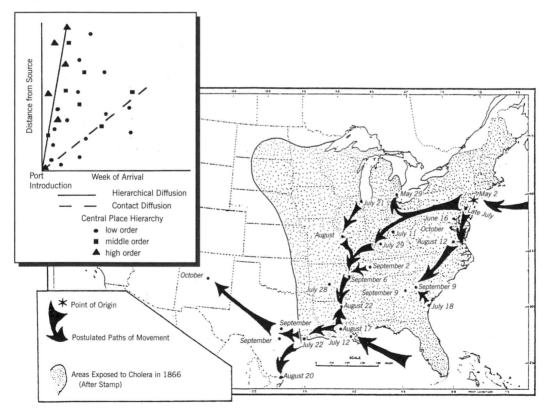

FIGURE 6.7b. Diffusion of the 1866 cholera epidemic in the United States. From Pyle (1969, p. 66). Copyright ©1969 Ohio State University Press. Reprinted by permission.

local newspapers. Pyle studied the diffusion of three major cholera epidemics in the United States by developing the methodology depicted in Figure 6.7. In the first epidemic, in 1832, cholera was introduced (relocated) at several points: the riverine ports of Canada, as well as New York City and New Orleans (Figure 6.7a). Diffusion occurred along the Atlantic coast and the riverine systems. Clearly the farther along the coast or a river a place was, the later the arrival of cholera, in a system dominated by water transport. The second epidemic, in 1849, occurred when the hierarchical structure of the urban system was emerging. Railroads were becoming important, although much of the interior was not yet connected to the East by rail. The epidemic was introduced from Europe in both New York City and New Orleans. From these cities it diffused up the rivers and along the coast in two separated regions; within each region, however, hierarchy (i.e., size of urban place) had emerged as the major predictor of how soon the epidemic would arrive. By the time of the third epidemic, in 1866, the rail system was well established but did not cross the continent (Figure 6.7b). Cholera was again introduced in New York City, but the epidemic now moved through the country as one system of urban hierarchy determining time of infection. Even the exceptions (residuals) to this served to emphasize the structure: Detroit was infected early for a city of its size, as explained by its proximity to nearby Canadian cities; San Francisco was infected more than a year late for a city of its size, but of course the transcontinental railroad was not yet built and pioneers still walked or sailed there; and Washington, DC, which then was almost deserted during summer, did not receive the epidemic until Congress convened in the fall.

The arrival of El Tor cholera in West Africa (1970–1975) in a "virgin" or totally susceptible population provided Stock with the chance to develop one of the most comprehensive and conceptually clear studies of diffusion. He examined the applicability of existing diffusion models. First he needed extensive fieldwork to identify, locate, and map the local timing of the outbreak of cholera throughout the region. He identified behavior at the village level that buffered or increased risk. This ecology was modeled as a local system. It involved such things as caring for the sick, preparing the dead for burial, washing hands, eating with hands out of common plates, use of latrines, and presence of local medical service. Then, addressing regional scales, he identified and developed four models (which are generically represented in Figure 6.6) to describe the overall diffusion pattern.

There were three models of contact diffusion: coastal, riverine, and radial. Fishermen carried the disease along the coast, and it spread from coastal villages to nearby towns. Diffusion along the Niger River was also dominated by the friction of distance. Larger towns received the infection no earlier than smaller towns closer to the origin. The radial pattern of contagious diffusion was identified around Lake Chad. Here the disease intensity was particularly virulent, and as people fled the lake area, there were no developed transportation routes or natural channels to structure the diffusion paths. People from a village fled to the nearest ones around, from which people fled outward to the next ones. Only in Nigeria did a clear example of hierarchical diffusion emerge. Coastal diffusion introduced cholera to the primate city of Lagos, from which river, bus, and rail routes emanated. A settlement's level in the urban hierarchy, rather than its distance from Lagos, determined when the epidemic arrived. Overland transportation routes channeled diffusion. The least affected areas of the country had low population density, little urbanization, and undeveloped transportation.

Stock's (1976) monograph emphasizes the complexity of the geographic structure of diffusion. All the basic models had some applicability. Immune populations, deserts, and sparsely populated areas were permeable barriers, but rivers, coast, and developed overland routes helped channel movement. Stock suggested that the linear pattern along the Niger River loosely resembled the 1832 cholera diffusion along the internal waterways of the United States discussed earlier, whereas the hierarchical diffusion in Nigeria more closely resembled the 1866 U.S. epidemic's structure. The difference between Nigeria and Mali in economic development in the early 1970s thus seems similar in spatial expression to the past structural development of the United States.

Local Diffusion by Chance: Stochastic Simulation

There have been few attempts to apply to disease diffusion the simulation methods formulated by Hagerstrand, which have otherwise been very useful in modeling innovation diffusion. This may be because the influences on the multiple waves of an epidemic are very complex. The aging but still the best example of the application of stochastic diffusion simulation to disease is a study of infectious hepatitis in Wollongong, Australia (Brownlea, 1972).

The infectious hepatitis virus is transmitted through fecal contamination. The virus is robust and tolerant of a wide range of environmental conditions. Water, fish, seafood such as oysters, and pet hair may be vehicles for disease transmission. Brownlea identified cyclical fluctuations in the number of cases reported. During epidemic years, incidence rates of 2 per 1,000 or more affected mainly children and showed an equal sex ratio, strong spring seasonality, and no spatial concentration. During interepidemic years, rates were rather uniformly 0.3 per 1,000, affected all ages, and showed little seasonality and no spatial concentration. Because hepatitis is often poorly reported, Brownlea aggregated cases over 15 years and used a Poisson probability test to determine just where active spread was present. He identified the center of this as a wave that moved spatially through time, and labeled it a "clinical front."

Brownlea developed several models for hepatitis behavior at different scales within this industrial–suburban, rapidly growing region. The basic model was based on Hagerstrandian-type stochastic simulation of a random walk. Assuming a closed population, an equal chance of diffusion in all directions, and essential community immunity as the epidemic passed, he simulated a random diffusion in which the clinical front would advance as a ring from Wollongong's initial node. Time periods for measurement were determined by incubation period. The bulges and bends in the actual, observed advance of the clinical front—that is, the deviation from the random—he attributed to ecological parameters he had identified, which operated as barriers or channels (Figure 6.8). Modifying the simulation to fit the actual disease behavior by adjusting the cell probabilities showed that concentration of

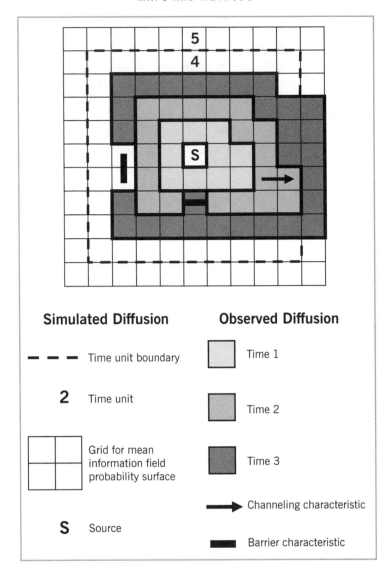

FIGURE 6.8 Simulation of the diffusion of infectious hepatitis.

young families, light sandy soils, concentration of older people, and a polluted lake used for fishing and swimming were important parts of the physical and sociodemographic surfaces that retarded or channeled epidemic diffusion. Recalibration of cell probabilities led to a close match between the simulated and real diffusion patterns.

These findings were applied to the movement of the clinical front among the settlement nodes in the study area. Brownlea found that the disease diffused from each population center, exhausting the susceptible population at the core and rippling out to the surrounding nodes. If a peripheral node developed a sufficient concentration of immigrants and young children, it became a new center of infection and node of diffusion. The first, exhausted node was occasionally reinfected from the surrounding, newly active nodes, which caused the in-filling epidemiological pattern of the interepidemic years. The model laid a basis for explaining the disease's rhythmicity and spatial patterning, and it provides a powerful example of using Hagerstrand's stochastic simulation for analysis of disease diffusion.

Approaches from the Family of Spatial Interaction Models

One of the most robust nomothetic models in geography is the **gravity model**, the prototype for geographic modeling of spatial interaction. It has been widely used to account for, and predict, an amazingly wide variety of flows: job migration, telephone traffic, airline passenger movements, mail delivery, commodity shipping, and the spread of information, among many others. Because it is about interaction, contact, flow, connection, and the spatial patterns that result, it is also about the diffusion of disease.

The basic concept is elegantly simple. The greater the number of people who live in two places, the greater the likelihood of contact between them; the farther apart the two places are, the less the likelihood of contact between them. We would expect more people to move, write, or telephone each other between Chicago and Philadelphia than between Chicago and farther Houston. We would expect more of such contact between New York City and Washington, DC, than between New York City and smaller Buffalo, New York. In an important way, even the hierarchical diffusion discussed earlier is "gravity model thinking," as the likelihood of interactions between the populations of the largest cities overwhelms the friction of the distance between them. If, furthermore, instead of the attraction between only pairs of places, we consider the influence of all places in a study region upon each other's contacts, we create a surface of potential interaction among people called "population potential."

The gravity model was derived from efforts in the late 19th century to analyze society "scientifically" and create a "social physics." The inverse relationship between the attraction of population size (i.e., mass) and the discouragement of distance seemed obviously analogous to Newton's gravity model. Some of the terminology used today, such as potential, is derived from Newtonian physics. Obviously, however, the surface of the Earth and the spatial arrangement of its settlements do not exist in a void, or on a perfect plane. Such things as direction, road routes, and mountain or sea barriers really matter. Different "frictions" of distance encountered by different modes of movement (whether by foot, car, plane, etc.) could be empirically adjusted in the model by varying the exponent on the distance (instead of dividing by distance squared, à la Newton) until the formula worked. After this initial model "calibration," more elaborate ways were developed, using constants to "constrain" the model so that the total flow predicted from an origin to a destination equaled the actual total flow observed. The computation of gravity model predictions of migration, for example, has become very complicated. The large literature on the gravity model and other models of spatial interaction can be conceptually approached for a general explanation (Alber, Adams, & Gould, 1971) or more advanced and mathematically sophisticated explanations (Cliff & Haggett, 1988, 1989; Thomas, 1992).

The family of spatial interaction models and their derivatives can provide a way to put real space into diffusion within a population. For example, the former Soviet Union used to collect remarkable incidence data (daily reports from more than 100 cities) and had developed detailed mathematical models of the diffusion of influenza within its population. These formulas ignored distance between the cities in that vast country, however. The models were adapted by British epidemiologists and others to the kinds of data available in the West, but they remained aspatial. When Pyle (1986) used the gravity model and U.S. data to put distance (between cities) into their predictions, he showed that distance as well as size (mass) was important by accounting for 70% to more than 80% of the variance among U.S. cities. He further added to his prediction by using population potential of the elderly population in his regression analysis.

The following two examples—of Gould (1993) using the gravity model as a "spatial filter" to transform the diffusion space of Ohio, and of Hunter and Young (1971) using a surface of population to explain the basic movement of an influenza epidemic through England and Wales—should clarify the gravity model and provide some sense of the power of these techniques in health geography.

Diffusion through Transformed Space

Cartographers have long transformed space for analytical purposes. For example, distances on roads may be represented by travel time instead of actual miles, so that areas get foreshortened or extended. Cartograms, such as demographic base maps, represent the content of a region as space. Indeed, it is impossible to represent the surface of our sphere on a flat sheet of paper without transforming space (Chapter 5). Gould (1993) used the gravity model as a "spatial filter" to transform the space of Ohio so that accessibility, in this case to the contagion of HIV/AIDS, could be illustrated and used predictively. Gould mapped the diffusion of HIV/AIDS across Ohio's 84 counties and in the three large cities of Cleveland, Columbus, and Cincinnati from 1982 on. He described an educational value in the dramatic impact of maps showing HIV/AIDS diffusing—and arriving. Instead of dismissing HIV/AIDS as spreading among some "other kind of people" or being someone else's problem, the reaction of Ohioans viewing the maps in sequence (and of course locating their own place first) was often "It's coming; . . . it's here!"

Figure 6.9 shows the 84 counties and three large cities in Ohio upon which Gould based his analysis. The three largest cities have a lot of interaction among their populations, of course, as well as contact with large cities outside the state. The gravity model was used to calculate all 3,486 potential interactions among the counties, which were then mapped into multidimensional "AIDS space." The rural counties isolated from other contacts, or places connected by proximity and population size, then showed the spatial logic of the mapped pattern of diffusion. Gould (1993) included a time series of maps from 1982 where the disease only existed in Cleveland to 1984 where it was in the three large cities and then in 1987 and 1990 filled in the rest of the state through a contagious diffusion process.

FIGURE 6.9 Diffusion of AIDS in Ohio and in space transformed by the gravity model. The gravity model can be used as a "spatial filter" to transform the space over which infectious agents spread.

Figure 6.10 depicts another population potential surface for England and Wales. After examining the uses of absenteeism at school and work, and various other estimates of infamously unreported influenza incidence, Hunter and Young (1971) used the incidence of acute pneumonia (primary and influenzal, and well reported) to map the Asian flu epidemic of 1961 week by week on areal and demographic base maps. They located the centroid (the mean center) of cases for each week. They calculated the population potential by county and mapped it. As the figure shows, the center of the epidemic advanced week by week along a ridge of population potential until it reached the "potential mountain" of London. Population potential and week of onset were highly correlated, but not total incidence.

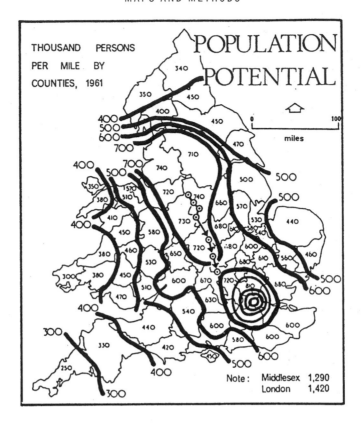

FIGURE 6.10 Population potential and diffusion of influenza in England and Wales in 1957, showing the movement of the epidemic center along the ridge of highest population potential. From Hunter and Young (1971). Copyright © 1971 the Association of American Geographers. Reprinted by permission.

Testing Models of Diffusion with Spatial Autocorrelation

The methodology of analyzing changes in relationship (or similarity) over space by correlating the values of geographic units, such as counties, at sequentially increasing distance for sequentially increasing time can provide a powerful and sensitive analysis of disease diffusion. This method of analysis, spatial autocorrelation, is described and developed in Chapter 5 (see especially Figure 5.16). In this section we look at how, in the hands of masters such as Andrew Cliff and Peter Haggett, it can illuminate the disease process and its human geography. The illustrations and models that follow are all from their work (Cliff & Haggett, 1982, 1983, 1988; Cliff et al., 1981, 1986, 1993; Haggett, 1976).

Measles

Measles is endemic throughout most of the world today, although from 2000 to 2013 the number of cases worldwide dropped by 75% because of vaccination programs. However, it is still one of the leading causes of death among children worldwide with 145,700 deaths in 2013. The virus is highly infectious, and nearly all susceptible individuals contract measles after close contact with an infectious person. Epidemics flare when a sufficiently large population of susceptible children are born in a population that is otherwise more than three-fourths immune as a consequence of previous infection (i.e., at the saturated end of the logistic curve). In small island populations the disease agent disappears (ceases to exist) when the virus that has killed or created immunity can no longer find a susceptible person to infect. There is then no measles in that place until population mobility introduces it again from outside. Numerous mathematical models and island studies have determined that a minimum population size of 250,000–500,000 frequently interacting people is needed to generate the new susceptibles and contact rate required to maintain the disease. The difference in these population estimates is undoubtedly related to the relative isolation or intensity of interaction of the people. The need for such a large size of interacting population is strong evidence that measles could not have existed as a human disease until the first cities were built in the trading nexus of today's Middle East.

Haggett (1976) first studied measles diffusion in the relatively isolated county of Cornwall in southwestern England. He posed seven possible models from what was understood at the time about the process of spread. Which was important: spreading in separate regional systems; spreading from cities to surrounding rural areas; contact diffusion spreading in waves across distance from the largest city; local contagion from one geographic unit to another; contact by people in their "journey to work"; or simply the population size of the places or the population density of the unit areas? He had a two-wave measles epidemic to study—a 222-week series from late 1966 to the end of 1970, across 28 political jurisdictions. The genius of this study was to redefine nearness to count joins in terms of his postulated models, not just spatial contiguity. Each unit could be correlated with its neighbors, but so could commuting origins and destinations, cities of the same class size, units of the same density, and so on. Haggett discovered that different spatial processes of diffusion were more important during different stages in the epidemic, as well as during endemic or epidemic periods. During endemic periods, the population size model showed the persistent cases of disease in an otherwise weakly structured low level of contagion. Early in the epidemic, population size became less important, and waves of contagion effects (shortest distance) and journey to work took over. During the epidemic peak, local contagion was most important. As the epidemic waned, it did not shrink in geographic extent but decayed in place, with some tendency for lower-level hierarchical spread to continue. These observations could not be generalized, of course, beyond this one epidemic in this singular place and time. A more ambitious study was needed, on a greater scale.

Cliff and Haggett (1983, 1986, 1988, 1989) found in Iceland a large island population system that could be "closed" for analysis as a region and that had the great advantage of meticulous records kept over more than a century. They followed the introduction, spread, and disappearance of multiple epidemics of both measles and influenza. Over time the population grew large enough to maintain endemic disease, and the infrastructure of transportation, medical care, schooling, and general circulation of information and people was also transformed. They were able to study how intervals, peaking, duration/frequency, age structure, and other dynamics were affected. Spatial autocorrelation methods were at the core of the most powerful of these analyses.

As shown in Figure 6.11, they used the method developed in the Cornwall study. They defined joins in various ways to test hypotheses of process. The geographic patterns of diffusion of contagious measles and influenza could be expected to reflect population movements for work, schooling, shopping, holidays, and such social activities as visiting relatives. An index case, or infected person, could theoretically travel to any one of the 50 medical districts (used for epidemiological data). The center of each district could be considered a node, and nodes were linked to each other by links of minimum distance (represented by the minimum spanning tree, or MST). When joins were defined not only in terms of minimum distance, but also in terms of airports, road connections, and nearest larger place with or without contiguity (i.e., urban hierarchy), the progression, duration, and concentration in time of successive epidemics of measles and influenza could be examined over decades of change. On each of the histograms in Figure 6.11, the vertical axis plots the number of occasions in 17 influenza and eight measles epidemics that were statistically significant at the 95% level in a one-tailed test for positive spatial autocorrelation. Time in months is on the horizontal axis. For "nearest larger place," the bar graph shows nearest larger contiguous place, and the line graph shows hierarchical size regardless of contiguity (i.e., pure population). The MST and the road network are said to be "spatially constrained"; that is, they represent contagious diffusion in which contiguity and distance are important. The air network and nearest larger place represent hierarchy, nodes, and population connections as more important than distance.

The graphs show the importance of contagious spread for both diseases. In the case of influenza (the b graphs), for example, the MST was important from 1 month before to 2–4 months afterward, and for measles (the c graphs) at all stages. The road network was most important during the peak of an epidemic. The influenza epidemics waned in place, not through constriction of territory. So did measles, but it showed some role for urban hierarchical diffusion in filling in the latter stage. Contiguous hierarchy was important for influenza only at the peak and immediately after; pure population size hierarchical diffusion (the line) was important in the initial spread, during the peak, and until the end. In contrast, unconstrained hierarchical diffusion was never important for measles. For both diseases, the air network was surprisingly insignificant. The researchers speculated that the period of infectivity is too short for both of these diseases for long-term travel to be as important as local contagion. The difference in the importance of population size is likely related to the fact that measles gives permanent immunity to the exposed population, whereas influenza continuously mutates and finds new susceptibles.

As the transport system, school locations, residence types, and commercial connections in Iceland changed over decades, so did the frequency, peakedness, duration, intensity, and other characteristics of the epidemics. From studies in spatial analysis such as this, a lot can be learned both about the characteristics of an infectious disease and about the connectivity and structure of the human geography.

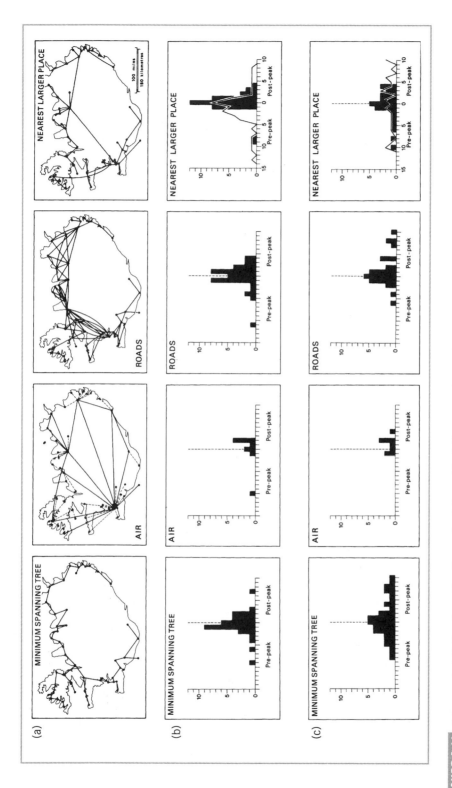

FIGURE 6.11 Join count statistics for estimating paths followed by influenza and measles epidemics in Iceland, 1945–1970. (a) Nodes and joins for diffusion models; (b) influenza; (c) measles. Spatial autocorrelation was used to study the links and pathways of diffusion of influenza and measles in Iceland. Number of significant deviates (alpha = .05 level, one-tailed test). From Cliff and Haggett (1988, p. 206). Copyright ©1988 Blackwell Science. Reprinted by permission.

Such knowledge is especially useful for developing strategies for vaccination (such as those used in the eradication of smallpox) and epidemic containment (Cliff et al., 1986).

Poliomyelitis

Geographers have recently looked at the epidemics of poliomyelitis that emerged in the early 20th century—in particular, the great epidemic of poliomyelitis (polio) in 1916 in New York City, and its spread (Trevelyan et al., 2005a, 2005b). Polio loomed as a horrific public health crisis until the late 1950s, when effective vaccines led to its retreat in the United States. The disease is now almost extinct in the world, hanging on in northern Nigeria and war-torn places like the Pakistani frontier, Afghanistan, and Somalia and adjacent areas of Ethiopia and Kenya. There have also been cases reported in Syria since the civil war broke out. The WHO has again studied the situation and resolved that it is possible to eradicate this terrible disease. Analysis of how it spreads spatially will be useful in this effort.

Using data from a seminal 1918 public health bulletin to examine the spread of polio in 1916 within New York City to its environs, Trevelyan et al. found a cluster of maximum activity with a 128-kilometer radius, and integrity as a cluster to 500 kilometers, with the level and rate of propagation of the wave (the logistic curve; see Figure 6.3) declining with distance. They determined that it "probably" followed a mixed contagious-hierarchical diffusion in the U.S. Northeast. Using data from the Mortality and Morbidity Weekly Report (Centers for Disease Control and Prevention [CDC]), and analyzing spatial autocorrelation and wave periodicity over 50 years, they found that polio activity increased in velocity of epidemic transmission over the decades (i.e., as U.S. transportation became increasingly networked, as in Iceland; see Figure 6.11) and became more spatially coherent. Over this half century the center of viral spatial activity shifted from the Northeast to the West Coast to occurrence across all states. After mass vaccine for polio was developed and implemented from the mid-1950s on, there was an increase in epidemic volatility and increased dependence upon endemic reservoirs in the most populous states, until in the end not only the incidence but the spatial structure collapsed.

These analyses of polio epidemics as they emerged and disappeared in the 20th century, using urban hierarchy, spatial autocorrelation, and infrastructure networking, seem to fit the process and pattern of the influenza and measles models above. There is something lacking in this basic spatial analysis, though, we think. Polio has a reversed disease ecology when compared with other 20th-century infections of childhood. It was a great puzzle. Polio, or infantile paralysis, was more common among the middle class and even the rich (e.g., Franklin D. Roosevelt) than among the poor. It was not a disease of filth and bad water; it was not vectored. In World War II it affected American soldiers in Egypt and/or the Middle East, but it did not seem to affect locals. It increased steadily as the U.S. infrastructure grew better, the population healthier, and the economy wealthier. The new postwar suburbs were its most fertile ground, and it was a terror there. Finally it was found that the polio virus was ubiquitous—a human infection that almost every child had previously contacted in infancy, everywhere, over the centuries. Babies received antibodies from their mothers' milk, which protected them until, constantly exposed, they developed their own antibodies. With improvements in sanitation and water supply, children were protected from the virus; some never contacted it. When children without any maternal antibodies did contact the polio virus, some developed infantile paralysis, or indeed died from the infection. Vaccination was needed as a cultural buffer to replace the mothers' antibodies, which safe water and sanitation (so protective against other enteric diseases) had removed, leaving children exposed. From reading the analyses of diffusion of polio epidemics above, and from thinking about how polio epidemics spread around the world (following the demographic transition) as India, Kenya, and Brazil improved their sanitation and water supply, it seems that what is diffusing is not the virus but protection from the virus—in other words, development, infrastructure, modernization. Leaving the disease ecology out of the spatial analysis would make the patterns of advance recorded above seem mere reflections in a mirror.

Emerging Methods

New computer models and geographic information system (GIS) simulations are opening up new methodological approaches. Old disease diffusion data, such as those on polio described above, can be approached in entirely new ways. One of the diseases being studied is the great 19th-century summer epidemic killer, yellow fever. Using data from public health reports cleverly applied to geocoded insurance lithographs of residences, Curtis, Mills, and Blackburn (2007) studied the great yellow fever epidemic of 1878 in New Orleans, officially reported as 27,000 cases with more than 4,000 deaths. They analyzed a new statistic, a spatial basic reproduction number (secondary infection), by putting 50-meter buffers keyed to *Aedes aegypti* mosquito activity range around each death as a "neighborhood." Thirty-three neighborhood index locations were finally identified, with age and nativity pattern for each, which followed the epidemic. The study highlighted the importance of multideath residences in numbers of deaths and as nodes in patterns of neighborhood diffusion sequences. In innovative hands, GIS applications are beginning to tease new insights out of even our oldest data.

A new method of analyzing diffusion uses the power of spatial data managed in a GIS to address the mobility in time and space. In particular, Hagerstrand's lifelines in space are "back" after decades of neglect and can be analyzed with GIS. Ling Bian (2003, 2007; Bian & Liebner, 2005) has challenged traditional epidemiological models based on uniform populations with uniform distributions interacting globally with equal contact rates. She has constructed a conceptual framework with network models of diffusion based on individuals interacting at multiple scales (see Plate 12 for an example). Time matters for both continuous processes and discrete events. Time periods in these models can be set to latency and incubation periods. Individual life paths intersect in two "layers," of homes and workplaces. These nodes are linked in a network of connections between groups, so that within-group and between-group interactions represent local and long-distance transmissions. This is stochastic modeling of discrete individuals moving in space, dispersing through networks (not waves). Data are gathered on occupations and workplaces; transportation routes/means and commuting; night and day residential patterns; shopping patterns; and so on to place individual people into multiple levels of spatial interactions, each level a medium of contagious diffusion. This form of modeling is only at its beginning, but holds enormous promise.

INFLUENZAS

The "swine flu epidemic" that did not happen in 1976 provides a major example of the need for geographic modeling of diffusion pathways. After the type of influenza isolated in an outbreak at Fort Dix, New Jersey, in January was thought to be the same as that which killed 20–40 million people in a world pandemic in 1918–1920, massive efforts were made to vaccinate the entire U.S. population. A particular characteristic of the deadly pandemic was that it killed not only elderly, very young, and other vulnerable people, but also healthy young adults. The entire population was at risk. There was no way to target vaccinations because little was known (or spatially considered, of course) about how such an epidemic would spread. In the end, there was no outbreak. Not only was a large amount of money spent, but because there are inevitably some people who have reactions to any vaccination, the serious side effects became more important than the danger of the swine flu epidemic that did not develop. The vaccination campaign created a popular backlash that impugned both science and government. How much better it would have been if the vaccine could have been produced and storehoused while a rational strategy for containment of any outbreak and spatially targeted vaccination coverage was implemented.

Basic Ecology

The ecology of influenza is intrinsically interesting. Influenza is notorious as an annual winter epidemic. The usual epidemiological explanation for winter infections has been that people gather together inside and so communicate contagion better. Even in places that are crowded all the time (e.g., prisons), however, influenza is a winter epidemic. It has been suggested that in lower humidity, or perhaps in lower ultraviolet radiation, the virus may survive longer in the air between people and so be able to infect more of them; still, nothing could be proved. Cliff and Haggett have addressed how influenza that is newly epidemic in the northern hemisphere in winter also becomes epidemic in the southern hemisphere in winter, but seems to occur at low rates in the tropics all year long (Cliff et al., 1986). Now that it has been found that the guinea pig gets the flu, and so can be used as the animal model that was previously lacking, the mystery has been solved (Lowen, Mubareka, Steel, & Palese, 2007; Lowen, Steel, Mubareka, & Palese, 2008). In subtropical climates measurements have shown the virus to survive longest in cold temperatures and to be most contagious in low humidity because the virus adheres to heavier water molecules and falls to the ground. However, curiously, while influenza peaks during the winter in the subtropics and polar regions, it peaks during the rainy season in the tropics (James et al., 2013).

Influenza is one of those few contagious diseases that could have occurred even when human population on earth lacked great cities. There are two reasons for this: humans share the influenza virus (or viruses) with several animals, especially swine and ducks, and the virus has an extraordinary ability to mutate in gradual (drift) and radical (shift) transformations that baffle the recognition and memory of the human immune system. Three major classes of influenza are known. Influenza C is least known, but does not seem to cause epidemic disease. Influenza B causes morbidity in youthful populations and has low case mortality. Human epidemics are caused by influenza A strains of virus. Influenza A strains have a coat, or envelope, of glycoprotein in the form of spikes of hemagglutinin (H) and neuraminadase (N), and are referenced by ratios and patterns of these spikes. The H is responsible for binding the virus to host cells. The N facilitates release of the virus from infected cells and the spread to others. As "the flu" spreads endemically in a population, continuing all year but spreading especially in winter, the virus changes the angle, bend, and other properties of its spikes. This gradual "drift" finds enough new susceptibles for the contagion to continue its presence. Then, with irregular periodicity, the virus rearranges its spikes into different HN ratios and patterns that are not recognized by the immune system ("shift"), and the whole population becomes susceptible to a new infection that can spread in a great pandemic wave. There were three pandemics in the 20th century. The so-called Spanish flu (which did not come from Spain) killed 20–40 million people in 18 months from 1918 to 1920, as noted above. It was caused by an H5 virus, which usually does not occur in people but in birds. The Asian flu of 1957–1958, H2N2, is considered mild, as it killed "only" a million people while causing massive absenteeism and misery. The Hong Kong flu of 1968–1969, H3N2, was another mild pandemic thought to have killed "only" a million people.

This rearrangement of H and N surface proteins seems to occur through genetic exchange as multiple varieties of the influenza virus infect pigs and perhaps ducks simultaneously. They emerge where the "three P's"—people, pigs, and poultry—reside together. Cultural processes of herding pigs and ducks and agricultural domestication and intensity of rearing have played a role in creating pandemics, but the recombination process probably also occurred occasionally in wild birds and mammals in prehistoric time. There has been a tendency for new varieties to emerge from the cultural hearth that domesticated pigs and ducks, southern China, which continues an especially numerous and intense association among the three P's. This has given Hong Kong special importance as an advance warning surveillance site for the development of appropriate vaccines for the future winter. It should be noted, however, that the U.S. Midwest also has an exceptionally intense historic focus of swine and people, and that the great North American flyways (of migrating waterfowl) may be implicated in the early emergence of epidemics. A recent (2015) bird flu in the Midwest resulted in the natural and unnatural (culling to prevent further infection) deaths of millions of poultry, though no human cases, as a result of a shifted influenza virus brought to the Americas by migratory water-fowl from Asia. Pyle (1986) noted the xenophobia implicit in naming influenza epidemics for other places. Indeed, the "Spanish flu" (or deadly "swine flu," in today's vernacular) of World War I was quite possibly brought by American farm boys from the Midwest to Europe.

Simulating U.S. 20th-Century Epidemics

Pyle (1986) tried to discover whether these drifted (endemically epidemic) and shifted (rapidly pandemic) forms of influenza have different and predictable spatial patterns. Such footprints would be useful in detection, but especially important for developing containment and intervention strategies. Using harmonic analysis to map the timing of epidemic arrival and peak, Pyle studied centuries of influenza epidemics in Europe and decades in the United States. He reconstructed diffusion pathways to identify patterns of great "frontal" assaults sweeping across continents; patterns of year-to-year continuous advance; and patterns of multiple-node, gradual fill-in. Having seen these footprints of shift and drift, he then combined many of the modeling techniques discussed in this chapter to simulate how the great epidemics of swine flu, Asian flu, and Hong Kong flu spread in the United States, compared with the endemic spread of intervening years. Pyle simulated how influenza would spread randomly if it began in the sites harmonic analysis had designated as "repeated events," and how it would spread if hierarchical diffusion dominated. He added map surfaces based on population potential for the elderly instead of city locations. He then used regression to test the prediction of these simulations for the spatial spread of drifting and shifting influenza. Clever though these models were, the results are somewhat muddied by the quality of data: even while the newly shifted strains spread, the old normal form also continued to circulate, and there was no realistic way to separate their infections.

Getting Sick with the Birds and the Pigs

We humans are now studying the origins of influenza and trying to prepare for what may well be the next great, and not mild, pandemic. Influenza is considered overdue for a shift. In 1997 an influenza virus that sickened chickens in Hong Kong managed to infect people and led to the immediate slaughter of all the millions of chickens in Hong Kong. It was only the beginning. The agent was and is H5N1, an avian influenza virus (AIV). There have now been more than a thousand human cases of H5N1, with a mortality of 60%. Cases have all resulted from contact with a sick bird, except for a handful of cases involving very close contact with an infected person (e.g., from a chicken to a grandmother cleaning the chicken for cooking to a granddaughter). There has been no "casual" transmission of H5N1. Although this AIV is thought to require only minor genetic changes to adapt to human cell receptors, it has not yet developed the ability to pass between human hosts through nasal aerosols. Production of a vaccine for the world's population is an especially difficult undertaking. Not only does it take months after isolating a strain to mass-produce enough of it to make vaccines, but the medium used to grow the virus has in the past been chicken eggs—and H5N1 is a chicken disease! New high-technology, 21st-century production systems need to be developed. This AIV has spread via bird migration routes and winter contacts, and probably some human (poultry) trade routes as well. It has now occurred—resulting in the mass slaughter of many hundreds of millions of chickens and ducks—from China, Vietnam, Thailand, and Indonesia to Nigeria, Saudi Arabia, Egypt, and Turkey, as well as to Germany, Poland, Russia, and England (see Plate 11a). The "bird flu" has moved out of the headlines, but there remain catastrophic possibilities when it crosses species. Much effort has gone into trying to develop a sentinel system, good reporting, and an emergency vaccination containment strategy. Geographers are using GIS to overlay environment, people, fowl, and many other variables to develop predictive models of where it will emerge.

QUICK REVIEW 2

✓ During the 1800s in the United States the diffusion of cholera took several forms. The first cholera epidemic was introduced at several points: the riverine ports of Canada, as well as New York City and New Orleans. Subsequent epidemics followed the hierarchical structure of the urban system or diffused along the rail system.

✓ During the 1980s the gravity model was used to predict the diffusion of HIV/AIDS in Ohio through the urban hierarchy. The gravity model is a geographic model of spatial interaction that takes into account distance and size of places.

✓ Measles diffusion is associated with population movements for work, schooling, shopping, holidays, and such social activities as visiting relatives.

✓ Influenza diffuses throughout the world each year and the virus evolves as it spreads. Whether or not there will be a global pandemic is dependent on many things including the number of susceptible people, human movement, and the ecology of places.

CONCLUSION

Maybe nothing is more inherently "geographic" than the study of the diffusion of disease at a variety of appropriate scales. This chapter has looked at how disease agents diffuse through settlement systems with social contexts. Many epidemic infections that need study have not been considered including the recent Ebola epidemic in West Africa or Middle East respiratory syndrome (MERS), a viral respiratory illness first reported in Saudi Arabia in 2012. Contagious diseases have an ecology, but their characteristics, impact, and prediction are especially appropriate for spatial analysis. Chapter 7 builds on this chapter by describing how changing ecology leads to emerging diseases and how genetics and space can be used simultaneously to understand how diseases diffuse and evolve.

REVIEW QUESTIONS

1. Figure 6.1 displays how innovation fades with distance and time and Figure 6.3 is a curve summarizing how the proportion of people accepting an innovation often occurs over time. How can these curves based on a theoretical understanding of diffusion processes in general be used to understand disease diffusion? Give an example of a communicable disease that you think would diffuse through a population in the way shown in these curves. Explain why.

2. Ebola diffused quickly through the population of three adjacent countries in West Africa in 2014–2015. Explain how you think this occurred using the SIR disease transmission-modeling framework. What do you think the R_0 is for the disease? Why do you think it spread so quickly in the West African context when all other epidemics in other places during the past 40 years have quickly fizzled out?

3. Chapter 6 describes types of diffusion including relocation, contagious (also called contact), and hierarchical. Using the historic examples of cholera and HIV/AIDS, compare and contrast the types of diffusion for these two diseases in the United States.

4. Figure 6.6 summarizes the hierarchical and contact diffusion of cholera in West Africa during the 1970s. Describe how urban hierarchy is related to diffusion of cholera in this setting. Explain the different types of contact diffusion over different geographic landscapes and features.

5. In 2015 there was a large outbreak of measles in the United States that started at Disneyland and some have blamed the lack of herd immunity in the population especially in California. Make an argument backing up the claim. Do you think there are any similarities in the diffusion patterns of the 2015 United States measles epidemic and the epidemic described by Haggett from 1966 to 1970 across southwest England? Explain.

6. What are some barriers that could stop the diffusion of some of the diseases mentioned in discussion questions 2–5: Ebola, measles, cholera, HIV/AIDS?

7. The gravity model assumes that there is an inverse relationship between population size and interaction between places. It also assumes an inverse relationship between distance between places and interaction of the population. Explain how the gravity model might have been useful in predicting the spread of Ebola in West Africa during the 2014–2015 epidemic.

8. Human and avian influenza diffusion is complex and still not completely understood. Cliff and Haggett described patterns of influenza diffusion in Iceland differentiating between patterns via suspected pathways including airplane travel, road travel, and urban hierarchy (nearest larger place). Explain how geographic context and genetic change in influenza viruses are factors that simultaneously control diffusion of this disease.

REFERENCES

Alber, R., Adams, J., & Gould, P. (1971). *Spatial organization: The geographer's view of the world*. Englewood Cliffs, NJ: Prentice-Hall.

Bian, L. (2003). A conceptual framework for an individual-based spatially explicit epidemiological model. *Environment and Planning B, 31*, 381–395.

Bian, L. (2007). A network model for dispersion of communicable diseases. *Transactions in GIS, 11*(2), 155–173.

Bian, L., & Liebner, D. (2005). Simulating spatially explicit networks for dispersion of infectious diseases. In D. Maguire, M. Goodchild, & M. Batty (Eds.), *GIS, spatial analysis and modeling* (pp. 245–265). Redlands, CA: ERSI Press.

Brownlea, A. (1972). Modeling the geographic epidemiology of infectious hepatitis. In N. McGlashan (Ed.), *Medical geography: Techniques and field studies* (pp. 279–300). London: Methuen.

Carrel, M., & Bitterman, P. (2015). Personal belief exemptions to vaccination in California: A spatial analysis. *Pediatrics, 136*(1), 80–88.

Cliff, A., & Haggett, P. (1982). Methods for the measurement of epidemic velocity. *International Journal of Epidemiology, 11,* 82–89.

Cliff, A., & Haggett, P. (1983). Changing urban–rural contrasts in the velocity of measles epidemics in an island community. In N. McGlashan & J. Blunden (Eds.), *Geographical aspects of health* (pp. 335–348). New York: Academic Press.

Cliff, A., & Haggett, P. (1988). *Atlas of disease distributions: Analytic approaches to epidemiological data.* Oxford, UK: Blackwell.

Cliff, A., & Haggett, P. (1989). Spatial aspects of epidemic control. *Progress in Human Geography, 13,* 315–347.

Cliff, A., Haggett, P., & Ord, J. (1986). *Spatial aspects of influenza epidemics.* London: Pion.

Cliff, A., Haggett, P., Ord, J., & Versey, C. (1981). *Spatial diffusion: An historical geography of epidemics in an island community.* Cambridge, UK: Cambridge University Press.

Cliff, A., Haggett, P., & Smallman-Raynor, M. (1993). *Measles: An historical geography of a major human viral disease from global expansion to local retreat.* Oxford, UK: Blackwell.

Currie, W. (1792). *Historical account of the climates and diseases of the United States of America.* Philadelphia: Dobson.

Currie, W. (1811). *A view of the diseases most prevalent in the United States of America.* Philadelphia: Humphreys.

Curtis, A., Mills, J., & Blackburn, J. (2007). A spatial variant of the basic reproduction number for the New Orleans yellow fever epidemic of 1878. *The Professional Geographer, 59*(4), 492–502.

Gould, P. (1993). *The slow plague: A geography of the AIDS pandemic.* Oxford, UK: Blackwell.

Gould, R. (1969). *Spatial diffusion.* Washington, DC: Association of American Geographers Commission on College Geography.

Hagerstrand, T. (1952). *The propagation of innovation waves.* Lund, Sweden: Gleerup.

Haggett, P. (1976). Hybridizing alternative models of an epidemic diffusion process. *Economic Geography, 52,* 136–146.

Hunter, J., & Young, J. (1971). Diffusion of influenza in England and Wales. *Annals of the Association of American Geographers, 61,* 627–653.

Lowen, A., Mubareka, S., Steel, J., & Palese, P. (2007). Influenza virus transmission is dependent on relative humidity and temperature. *PLoS Pathogens, 3.*

Lowen, A., Steel, J., Mubareka, S., & Palese, P. (2008). High temperature (30° centigrade) blocks aerosol but not contact transmission of influenza virus. *Journal of Virology, 82*(11), 5650–5652.

Loytonen, M. (1991). The spatial diffusion of human immunodeficiency virus type 1 in Finland, 1982–1997. *Annals of the Association of American Geographers, 81,* 127–151.

Pyle, G. (1969). The diffusion of cholera in the United States in the nineteenth century. *Geographic Analysis, 1,* 59–75.

Pyle, G. (1984). Spatial perspectives on influenza inoculation, acceptance, and policy. *Economic Geography, 60,* 273–293.

Pyle, G. (1986). *The diffusion of influenza.* Totowa, NJ: Rowman & Littlefield.

Smallman-Raynor, M., Cliff, A., & Barford, A. (2005). Geographical perspectives on epidemic transmission of cholera in Haiti, October 2010 through March 2013. *Annals of the Association of American Geographers, 105*(4), 665–683.

Stock, R. (1976). *Cholera in Africa.* Plymouth, UK: International Africa Institute.

Thomas, R. (1992). *Geomedical systems: Intervention and control.* London: Routledge.

Trevelyan, B., Smallman-Raynor, M., & Cliff, A. (2005a). The spatial dynamics of poliomyelitis in the United States: From epidemic emergence to vaccine-induced retreat, 1910–1971. *Annals of the Association of American Geographers, 95*(2), 269–293.

Trevelyan, B., Smallman-Raynor, M., & Cliff, A. (2005b). The spatial structure of epidemic emergence: Geographical aspects of poliomyelitis in north-eastern USA, July–October 1916. *Journal of the Royal Statistical Society A, 168*(Pt. 4), 701–722.

Webster, N. (1799). *A brief history of epidemic and pestilential diseases.* Hartford, CT: Hudson & Goodwin.

📖 FURTHER READING •

Adesina, H. (1984a). The diffusion of cholera outside Ibadan City, Nigeria, 1971. *Social Science and Medicine, 18,* 421–428.

Adesina, H. (1984b). Identification of the cholera diffusion process in Ibadan, 1971. *Social Science and Medicine, 18,* 429–440.

Althaus, C. (2014). Estimating the reproduction number of Ebola virus (EBOV) during the 2014 outbreak in West Africa. *PLoS Currents, 6.*

Anderson, R,, & May, R. (1991). *Infectious diseases of humans: Dynamics and control.* Oxford, UK: Oxford University Press.

Angulo, J. (1987). Interdisciplinary approaches in epidemic studies: II. Four geographic models of the flow of contagious disease. *Social Science and Medicine, 24,* 57–69.

Angulo, J., Haggett, P., Meghale, P., & Perderneiras, C. (1979). Variola minor in Braganca Paulista County, 1956: A trend–surface analysis. *American Journal of Epidemiology, 105,* 272–280.

Auchincloss, A., & Diez-Roux, A. (2008). A new tool for epidemiology: The usefulness of dynamic-agent models in understanding place effects on health. *American Journal of Epidemiology, 168,* 1–8.

Bailey, N. (1975). *The mathematical theory of infectious diseases and its applications.* High Wycombe, UK: Charles Griffin.

Burnet, M., & White, D. (1974). *Natural history of infectious diseases* (4th ed.). Cambridge, UK: Cambridge University Press.

Cohn, S., Klein, J., Mohr, J., van der Horst, C., & Weber, D. (1994). The geography of AIDS: Patterns of urban and rural migration. *Southern Medical Journal, 87,* 599–606.

Dutt, A., Monroe, C., Dutta, H., & Prince, B. (1988). Geographical patterns of AIDS in the United States. *Geographical Review, 77,* 456–471.

Elmore, K. (2006). The migratory experiences of people with HIV/AIDS in Wilmington, North Carolina. *Health and Place, 12*(4), 570–579.

Fairclough, G. (1995). A gathering storm. *Far Eastern Economic Review, 158,* 26–30.

Ferguson, N., May, R., & Anderson, R. (1997). Measles: Persistence and synchronicity in disease dynamics. In N. McGlashan & J. Blunden (Eds.), *Spatial ecology* (pp. 137–157). Princeton, NJ: Princeton University Press.

Gardner, L., Brundage, J., McNeil, J., Visintine, R., & Miller, R. (1989). Spatial diffusion of the human immunodeficiency virus infection epidemic in the United States, 1985–87. *Annals of the Association of American Geographers, 79,* 25–43.

Gould, P., & Wallace, R. (1994). Spatial structures and scientific paradoxes in the AIDS pandemic. *Geografiska Annaler, 76B,* 105–116.

Greenwood, B., & DeCock, K. (1998). *New and resurgent infections: Detection and management of tomorrow's infections.* Chichester, UK: Wiley.

Grenfell, B., Bjornstad, O., & Kappey, J. (2001). Travelling waves and spatial hierarchies in measles epidemics. *Nature, 414,* 716–723.

Hagerstrand, T. (1970). What about people in regional science? *Papers of the Regional Science Association, 24,* 7–21.

Johnson, N., & Muller, J. (2002). Updating the accounts: Global mortality of the 1918–1920 "Spanish" influenza pandemic. *Bulletin of the History of Medicine, 76*(1), 105–115.

Kwofie, K. (1976). A spatio-temporal analysis of cholera diffusion in western Africa. *Economic Geography, 52,* 127–135.

Loytonen, M. (1998). Time geography, health and GIS. In A. Gatrell & M. Loytonen (Eds.), *GIS and Health* (pp. 97–110). London: Taylor & Francis.

MacKellar, L. (2007). Pandemic influenza: A review. *Population and Development Review, 33*(3), 429–451.

Nijkamp, P., & Reggiani, A. (1996). Space–time synergetics in innovation diffusion: A nested network simulation approach. *Geographic Analysis, 28,* 18–37.

Pan American Health Organization. (1996). New emerging and re-emerging infectious diseases. *Bulletin of the Pan American Health Organization, 30,* 176–181.

Pollitzer, R. (1959). *Cholera.* Geneva, Switzerland: World Health Organization.

Pyle, G., & Furuseth, O. (1992). The diffusion of AIDS and social deprivation in North Carolina. *North Carolina Geographer, 1,* 1–10.

Shannon, G., & Pyle, G. (1989). The origin and diffusion of AIDS: A view from medical geography. *Annals of the Association of American Geographers, 79,* 1–24.

Shannon, G., Pyle, G., & Bashshur, R. (1991). *The geography of AIDS.* New York: Guilford Press.

Smallman-Raynor, M., & Cliff, A. (1990). Acquired immunodeficiency syndrome (AIDS): Literature, geographical origins and global patterns. *Progress in Human Geography, 14,* 157–213.

Smallman-Raynor, M., Cliff, A., & Haggett, P. (1992). *London international atlas of AIDS.* Oxford, UK: Blackwell.

Smyth, F., & Thomas, R. (1996). Preventative action and the diffusion of HIV/AIDS. *Progress in Human Geography, 20,* 1–22.

Stephenson, J. (1995). AIDS data animation maps evolving US epidemic. *Journal of the American Medical Association, 274*(10), 784–785.

Thomas, R. (1996). Modeling space–time HIV/AIDS dynamics: Applications to disease control. *Social Science and Medicine, 43,* 353–366.

Westvelt, J., & Hopkins, L. (1999). Modeling mobile individuals in dynamic landscapes. *International Journal of Geographic Information Science, 13,* 191–208.

Yang, Y., & Atkinso, P. (2008). Individual space–time activity-based model: A model for the simulation of airborne infectious-disease transmission by activity-bundle simulation. *Environment and Planning B: Planning and Design, 35,* 80–99.

CHAPTER 7

Emerging Infectious Diseases and Landscape Genetics

If you list the top 10 causes of death in the world, stratified by the wealth of countries, you observe that in wealthy places the causes of death are chronic or noninfectious: cancers and heart disease and stroke. A single infectious cause may be present in this top-10 list for the wealthy populations of the world: lower respiratory infections, caused by pneumonia or bronchitis and most often afflicting the elderly. In the middle-income countries of the world, where the bulk of humanity resides, the leading causes of death are a mixture of chronic/noninfectious and infectious diseases, things such as diabetes and heart disease alongside road traffic accidents (see Chapter 10) and diarrheal diseases and HIV. The leading causes of death in poor countries, in sharp contrast to those of wealthy countries, are still primarily infectious: malaria, neonatal infections, diarrheal disease, HIV, and so on. While the impacts of chronic noninfectious diseases are increasingly felt in the poor places of the world, infectious diseases still make up a large share of the overall disease burden.

This shift from infectious to noninfectious diseases over time was thought, in the mid-20th century, to be a natural progression, and that science and technology would eventually overcome nature's viruses and bacteria. William Stewart, surgeon general of the United States in the late 1960s, is often credited with saying that it was time to "close the book" on infectious diseases, that the war against pestilence had been won (Avila, Saïd, & Ojcius, 2008; Cohen, 2000; Francis, Battin, Jacobson, Smith, & Botkin, 2005; Nesse & Williams, 1998; Upshur, 2008). The quote is so famous it even appeared in his obituary in the preeminent medical journal *The Lancet* (Bristol, 2008). The quote is held up as an example of medical hubris, thinking that infectious diseases would no longer threaten human health. Problematically,

Dr. Stewart never actually said this, which might explain why the dozens of publications that use this quote always have different wording and different years of attribution (Spellberg & Taylor-Blake, 2013). Despite the false tarnishing of poor Dr. Stewart's name, the idea that infectious diseases had been conquered, or were conquerable, was not unheard of in the 1960s and 1970s. In 1962, Sir Frank MacFarlan Burnet, a cowinner of the Nobel Prize in Medicine in 1960, said, "One can think of the middle of the twentieth century as the end of one of the most important social revolutions in history, the virtual elimination of the infectious diseases as a significant factor in social life." His optimism is reflective of advances in medicine at the time: the introduction of penicillin in 1942 and then other antibiotics in subsequent years; the development of a polio vaccine and the first countrywide elimination of polio in Czechoslovakia in 1960; the success in using DDT to eliminate malaria from many parts of the world; and the promise of smallpox eradication on the horizon.

Instead of this dream of a world free, or virtually free, from the effects of infectious disease, we live in a world where infectious diseases are still very much a part of everyday life for everyone, although their role in life is much larger for the poor than for the wealthy. The first part of this chapter will focus on why, where, and how infectious diseases emerge, reemerge, or remain persistently intractable. Then, building on landscape epidemiology (see Chapter 2), the last part of the chapter will explore how advances in genetic sequencing and computational capability, combined with geographic understanding of landscapes, can allow us to understand infectious diseases in ways that were not previously possible.

WHAT'S IN A NAME? EMERGING, REEMERGING, OR ALWAYS THERE

Research by Jones et al. (2008) attempts to quantify the emergence of infectious diseases over time, from the 1940s through 2000. This and other research is part of a larger agenda in public health, epidemiology, health geography, and other fields: to understand where emerging infectious diseases originate, why they originate, and what threat they pose to human health. Although Jones and coauthors document new infectious diseases in the decades between 1940 and 1980, it was not until the emergence of HIV on a global scale in the 1980s that the world's public health and medical communities realized the need to refocus energy on the surveillance and prevention of infectious diseases (Figure 7.1). Those advances in medicine listed above, antibiotics and vaccines, had meant that many resources had shifted away from infectious disease toward chronic, noninfectious outcomes. HIV, and then other infectious diseases such as Ebola, changed all that.

There is no single agreed-upon definition of an emerging infectious disease, but what is generally agreed upon is that a disease is emergent/emerged if it is observed in humans for the first time, if it is caused by a newly evolved pathogen, or if it has changed and taken on more severe symptomology (Funk, Bogich, Jones, Kilpatrick, & Daszak, 2013). A reemerging infectious disease, in contrast, is one that is not new but rather is moving into areas where it has never previously been

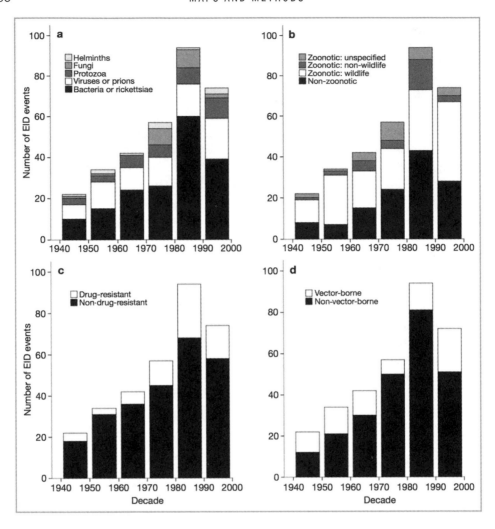

FIGURE 7.1 Emerging infectious disease (EID) events over time according to (a) pathogen type, (b) transmission type, (c) drug resistance status, and (d) transmission mode (vectored or nonvectored). From Jones et al. (2008). Copyright ©2008 Nature Publishing Group. Reprinted by permission.

observed or from where it had been eradicated or eliminated, or a disease whose numbers had fallen dramatically but now causes high levels of morbidity and mortality, more a matter of geography and scale than novelty. Finally, billions of people in the world continue to experience infectious diseases that never, for them, went away, diseases that are neither emerging nor reemerging, the persistent and intractable diseases of the world. Examples of each are seen in Table 7.1. The distinctions between each category can be a bit blurry. For instance, at what point in time do we draw a line to say diseases after are newly emerging but diseases before are reemerging? Or what if a disease has been persistent in one place but is reemerging elsewhere? It can be argued that these distinctions are subjective but they have major impacts on how funding is allocated.

TABLE 7.1. Examples of Emerging, Reemerging, and Persistent Infectious Diseases		
Emerging	**Reemerging**	**Persistent**
HIV/AIDS	Malaria	Onchocerciasis (river blindness)
SARS	Chikungunya	Dracunculiasis (Guinea worm)
Nipah	West Nile virus	Buruli ulcer
Hendra	Dengue	Rotavirus
MERS	Ebola	Schistosomiasis
H5N1 Avian influenza	Cholera	
H1N1 Influenza (2009)	Lymphatic filariasis	
Methicillin-resistant *Staphylococcus aureus* (MRSA)	Tuberculosis	
	Chagas disease	

Chagas disease is listed in the table as reemerging, given its range expansion both into urban areas and into the United States with migrants from Central and South America. The WHO, however, lists Chagas as a neglected tropical disease (NTD) that is persistent and intractable in those populations. Similarly, lymphatic filariasis is considered a NTD and has persisted in affected populations for decades but is also reemerging into new urban settings (see below). Finally, drug-resistant organisms, such as MRSA, are included in the emerging category even as their base causal agent, in this case *Staphylococcus aureus,* has existed for hundreds of years and as their range is expanding, both arguments for a persistent or reemerging categorization. Despite the problems with these categories, however, they form the basis for how the public health community talks about infectious diseases.

WHY DO DISEASES EMERGE, REEMERGE, OR PERSIST?

A variety of environmental, social, behavioral, and biological factors influence the process of the introduction and diffusion of disease within a population. Jonathan Mayer, a health geographer at the University of Washington, attributes the emergence and reemergence of infectious diseases to five factors: cross-species transfer, spatial diffusion, pathogenic evolution, new recognition of disease, and changing human–environment relationships (Mayer, 2000). A sixth factor, poverty, combines with these other five and escalates the likelihood of emergence/reemergence and is a primary driver for the persistence of intractable and neglected diseases. The Ebola epidemic in West Africa is a recent example of an emerging infectious disease that has been neglected in past African epidemics when most of the people who were infected were very poor. Emch and Root (2009) frame these changing human–environment relationships as a series of drivers that reflect the underlying forces or processes that lead to emergence or reemergence: ecological change, evolutionary forces and microbial adaptation, human behavior and demographic change, political conflict and public health infrastructure collapse, and increased mobility and globalization. Table 7.2 provides examples for each of these potential

TABLE 7.2. Underlying Drivers of Emerging and Reemerging Infectious Diseases

Driver	Example	Disease
Evolutionary forces and microbial adaptation	Drug resistance (e.g., overuse of antibiotics); selective pressures in the environment; microbial evolution	Influenza virus; *Staphylococcus aureus;* multidrug-resistant tuberculosis
Ecological change	Deforestation/reforestation; flood/drought; climate change; dams/irrigation	Schistosomiasis; hantaviruses; Ebola; Lyme disease; Rocky Mountain spotted fever
Human behavior and demographic change	Population growth; urbanization and rural-to-urban migration; aging population structure; dietary changes; sexual behaviors; intravenous drug use; poverty and income inequality	Dengue; HIV
Mobility and globalization	Globalization of food supply; air travel; worldwide movement of goods and people	Influenza virus; Dengue; SARS
Political conflict and public health infrastructure collapse	Inadequate sanitation and potable water sources; insufficient number of trained public health personnel; insufficient coverage of hospitals/clinics; curtailment of vector control measures; reduction of immunization campaigns; elimination of disease surveillance and prevention programs	Trypanosomiasis; *Vibrio cholerae;* yellow fever; measles; polio

From Emch and Root (2010). Copyright © 2010 Blackwell Publishing Ltd. Reprinted with permission from John Wiley & Sons, Inc.

drivers. Many of these factors are discussed in detail in other chapters in this book; for example, Chapter 10 examines the process of urbanization and unique health concerns of urban populations while Chapter 12 examines the impact of climate change on ecological systems and population health. Emch and Root (2009) suggest that "some factors influence the introduction of a disease while others primarily affect the spread or changing geographic distribution of the disease" (p. 155). Some diseases emerge because of interactions between several factors.

Lymphatic Filariasis and Urbanization

Lymphatic filariasis (LF) is caused by three species of parasitic filarial worm. LF is also known as elephantiasis: one of the symptoms of the disease is a buildup of worms in various parts of the lymph system that causes swelling. The worms are transmitted via mosquito bites, and several types of mosquitoes can transmit LF. While infection is usually acquired in childhood, the major symptoms often don't appear until adulthood since it takes time for the worm populations to build up in the body.

LF is endemic in many rural areas of the world. However, rapid and unplanned urbanization in low-income countries in the decades following World War II has changed the ecology of LF, increasing urban-based transmission of the worms. In particular, the ecology of one of the types of mosquitoes that can transmit LF, the *Culex* genus, has helped to urbanize this previously rural disease and to cause a reemergence and an increase in the number of individuals infected with and at risk of LF. *Culex* mosquitoes prefer biting humans to other animals, and are night biters, taking blood meals when the worms in an infected body are at their most active. *Culex* mosquitoes like to lay their eggs and hatch their larvae in water supplies contaminated with organic matter. The polluted drainage and septic systems of poor areas of tropical cities provide a ready breeding habitat, and the cities themselves provide a source of hosts to feed on, transmitting worms from infected to susceptible. The WHO estimates that 120 million people worldwide are infected with LF, and a further 1.4 billion are at risk of infection, particularly in low- and middle-income urbanized areas of the world. Efforts are under way to eliminate LF, via mass administration of antiworming drugs and increased usage of bed nets. By reducing the worm burden in infected individuals the transmission chain can, hopefully, be broken.

Nipah and Cross-Species Transfer

In late 1998 and early 1999 a strange outbreak of respiratory and neurological disease occurred on Malaysian pig farms. Pigs were exhibiting fever and respiratory symptoms, and people were experiencing high fever, disorientation, and even comas. This initial epidemic resulted in the deaths of over 100 people and the culling (killing) of over 1 million pigs in an attempt to stop the disease. Following exposure to Malaysian pigs in Singaporean abattoirs (slaughterhouses), a further 11 people became sick and one died. The cause of this epidemic was unknown, though it was initially thought to be Japanese encephalitis (JE), a mosquito-borne disease that in humans manifests with similar symptoms. The age distribution of the epidemic was wrong, however, for JE, since JE usually impacts children, and a number of individuals who were vaccinated against JE showed no protection against this new disease.

In reality the outbreak was caused by a previously unknown virus, named Nipah virus for the location of the outbreak in Malaysia. Nipah was molecularly similar to another recently discovered virus, Hendra virus, which had emerged for the first time in Australia in 1994 and caused mortality in horses, horse trainers, and veterinarians. Hendra virus had been associated with exposure of horses to flying foxes (fruit-eating bats).

Malaysia is a predominantly Muslim country, a faith that prohibits the consumption of pork. Why then were Malaysians raising pigs? Malaysia hosts a large minority population of ethnic Chinese, the main consumers of pork in the country, and also exports pork to its wealthy neighbor, Singapore, which is majority ethnic Chinese. Increased demand for meat is associated with increased wealth (see Nutrition Transitions in Chapters 4 and 8), and Malaysian pig farmers expanded their production capacity by establishing new farms in previously forested areas. This **agricultural extensification** of agricultural production into previously unfarmed areas brought humans and pigs into close and frequent contact with species that they would not before have encountered: fruit bats.

Pigs can often act as a bridge for pathogens between humans and other species; they are a natural mixing vessel for influenza viruses, for instance. Their immunological and systemic similarity to humans makes them a suitable environment for pathogens to adapt to a human system. While influenza viruses often have to change substantially for transmission within mammals, the reservoir of Nipah was also a mammal, so fewer changes were needed. Pigs raised under trees where bats were feeding or roosting became sick and infected people. There is also evidence that humans who directly consumed fruit or juice that had been contaminated with bat urine or saliva contracted the virus.

Nipah virus died out in Malaysia, but has reappeared multiple times since then, causing relatively small and isolated outbreaks in India and Bangladesh. Retrospective studies indicate that Nipah was present in pigs in 1996, years before the first large outbreak in pig and human populations. Most countries of South and Southeast Asia are within the ranges of multiple fruit bat/flying fox species, the primary reservoirs for Nipah, Hendra, and, potentially, other as yet unknown viruses in the same family. The loss of habitat and food as humans encroach on previously forested areas brings bat populations into contact with humans facing population pressure and increased demand for resources. Given changing diets across South and Southeast Asia and still-growing populations, the continued **intensification** of contact between wild animal species and humans and domesticated animals is likely, as is the emergence of previously unknown infectious diseases of wildlife when formerly silent zones of infection are intruded upon by humans and domesticates.

Staphylococcus Aureus and Pathogenic Evolution

In the case of Lymphatic filariasis, a changing human ecology (e.g., urbanization) is interacting with an adaptable vector ecology to produce reemergent disease. Our changing human ecology also works to encourage the adaptation of pathogens themselves. Pathogens are evolutionary beings: they will evolve new strategies for success when given ample opportunities to reproduce and transmit to new hosts. The strong will outcompete the weak, and the use of antibiotics to treat infections kills only those bacteria that are susceptible, leaving the strong alive.

Penicillin was first discovered by Alexander Fleming in 1928, and it was found to kill dozens of different types of bacteria. During World War II, penicillin was manufactured in mass quantities by the U.S. government as part of the war effort: it was hailed as the miracle drug of the war, that because of penicillin soldiers were not left dead on the battlefield but survived to come home. In the decades following the war, multiple types and classes of antibiotics were discovered. Broad-spectrum antibiotics, such as penicillin, were and are particularly favored by physicians because they can target many types of bacteria simultaneously and can take effect before lab work comes back providing a definitive diagnosis. By the mid-1940s Fleming and others warned of evolutionary selection of resistance to penicillin and other antibiotics, cautioning that bacteria are highly adaptable and will overcome the effectiveness of antibiotics. This describes the **antibiotic paradox**: any treatment with antibiotics triggers selection in bacteria for those with resistance to the antibiotic, so the only way to prevent antibiotic resistance is to not use antibiotics. By 1946 there were reports of penicillin-resistant *Staphylococcus aureus* (SA).

The use of broad-spectrum antibiotics allows for multiple types of bacteria to develop resistance to the same drug. The preference of patients for immediate relief also means that antibiotics are sometimes prescribed for infections that are minor and would be cleared by the body's immune system. Simultaneously, hygiene and sanitation practices have suffered with the arrival of antibiotics: the safety net provided by antibiotics leads some health care workers to rely on medical treatment to keep patients healthy rather than hand hygiene and other practices. In the increasingly crowded health landscape, an overworked health care workforce and steadily aging population also contribute to an overreliance on pharmaceuticals and other medical treatments rather than hygienic prevention.

The aging of the U.S. and other populations in the final stages of the demographic transition leads to the concentration of large numbers of elderly, sick patients into hospital intensive care units (ICUs), long-term care facilities, outpatient care centers, nursing homes, and the like. These environments become places where infections circulate among people and contaminate surfaces. Present in these environments are a triple threat to human health: sick people, invasive devices (e.g., catheters and IVs), and treatment protocols such as antimicrobial therapy or others that cause **immunosuppression**. As these environments and ecologies have expanded, so too has the threat posed by **nosocomial** infections. Nosocomial infections are those that are the result of treatment in a hospital or other health care setting, infections acquired by patients admitted to or interacting with a health care setting for some other reason. The CDC estimates that these hospital-associated infections, from all types of bacteria, cause close to 2 million infections and 100,000 deaths each year in the United States.

SA is a bacteria often found colonizing the nose, throat, and skin of humans, typically not causing any sort of problem; some estimates indicate that 20–30% of the population is colonized at any given point in time. Most individuals who do develop a SA infection do not experience any symptoms, or experience minor symptoms such as pimples on the skin. SA becomes a problem when the bacteria is able to breach the skin or mucous membranes of the nose and throat and enter the bloodstream or other internal organs, causing sepsis, endocarditis, or other life-threatening conditions. SA can spread via contact with an infected wound, via skin-to-skin contact with an infected or colonized individual, or via contact with objects such as sheets or towels that were used by a colonized or infected individual.

The ubiquity of SA meant that it was one of the first bacterial infections to benefit from the introduction of antibiotics. Penicillin in particular was the first choice to treat an SA infection. By 1950 some 40% of hospital SA bacteria were penicillin-resistant, the number by 1960 was 80%. Over the subsequent decades SA developed resistance to a string of antibiotics: the term methicillin-resistant SA (MRSA) describes SA bacteria that have developed resistance to methicillin, penicillin, oxacillin, and other members of that antibiotic family. This drug resistance is based upon the *mecA* gene that allows SA to produce an enzyme (beta-lactamase) that breaks down the key mechanism in this family (known as the beta-lactams).

MRSA is not more virulent than its drug-susceptible brethren. What makes MRSA so much riskier is that our treatment options against it are extremely limited. Because treatment options are limited and because SA is so well adapted to spreading in health care settings, the number of hospital-associated MRSA infections in the United States was estimated at 70,000 in 2012. While this number is high, it actually represents a major decrease in nosocomial MRSA infections, part of intensive efforts by hospital staff to decrease contamination and infection.

When methicillin, penicillin, and other beta-lactam antibiotics became ineffective against infections caused by MRSA, doctors turned to another antibiotic, vancomycin, to treat their patients. Unlike beta-lactams which can be taken in pill form, vancomycin cannot be absorbed by the body via capsules. Instead, vancomycin must be delivered via intravenous methods. This breaching of the skin potentially exacerbates a patient's MRSA infection and increases the risk that he or she will pass the infection on to nearby surfaces, equipment, health care workers' hands or clothes, and other patients. While vancomycin was originally very effective in treating MRSA, over time the dosage required to kill MRSA has doubled and tripled and now there are SA bacteria that are resistant to vancomycin altogether (VRSA). Problems arise when VRSA bacteria also carry the *mecA* gene that encodes for methicillin resistance because treatment options then become even more limited.

While drug-resistant *S. aureus* is highly problematic in health care (and community) settings, the problem is bigger than the drug resistance observed in a single bacterial species. All the procedures of modern medicine depend upon antibiotics. When the effectiveness of antibiotic treatments, **prophylactic** or **therapeutic**, diminishes or disappears, then so too do options for medical care. Dialysis, an increasingly important treatment given kidney failure associated with type II diabetes, becomes enormously risky. Elective surgeries would virtually cease. Every trip to the ICU and every IV or blood draw could transmit drug-resistant bacteria with no options for treatment. While the epidemiological transition began in the wealthy world because of sanitation and hygiene, it has increasingly been the result of antibiotic treatment protocols. Antibiotics are not an infinite resource, however, and drug discovery focused on finding new antibiotics has languished in the last decades. Antibiotics are cheap, relatively speaking, and are likely to be generic rather than branded, so there is little money to be made in that business. The rise in antibiotic resistance has been, unfortunately, paralleled by a decline in new antibiotics entering the market (Figure 7.2).

As we are reminded by the Infectious Diseases Society of America:

> We need to remember that human beings did not invent antibiotics; we merely discovered them. Genetic analysis of microbial metabolic pathways indicates that microbes invented both beta-lactam antibiotics and beta-lactamase enzymes to resist those antibiotics >2 billion years ago (Hall, Salipante, & Barlow, 2004). In contrast, antibiotics were not discovered by humans until the first half of the 20th century. Thus, microbes have had collective experience creating and defeating antibiotics for 20 million times longer than *Homo sapiens* have known that antibiotics existed. (Spellberg et al., 2008)

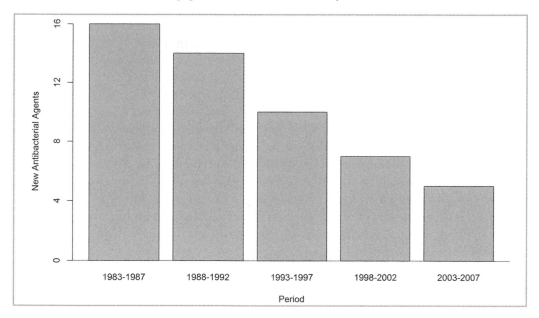

FIGURE 7.2 Systemic (i.e., nontopical) antibacterial new molecular entities approved by the U.S. Food and Drug Administration, per 5-year period.

SARS and Trade

In early 2003, a new infectious disease was observed, first in Asia and then worldwide. It was termed SARS, for severe acute respiratory syndrome. A **syndrome** is a set of signs and symptoms that usually occur together. AIDS, for instance, is a syndrome of associated symptoms that we now know is caused by the human immunodeficiency virus (HIV). Just as with AIDS, the cause of SARS was unknown when it originally appeared. SARS first came to the attention of the world in February and March of 2003 when individuals in Hong Kong, and individuals who had traveled to Hong Kong and then returned home, developed very high fevers and flu-like symptoms. From Hong Kong, SARS radiated outward with airline passengers to North America, Europe, and other parts of Asia. Health care workers treating sick travelers contracted the disease themselves and passed it along to their families and communities. In April 2003 the cause of SARS, a novel, never-before-seen coronavirus (SARS-CoV), had been isolated.

While SARS had come to the attention of the world and the WHO in early 2003, it had been circulating in southern China since 2002. The Chinese government received a significant amount of negative press in the wake of the SARS epidemic for not notifying the WHO about reports of a new, deadly disease as soon as it was observed. Guangdong Province is the most populous province in China, with over 100 million residents. It is a primary destination for internal migrants in China because it is a production and export hub. This export industry makes Guangdong a site of massive income inequality, with incredible wealth being generated by millions of low-paid workers.

In November 2002 an individual was admitted to the intensive care unit of a hospital in Guangdong Province. The patient had a high fever, respiratory failure, and lung damage. Four family members who had direct contact with the patient were also sick. The patient had a history of direct contact with and consumption of wild cat meat. A month later a chef working in Shenzhen, a city in Guangdong Province directly across the border from Hong Kong, also reported to the hospital with high fever, respiratory symptoms, and lung damage. He infected three family members and hospital staff who all developed the same symptoms. As a chef the patient had come into frequent contact with live caged animals being served as food in the restaurant.

Scientists began to examine blood samples from previous years to see if the virus appeared in stored blood, an indication that the virus had been cycling for some time but not caused an epidemic. Retrospective studies find no evidence of the SARS coronavirus in the population, unlike other diseases such as HIV and Legionnaires' disease, which have been found in stored blood samples. However, analysis of new blood samples indicated that antibodies against SARS-CoV or related viruses were present at a higher ratio in animal traders than control populations. Epidemiological studies also indicated that early case-patients were more likely than later case-patients to report living near a produce market but not near a farm, and almost half of them were food handlers with probable animal contact. Scientists began testing local wildlife to find evidence of SARS-CoV in an animal population.

The original suspect for SARS-CoV was the masked palm civet, a type of cat described above as being consumed by one of the first SARS patients. SARS-CoV found in humans was the same as SARS Co-V isolated in civet cats. People were exposed to civets via live animal markets and in restaurants, as the small animal is considered a delicacy in southern China. As wealth in China increases and the country passes through the nutrition transition, consumption of civets and other meats has increased. Civets are raised on farms in Vietnam, Laos, and Cambodia, as well as in Southern China, and are traded across borders to satisfy increased demand for their meat. In response to civets being linked to SARS Co-V, thousands were culled in Guangdong Province in an attempt to stop the epidemic.

While civets tested positive for SARS Co-V, they also showed the same symptoms of infection as people: high fever, respiratory distress, etc. Additionally, there were no widespread infections detected in civet farms. This suggested that, in fact, civets were not the reservoir host of SARS Co-V, since reservoir hosts usually exhibit **asymptomatic** or low-grade infections. Instead, the horseshoe bat emerged as the likely reservoir of SARS Co-V. Horseshoe bats are consumed as food and their manure is used in traditional Chinese medicine. While horseshoe bats are not farmed in the same way that civets are, they often overlap with farmed or other wild caught animals in warehouses, trucks, or live markets where the virus can be exchanged in feces, urine, blood, or aerosols. These close ecological interactions between humans, civets, and bats allowed SARS to jump the species barrier. Bats, in addition to their role in SARS and Nipah and Hendra, have also been implicated in the emergence of Ebola.

SARS disappeared in 2003 and has not been observed since that time. Public health interventions, such as isolating sick patients, closing schools and live markets, installing temperature scanners in airports to detect feverish patients before they get on planes or enter the country, helped to stop the epidemic. The potential remains, however, for SARS Co-V or another novel virus to emerge from the same ecological system. Population growth and mobility, economic change, the farming of animals, the harvesting of wild animals, cross-border trade—all this creates an ecology that is conducive to the emergence of previously unknown infectious diseases.

WHERE CAN WE EXPECT THESE DISEASES TO EMERGE/REEMERGE?

The causes of infectious disease persistence, emergence, and reemergence are complex and operate differently in different places. However, if we know that phenomena such as increased human/animal contact during agricultural extensification or increased human mobility or poverty are the underlying drivers of infectious disease, and if we know where previous emerging infectious diseases (EIDs) originated, then we can start to predict where, if not when, the emergence or reemergence of disease is most likely to take place.

Jones and coauthors utilized data on previous EID events and spatial analysis of variables such as human population growth and density, animal species richness, and climate variables (temperature and rainfall) to produce maps predicting where EID events are most likely to occur (Plate 13a). They predict, based upon their analysis, that infectious diseases that are transmitted into humans from animal species are most likely to emerge in places such as the Himalayan region, eastern China, and parts of sub-Saharan Africa. These are places in the world where high human populations are forcing increased interactions with animal populations, either through the hunting of bush meat, the expansion of human agricultural activity into forests, the wild pet trade, or via other pathways.

HOW WILL THESE DISEASES BEHAVE?

As has previously been discussed (Chapter 2 and Chapter 6), the general pattern of disease seems to indicate that they evolved toward benignity, that new diseases are highly pathogenic, and that less virulent diseases are those that are older and more adapted to humans. This phenomenon, known as commensalism, holds true for many pathogens but not for others. An evolutionary reasoning underlies commensalism: if a pathogen is too deadly, then it will kill off its human hosts too quickly, before they are able to pass the pathogen on to the next susceptible individual. To increase chances of survival in host populations, a pathogen will evolve to be less virulent, to maintain infection for longer, and to increase reproductive chances. Examples of diseases following this pattern include syphilis, which was highly deadly when it first appeared in Europe in the early 16th century but gradually became a chronic disease, and cholera, which initially caused pandemics with extremely high mortality in Europe but subsequent pandemics were much less deadly.

Commensalism, as described in Chapter 2, is not a given, not a predetermined outcome that all pathogens obey. It may happen, it may not. When it is not in the pathogen's best interest, evolutionarily speaking, to become more benign, it will maintain or even increase virulence. The goal of pathogens is the same as the goal of all life on Earth: to reproduce and pass genetic material on to the next generation. Natural selection will favor those strains or variants of a pathogen that are able not only to survive but to generate offspring. When high levels of virulence will not limit transmission and reproduction in other susceptible hosts, pathogens have little incentive to become less virulent. Situations where high virulence does not limit transmission and reproduction include when there are large numbers of susceptible individuals interacting at high levels, or when a disease is vectored by a fly or mosquito and does not depend on direct exchange between a sick individual and a susceptible individual. Examples of diseases that have not moderated virulence over time include HIV and dengue.

HIV

HIV is a retrovirus that spreads via body fluids such as blood and semen. HIV evolved from simian immunodeficiency viruses (SIVs), which circulate in nonhuman primates of Central and West Africa. While SIVs can be present at high levels in hosts such as green monkeys and chimpanzees, the animals typically exhibit only a mild immune response. This high viral load but lack of symptomology indicates that SIVs are commensal in nonhuman primates, that is, they have evolved to a relatively benign state. The same cannot be said for HIV.

HIV transferred from nonhuman primates into humans sometime in the early 20th century. Genetic evidence suggests that SIVs in chimps from Cameroon are most similar to HIV (Keele et al., 2006). This transmission event was likely one of several times that SIVs passed into people during hunting and butchering of nonhuman primates, but the small, isolated populations in forests caused viruses to extinguish before finding enough susceptibles to sustain transmission.

While the SIV that became HIV may have originated in Cameroon, it is the Democratic Republic of the Congo that is likely the site of the modern HIV epidemic (Faria et al., 2014; Worobey et al., 2008). Economic connections between Cameroon and the Congo, the city of Kinshasa in particular, meant that HIV moved with people and goods. HIV traveled from the jungle to villages to the cities of sub-Saharan Africa along with economic migrants. From Kinshasa it radiated outward along rivers, roads, and rail lines, into cities and plantations and mining operations. The increasing size of sexual networks present in these places meant that HIV was no longer confined to a village of a few dozen people with only one or two sex partners. The decades surrounding independence, the 1960s and 1970s, in the Congo and other Central African countries coincide with a spike in HIV infections and genetic diversity of HIV. Independence meant several things: that colonial government controls on population movement ended, that colonial plantation and mining jobs disappeared and people moved to cities, that there were more opportunities for people in the emerging cities of independent countries than there had been in colonial outposts. Simultaneously, the demographic transition meant there were more people than ever before in sub-Saharan Africa.

In addition to radiating outward from Kinshasa and other cities of Central Africa, HIV jumped to Haiti. The newly independent Democratic Republic of the Congo invited French-speaking, educated professionals from Haiti to fill gaps in teaching and civil service and other occupations left by the departing Belgians. When Haitians returned home they took HIV with them: HIV in the Western Hemisphere can be traced to the mid-1960s. In both Haiti and sub-Saharan Africa HIV was predominantly transmitted via heterosexual sex. Haiti in the 1970s became a sex tourism destination for American men, seeking either heterosexual or homosexual encounters. Retrospective analysis of stored blood samples from the United States detects HIV as early as 1969.

HIV remained a simmering epidemic in the United States until the 1980s, even as it was ravaging through urban populations in Africa. CDC doctors dispatched to Kinshasa after HIV emerged in the United States in the early 1980s describe finding hospital wards full of individuals dying of AIDS, a larger and more diverse epidemic than was observed in the United States at the same time. In the United States, HIV began to spread within the men who have sex with men (MSM) community, a population experiencing new levels of openness and freedom, and a mobile and often wealthy community, and a community that was not worried about sexually transmitted diseases since so many were cured by antibiotics. Simultaneously HIV made its way into the injecting drug user (IDU) communities; the 1980s were a time of exploding crack and heroin epidemics with needle sharing and needle renting in the shooting galleries of major cities. HIV found in the United States what it had found decades before in Congo and elsewhere: a population of well-connected susceptible individuals. These connected, unprotected, and susceptible populations drove the emergence of a worldwide epidemic, and created an environment wherein HIV faces little or no pressure to decrease its virulence.

Dengue Fever

Dengue fever is caused by viruses belonging to the family *Flaviviradae*. There are four main types of virus, known as DENV 1-4. Dengue viruses are transmitted primarily by two vectors in the *Aedes* genus of mosquitoes: *Aedes aegypti* and *Aedes albopictus*. Known colloquially as "breakbone fever," the primary symptoms of a dengue infection are severe aches and pains, fever, and rash. Historically, dengue was a childhood disease that was rarely fatal, primarily found in the jungles of southeast Asia. Dengue was able to maintain a sylvatic enzootic cycle, transmitting from nonhuman primates into mosquitoes and back again (Vasilakis & Weaver, 2008). Humans were often incidental hosts, bringing infections back into human settlements after being bitten by mosquitoes in the forest. The primary vector responsible for this forest cycle and low infection rates in people was *Ae. albopictus*, the tiger mosquito, which will bite people but prefers to bite other animals.

As population exchange increased and travel time decreased in the 16th to 18th centuries, dengue fever emerged from the jungles of southeast Asia and was transported both to Africa and the Americas. It appeared in the Caribbean in the 17th century, and a pandemic was described in Philadelphia in 1780. In the Americas and Africa dengue encountered a new vector, *Ae. aegypti*. This African mosquito was introduced to the Americas via the slave trade and to Asia via merchant trade in the 17th and 18th centuries, stowing away on sailing ships. *Ae. aegypti* is the vector for yellow fever, a virus in the same family as dengue that originates in Africa. Wherever *Ae. aegypti* spread, so too did yellow fever.

Dengue fever control benefitted greatly from yellow fever research. Yellow fever was a much bigger threat to human health in the 18th and 19th centuries, and control programs to decrease yellow fever epidemics had the side effect of also decreasing dengue incidence. In particular, the use of DDT in mosquito eradication efforts led to the near eradication of dengue fever by the 1960s and 1970s, particularly in the Americas. By getting rid of the vector species, the transmission cycle was cut. The recognition that DDT had harmful and long-lasting impacts on other species, however, discontinued its use in vector control and *Ae. aegypti* populations came roaring back.

At the same time as vector species were rebounding, the world was experiencing a significant shift in the speed and scale of population movement. The huge increase in the global exchange of people and goods in the 16th and 17th centuries that had enabled dengue fever and *Ae. aegypti* to meet was paralleled in the growth of population exchange in the 1960s and subsequent decades. Populations in low- and middle-income countries began to grow rapidly as they entered the demographic transition, cities expanded as rural to urban migration increased, increasing wealth and consumerism meant more goods were being shipped around the world and more people were moving for labor. *Ae. aegypti* found a comfortable home in the cities of the developing world, while *Ae. albopictus* preferred the suburbs or countryside. *Ae. aegypti* is a container breeder, laying its eggs in any small pool of water it can find, and prefers to bite people over other animals. Cities are full not only of *Ae. aegypti*'s preferred meal but also of places to lay eggs: standing water in poorly drained areas, buckets or containers of water kept in households without indoor plumbing, tires or other abandoned trash that hold rainwater, and the like. As the number of people living in cities grew, so too did populations of *Ae. aegypti*, a "super vector" for dengue.

When people left their villages for the city, or left their countries to seek their fortunes overseas, they took with them their dengue viruses. The four types, DENV 1–4, were no longer geographically isolated in their areas of primary circulation. Until the 1980s most places where dengue existed only experienced one type (Messina et al., 2014). Beginning first in southeast Asia, where the four types have their ancestral home, where the four types were more likely to have overlapping ranges, where population growth and mobility was particularly high, a new type of dengue fever was observed: dengue hemorrhagic fever (DHF). Initially DHF was undiagnosed, and there was confusion about what pathogen could be causing the new epidemic with its high mortality rate. Dengue was not suspected, as it had never before caused hemorrhaging of the brain. Eventually it was recognized that DHF was the result of infection with multiple types of dengue viruses. Infection with one type confers lifelong immunity to that type. For a long time that meant that individuals would experience the type in their region and recover. With the mixing of peoples, mosquitoes, and types, however, increasingly people were experiencing infection with a second type after recovering from infection with another. DHF is caused by repetitive infections with different strains of dengue viruses, when the body's immune system overresponds and sends the body into hemorrhage and shock. Increasing transmission in cities by *Ae. aegypti* results in 50–100 million new infections annually, of those approximately 50,000 are DHF. As urbanization and migration continue, as people mix and bring their dengue viruses with them, as *Ae. aegypti* continues to find a home in cities, dengue and DHF will continue without decreasing virulence.

Recently a new type of dengue virus, DENV 5, has been found. This strain was discovered during a dengue outbreak in Malaysian Borneo in 2007 and appears to be most genetically similar to DENV 4 (Normile, 2013). While sustained transmission of DENV 5 in humans is not yet observed, it is likely circulating in nonhuman primates in Borneo. The potential for new types of dengue to emerge from the forests and infect people complicates treatment and prevention efforts such as vaccine development and makes total eradication of dengue unlikely if not impossible.

Zika Virus Disease

As this book was going to print the WHO had just declared Zika virus a "public health emergency of international concern" and it had spread to many countries in Latin America and it continues to spread. Zika was first discovered in Uganda in the early 1950s and is known to circulate in many other African countries as well as Asia, many Pacific Islands, and most recently in the Americas. Like dengue fever, Zika is a mosquito-borne virus that spreads when bitten by *Ae. aegypti* and *Ae. albopictus*. Zika disease is usually mild and many infections are subclinical; however, when a pregnant woman is infected with the virus it can cause a birth defect called "microcephaly." At the time of printing it was still unclear how prevalent these birth defects are among mothers infected with Zika virus. It was also unclear whether the virus would spread to the southern United States where both vector species, *Ae. aegypti* and *Ae. albopictus,* are present. While there could be some isolated transmission it is unlikely to spread widely compared to other countries such as Brazil because the U.S. climate and built environment are not as conducive to transmission. That being said, the U.S. government is so very concerned about the disease that the Obama administration requested that $1.8 billion in emergency funds be allotted for research into a vaccine and other prevention measures. This is the latest in a sequence of emerging viruses that involve large government response for treatment and prevention.

LANDSCAPE GENETICS

The emergence and reemergence of infectious diseases are due largely to the development of drug resistance or to pathogens jumping species (Figure 7.1; Jones et al., 2008), both of which require genetic changes on the part of the pathogen.

Therefore, incorporating a genetic perspective into the study of emerging infectious diseases is useful.

Humans exist in a state of dynamic equilibrium with their environments, with both people and places changing in response to the other. To understand disease, then, you have to understand both the person and the place (Hunter, 1974; May, 1958). However, this person/place duality overlooks a major player in emerging infectious diseases: the pathogen itself. Pathogens are evolving and changing just as people and places are, and those changes are often the result of human–environment interactions, an extension of the idea of commensalism (or noncommensalism): successful pathogens act in their own best interest, and that interest is often constrained or influenced by human–environment interactions. Animals and human hosts making contact for the first time, or animals and humans experiencing greater intensity of contact, or pathogens experiencing selective pressure from drug treatments—all of these opportunities for pathogenic evolution to be more fit or to better reproduce indicate that we need to consider the *pathogen* in the *person* in the *place* (Carrel, 2015). Disease in place A might not equal disease in place B, or person A and person B; just because two people and two places have influenza doesn't mean they have the *same* influenza, and these differences can shed light on how the virus is adapting to new environments or hosts. Or they can show how the virus is responding to drug pressure. Alternately, if these two people or two places *do* have the same influenza, then this tells us something about interactions between people over space.

The theory and methods for integrating a pathogen component into the traditional disease ecology approach of people and places are relatively new. It is only in the last decade or so that the speed and cost of genetic analysis has decreased to the point of being available on a wide scale. Genetic analysis enables detection of characteristics such as drug-resistant genes or calculation of genetic relatedness, an indicator of time since divergence from a single ancestor. **Phylogenetic** analysis, wherein family trees of viruses or other pathogens are estimated, is one way of measuring genetic relatedness or genetic proximity (see Figure 7.3). In this example, eight influenza viruses have been taken from eight infected individuals and had one of their genes, the surface gene hemagglutinin (HA), which controls binding of the virus to host cells, sequenced. While these are all sequences from the same gene of the same species, they have evolved differently over time and space and within hosts. Viruses 7 and 8 are quite close to each other in the phylogenetic tree, indicating that they were likely found in individuals that were close geographically or individuals who became infected around the same time. Viruses 7 and 8 are most closely related to virus 6 and have the greatest genetic distance from viruses 1, 2, and 3.

Landscape genetics is a subfield of landscape ecology that combines that field with population genetics. Landscape ecology is focused on understanding how patterns in the landscape influence processes of the organisms in that landscape (Turner, 1989). "Pattern" in this sense could describe the arrangement of different types of land cover (scrub vs. forest vs. water), how well connected or fragmented patches of the same type are (can a species that only lives in forest move across the landscape without leaving forested areas), or how borders or edges are arranged

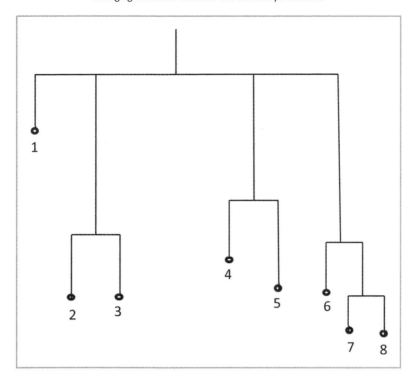

FIGURE 7.3 Hypothetical phylogenetic tree representing genetic relatedness between viruses.

in a landscape. The processes that landscape ecologists are interested in include how species are distributed across the landscape, how species migrate within a landscape, and how landscapes respond to disturbances (e.g., fire, tornadoes, or invasive species). With the addition of a genetic component, landscape ecologists have been able to determine how mountain ranges impact the mixing of populations on either side of the range, how human alteration of landscape impacts the reproductive success of species, and how highways act as barriers to the exchange of wildlife illness.

Utilizing landscape genetics to consider the evolution of human pathogens in a landscape is a natural extension of its typical application to plant and animal species. Integrating disease ecology theory on population and environment interactions (such as is suggested in the triangle of human ecology) with outcome data from population genetics and methods from landscape ecology allows us to answer questions that were not before possible to pose or answer (Carrel & Emch, 2013). For instance, we can start to understand how and why pathogens spread and evolve across landscape, what types of places are conducive to pathogens evolving quickly or developing drug resistance, and what barriers or interventions might slow or prevent disease emergence. Taking the example above, we can associate spatial and temporal information with each of our eight influenza viruses and also determine the population and environment characteristics associated with their place of isolation. We can see if viruses 7 and 8 share time and space, and what characteristics of the places where viruses 1, 2, and 3 are different from where 7 and 8 were found. If virus 1 represents a branch of the influenza family tree that is now extinct, we could determine if the place where virus 1 was found is not conducive to the continuous circulation of influenza and why. Alternately, if viruses 2–8 were all found in the same place, this could suggest an environment that is conducive to rapid evolution of influenza viruses or a place where influenza viruses are continually being introduced.

Landscape Genetics of Avian Influenza Evolution

Highly pathogenic H5N1 avian influenza has persisted in poultry and human populations in Asian countries since 2003. While the total number of human cases since then has remained relatively low, only a little over 600, the fatality rate in these cases has been quite high, around 60%. The morbidity and mortality in domestic poultry has been much higher, in the hundreds of millions, either due to infection or to culling to prevent or contain outbreaks. H5N1 poses a huge economic risk in addition to a health risk, as the infection and destruction of domestic poultry populations threatens the livelihoods of farmers and decreases the food supply. The continuous circulation of H5N1 despite containment and eradication efforts, and the potential that H5N1 has to develop efficient and sustained human–human transmission, as has happened with so many other influenza strains, has led to a great deal of research into the ecology and genetics of H5N1.

Combining these types of studies into a landscape genetics analysis of the ecology of H5N1 takes previous research attempts a step further. Building on research that explored the types of populations and environments that were conducive to incidence, a landscape genetic analysis revealed that places in Vietnam with high amounts of aquaculture surfaces (i.e., lots of ponds), with low high school graduation rates (i.e., poor and uneducated populations), and with high human population densities were places where H5N1 avian influenza viruses demonstrated the highest rates of evolution (Carrel, Emch, Nguyen, Todd Jobe, & Wan, 2012). The combination of susceptible hosts, spaces of interaction between hosts, and low education/SES drive genetic variation. High rates of evolution equate to a greater number of opportunities for the virus to mutate and develop traits that make more efficient human–human transmission likely.

Landscape Genetics of Drug-Resistant Malaria

Malaria, like HIV and dengue, is a disease whose virulence has not lessened over time, even though it is an old disease, so old that some humans have evolved protection against malaria via sickle cell traits (which also cause sickle cell anemia). Malaria is caused by protozoan species of the genus *Plasmodia,* the most prevalent and deadliest of which is *Plasmodium falciparum.* Malarial parasites are spread exclusively by mosquitoes in the genus *Anopheles,* of which there are hundreds of species although only a few dozen are responsible for the majority of malarial transmission. The ecology of malaria is thus more complex than that of flu, which can be directly transmitted between humans (although in the case of avian influenza, discussed above, was more commonly transmitted between birds). Malaria genetics is reliant on the transmission of a malaria protozoa between two people via an *Anopheline* mosquito. Places where either mosquitoes or infected individuals are not present, then, represent places where malaria faces barriers to transmission. Research into malaria landscape genetics in the Democratic Republic of Congo found that malarial parasites found far apart were also far apart genetically but that there were no significant spatial barriers to malaria transmission (Carrel et al., 2015).

The evolution of drug resistance, discussed earlier in the context of *Staphylococcus aureus* and antibiotics, is also problematic for global efforts to reduce or eliminate malaria. One gene that confers drug resistance in malaria, the *dhps* gene, has been shown to have distinct regional groupings and lineages in sub-Saharan Africa (see Plate 13b; Pearce et al., 2009). This phylogeographic analysis can suggest landscape genetic studies exploring why these geographic differences exist and what they tell us about how malaria diffuses across the landscape.

Challenges Facing Landscape Genetics of Infectious Disease Studies

While the application of landscape genetics methods to the study of human pathogens has been made possible by the increased efficiency and decreased cost of computing and molecular typing, many challenges face the field. For instance, while genetic sequencing costs and times have fallen dramatically in the past decade, it is still often prohibitively expensive to sequence enough samples of a pathogen to generate an adequate sample size. Knowing whether a sample is an accurate representation of the pathogen is also difficult. HIV, for instance, evolves so quickly that a single individual can have vastly different viruses circulating in the body. Influenza viruses have eight gene segments and mutations can occur on each, though often researchers are limited to sequencing only one or two of the gene segments. In the case of more complex pathogens, such as malarial parasites, testing different parts of the genome can give different indications as to the speed or degree of molecular change.

Once genetics have been measured, interpreting those findings is also complex. Increased genetic distance or genetic change between two samples could mean two different things. It might indicate a degree of isolation or separation between the samples and their associated locations or populations: populations that are isolated are different genetically. Alternately, it might indicate a high level of mixing, that lots of infected individuals are sharing infections and providing greater opportunities for evolution and change.

Issues of time and mobility are inherent in many health geographic analyses and are present in the application of landscape genetics as well. When a sample of a pathogen was taken from an ill patient or sick animal is important as different diseases have differing lengths of time between infection and symptoms. Knowing how long the pathogen has had to evolve inside the body from when the pathogen first entered the body is important if there are questions about how pathogens are evolving across space and time. Landscape genetics studies have typically focused on plant and animal species, many of which have highly limited or circumscribed mobility, but humans are a very different type of animal with very different mobility patterns. The application of methods that were developed for plants, which don't move but whose offspring may be distributed via wind or water, to humans who move frequently and over great distances in short time periods, is potentially problematic.

The integration of genetics and geography has huge potential for understanding the ecology and evolution of infectious diseases, either emergent, reemergent, or persistent. The convergence of spatial data and molecular data allows for the addition of another layer of complexity to the analytic process, but it is not without complication.

QUICK REVIEW

✓ The burden of infectious diseases remains high due to the emergence, reemergence, and persistence of infectious diseases, particularly in low-income settings.

✓ Infectious disease emergence and reemergence is the result of several factors, including increased human–animal contact through agricultural extensification and increased human–human contact through migration and urbanization.

✓ Infectious disease emergence is also due to changes in infectious pathogens themselves, such as the development of resistance to antibiotics and vaccines.

✓ The development of new antibiotics has declined in recent decades in favor of more lucrative medicines.

✓ Combining genetic analysis with health geography allows researchers to understand why and where infectious disease pathogens evolve and emerge.

CONCLUSION

The morbidity and mortality burden of infectious diseases is low in wealthy countries and continues to be high in low- and middle-income countries. The double burden of disease experienced by much of the world will continue: infectious diseases will never be absent from the disease landscape. The optimism experienced by the public health community in the 1960s and 1970s, the belief that infectious diseases could be virtually eliminated, is no longer the prevailing sentiment. Instead, even as public health makes headway reducing malaria (via bed nets, for instance) or polio (via vaccination), other infectious diseases that emerge or reemerge, or persist, maintain or increase the global burden of infectious disease.

There are many causes of infectious disease emergence/reemergence/persistence, typically operating together to produce conditions ripe for infectious disease. These include, but are not limited to, the expansion of humans into animal habitats via deforestation for agricultural production or resource extraction, the increased migration and circulation of humans, the crowding of susceptible populations into urban environments, the evolution of pathogens due to antibiotic treatments, increasing elderly populations in health care settings, and the persistence of poverty. Put succinctly, high population densities, well-connected populations, and changing interactions with environments drive infectious disease patterns. As the world's population lives in closer quarters (see Chapter 10) and mobility increases, the chances of a disease not only emerging but reaching a large portion of the world increases as well; no longer are emergent diseases likely to be geographically confined. Some emergent and reemergent infectious diseases appear as though they are here to stay (e.g., HIV), while others flare up and die out (e.g., SARS). Previous disease patterns suggest that when less-virulent strains are able to outcompete more virulent strains, such as when higher virulence kills infected individuals too quickly for them to be able to pass the pathogen onward, pathogens tend toward commensalism, becoming minor infections that result in morbidity but not mortality. As population interactions increase, however, it is less likely that this commensalism will take place and that less-virulent strains will not outcompete their more virulent brethren. Containment of infectious disease will thus depend not on treatment but on prevention, on quarantine and school closings and airline cancellations.

REVIEW QUESTIONS

1. Choose two recent emerging infectious disease events and compare and contrast the underlying drivers of their emergence. Did they emerge from where Jones et al. (2008) predicted that they would?

2. Thinking about the list of underlying drivers of disease that was put forth by Mayer, do you think anything is missing? Which of the factors from the list is most important? Which is the most modifiable (i.e., Can we do anything about these factors)?

3. The Centers for Disease Control and Prevention publishes a journal called *Emerging Infectious Diseases*. Visit the website of the journal (*www.nc.cdc.gov/eid*). Do the papers address diseases that you would have expected? Are they focused on domestic disease incidence (in the United States) or internationally?

4. Visit *healthmap.org* and explore what infectious diseases are currently being reported. Zoom in on your location. Is anything being reported in your area?

5. How do your own behaviors influence genetic change of pathogens? Think about your use of antibacterial soaps or other products, your consumption of meat produced with antibiotics, your uptake of seasonal influenza vaccines, and the like.

6. If infectious diseases have always been a threat to human health, what is different about this time for our species? Why are we so worried about emerging and reemerging infectious diseases?

7. Given that poverty is the underlying risk factor for most infectious disease, how should this fact inform global funding for health?

REFERENCES

Avila, M., Saïd, N., & Ojcius, D. M. (2008). The book reopened on infectious diseases. *Microbes and Infection, 10*(9), 942–947.

Bristol, N. (2008). William H. Stewart. *The Lancet, 372*(9633), 110.

Carrel, M. (2015). Disease at the molecular scale: Methods for exploring spatial patterns of pathogen genetics. In P. D. Kanaroglou, E. Delmelle, & A. Paez (Eds.), *Spatial analysis in health geography* (pp. 101–117). Surrey, UK: Ashgate.

Carrel, M. A., & Emch, M. E. (2013). Genetics: A new landscape for medical geography. *Annals of the Association of American Geographers, 103*(6), 1452–1467.

Carrel, M. A., Emch, M., Nguyen, T., Todd Jobe, R., & Wan, X. F. (2012). Population-environment drivers of H5N1 avian influenza molecular change in Vietnam. *Health and Place, 18*(5), 1122–1131.

Carrel, M. A., Patel, J., Taylor, S. M., Janko, M., Mwandagalirwa, M. K., Tshefu, A. K., et al. (2015). The geography of malaria genetics in the Democratic Republic of Congo: A complex and fragmented landscape. *Social Science and Medicine, 133,* 233–241.

Cohen, M. L. (2000). Changing patterns of infectious disease. *Nature, 406*(6797), 762–767.

Emch, M., & Root, E. D. (2009). Emerging and re-emerging diseases. In T. Brown, S. McLafferty, & G. Moon (Eds.), *A Companion to Health and Medical Geography* (pp. 154–172). Oxford, UK: Wiley-Blackwell.

Faria, N. R., Rambaut, A., Suchard, M. A., Baele, G., Bedford, T., Ward, M. J., et al. (2014). The early spread and epidemic ignition of HIV-1 in human populations. *Science, 346*(6205), 56–61.

Francis, L. P., Battin, M. P., Jacobson, J. A., Smith, C. B., & Botkin, J. (2005). How infectious diseases got left out—and what this omission might have meant for bioethics. *Bioethics, 19*(4), 307–322.

Funk, S., Bogich, T. L., Jones, K. E., Kilpatrick, A. M., & Daszak, P. (2013). Quantifying trends in disease impact to produce a consistent and reproducible definition of an emerging infectious disease. *PLoS ONE, 8*(8), e69951.

Hall, B. G., Salipante, S. J., & Barlow, M. (2004). Independent origins of subgroup Bl+ B2 and subgroup B3 metallo-β-lactamases. *Journal of Molecular Evolution, 59*(1), 133–141.

Hunter, J. M. (1974). The challenge of medical geography. In J. M. Hunter (Ed.), *The geography of health and disease* (pp. 1–31). Chapel Hill: Department of Geography, University of North Carolina at Chapel Hill.

Jones, K. E., Patel, N. G., Levy, M. A., Storeygard, A., Balk, D., Gittleman, J. L., et al. (2008). Global trends in emerging infectious diseases. *Nature, 451*(7181), 990–993.

Keele, B. F., Van Heuverswyn, F., Li, Y., Bailes, E., Takehisa, J., Santiago, M. L., et al. (2006). Chimpanzee reservoirs of pandemic and nonpandemic HIV-1. *Science, 313*(5786), 523–526.

May, J. M. (1958). *The ecology of human disease: Studies in medical geography.* New York: MD Publications.

Mayer, J. D. (2000). Geography, ecology and emerging infectious diseases. *Social Science and Medicine, 50*(7–8), 937–952.

Meade, M. S., & Emch, M. (2010). *Medical geography* (3rd ed.). New York: Guilford Press.

Messina, J. P., Brady, O. J., Scott, T. W., Zou, C., Pigott, D. M., Duda, K. A., et al. (2014). Global spread of dengue virus types: Mapping the 70 year history. *Trends in Microbiology, 22*(3), 138–146.

Nesse, R. M., & Williams, G. C. (1998). Evolution and the origins of disease. *Scientific American–American Edition, 279,* 58–65.

Normile, D. (2013). Surprising new dengue virus throws a spanner in disease control efforts. *Science, 342*(6157), 415.

Pearce, R. J., Pota, H., Evehe, M. S. B., Bâ, E. H., Mombo-Ngoma, G., Malisa, A. L., et al. (2009). Multiple origins and regional dispersal of resistant dhps in African plasmodium falciparum malaria. *PLoS Medicine, 6*(4), e1000055.

Spellberg, B., Guidos, R., Gilbert, D., Bradley, J., Boucher, H. W., Scheld, W. M., et al. (2008). The epidemic of antibiotic-resistant infections: A call to action for the medical community from the Infectious Diseases Society of America. *Clinical Infectious Diseases, 46*(2), 155–164.

Spellberg, B., & Taylor-Blake, B. (2013). On the exoneration of Dr. William H. Stewart: Debunking an urban legend. *Infectious Diseases of Poverty, 2*(3).

Turner, M. G. (1989). Landscape ecology: The effect of pattern on process. *Annual Review of Ecology and Systematics, 20*, 171–197.

Upshur, R. (2008). Ethics and infectious disease. *Bulletin of the World Health Organization, 86*(8), 654.

Vasilakis, N., & Weaver, S. C. (2008). The history and evolution of human dengue emergence. *Advances in Virus Research, 72*, 1–76.

Worobey, M., Gemmel, M., Teuwen, D. E., Haselkorn, T., Kunstman, K., Bunce, M., et al. (2008). Direct evidence of extensive diversity of HIV-1 in Kinshasa by 1960. *Nature, 455*(7213), 661–664.

FURTHER READING •

Desowitz, R. S. (1981). *New Guinea tapeworms and Jewish grandmothers.* New York: Norton.

Drexler, M. (2002). *Secret agents: The menace of emerging infections.* Washington, DC: National Academy Press.

Garrett, L. (1994). *The coming plague: Newly emerging diseases in a world out of balance.* New York: Farrar, Straus & Giroux.

Hubbert, W. T., McCulloch, W. F., & Schnurrenberger, P. R. (Eds.). (1975). *Diseases transmitted from animals to man* (6th ed.). Springfield, IL: Chales C Thomas.

Karlen, A. (1995). *Man and microbes: Diseases and plagues in history and modern times.* New York: Putnum Press.

McMichael, T. (2001). *Human frontiers, environments and disease.* Cambridge, UK: Cambridge University Press.

Quammen, D. (2012). *Spillover: Animal infections and the next human pandemic.* New York: Norton.

Sharma, V. P. (1996). Re-emergence of malaria in India. *Indian Journal of Medical Research, 103*, 26–45.

Singhanetra, R. A. (1993). Malaria and mobility in Thailand. *Social Science and Medicine, 37*, 1147–1154.

Stanley, N. F., & Alpers, M. P. (Eds.). (1975). *Man-made lakes and human health.* New York: Academic Press.

Wills, C. (1996). *Yellow fever, black goddess: The coevolution of people and plagues.* Reading, MA: Addison-Wesley.

World Health Organization (WHO). (1976). *Water resources development and health: A selected bibliography* (Document No. MPD/76.6). Geneva, Switzerland: Author.

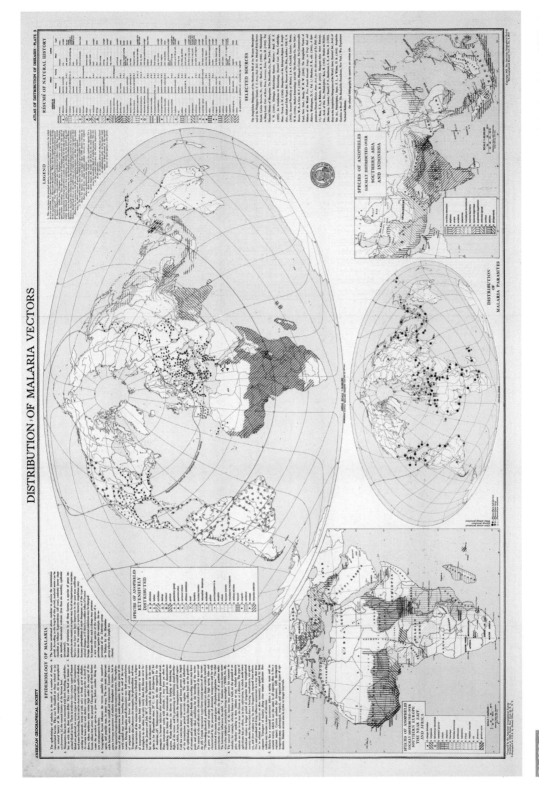

PLATE 1 May's map of the distribution of species of *Anopheles* mosquito vectors of malaria, inserted into the *Geographical Review* in 1951. Copyright ©1951 the American Geographical Society.

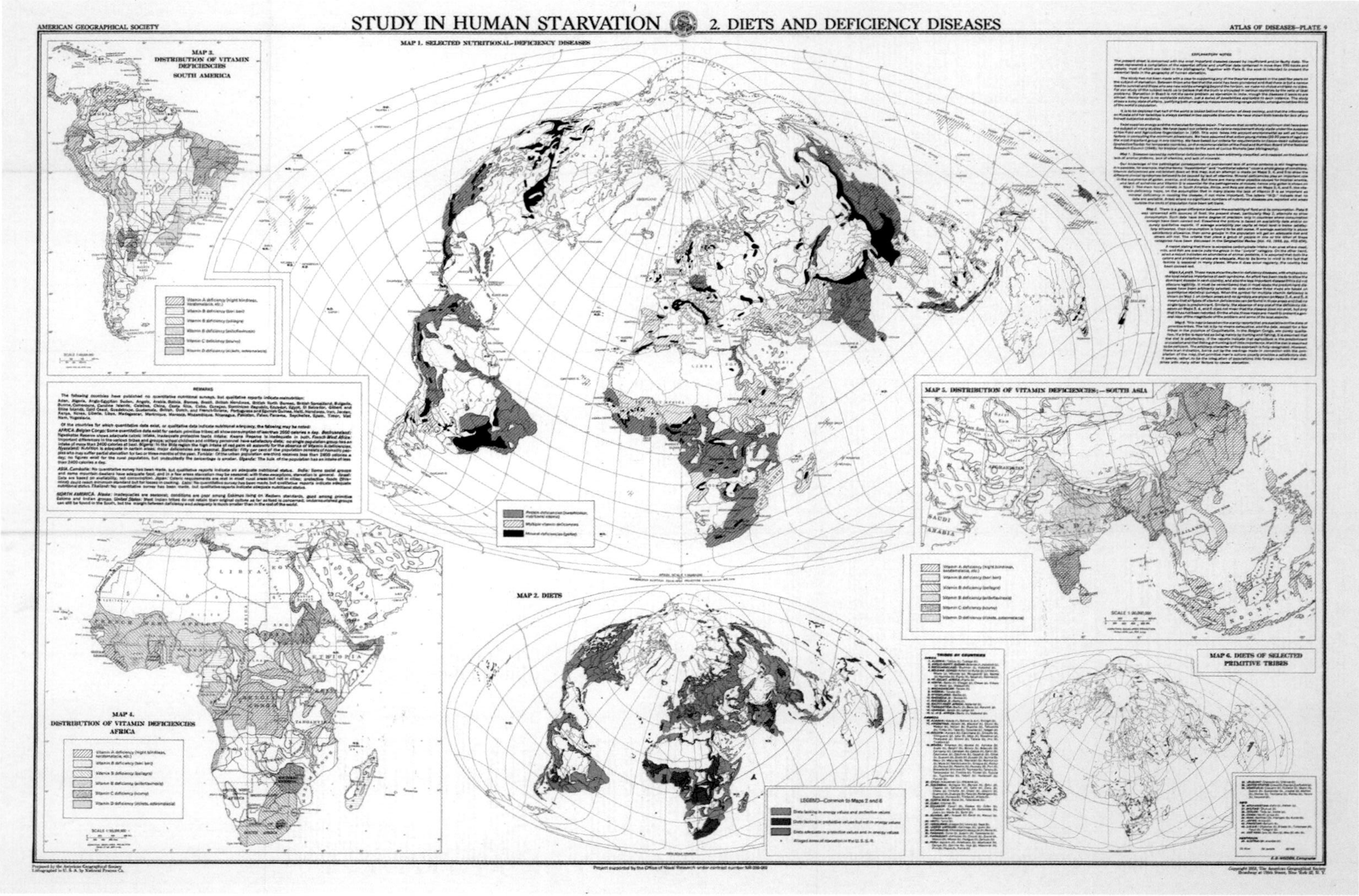

PLATE 2 May's map of starvation, diet, and deficiencies, inserted into the *Geographical Review* in 1953. Copyright ©1953 the American Geographical Society.

PLATE 3 Household type and vulnerability. *Upper left:* Rural Bangladesh (2008). Houseboats where people live permanently. The river is a reservoir for cholera. Cooking is done on small stoves fueled with wood or leaves; these stoves give off a lot of smoke, which is inhaled. *Upper right:* Northwest Frontier Province, Pakistan, near Afghanistan border (1988). Mud and stone house where cracked walls could be habitat for sand flies that transmit visceral leishmaniasis (*kalaazar*). *Middle left:* Northeast Thailand (1968). Local (*lao*) houses with typical architecture—open porch/house side, on stilts (to shade animals and weaving, etc.), now with a corrugated iron roof; burning garbage pit; water jar on porch. *Middle right:* Southern Belize (2005). A virtually identical house to the northeast Thailand house on stilts, but almost 40 years later and halfway across the world. *Lower left:* Southern Belize (2005). Mayan house with thatched roof, pigs in courtyard. *Lower right:* Chapel Hill, North Carolina (2008). House with screens, central heating, central air conditioning.

PLATE 4 Resources for drinking water. *Upper left:* Guanxi, China (2006). Dug well with hand pump. Cement cover to protect the water. *Upper right:* Rural Malaysia (1974). High school lesson for making a safe well: common dipper, so individual pails don't contaminate; protected by fence, so chickens, other animals, and small children don't fall in. But really useless, with water table so high and contaminated water so near. *Lower left:* Rural Punjab, India (1984). Residential tube well in Punjab as Green Revolution prosperity transformed water safety. *Lower right:* Rural Bangladesh (2007). Shallow tube well painted red because of high arsenic concentrations. Arsenic occurs naturally in the shallow aquifer in much of rural Bangladesh and causes cancer when water from these wells is drunk over long periods.

PLATE 5 Sanitation systems. *Upper left:* Guanxi, China (2006). Latrines at a school where a typhoid fever outbreak occurred. *Upper right:* Rural Malaysia (1974). This banana-shaded private place was designated in the national census as a latrine shared by three households. It drains right into the water! *Middle left:* Rural Bangladesh (2008). Latrine draining into pond that is often used for bathing or washing dishes. *Middle right:* Thailand (1974). Thai "squat toilet," a common type of Asian floor toilet. It's water-sealed, but jar must be filled with water from well to fill bowl for flushing; jar of water breeds *Aedes aegypti*. *Lower left:* Chapel Hill, North Carolina (2008). Water-sealed, water-saving toilet. *Lower right:* Jaipur, India (1984). Alley designated locally for defecation.

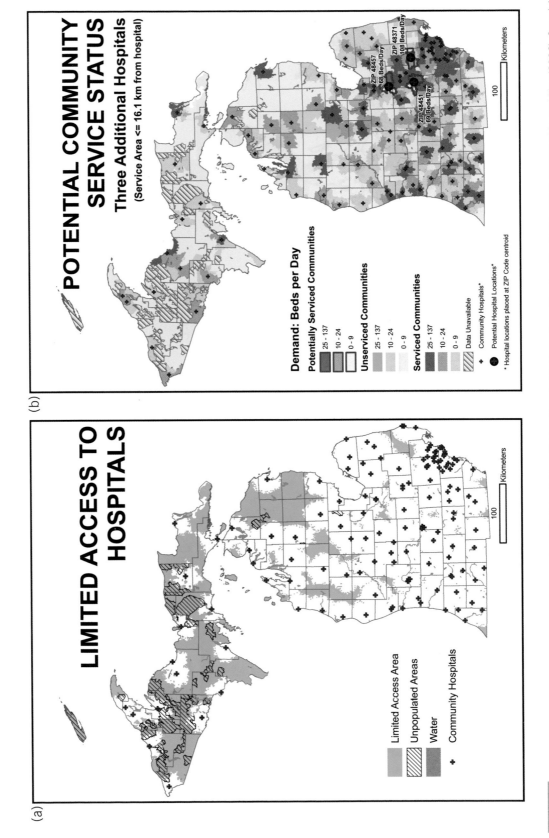

PLATE 16 (a) Areas with limited access to hospitals in Michigan. (b) Potential hospital locations in Michigan. From Messina, Shortridge, Groop, Varnakovida, and Finn (2006). Copyright ©2006 BioMed Central Ltd., part of Springer + Business Media. Reprinted by permission.

PART III

WHAT WE EAT
and WHERE WE LIVE

The ecology of humanity is undergoing a radical shift in terms of diet and habitat. More people are eating more, and different, calories. Our physical environment is changing, as more of us live in cities than in rural areas. This section explores how our changing food behaviors and living environments will impact human health in a number of ways.

Chapter 8 comprehensively discusses the geography of food, diet, and nutrition. The emergence of obesity, diabetes, and other diet-related diseases is explored in this chapter, in light of the revolutionary changes in modern food production, marketing, and distribution. The context for these discussions is the nutrition transition, in terms of global movement toward a "Western" diet and the reasons for this change. Problems of overnutrition, and the spatial distributions of food outlets and recreational facilities (food deserts), are also examined. This chapter further explores the health impacts associated with intensive production of crops, such as the depletion of groundwater sources, and intensive production of animals, such as antibiotic resistance from subtherapeutic dosing of confined swine and cattle.

Chapter 9 explores the impact the environments we live in (neighborhoods) have on health.

Neighborhoods are broadly defined to include both the built and the social environments. We explore how the social connections that people make within neighborhood environments can either support good health or contribute to poor health behaviors and stress, which impacts both mental and physical health outcomes. Structural factors operating at larger scales but playing out in individual neighborhoods are also discussed. Studying the effect of neighborhood environments (both social and structural) is difficult, especially with complex health problems such as cardiovascular disease, and so we explore many of the challenges that geographers have encountered in this work. Methodologically, the chapter describes how to separate compositional effects (i.e., who lives in a place) from contextual effects (i.e., what the place is like) and how to define the spatial bounds of a neighborhood.

Chapter 10 focuses on how the movement of the global population into urban areas is impacting human health. Urban areas represent unique ecologies for human health, with massive population densities and interactions with built environments. How a city can support itself, via provision of water, food, sewer, electricity, trash pickup, education, health care, and other services, is explored in the context of human health.

The chapter discusses how cities struggle to provide these public services in the face of rapidly increasing populations, with particular reference to the disease experience of rapidly growing cities historically (i.e., the tenements and rookeries of Industrial Revolution London and New York), and how modern megacities face a double burden of infectious and noninfectious disease. The chapter includes an examination of traffic, in the context of morbidity and mortality due to traffic accidents and air pollution in low- and middle-income countries, where car and motorbike ownership is increasing but traffic safety measures are scarce. Finally, the chapter discusses the issue of shrinking cities, where population growth is slowing or has already peaked and where cities have to scale back on the services that they provide.

CHAPTER 8

Food, Diet, and the Nutrition Transition

Humans are the only animals that cook their food. Cooking food increases the efficiency of eating by breaking plants and animals down into more easily absorbed components. Increased absorption efficiency meant that our digestive tract could be smaller, that more energy and space could go to developing our brains, and that humans could spend less time eating and more time doing other things, like developing language, tools, and art. Humans have taken *food,* that most basic of needs in **Maslow's hierarchy of needs**, alongside breathing, sleep, and sex, and have transformed it into something much more: *cuisine.* Our food exemplifies "high culture," meals that can cost up to thousands of dollars at restaurants where you have to make reservations years in advance, techniques and rules that are passed from chef to chef; and our food exemplifies "deep culture," food that is closely and emotionally associated with family and social history, homestyle cooking, and soul foods. Humans use food to express culture: most religions of the world have rules or traditions regarding food consumption or food abstinence at certain times of the year, certain foods are held up as examples of national or regional dishes, cultural membership is bound up in whether you follow food conventions, and so on. Even deciding what *is* and what *is not* food is deeply cultural, as food taboos are seldom universal. Cows to a devout Hindu are not a source of food, just as horses and dogs are not sources of food to most Americans, though they may be eaten elsewhere.

Just as food and cuisine are intensely cultural, they are also heavily influenced by the environment. This is where geography enters, given the focus of geographers on interactions between people and their environments. Often what has become or appears to be cultural about a regional cuisine is, in fact, a factor of the environment. What we eat was, for much of human history, limited by what was locally available. Regional variation in plants and animals determined what became regional cuisine: What are the agricultural limits of corn? Where did the ancestors of cattle live? How safe was the water to drink in a region? Tea consumption in Asian countries and wine and beer consumption in European countries may appear cultural in modern contexts, but these traditions have their roots in unsafe drinking water sources: boiling water for tea or fermenting wheat, grapes, barley, or hops in water removes the threat of bacteria. The practice of eating with your right hand rather than the left in places such as Bangladesh, or of consuming kosher or halal meats by Jews and Muslims, are reflections of historical environmental threats from lack of sanitation and refrigeration. The spices that we associate with specific cuisines, such as garlic and ginger and turmeric, are used today for their flavoring but at one point were also used for their antibacterial and antiworming properties.

Even today, when our local food environment is used to describe not the natural growing patterns of other species but rather the food outlets that we can access, this interaction between environment and culture is highly influential on our cuisine and our health. This chapter explores humans' history with food and agriculture, focusing in particular on how our modern food production system influences human health in direct and indirect ways. The first section of the chapter reviews the transition that humans made from collecting their food to purposely growing it, and the health consequences associated with this transition. The second portion of the chapter examines major shifts in agriculture that have taken place since humans began growing their own food, namely, the Columbian exchange and the Green Revolution. The final portion of the chapter explores the major changes in food consumption that have taken place in the last few decades and the direct and indirect health consequences of these changes, made possible by commercial agriculture.

FROM HUNTER–GATHERERS TO FARMERS

Modern *Homo sapiens* evolved ~200,000 years ago (conservatively), while agriculture is only ~10,000 years old. For the vast majority (95%) of human history we have been **foragers**, hunting and gathering from the land. Our bodies are adapted to consume an **omnivorous**, hunter–gatherer diet, one that includes fruits, nuts, seeds, tubers and root vegetables, insects, amphibians, reptiles, and lean wild meats. While the plant and animal species comprising this hunter–gatherer diet varied widely around the world, all diets must satisfy the same basic nutritional needs, regardless of culture or environment: energy, protein, vitamins, and minerals. Most plants, and some animals, contain toxins and are unsuitable for human consumption. Sweet foods are usually safe, bitter flavors can signal toxicity. This may be why young children will readily eat sweet fruits but disdain vegetables—some evolutionary protection is kicking in. Because of the potential for strange or untested plants to harm us, humans tend to be conservative as a species—we find what works and we stick with it. Populations around the world found what worked in their region and, unwilling to risk the threats to exploring consumption of new species, transformed those safe species into cuisines and cultures.

The amount of territory needed to support hunter–gatherer populations varies. Water-based hunting and foraging requires less space than does land-based foraging, while diets that are composed of mainly gathered plant species require less range than do diets higher in hunted animal products (Hamilton, Milne, Walker, & Brown, 2007). Overall population size of these hunter–gathering *Homo sapiens* remained small. Life expectancy was very low, in the 20s, and deaths among adults were mainly due to violence, trauma, or childbirth.

At several points in time in several places around the world, groups of *H. sapiens* made the transition from foragers to **agriculturalists**. Exactly how this happened is still unclear; a popular explanation known as the "dump heap" theory suggests that people noticed plants growing from seeds discarded as waste from previous meals and began to grow these desired plants rather than forage for them. Agriculture, somewhat counterintuitively, actually requires more labor to produce food than does foraging, so why would hunter–gatherer groups make the transition to farming? It has been suggested that as hunter–gatherer populations grew there was less range available to groups for foraging, and fewer plants and animals to be hunted and gathered, so there was greater incentive for groups to invest their energies in growing food. Better tools for food preparation and storage as well as changing environmental conditions may also have encouraged groups to become **sedentary agriculturalists.** How and why it happened aside, we cannot understate the importance of this change for the history of our species.

The overall health of agriculturalists was poorer than that of hunter–gatherers; **paleopathological** evidence indicates shorter stature, for instance, due mainly to their narrower diets that were heavily based on grains rather than vegetables. However, as groups began to grow their own food, their population sizes increased. Increased population size among agriculturalists has been attributed to decreased birth spacing (shorter intervals between babies) as a result of women in farming groups weaning their children earlier than in hunter–gatherer groups (breastfeeding can suppress ovulation), or to children being helpful sources of labor on farms rather than a burden in a hunter–gatherer context. As agriculturalist populations grew there were opportunities for the division of labor, the emergence of organized religions, a need for bureaucracies, and the development of a leadership structure to protect agriculturalists in exchange for a portion of their crop, in a word: cities. The oldest cities in the world are located in the original agricultural hearths of the world: Mesopotamia (the Fertile Crescent), Mesoamerica, the Andes, the eastern United States, and China.

Agricultural hearths are regions where agriculture developed independently, without the introduction of ideas or technology from elsewhere. Mesopotamia, Mesoamerica, the Andes, the eastern United States, and China are five definitive hearths. Other areas, such as Amazonia, Ethiopia, and West Africa, are definite sites of early agriculture but it remains unclear whether the techniques were introduced from other definitive hearths. What also remains unclear is the exact timing of these independent emergences of agricultural practices. It was long thought that the Mesopotamian agricultural hearth was the oldest, but evidence from squash and other seeds found in Mesoamerican archaeological sites suggests that agriculture emerged there much earlier than previously thought.

Each agricultural hearth is associated with a unique set of plant and animal domesticates, shaped by what species were locally available. Mesopotamian agriculturalists domesticated wheat, barley, peas, sheep, and goats, among other things. Chinese agriculturalists domesticated rice, millet, pigs, and ducks. Mesoamericans domesticated maize (i.e., corn), beans, and squash. Andeans domesticated potatoes, manioc, llamas, and guinea pigs. The **staple crops** that we still associate with regional cuisines today (rice in China, maize in Mexico, potatoes in Peru) have their origins thousands of years ago.

Domestication means bringing a wild plant or animal under human management. Humans did not stop at domestication, however, they also practiced **selection**. The domesticated maize, wheat, and rice that we grow today bear little resemblance to their wild relatives. Over thousands of plant and animal generations humans preferentially selected which offspring to reproduce, which "children" of maize and wheat would go on to be planted and harvested. We selected for traits that benefitted humans and our need for enhanced food production, such as a loss of bitter or toxic substances, loss of defensive or protective structures to better allow us to get at the grains or fruits of the plant, for early and simultaneous ripening, for shorter and faster growth times, and for larger grain or fruit sizes. In animals we selected for animals that were smaller and more manageable than their wild ancestors, for greater fur production, for higher milk or meat yields, and for smaller brains and less acute sense organs.

Plant and animal domestication and selection enabled humans to produce larger amounts of food and spurred population growth. At the same time, humans were now in much closer contact with animals than they had been previously, encountering domesticated cattle and pigs and sheep on a daily basis, sometimes even sharing houses with them. Hunter–gatherer populations were shielded from exposure to infectious diseases by not coming into repeated and close contact with animals, and by living in small and isolated groups where disease transmission could not be sustained. Many of our oldest infectious diseases transferred to us from domesticated species, such as measles from the rinderpest virus of cattle and influenza from poultry, and we share the greatest number of pathogens with those species that have been domesticated the longest (McNeill, 1976; Morand, McIntyre, & Baylis, 2014).

THE COLUMBIAN EXCHANGE

Modern humans first migrated to the western hemisphere from Asia, crossing a land bridge known as Beringia, territory between contemporary Alaska and Russia. While this area is now underwater, the Bering Strait, during the migration of humans from the eastern to the western hemisphere, sea levels were lower as water was trapped in massive ice sheets. Though archaeological evidence is contradictory as to exactly when this migration took place, and precisely the route that migrants took from Alaska southward into South America, what is agreed upon is that by 13,000 years ago there was widespread human habitation of the Americas. As the ice sheets melted and sea levels rose, this land route between the eastern and western hemispheres disappeared and then no (substantial) interaction took place for thousands of years. In the eastern hemisphere, trade routes connected Asia and Africa and Europe, meaning that the disease experience of one continent was not confined, that infectious diseases such as smallpox, measles, influenza, and plague were shared by the entire hemisphere.

In the Americas, few large animals, particularly mammals, were candidates for domestication. According to Jared Diamond, an evolutionary biologist, there are six traits or characteristics that animals must satisfy in order to be good candidates for domestication (Diamond, 1999, 2002). They must have a fairly flexible diet, in particular a diet that does not utilize food that a human could consume. Our major domesticates are herbivores or, in the case of pigs, omnivores, rather than carnivores, and are primarily herbivores that can forage on grass or other cellulosic substances that humans cannot consume. Domestication candidates must grow relatively quickly, reproduce in captivity, lack aggression toward humans, have a tendency to herd and not flee when panicked, and to recognize humans as leaders (i.e., have a social hierarchy). In the eastern hemisphere, animals such as zebras, warthogs, and gazelles lacked one or more of these characteristics and were not domesticated. In the western hemisphere, only a very few animals met these requirements and were domesticated by Amerindians: turkeys, llamas, alpacas, and guinea pigs. Many cultures in the Americas relied on fish and aquatic mammals as the major source of protein in their diet. This relative lack of domesticated animals, and the isolation from the western hemisphere prior to the emergence of agriculture, meant that Amerindians were an **immunologically naïve** population: no individual had prior experience with smallpox or influenza or measles infection.

When Europeans arrived in the Caribbean and then South, Central, and North America at the end of the 15th century, they initiated one of the greatest exchanges in the history of our species: the **Columbian exchange** (Crosby, 2003). The renewal of contact between the eastern and western hemispheres resulted in phenomenal transfers of human populations (forced and voluntary), plant and animal populations, and the exchange of human pathogens. Plant species from the Americas, such as manioc (cassava), peanuts, and maize (corn), were introduced into Africa and became staples in diets there. Potatoes from the Andes were introduced into Europe and became a major source of calories in countries such as Ireland. Sugar, onions, bananas, apples, and citrus fruits were all introduced to the western hemisphere. While the exchange of plants was a two-way street, the exchange of animals and pathogens was more heavily tipped toward introduction from the Old World (the eastern hemisphere) to the New World. Chickens, horses, sheep, cows, and pigs were all transferred to the New World. While turkeys and llamas were sent back to the Old World, their importance is far smaller than the animals that arrived in the New World.

The populations of the Americas were not pristine and disease-free. Infectious diseases from pathogens whose origins are not from domesticated animals, such as staphylococcus, streptococcus, and treponemiasis, did exist. The infectious disease pathogens of the Old World, however, were overwhelming to the populations of the New World. While the exact number of Amerindians who were living in the Americas at the time of contact is unknown and disputed, what is known is that smallpox and other infectious diseases of the Europeans, and later African slaves, decimated indigenous populations. Cortes invaded Mexico with 500 men and 23 cannons, but his deadliest weapon was a smallpox-infected slave. In 1519 there were 25 million Aztecs, by 1605 there were less than 1 million. Smallpox arrived in Peru ahead of Pizarro, and 93% of the 16 million Incans died. Smallpox was accompanied by other diseases of the eastern hemisphere, including measles and typhus. Often disease would arrive ahead of the Europeans, traveling along the trade routes that connected Amerindian populations across North and South America. Most affected were agricultural tribes, those with large cities or well-connected settlements, such as those of the Mississippi River Valley and East Coast of the United States. Hunter–gatherer groups of the Great Plains did not suffer losses on the same scale, though they would in later centuries.

MODERN AGRICULTURAL SYSTEMS

Despite the massive transfer of plants and animals during the Columbian exchange, and the major impact of this transfer on diets in sub-Saharan Africa and Europe, for centuries after that the diets of most people remained fairly limited. The same foodstuffs were and are consumed day after day, sometimes meal after meal. In many countries of sub-Saharan Africa, for instance, meals rely heavily on caloric intake from cassava or maize supplemented with small amounts of fruits and vegetables. Figure 8.1 shows a typical day's food for a lower-middle-class family in rural Malawi; the picture was taken as part of a honor's thesis by a University of Iowa geography major looking at patterns of food consumption and production. The

FIGURE 8.1 Daily diet for a family of five in rural Malawi, consisting of maize porridge (*nsima*) and cooked pumpkin leaves (*ndiwo*). Photo copyright © Austin Dunn. Reprinted by permission.

family will consume the majority of their calories in the form of a maize porridge, accompanied by salads of greens and pumpkin. Cassava is relatively easy to grow, and produces well even in poor soils or in drought conditions. The introduction of cassava from its native home in South America to Africa following the Columbian exchange radically changed diets. An overreliance on cassava, however, with limited protein intake, can result in what is known as kwashiorkor, or protein-energy malnutrition. Individuals, particularly children, consume enough calories but not enough protein, and suffer either death or long-term restrictions on physical growth and mental development.

The practice of farming remained much the same in the hundreds of years following the Columbian exchange, also despite the massive transfer of plant and animal species. Most farms were small, supporting a single family, and labor demands were high. Production was relatively low, with small surpluses going to support towns and cities. Major changes in the way food is produced began to take place in the United States in the decades surrounding World War II. There are three major nutrients that are important to plant growth and production: nitrogen (N), phosphorus (P), and potassium (K). Bags of fertilizer or potting soil will list their NPK ratios. Natural sources of these three elements have been used by farmers for thousands of years, well before they were discovered and named by science. Nitrogen, for instance, can be found in human or animal manure, or is deposited in soils by nitrogen-fixing plants such as soybeans and clover. Crop rotation with these nitrogen fixers or the use of manure as fertilizer spurred growth and production of agricultural crops for thousands of years.

In the late 19th and the early 20th centuries, countries in Europe and North America entered the demographic transition (discussed in Chapter 4). Meeting growing demand for food was problematic, as extensification of agricultural land was no longer possible: the midwestern states of the United States, the Canadian prairies, the Russian steppes, and the Argentinian pampas were all farmed to their full extent. **Intensification** of agricultural production was necessary, to generate higher yields in the same acreage, and would rest on being able to provide more NPK to plants. Increasing yields without increasing land under acreage was no easy task; yields per acre had been steady for decades. Phosphorus and potassium deposits were relatively easy to access, but finding enough nitrogen was problematic. While nitrogen is abundant in the air, atmospheric nitrogen cannot be used by plants. Premodern sources of nitrogen included bird guano, found in large amounts on islands where aquatic birds roost. So important were these islands to the mining and production of nitrogen for agricultural production that the United States passed the Guano Islands Act in 1856, which empowered U.S. citizens to take possession of islands containing guano deposits and empowered the U.S. military to protect those claims of possession. The synthetic fixation of nitrogen, that is, development of a means to create nitrogen in sufficient quantity for agriculture, was a major concern at the time: "The fixation of nitrogen is vital to the progress of civilized humanity," claimed William Crookes, of the British Royal Academy of Science in 1898.

The fixation of nitrogen was a goal not only for agricultural science but also for militaries of the world. Nitrates are a main ingredient in explosives, and being able

production. These new varieties of wheat were also highly responsive to nitrogen fertilizer, responding to increased fertilizer input with increased yield.

The Indian government invited Borlaug to introduce his high-yield varieties of wheat into the Indian agricultural system in the early 1960s. Wheat is a primary staple crop of northwest India, while rice agriculture dominates in the east and south. The Rockefeller and Ford Foundations invested in developing new high-yield varieties of rice via the establishment of the International Rice Institute. The result was IR8, a cross between a short Taiwanese variety of rice and a tall Indonesian variety. It matured quickly, enabling a double or triple crop in a year; it had a strong stem that would not fall over and destroy the rice grain; and it responded exceptionally well to nitrogen fertilizer. This "miracle rice" was adopted by India alongside Borlaug's Mexican high-yield varieties of wheat.

Producing enough food to meet the needs of India's growing population required not only new high-yield varieties of wheat and rice, however. These seeds were only one element of the "miracle." In addition to seeds, farmers needed fertilizer to generate maximum yields, since without fertilizer the seeds would produce yields similar to traditional varieties. They also needed reliable water, which meant greater dependence on wells or rivers for irrigation, since you cannot double- or triple-crop if you depend on the monsoon to arrive on time. Maximum yields are generated through the application of pesticides, to crowd out competition from undesired plants, and through mechanization. As Borlaug said to Prime Minister Indira Ghandi, the need is for "fertilizer, fertilizer, fertilizer, credit, credit, credit." Farmers need access to credit because inputs were now coming from off-farm, there was no seed saving from year to year, and farmers could not generate fertilizer with the same nitrogen load as agribusinesses could.

These major changes in agricultural production in India, and elsewhere in Asia such as Pakistan, Bangladesh, Vietnam, and the Philippines, are known as the **Green Revolution**. In 1968, William Gaud of the U.S. Agency for International Development (Gaud, 1968) said: "These and other developments in the field of agriculture contain the makings of a new revolution. It is not a violent Red Revolution like that of the Soviets, nor is it a White Revolution like that of the Shah of Iran. I call it the Green Revolution." This cold war-era policy of the United States focused on increasing food production in low-income countries under the idea that well-fed populations would be less likely to be seduced by communism, that well-fed populations would remain happy with their pro-United States governments.

While the global area cultivated has remained fairly constant since the 1960s, fully extensified, the yields from this land have increased dramatically: they have intensified. World grain production increased by 250% (Kendall & Pimentel, 1994). Global daily caloric consumption in poor countries has increased by 25% (Conway, 1998). India is now not only the second largest consumer of grain in the world, it is also the second largest producer. While malnutrition is still a concern in India, widespread famine is not. World population has grown from 2 billion to over 7 billion people since the 1960s and the Green Revolution averted the starvation of millions of people. Borlaug received the Nobel Peace Prize in 1970 for his contribution to world peace through increasing the world food supply. While the agricultural system that Borlaug fostered globally is not without its problems, problems that mirror those of commercial agricultural production in the United States, Europe, and China, to his critics Borlaug responded that

> Some of the environmental lobbyists of the Western nations are the salt of the earth, but many of them are elitists. *They've never experienced the physical sensation of hunger.* They do their lobbying from comfortable office suites in Washington or Brussels. If they lived just one month amid the misery of the developing world, as I have for fifty years, *they'd be crying out for tractors and fertilizer and irrigation canals* and be outraged that fashionable elitists back home were trying to deny them these things. (Easterbrook, 1997, emphasis added)

THE NUTRITION TRANSITION

The massive change in agricultural production from hunter–gatherers to industrial agriculture has led to shifts in food consumption. The types of food we eat now, and the amount we consume, is vastly different than hunter–gatherer or preindustrial populations. Chapter 4 examined three types of transitions: the demographic, the mobility, and the epidemiological transitions. Over time and across countries, populations appear to behave in similar ways. Populations grow as first death rates and then birth rates fall, then population growth slows as total fertility rates fall to around replacement value. People begin to move farther than in generations before, migrating to cities or the agricultural frontier, until populations at the end of the mobility transition are primarily urban. At the same time, the causes of morbidity and mortality in populations shift, from infectious to chronic diseases or a mixture of the two. The simultaneous changes in agricultural production and food consumption are deeply interconnected with these changes in population growth, mobility and urbanization, and major causes of morbidity and mortality. The shift in patterns of food consumption is referred to as the nutrition transition. There are interactions and feedback loops across all four types of transitions. As death rates fall in the early stages of the demographic transition, and as more children survive to adulthood, there is an overabundance of labor on farms in rural populations. Now the farm had to be divided and shared among surviving sons, or some sons had to leave the farm and make their way to the agricultural frontier or the city to make their fortune. Population growth in Europe during the 19th century was influenced, in part, by improved nutrition from crops, such as the potato, that were introduced after the Columbian exchange. Ireland's population grew enormously after the adoption of potato agriculture, as a small plot of potatoes, combined with milk from a cow, provided all the calories, protein, vitamins, and minerals that families needed.

Over time, the number of people engaged in agricultural labor has decreased as farms have gotten larger and more efficient worldwide and as human labor has been replaced with technology. Generally, decreasing global fertility is associated with an aging population and a population that has more disposable income. The rise of a middle class in countries such as India and China is particularly striking: people can now afford to spend more on food. In many countries there have been large shifts in physical activity patterns, as individuals work in offices or other low-activity settings and as people rely on motorized transport and spend their leisure time in ways that are not physically active. Generally speaking, wealthier and urban populations are consuming a diet that is "Western" in nature, high in animal products like meat and dairy and in processed foods. The gradual transition of populations to this type of diet, in country after country, is known as the nutrition transition (Popkin, 1994).

The first stage of the nutrition transition characterizes all humans prior to the development of agriculture. Hunter–gatherer populations have diets low in fat and high in fiber, they live lives of physical activity, and they experience little to no obesity. The second stage is known as the famine stage, after humans become sedentary farmers. Their diets, as previously mentioned, lack variety, depend heavily on staple grains, and there are times of severe food scarcity when crops fail. Physical activity in this stage is high. Very few populations are left in either of these stages. In the third stage of the nutrition transition diets remain much as they did in stage 2, high in starch and low in fat, with little day-to-day variety in what people are eating. The threat of famine recedes, however, as countries progress through industrialization and there is greater governmental organization. Rural families in low- and middle-income countries are primarily in this stage of the nutrition transition, while urban population in low- and middle-income countries are moving out of this stage into the next phase of the transition. In the next stage, major changes to the diet take place. The intake of fats (particularly saturated ones), sugar, and salt increases as individuals consume greater amounts of caloric beverages (e.g., sodas), other processed foods, meats, eggs, and dairy products. This stage of the nutrition transition is also associated with declines in physical activity and the emergence of health effects related to **overnutrition**: obesity, type II diabetes, heart disease, cancer, and the like. These are the diseases found at the end of the epidemiological transition. Wealthy countries such as the United States are solidly in this stage of the nutrition transition and have been for decades. The emerging middle class of the Global South is joining us in this stage. The fifth and final stage of the nutrition transition is associated with behavioral change. In this stage, people begin to deliberately increase their consumption of fruits, vegetables, and whole grains and to decrease their consumption of salt, sugar, and fat. They increase their physical activity and experience less obesity and other chronic diseases of overnutrition. No countries are fully in this stage of the nutrition transition, although individuals within countries may be.

The nutrition transition, in this case the movement from stage 3 to stage 4, happens first in the urban areas of low- and middle-income countries, before spreading outward to rural areas. It is urban populations who first have the income levels necessary for increased purchases of Western types of food, and it is urban areas that provide a large enough customer base for global food chains. Globally, a higher percentage of population living in urban areas is associated with higher body mass index (BMI), a commonly used metric for measuring overweight/obesity (Figure 8.3).

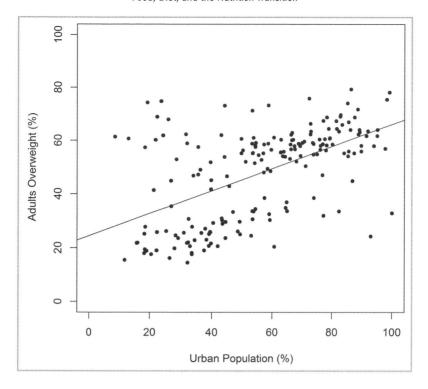

FIGURE 8.3 Percentage of adults who are overweight in countries versus the percentage of residents in those countries who are urban dwellers in 2014. Data from World Health Organization.

COMMERCIAL AGRICULTURE AND THE NUTRITION TRANSITION

A primary characteristic of populations in or entering the fourth stage of the nutrition transition is increased caloric intake from meat and sweetened beverages. Global meat consumption between 1961 and 2011 increased dramatically, most especially in countries like China, Brazil, and Indonesia, countries with growing middle-class populations (Figure 8.4). No longer confined to special days of the week or for celebration of special events, meat is now consumed at most meals in these countries. To meet growing consumer demand, global meat production has had to increase in parallel.

Pork is the world's most-consumed meat, followed by poultry (primarily chicken), though beef remains a popular product globally. A primary characteristic of these domesticates, as previously discussed, is that their diets did not rely on food that could otherwise go to human consumption. Pigs are omnivores and can be fed table scraps or scavenge for food. Chickens in backyard flocks similarly scavenge, requiring only a small amount of grain to supplement a diet of insects and foraged seeds. Cows consuming grass and other pasture forage turn cellulose that is inedible to humans into usable energy: meat and milk. There is simply no way, however, for the production of pigs, chickens, and cows to remain dependent on scavenging or foraging for food and still meet the 200% increase in consumption

CAFO technology originated in the United States, it is now the primary means by which livestock is produced globally. China in particular has invested heavily in the CAFO production of hogs.

Just as global meat consumption has increased since the 1960s, so has consumption of sugar-sweetened beverages (SSBs) such as sodas, juices, and energy drinks. Calling these drinks "sugar"-sweetened is a bit of a misnomer, however, since the primary sweetener in many cases is not sugar but **high fructose corn syrup (HFCS)**. HFCS is the result of a technologically intensive process that transforms field corn into a sweet, sticky liquid. The majority of maize grown in the United States and globally is not sweet corn that can be eaten directly by humans, it is a raw material for other types of production, such as meat or HFCS. In particular, the United States grows a type of corn known as yellow dent, so-called for the little dent on each kernel. Yellow dent has been specifically developed to be high in sugar content, ideal for HFCS production as well as for putting weight onto livestock.

While there was a capability to make HFCS before the 1970s, for the process by which corn could be transformed into this syrup was well understood, HFCS was virtually absent from the American and world diet before this time. It was only when corn production in the United States was greatly expanded, thanks in part to better seeds, better fertilizer, and mechanization, but also to a shift in federal agricultural policy that encouraged the production of as much corn as possible, that the raw ingredient for HFCS became cheap enough for the final product to be economically viable (Figure 8.6).

Without cheap and abundant corn, HFCS becomes too expensive to manufacture. But with cheap and abundant corn, HFCS also becomes cheap and abundant. HFCS now occupies a central role in the global food system, used to sweeten beverages such as sodas and juices. It is also in many other foods: it browns well so is used in the production of breads and other snack pastries, it cuts the sharp tang that jarred spaghetti sauce may have. Because it is manufactured in secure facilities and is exposed to high heat, HFCS is a sterile product and makes items shelf stable for much longer than would traditional sugar sources such as cane or beet sugar or honey. Because HFCS is cheap and abundant, the products that it goes into can be sold for lower prices than if they were sweetened with other sources; sweetened processed foods thus become some of the lowest cost calories in the marketplace.

DIRECT AND INDIRECT HEALTH EFFECTS OF AGRICULTURAL AND DIETARY CHANGES

A hallmark of the nutrition transition is that the world now has problems of both undernutrition and overnutrition (Figure 8.7).

For much of human history it was a *lack* of calories and protein that influenced human health. As millions of people move from stage 3 to stage 4 of the nutrition transition, joining Americans and Europeans and others who have been there for decades, increasingly negative health effects are associated with an *abundance* of calories and protein. Cheap meat and dairy which, along with HFCS and processed foods, is built on a foundation of cheap and abundant corn and soy, have added

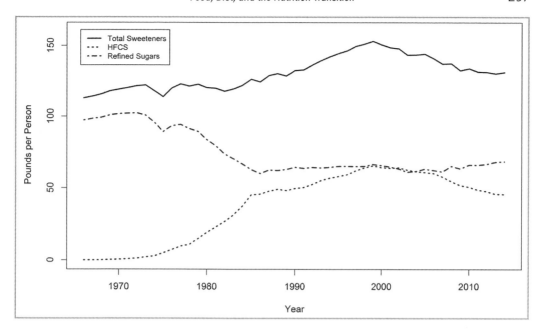

FIGURE 8.6 Per capita consumption of sweeteners from 1966 to 2014 in the United States. Data from U.S. Department of Agriculture Economic Research Service *Sugar and Sweeteners Outlook.*

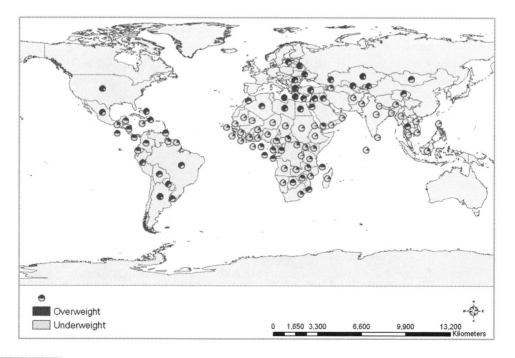

FIGURE 8.7 Proportion of underweight and overweight children under age 5. Data from World Health Organization.

hundreds of calories to global daily caloric intake. According to the Food and Agriculture Organization (FAO), in 1961 there were only a handful of countries whose daily caloric intake was greater than 2,500 calories, and whose protein intake classified them as protein-rich. By 1999 nearly half of the world's countries were consuming more than 2,500 calories per day, and by 2030 there are projected to be only a very few countries with fewer than 2,500 calories per day per capita, countries almost exclusively located in sub-Saharan Africa and the Middle East. The increase in caloric intake will be more rapid than the increase in protein intake, but global protein intake (i.e., meat and dairy) will increase substantially by 2030.

Why do these changing diets matter? There are both direct and indirect consequences to human health associated with the global nutrition transition. *Direct health effects* are associated with differences in the types and amounts of food being consumed. *Indirect health effects* are associated with the manner and means of food production.

Direct Health Effects

The World Health Organization (WHO) estimates that millions of deaths occur each year due to unhealthy diet and lack of physical activity. WHO classifies these as modifiable risk factors for chronic diseases because, unlike age or heredity, they are the result of choices made by individuals. The WHO is particularly interested in these modifiable risk factors because they indicate potential areas for intervention or prevention. The WHO estimates that if risk factors (i.e., poor diet and physical inactivity) were eliminated, the majority of heart disease, stroke, and type II diabetes would be preventable.

The biggest emerging health problems by far are diabetes and obesity. Both today are prevalent in both developing and developed countries; they affect children and adolescents as well as adults. Among the more affluent, late in the transition and after their completion, diets, occupations, leisure activities, and means of transportation "modernize" to parallel Western lifestyles. Diets high in fats and sugars, and lives low in activity, result in obesity. Popkin (2002a, 2002b, 2003; Popkin & Gordon-Larson, 2004) has found that this shift in the burden of the malnutrition of obesity to the poor occurs when countries reach a gross national product per capita of about U.S.$1,700. Not only is it possible (as Figure 8.7 shows), but it is becoming increasingly common, for developing countries to have both a high percentage of children extremely underweight and malnourished, and at the same time a high percentage of children very overweight and malnourished. Obesity is a major risk factor for serious noncommunicable diseases, such as cardiovascular disease, hypertension, stroke, diabetes mellitus, and various forms of cancer.

Obesity was, as of 2012, prevalent among all groups of Americans. One-third (34.9%) of American adults were obese, which was defined as having a body mass index (ratio of weight to height) of 30 or more. As recently as 1995, the portion of the population with that BMI was only 15%. Not one U.S. state met the new 15% target established by the Centers for Disease Control and Prevention's (CDC's) Healthy People 2010 initiative (Colorado was closest, at <20%). Two states, Mississippi and

West Virginia, had obesity rates exceeding 35% of the population (Centers for Disease Control, 2013). Even as U.S. public health agencies struggle to change the conditions promoting this epidemic, it is spreading throughout the world.

Since 1980, worldwide obesity has more than doubled; 1.9 billion adults, or 39% of the world population over 18 years of age, were overweight in 2014 (World Health Organization, 2015). Six hundred million of these adults, or 13%, were obese. Globally, the majority of the world population lives in a country where mortality due to overweight- and obesity-related conditions kill more people than do conditions related to being underweight. In the last two decades we have seen a radical shift, unprecedented in human history, in which morbidity and mortality are now more related to an overabundance of food than to a lack of food. Agricultural developments at the end of the 20th century have allowed us to produce so much food, and so many calories, that we are killing ourselves by overeating.

While HIV receives huge amounts of attention and funding (deservedly so), there are far more individuals living with type II diabetes than there are with HIV. The WHO estimates that there are 347 million diabetics worldwide, and that 80% of diabetics live in low- and middle-income countries. Diabetes is the result of a problem in the way the body makes or uses insulin. Insulin is necessary to move blood sugar (glucose) into cells, where it is stored and later used for energy. In type II diabetes the fat, liver, and muscle cells of the body do not respond correctly to insulin (insulin resistance) and blood sugar does not move into cells for storage. When sugar cannot enter these cells, abnormally high levels of sugar build up in the blood, a condition known as hyperglycemia. People who are overweight are more likely to have insulin resistance because fat interferes with the body's ability to use insulin. Long-term hyperglycemia can lead to blindness, kidney failure, and nerve damage, as well as to heart disease and stroke.

There are distinct spatial patterns to overweight and obesity. These spatial patterns are influenced by factors beyond the control of individuals but which directly influence personal choices regarding food consumption. How a person's environment or neighborhood influences his or her health will be described in greater detail in the next chapter, but here we will briefly outline how proximal and distal factors influence the spatial patterns we observe in the United States and elsewhere for obesity and overweight and their associated health conditions.

Proximal factors are those that are close to an individual, directly influencing what he or she chooses to eat, in what amounts, and with what frequency. *Distal factors* are those that are far away, metaphorically speaking, from an individual, operating at higher spatial scales and beyond the control or modification of an individual. Oftentimes people are unaware of how distal factors are influencing their choices. Distal influences usually condition or change the choices that are available to individuals, so people feel as though they have free choices and personal responsibility, but in fact the options available to them have been chosen by others.

Proximal factors affecting food choice include psychological aspects, such as emotions and body image, as well as food knowledge, and social/cultural characteristics such as social norms, acculturation, and peer and family influence. In other words, how you feel about yourself and about food, and the food expectations that you and your family and peers have, directly influence your food choices. Operating distally are factors that impact the physical, policy, and organizational environment in which you live, creating what are known as **obesogenic** environments (Popkin, Duffey, & Gordon-Larsen, 2005).

Obesogenic, mimicking the word *carcinogenic,* indicates an environment that is conducive to obesity (this is discussed in more detail in Chapter 9). The physical safety and walkability of neighborhoods, as well as whether it has parks and playgrounds for children to play, is beyond the control of an individual and is likely the result of decades of choices by local politicians for police and social funding. The availability of grocery stores or other food outlets is determined by the economic decisions of the companies that own those outlets, whether they believe they can make a profit locating in a neighborhood, as well as what food items they choose to carry. You cannot purchase fresh fruits and vegetables if your only food outlet is a corner store that stocks exclusively packaged and processed foods. If there is no full-service grocery store in a neighborhood, individuals must travel outside of their immediate home environment to purchase the range of groceries needed for a healthy diet. Many cities lack efficient public transit systems, however, and individuals without access to cars are left with the options that they can walk to. Some people may have to take several buses or trains, and several hours round-trip, to do their grocery shopping.

The economic vitality of a neighborhood and its residents is the product of decades of decisions at the federal, state, county, and local levels regarding education, health care, and investment for development. This can result in a majority of local residents relying on federal food and other forms of assistance. Many individuals on the Supplemental Nutrition Assistance Program (SNAP, formerly Food Stamps) indicate that their benefits are not supplemental but comprise the total of their food budget, and often only last 3 out of 4 weeks in a month. This leads people to purchase foods that have the highest calorie count for the lowest cost, and foods that will last as long as possible. Decades of federal agricultural policy have made corn- and soy-based food items, highly processed frozen or shelf-stable food items, much cheaper on a per-calorie basis than fresher, unprocessed fruits, vegetables, and whole grains. Some research also suggests that SNAP participation is correlated with obesity, over and above the effects of age, race/ethnicity, and SES. Reasons for this could be related to the "boom and bust" or "feast and famine" cycles that SNAP participants report, running out of food at the end of the month and then overconsuming when the benefits are received at the beginning of the month. The human body responds to times of famine, even short times of food shortages, by holding on to calories, by transforming food into fat more so than it otherwise would as a protection against future lack of food. *Epigenetics,* the study of how environmental factors regulate gene expression, suggests that food shortages or food insecurity when an individual is in the womb may affect his or her health later in life, and that children born to mothers who experienced famine are at greater risk of obesity (Bruce & Cagampang, 2011; Youngson & Morris, 2013).

Food deserts are a designation, now formalized by the U.S. Department of Agriculture (USDA), to describe an area with restricted access to healthy food sources. Under the USDA definition, a food desert is a census tract where 20% or more of residents live below the poverty line and where a third or more of residents live more than 1 mile from a grocery store or supermarket. According to this definition, the USDA Economic Research Service estimated that 23.5 million Americans lived in a food desert in 2010 (Figure 8.8). Food deserts are not confined to urban areas: many food deserts are in rural areas of low population density where grocery stores are sparsely located. In both urban and rural food deserts, the ability to access food is heavily reliant on having access either to a car or to public transit systems. The sprawl of American cities, especially in the South, and then the cutback to city services that came with the 2008 economic collapse, means that there are increasing numbers of suburbs that can also qualify as food deserts. People are left stranded in highly walkable subdivisions, but with sidewalks that go nowhere. The USDA definition of food deserts is but one way to define and understand the food environment, and the boundaries of food deserts may shift or evolve over time or may not conform to administrative boundaries such as census tracts.

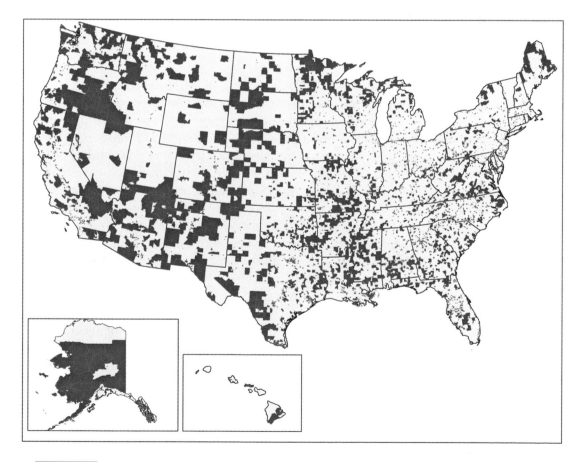

FIGURE 8.8 Census tracts defined as food deserts (shaded) by the U.S. Department of Agriculture in 2010. Data from U.S. Department of Agriculture Research Service.

While overweight and obesity are negatively associated with wealth in the United States, with poorer people having a higher prevalence of both, the opposite is true in the low- and middle-income countries moving through the nutrition transition. In these places, it is the urban middle class who are newly able to purchase the items that characterize stage 4 of the transition: meats, dairy, eggs, processed foods, and the like. In the United States food is cheap, even for those who require SNAP and other federal assistance such as Women, Infants, and Children (WIC) and the National School Lunch Program. The proportion of income that Americans and residents of other wealthy countries spend on food is very low. In contrast, people in low- and middle-income countries spend large portions of their income, sometimes as much as 50–75%, on food. Generally speaking, countries that eat the most spend the least money on food. As the SES of a country improves, its absolute spending on food increases, but the proportion of total income spent on food falls. The same amount of money is spent on fruits and vegetables and whole grains, but much more money goes toward purchasing meat, dairy, and processed foods, and on eating out at restaurants. We thus observe increases in BMI and the rates of obesity and overweight individuals first in wealthier individuals of urban areas of low- and middle-income countries.

Indirect Health Effects

The range of health outcomes associated with food is not limited to those that are the direct result of the types and amounts of food that are being consumed. Food production, from what is grown where and how, to the ways in which food is processed, distributed, and prepared, has a wide array of impacts on human health.

While the diets of the wealthy and emerging middle class in low- and middle-income countries are shifting, for billions of poor people in rural and urban areas diets and the way in which food is prepared have remained virtually unchanged. The WHO estimates that 3 billion people still use traditional fuel sources, such as wood, dung, or crop waste, for cooking. Using biomass for fuel rather than electric or natural gas or solar stoves produces large amounts of indoor air pollution on a daily basis. When these pollutants are trapped in small, poorly ventilated spaces they can do severe damage to the health of individuals inside those spaces, typically women who are cooking and their children (see section on Indoor Air Pollution in Chapter 11 for further discussion). The WHO, based on 2012 data, estimated an excess of 4 million deaths per year are attributable to this type of indoor air pollution, with higher numbers of people experiencing decreased health through pollution-associated morbidity. Causes of morbidity and mortality due to household air pollution are diverse, but include chronic obstructive pulmonary disease (COPD), stroke, pneumonia, lung cancer, and low birth weight.

Other health challenges posed by agriculture and food production are shared between populations in high-, middle-, and low-income countries. Pesticides, fungicides, and herbicides are a primary component of modern industrial food production. The goal of farmers since the development of agriculture has been to grow only those plants that are desired, and to protect those plants from insects and other pests. **Ecological simplification** means that the system is no longer capable of protecting itself through diversity and systemic regulation; it is now up to the farmer to provide protection to the vulnerable monoculture in the field. These ecologically simple agricultural landscapes have few species (e.g., one type of corn plant) but many members of that species, while a mature ecosystem such as a forest has a large variety of species but fewer members of each (see Table 8.1). A pest that invades a mature ecosystem will find only a few susceptible individuals. In contrast, a pest that can kill one corn plant can kill the entire field if the farmer is not there to intervene, usually in the form of application of chemicals.

Research suggests that climate change will shift the geographic range of many agricultural pests and create new microclimates that are dryer or wetter than current conditions. Simplified agricultural systems are less resilient to these types of changes, and there is significant potential for crop failure as the earth's climate shifts. The application of chemicals to fields to kill weeds, fungi, and insects can be dangerous to human health. Multiple studies have indicated that increased exposure to pesticides are associated with cancers, such as prostate cancer (Koutros et al., 2013; Parrón, Requena, Hernández, & Alarcón, 2014). Childhood cancer is also linked to pesticide exposure, both directly experienced by the child and by parental exposure while the child is *in utero* (Vinson, Merhi, Baldi, Raynal, & Gamet-Payrastre, 2011; Zahm & Ward, 1998). While agricultural workers have the most direct exposure to pesticides, evidence suggests that children whose houses and schools are close to agricultural fields also have high exposure to pesticides (Lu, Fenske, Simcox, & Kalman, 2000).

TABLE 8.1. Some Generalized Characteristics of Ecosystems

	Youthful	Mature
Number of species	Few	Many
Number of species individuals	Many	Few
Niche	Generalists	Very specialized
Growth	Rapid	Slow
Reproduction	Frequent, prolific	Infrequent, limited
Dissemination	Wide, fast	Local, slow
Size tendency	Small	Large
Information, feedback	Low	High
Food chains	Linear	Complex
Nutrient conservation	Low	High
Entropy	High	Low
Overall stability	Low	High
Prototype animals	Mice, sparrows	Elephant, chimpanzee
Prototype plants	Dandelion, pine	Mahogany tree

Note. Information derived, for the most part, from Odum (1978).

A major boost to the production of crops such as corn and soy in the United States and globally was the development of genetic modification of crops in the final decades of the 20th century. The majority of corn and soy planted in the United States are now genetically modified (USDA). A defining feature of most industrial production of corn is the use of seeds that have been genetically modified to be resistant to spraying with herbicides, allowing farmers to blanket entire fields of corn with herbicides that kill weeds while the corn plant remains healthy. Two pairings of genetically modified seed and herbicide exist, Bayer's Liberty herbicide and LibertyLink corn (and soy, cotton, canola, and sugarbeets) and Monsanto's Roundup herbicide and RoundupReady corn (and soy, cotton, canola, and alfalfa). The primary chemical agent in Roundup is glyphosate, which in 2015 was declared a probable carcinogenic agent by the WHO upon the basis of evidence from *in vitro*, epidemiological, and animal studies. The Environmental Protection Agency estimates that in 2007 (the latest data available) more than 180 million pounds of glyphosate were used in agricultural applications in the United States, with a further 5–8 million pounds used in homes and gardens.

It is estimated that of all the antibiotics used in the United States 80% are used in animal production. Pigs and cows and chickens raised in CAFOs represent a type of monoculture, one in which susceptibility to infection is shared by all individuals within a barn or feedlot. To protect animals from infection, and to help them gain weight as quickly as possible, and to combat the negative effects of a primarily corn- and soy-based diet, antibiotics make up a regular part of the feed. Only estimates of antibiotic usage in livestock production exist because livestock and pharmaceutical companies and farmers are not required to keep or divulge records of sales and dosing. A lack of data about what antibiotics are being used in what amounts in what places means that it is difficult, if not impossible, to detect when and where antibiotic resistance emerges and to further link this to antibiotic resistance in human infections. Antibiotics commonly used to treat human infections, such as tetracycline, are also routinely used in livestock production. The concern to public health is that human infections will become untreatable with available antibiotics because bacteria have been able to develop drug-resistant capabilities within a livestock production setting. Millions of animals are receiving millions of subtherapeutic doses of antibiotics each day, providing millions of opportunities for bacteria that have developed resistance genes to thrive and transmit them. Research has indicated that antibiotic-resistant bacteria in livestock animals are colonizing and infecting humans that come into contact with these animals, and that drug-resistant bacteria can be isolated from the air and soil surrounding farms. The FDA, USDA, and the CDC all have testified before Congress that there is a definitive link between the routine, nontherapeutic use of antibiotics in food animal production and the crisis of antibiotic resistance in humans, and in 2013 the FDA took steps to address these concerns. Antibiotics are a relatively new weapon in the arsenal of public health, and antibiotic discoveries have stalled as pharmaceutical companies focus on higher added-value drug discovery. When our currently available antibiotics are no longer effective, currently minor surgeries or infections will once again become as dangerous as they were in the first half of the 20th century.

For individuals working in livestock production a primary concern is occupational health, particularly in terms of the air quality inside barns. A single barn can hold thousands of animals, often with minimal ventilation. Sunlight from open windows would cause chickens and other birds to be more active, burning calories and increasing the time birds took to achieve market weight. In hog barns during the winter all windows and doors are kept closed so that the heat generated by the animals prevents the need for costly electric heating. Barns are filled with dust, and few workers wear respirators and masks due to the heat inside the barns. Twenty-five percent of all swine confinement workers have one or more respiratory conditions (Donham, Haglind, Peterson, Rylander, & Belin, 1989). Poultry workers have reduced pulmonary function (Donham, Cumro, & Reynolds, 2002). High airborne concentrations of dust, ammonia, and endotoxins are found inside CAFOs and are correlated to a drop in lung function. Other items inhaled by workers include fecal matter, bacteria and bacteria components, endotoxins, fungi, and hair and skin particles from the animals.

While the dust and particulates that are inhaled by livestock workers will not affect the majority of Americans, the impact of CAFOs on local waterways is also a potential health concern. CAFOs are regulated by the EPA under the Clean Water Act. The limiting factor for where CAFOs can be located and how many animals they can raise is the effect of waste production on surrounding ground and surface water supplies. Individuals wanting to operate CAFOs need to demonstrate to state departments of natural resources or environments that they will be able to safely store waste in perpetuity or, after ensuring that the waste is free of contaminants such as bacteria, deposit it somewhere on the surrounding landscape. Most hog CAFOs hold the waste produced by animals in storage pits or lagoons near the swine barns, allowing anaerobic decomposition to make the sludge safe enough to be sprayed onto or injected into nearby fields. Problems arise when the storage lagoons leak or when they are flooded, allowing the sludge to make its way into ground or surface water before contaminants are removed. In 1995, a 120,000-square-foot lagoon ruptured in North Carolina, releasing 25.8 million gallons of effluvium into the New River. The spill resulted in the killing of 10 million fish in local water bodies. It also contributed to an outbreak of *Pfiesteria piscicida,* which caused health problems for humans in the area including skin irritations and short-term cognitive problems.

The increase in meat and dairy consumption in the United States and globally has ripple effects beyond obesity and heart disease and cancer. Given that the foundation for the production of beef and pork and chicken and milk is readily available corn and soy and alfalfa, increasing the amount of meat and dairy available for consumers necessitates increases in corn and soy and alfalfa harvests. Reliable production of livestock feed crops depends on reliable rainfall or access to ground or surface water supplies. Iowa produces the most corn and soy of any U.S. state, and depends primarily on rainfall to do so. Other corn- and soy-producing states, such as Kansas and Nebraska, used to be too dry to grow corn and soy. Starting in the 1960s, improved well and irrigation technology enabled farmers in these and other states that did not receive enough rainfall to instead tap into the large Ogallala

Aquifer (Gutentag, 1984). This **aquifer** is one of the largest in the world and now supports a large percentage of agricultural production in the United States. Usage of water from the aquifer far outpaces the replenishment rate, however. The U.S. Geological Survey (USGS) reports that in many places the level of groundwater has dropped by over 50 feet, leaving farmers to either drill deeper or switch to less water-intensive crops (Konikow, 2013). The same phenomenon, drops in the level of groundwater due to overusage by agriculture, is observed elsewhere in the United States, such as in California, and in countries such as India and China. California produces huge amounts of alfalfa, a crop that increases milk production and milk quality in dairy cows. Alfalfa is a thirsty crop that is responsible for a large share of total water used in California agriculture. In India, the northwest of the country, where Borlaug's new strains of wheat were first introduced, has millions of acres where groundwater supplies are either in critical danger or already overexploited. This overuse critically endangers India's ability to produce enough wheat for its population. The government of India encouraged farmers to drill wells during the Green Revolution, providing free electricity to run water pumps as a cheaper alternative to building dams and irrigation canals to divert river water. Countries around the world face water shortages in the coming decades as current agricultural production is based on unsustainable usage of groundwater supplies; this artificially high agricultural production is known as "water-based food bubbles."

The water it takes to produce the corn, soy, and alfalfa fed to meat and dairy animals is just one aspect of the hidden **water footprint** associated with meat production. A water footprint is a way to describe the total amount of water that it takes in all stages of meat production, from growing feed to the water consumed by the animal during its lifetime to the water used in processing and transport of the product to the consumer (Hoekstra, 2010, 2012). It takes approximately 750 gallons of water to produce a pound of pork (~6,000 liters/kg), and almost 2,000 gallons of water to produce a pound of beef (15,400 liters/kg) (Hoekstra, 2012; Mekonnen & Hoekstra, 2012). Increasing consumption of animal products in stage 4 of the nutrition transition requires increased water usage both for the growth of feed crops and to sustain the animals themselves, a problematic increase when groundwater supplies are already stressed or depleted and when rainfall patterns are predicted to become less reliable with global climate change.

One final large-scale and long-term way in which our diets impact human health is how the nutrition transition is helping to drive global climate change. In addition to requiring huge amounts of water, the production and consumption of meat and dairy products results in much larger carbon dioxide and carbon dioxide equivalents (i.e., methane) emissions than does a primarily plant-based diet. The Environmental Working Group estimates that 1 kilogram of milk produces 1.9 kilograms of greenhouse gases (GHGs), while 1 kilogram of eggs produces 4.8 kilograms of GHG. A kilo of pork produces less GHG, 12.1 kilograms, than does a kilo of beef, 27 kilograms. Cows, and other ruminants such as sheep, release huge amounts of methane from their breath, a process known as enteric fermentation. While beef is not as popular a meat as pork or chicken, beef production has increased as countries have moved through the nutrition transition and an estimated one-third of arable land globally now goes to producing food to feed cattle. This results not only in large water footprints but also increased GHG emissions.

The Case of *E. coli* O157:H7

Ground beef is one of the most popular meat items consumed in the United States. The majority of ground beef in the United States comes from cattle raised in extremely large feedlots and goes into the production of hamburgers. Cattle in the United States, indeed all cattle, used to live their lives eating a range of plants found in forage pastures, or hay in the winter. Since the 1950s, however, the share of a beef cow's diet that is made up of corn has dramatically increased. Unlike pigs and chickens, beef cows still begin their lives freely ranging in pastures. The time that a calf spends grazing on pasture has gotten shorter and shorter, however. Animal science and economic pressure have increased the speed with which calves are transferred to feed lots for fattening to market weight. American cows now consume several types of corn product: the corn kernels themselves, corn gluten feed, a by-product of ethanol and HFCS production, and silage, the ground-up stalks and corn cobs from the plant. While cows will quickly gain weight, in terms of both meat and fat, on a diet composed of higher amounts of corn and corn products, this is not a diet to which the cow is evolutionarily adapted. Feeding cows corn changes the pH of their stomach, making it more acidic and encouraging the growth of acid-producing bacteria, and over time this can lead to a condition known as *acidosis*. Left untreated, acidosis will eventually kill a cow. For livestock producers, the goal is to stave off acidosis for as long as possible to allow a cow to gain weight before slaughter. One way to delay the effects of acidosis is to provide the cow with low doses of antibiotics.

Escherichia coli (*E. coli*) is a bacteria that lives in the gastrointestinal tract of humans and animals. Most types of *E. coli* are benign or cause only minor diarrhea. One type, however, is particularly harmful to humans. *E. coli* O157:H7 is a form of the bacteria that at some point in its evolutionary history incorporated the DNA for toxin production from *Shigella* bacteria. The primary symptom of an *E. coli* O157:H7 infection is bloody diarrhea and painful cramping. Some individuals develop hemolytic uremic syndrome (HUS) when Shiga toxins produced by the *E. coli* affect the cells of the intestines. Bleeding of the intestinal walls allows the bacteria to enter the bloodstream, where the toxin destroys red blood cells. The kidneys begin to fail and the body swells with fluid as the body stops filtering and excreting fluids. Eventually heart failure can kill the person. HUS was first described in the 1950s, and retrospective analysis has identified *E. coli* O157:H7 in a sample of bloody diarrhea from the 1970s. This deadly strain of *E. coli* first entered the public health consciousness in the early 1980s, however, when it caused an epidemic associated with consumption of undercooked hamburger patties in Michigan and Oregon. A larger outbreak followed a decade later when individuals eating at Jack in the Box restaurants consumed hamburgers whose advertising tagline was "So good it's scary." The large, lightly grilled patties were not exposed to high enough heat to kill the bacteria; 700 people became sick and four children died.

E. coli is an asymptomatic resident of the gastrointestinal tracts of feedlot cattle; an estimated 30% of feedlot cattle shed the *E. coli* O157:H7 bacteria in their feces (Callaway, Carr, Edrington, Anderson, & Nisbet, 2009). A cow can produce 50–60 pounds of manure per day, so feedlots with tens of thousands of cows are producing more waste than many cities . . . with no sanitation system. Cattle stand in their own and the manure of others, meaning that *E. coli* shed from one cow can easily infect nearby cows, or can be found in dust on the hides of uninfected cattle. By changing the pH of cows' stomachs via feeding them corn and corn products, we are making the stomach environment more similar to the pH of human stomachs. The *E. coli* bacteria that survive in the newly acidic environments of cow stomachs are thus well adapted to the pH of the human gut. They are also potentially drug-resistant, as the antibiotics we feed to cattle are the same as we use in therapeutic treatment of people.

Bacteria is shed in cattle feces, but crowded conditions in feedlots and lack of hygiene means that fecal material sometimes coats animal hides. When cows are taken to be slaughtered their hides are stripped off the carcass. While steps are taken to prevent any contaminants on the animals' skin from making it into the finished meat supply, fecal dust can fly into the air and land on carcasses whose hides have already been removed. Additionally, gutting of the animals takes place by hand and a worker's knife may slip and the contents of the gastrointestinal tract (i.e., the *E. coli* bacteria) may make their way onto the meat of the animal.

Consolidation of livestock production into fewer but larger CAFOs is mirrored by consolidation in the slaughter/processing/packing industry. A few large firms control the majority of the market share for beef, pork, and poultry processing. This means that a single slaughterhouse can be producing hundreds of thousands, even millions, of pounds of meat a day. Thus, one infected cow can infect millions of pounds of ground beef.

Ground beef is an ideal environment for bacteria to grow: it has a much larger surface area than does an intact steak. Ground beef is also an ideal vector for transmission of bacteria: if a burger is not cooked to an internal temperature that can kill bacteria, the infection is easily transmitted, whereas the outside of steaks are always exposed to heat and the bacteria are not located inside the steak itself. Increasing the risks of bacterial contamination of ground beef are the economic realities of food production in the United States. One hamburger sold in the United States could contain beef products from multiple processing facilities located in multiple countries, representing many points of entry for *E. coli* infection. An investigation by the *New York Times* into the ground beef that was responsible for a 2007 *E. coli* O157:H7 outbreak indicated that the frozen beef patties the infected individuals consumed contained fat from a packing plant in Nebraska, fresh lean ground beef from a processor in Texas, frozen lean ground beef from a slaughterhouse in Uruguay, and lean finely textured beef (aka "pink slime") from either Iowa or South Dakota. These meat products were combined with bread crumbs and spices in a Cargill facility in Wisconsin. The contamination of the final product with bacteria could have occurred at one or many points on the production chain, given that the four facilities producing the ingredients for the final burger patty were themselves drawing in beef from widely distributed feedlots.

The problem of diarrheal diseases in low-income countries, where they are the second most common cause of mortality, is discussed further in Chapter 11. Diarrheal transmission is both water-borne and food-borne in these low-income countries where the water, sanitation, and hygiene infrastructure and practices are underdeveloped.

QUICK REVIEW

✓ Food is a product of both culture and environment.

✓ The development of agricultural techniques ~10,000 years ago fundamentally altered the health and size of human populations, and began independently in multiple agricultural hearths around the world.

✓ Agriculture results in ecological simplification, with fewer species grown in fields, and necessitates that farmers provide protection to crops because the system is no longer internally regulated.

✓ Modern commercial agriculture relies on inputs such as genetically modified seeds, fertilizers, pesticides, and irrigation water, to generate phenomenal amounts of food.

✓ Human health is now affected as much or more by diseases of overnutrition than it is by malnutrition, and this will increasingly be the case as countries move through the stages of the nutrition transition and increase consumption of fats and sugars.

CONCLUSION

The emergence of agriculture has fundamentally altered human history, allowing for food production and population growth far beyond the capacity of hunter–gatherers. The transition from foraging to sedentary agriculture altered the physiological structure of humans, decreasing stature and overall health, and also changed fertility patterns and family size. Major changes to agricultural production that took place in the second half of the 20th century, including scientific and technological innovations such as high-yield varieties of seeds, synthetic nitrogen manufacture, and mechanization of farm equipment, decreased the labor to grow food while increasing the productivity of the land. The introduction of these techniques from the wealthy world to low- and middle-income countries in the 1960s and 1970s, during what is known as the Green Revolution, averted the starvation of millions of individuals. The phenomenal production of corn and soy and other crops is not without health and environmental consequences, however, as it has resulted in increased caloric and fat consumption, unsustainable groundwater usage, pollution of soil and water with carcinogens and animal waste, an increase in the speed and scale of antibiotic resistance, and increased carbon dioxide emissions. One way to think about these interrelated phenomena is as internalities versus externalities.

Internalities are outcomes directly experienced by customers, while externalities are those that are felt by society at large. The price paid at the grocery store for food, an internality, does not reflect its true costs, or externalities. Prices are felt directly by individuals and are immediate. Costs are shared by groups of people and are long term. Costs of food production and consumption include those listed above, as well as things like the stresses placed on health care systems by millions of diabetic individuals and the economic consequence of an unhealthy population.

Estimates of future population growth project that another 2 billion people will join the planet in the coming decades. Given the instability and unsustainability of our current diets, and the race into stage 4 of the nutrition transition that we currently observe in low- and middle-income countries, there is no way to support these new members unless we progress to stage 5, behavioral change.

REVIEW QUESTIONS

1. Where are the major agricultural hearths of the world? What is the relationship between plant and animal domestication, food production, and cities?

2. What happened during the Columbian exchange? What are the positive and negative results of this exchange? How would the history of humanity be different if the eastern and western hemispheres had never interacted?

3. What are the benefits and consequences of the Green Revolution? And of the changes to agriculture that occurred in the United States during the 1950s/1960s? How can we learn from the past to feed the extra 2 billion people who will join the planet in the coming decades?

4. The USDA defines food deserts based on a combination of low-income populations and distance to a grocery store within a census tract. Do you think this is a good way to define a food desert? What other variables or characteristics would you include in the definition? How might what constitutes a food desert vary between urban and rural areas?

5. Visit the site for the USDA Food Access Research Atlas (*www.ers.usda.gov/data-products/food-access-research-atlas/go-to-the-atlas.aspx*). Explore the different ways you can think about food access (e.g., varying distances to grocery stores) and how this changes the map of food deserts. Zoom in on the place you live, are there any census tracts that are considered food deserts? Do these places match up with how you would define a food desert in your area?

6. What is a CAFO? What are health consequences, both direct and indirect, associated with the increase in meat production taking place in CAFOs?

7. How does the nutrition transition interact with the demographic, mobility, and epidemiological transitions?

8. Record your food consumption over the course of a week, then categorize your food intake into categories such as grains, meat, dairy, and processed food. What phase of the nutrition transition are you in? How do you interact with the industrial food chain?

9. How does the price of food differ from its cost? How can you internalize the externalities?

REFERENCES

Bruce, K. D., & Cagampang, F. R. (2011). Epigenetic priming of the metabolic syndrome. *Toxicology Mechanisms and Methods, 21*(4), 353–361.

Callaway, T. R., Carr, M., Edrington, T., Anderson, R. C., & Nisbet, D. J. (2009). Diet, Escherichia coli O157: H7, and cattle: A review after 10 years. *Current Issues in Molecular Biology, 11*(2), 67–80.

Centers for Disease Control. (2013). *Prevalence of self-reported obesity among U.S. adults by state and territory.* Retrieved August 30, 2015, from *www.cdc.gov/obesity/data/prevalence-maps.html*.

Conway, G. (1998). *The doubly green revolution: Food for all in the twenty-first century.* Ithaca, NY: Cornell University Press.

Crosby, A. W. (2003). *The Columbian exchange: Biological and cultural consequences of 1492* (Vol. 2). Westport, CT: Greenwood Publishing Group.

Diamond, J. M. (1999). *Guns, germs, and steel.* New York: Norton.

Diamond, J. M. (2002). Evolution, consequences and future of plant and animal domestication. *Nature, 418*(6898), 700–707.

Dimitri, C., Effland, A. B., & Conklin, N. C. (2005). *The 20th century transformation of US agriculture and farm policy. Economic Information Bulletin 3* (pp. 1–17). Washington, DC: U.S. Department of Agriculture, Economic Research Service.

Donham, K. J., Cumro, D., & Reynolds, S. (2002). Synergistic effects of dust and ammonia on the occupational health effects of poultry production workers. *Journal of Agromedicine, 8*(2), 57–76.

Donham, K. J., Haglind, P., Peterson, Y., Rylander, R., & Belin, L. (1989). Environmental and health studies of farm workers in Swedish swine confinement buildings. *British Journal of Industrial Medicine, 46*(1), 31–37.

Easterbrook, G. (1997, January). Forgotten benefactor of humanity. *The Atlantic.*

Ehrlich, P. (1970). The population bomb. *New York Times,* p. 47.

Gaud, W. S. (1968, March 8). The Green Revolution: Accomplishments and apprehensions. Address before The Society for International Development, Washington, DC. Retrieved August 30, 2015, from *www.agbioworld.org/biotech-info/topics/borlaug/borlaug-green.html*.

Gutentag, E. D. (1984). Geohydrology of the High Plains aquifer in parts of Colorado, Kansas, Nebraska, New Mexico, Oklahoma, South Dakota, Texas, and Wyoming: High Plains RASA Project [Western States (USA); South Central States (USA)]. *Geological Survey Professional Paper (USA).* Washington, DC: U.S. Government Printing Office.

Hamilton, M. J., Milne, B. T., Walker, R. S., & Brown, J. H. (2007). Nonlinear scaling of space use in human hunter–gatherers. *Proceedings of the National Academy of Sciences USA, 104*(11), 4765–4769.

Hoekstra, A. Y. (2010). The water footprint of animal products. In J. D'Silva & J. Webster (Eds.), *The meat crisis: Developing more sustainable production and consumption* (pp. 22–33). London: Earthscan.

Hoekstra, A. Y. (2012). The hidden water resource use behind meat and dairy. *Animal Frontiers, 2*(2), 3–8.

Kendall, H. W., & Pimentel, D. (1994). Constraints on the expansion of the global food supply. *Ambio, 23*(3), 198–205.

Konikow, L. F. (2013). *Groundwater depletion in the United States (1900–2008).* U.S. Department of the Interior, U.S. Geological Survey. Available at *http://pubs.usgs.gov/sir/2013/5079*.

Koutros, S., Freeman, L. E. B., Lubin, J. H., Heltshe, S. L., Andreotti, G., Barry, K. H., et al.

(2013). Risk of total and aggressive prostate cancer and pesticide use in the Agricultural Health Study. *American Journal of Epidemiology, 177*(1), 59–74.

Lu, C., Fenske, R. A., Simcox, N. J., & Kalman, D. (2000). Pesticide exposure of children in an agricultural community: Evidence of household proximity to farmland and take home exposure pathways. *Environmental Research, 84*(3), 290–302.

McNeill, W. H. (1976). *Plagues and peoples.* Garden City, NY: Anchor Press/Doubleday.

Mekonnen, M. M., & Hoekstra, A. Y. (2012). A global assessment of the water footprint of farm animal products. *Ecosystems, 15*(3), 401–415.

Morand, S., McIntyre, K. M., & Baylis, M. (2014). Domesticated animals and human infectious diseases of zoonotic origins: Domestication time matters. *Infection, Genetics and Evolution, 24,* 76–81.

Parrón, T., Requena, M., Hernández, A. F., & Alarcón, R. (2014). Environmental exposure to pesticides and cancer risk in multiple human organ systems. *Toxicology Letters, 230*(2), 157–165.

Popkin, B. M. (1994). The nutrition transition in low-income countries: An emerging crisis. *Nutrition Reviews, 52*(9), 285–298.

Popkin, B. M. (2002a). An overview on the nutrition transition and its health impacts. *Public Health Nutrition, 5*(1A), 93–103.

Popkin, B. M. (2002b). The shift in stages of the nutrition transition in the developing world differs from past experiences. *Public Health Nutrition, 5,* 205–214.

Popkin, B. M. (2003). The nutrition transition in the developing world. *Development Policy Review, 21*(5–6), 581–597.

Popkin, B. M., Duffey, K., & Gordon-Larsen, P. (2005). Environmental influences on food choice, physical activity and energy balance. *Physiology and Behavior, 86*(5), 603–613.

Popkin, B. M., & Gordon-Larsen, P. (2004). The Nutrition Transition: Worldwide obesity dynamics and their determinants. *International Journal of Obesity, 28*(Suppl. 3), S2–S9.

Vinson, F., Merhi, M., Baldi, I., Raynal, H., & Gamet-Payrastre, L. (2011). Exposure to pesticides and risk of childhood cancer: A meta-analysis of recent epidemiological studies. *Occupational and Environmental Medicine, 68*(9), 694–702.

World Health Organization (WHO). (2015). Obesity and overweight: Fact sheet number 311. Retrieved from *www.who.int/mediacentre/factsheets/fs311/en*.

Youngson, N. A., & Morris, M. J. (2013). What obesity research tells us about epigenetic mechanisms. *Philosophical Transactions of the Royal Society B: Biological Sciences, 368*(1609), 20110337.

Zahm, S. H., & Ward, M. H. (1998). Pesticides and childhood cancer. *Environmental Health Perspectives, 106*(Suppl. 3), 893.

FURTHER READING ••

Bell, A. C., Ge, K., & Popkin, B. M. (2002). The road to obesity or the path to prevention?: Motorized transportation and obesity in China. *Obesity Research, 10,* 277–283.

Brown, L. (2004). *Outgrowing the earth: The food security challenge in an age of falling water tables and rising temperatures.* Washington, DC: Earth Policy Institute.

Cook, C. D. (2004). *Diet for a dead planet: How the food industry is killing us.* New York: New Press.

Drexler, M. (2002). *Secret agents: The menace of emerging infections.* Washington, DC: National Academy Press.

Fitting, E. (2006). Importing corn, exporting labor: The neoliberal corn regime, GMOs, and the erosion of Mexican biodiversity. *Agriculture and Human Values, 23,* 15–26.

Gade, D. W. (1979). Inca and colonial settlement, coca cultivation and endemic disease in the tropical forest. *Journal of Historical Geography, 5,* 263–279.

Gopalan, C. (1992). *Nutrition in developmental transition in South-East Asia.* New Delhi: World Health Organization.

Mann, C. (2006). *1491: New revelations of the Americas before Columbus.* New York: Vintage Books.

Menzel, P., & D'Aluisio, F. (2007). *Hungry planet: What the world eats.* Danvers, MA: Crown.

Millstone, E., & Lang, T. (2013). *The atlas of food: Who eats what, where and why.* Oakland: University of California Press.

Nestle, M. (2007). *Food politics: How the food industry influences nutrition and health.* Oakland: University of California Press.

Patel, R. (2007). *Stuffed and starved: The hidden battle for the world food system.* Brooklyn, NY: Melville House.

Pollan, M. (2002). *The botany of desire.* New York: Random House.

Pollan, M. (2007). *The omnivore's dilemma.* New York: Penguin Books.

Powell, K. E., Martin, L. M., & Chowdhury, P. P. (2003). Places to walk: Convenience and regular physical activity. *American Journal of Public Health, 93*(9), 1519–1521.

Reardon, T., & Berdegue, J. A. (2002). The rapid rise of supermarkets in Latin America: Challenges and opportunities for development. *Development Policy Review, 20*(4), 371–388.

Saelens, B. E., Sallis, J. F., Black, J. B., & Chen, D. (2003). Neighborhood-based differences in physical activity: An environment scale evaluation. *American Journal of Public Health, 93*(9), 1552–1558.

Sage, C. (2011). *Environment and food.* New York: Routledge.

Schlosser, E. (2001). *Fast food nation: The dark side of the all-American meal.* Boston: Houghton-Mifflin.

Weil, C. (1981). Health problems associated with agricultural colonization in Latin America. *Social Science and Medicine, 15D,* 449–461.

Williams, O. (2005). Food and justice: The critical link to healthy communities. In D. Naguib Pellow & R. J. Brulle (Eds.), *Power, justice and the environment: A critical appraisal of the environmental justice movement* (pp. 117–130). Cambridge, MA: MIT Press.

CHAPTER 9

Neighborhoods and Health

Investigations of geographic variation in health have a long history; John Snow tallied cholera deaths by water district in the late 1800s in order to explain why certain areas in London had higher rates of mortality. But until recently for more than a century the field of public health was largely dominated by studies that investigate individual risk factors for health and disease. Increasing acknowledgment that this research failed to fully explain differentials in health has led to a broadening of the field. In particular, some of the focus has shifted toward examining the social determinants of health (even giving rise to a new field of social epidemiology!) and the importance of neighborhoods as contexts where social relations and the physical environment interact to modify individual health. A growing number of studies consider neighborhood-level determinants of health (Curtis, 1990; Diez-Roux, 1998, 2001; Kawachi & Berkman, 2003; Macintyre, Ellaway, & Cubbins, 2002; Schwartz, 1994; Susser, 1994). Risk and protective factors may inherently reside at the community level, and therefore measurement and intervention should take place at that same level. Spatial variations in health outcomes are caused by differences in the kinds of people who live in these places and differences in the physical or social environment. The goal of these studies is to differentiate between **contextual** (area) effects and **compositional** (population) effects of a neighborhood on health. *Contextual effects* occur when factors in an individual's social and physical environment have a direct impact on health. *Compositional effects* occur when individuals with the same set of risk factors live in the same area, giving rise to spatial patterns of disease. These spatial patterns are not directly related to environmental risk; rather, they are an artifact of the clustering of people with similar characteristics.

The remainder of this chapter has four major sections. The first explores how social context and neighborhood social environments affect health; this includes a discussion of human stress response and the biological mechanisms that link social environments and health. The second section summarizes research on how the built environment impacts health behaviors and health outcomes. Though there is also an extensive literature on how environmental toxicants lead to adverse health, this topic is covered in Chapter 11. The third section is concerned with understanding the methods by which researchers have explored links between neighborhoods and health, including statistical methods and experimental studies. The final section addresses the major challenges neighborhoods and health researchers face today.

THE CONCEPT OF NEIGHBORHOOD HEALTH

Influences on the health of individuals are often explained through an analogy to a stream. "Upstream" influences may include a political decision at the national level, such as corn subsidies that may indirectly lead to increased obesity rates. "Downstream" influences may include individuals' choices to "supersize" their meals at the local fast-food restaurant. These upstream and downstream influences on health can also be conceptualized spatially. Figure 9.1 shows how individuals are part of neighborhoods, communities, and regions. Individual health can be influenced by neighborhood-level processes, such as whether or not there is a fast-food restaurant near a person's home, or further upstream at the community level. For instance, the towns (communities) that two of us (Emch and Root) grew up in did not have any fast-food restaurants at all because the local government did not allow them.

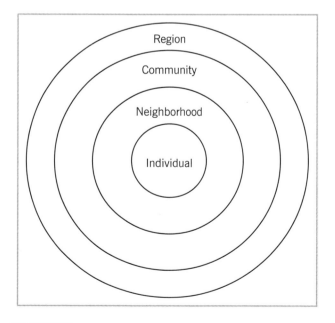

FIGURE 9.1 Neighborhoods within a nested spatial hierarchy.

A growing body of literature has demonstrated that overall living environments, including the neighborhoods in which people reside, have important effects on health. In studies using multilevel statistical methods to disentangle community-level and individual-level effects, neighborhood characteristics have been shown to independently influence a wide variety of health outcomes, including cardiovascular disease, mental health, overall mortality, adverse birth outcomes, self-rated health, asthma, and Lyme disease (Jones & Duncan, 1995). Neighborhood context has also been shown to affect behaviors and health outcomes, such as physical activity (Gordon-Larsen, Nelson, Page, & Popkin, 2006; Gordon-Larsen & Reynolds, 2006), food choices (Popkin, Duffey, & Gordon-Larsen, 2005), crime (Morenoff, Sampson, & Raudenbush, 2001; Sampson, Raudenbush, & Earls, 1997), hypertension (Morenoff et al., 2007), heart disease (Diez-Roux et al., 2001), smoking (Barnett, 2000; Barnett, Moon, & Kearns, 2004; Barnett, Pearce, & Moon, 2005; Moon & Barnett, 2003; Pearce, Hiscock, Moon, & Barnett, 2009), and self-perceived health (Dunn, Veenstra, & Ross, 2006). There is growing awareness of the role the human-made environment plays in health. For example, the layout of a community affects the occurrence of cardiovascular disease, mental health, and Lyme disease among residents.

The **neighborhood approach** to analyzing disease outcomes is similar to the health geographic approach of disease ecology (discussed in Chapter 2) and its analytical approaches. Contextual effects are included in the environment and behavior vertices, which include aspects of the physical and built environments as well as the larger social context. Compositional effects are represented in the population and behavior vertices, which include individual demographic and economic characteristics. All of these factors exist in a certain place (the "neighborhood") at a certain point in time. The neighborhood approach is a conceptual framework that is spatial in nature and can be used to address the complexities of social and environmental correlates of disease over time and space. The health geographic tradition of disease ecology is not only spatial, but holistic; it includes the integration of many different types of variables responsible for disease. The theory of disease ecology fits into both the spatial organization and human–environment traditions of geography. The fundamental question asked in studies using this approach is "What factors are associated with the time–space variation of disease?" It emphasizes the complex sets of interactions between people and their environments.

Figure 9.2 summarizes these interactions between people and their social and physical neighborhood environments. Previous research has distinguished between two types of neighborhood characteristics—structural and social processes (Shaw & McKay, 1942; Sampson, Rosenoff, & Gannon-Rowley, 2002). Structural characteristics refer to sociodemographic attributes, in particular, poverty, racial/ethnic makeup, and residential mobility. These structural characteristics of a neighborhood are the result of a variety of larger social and economic forces. For example, historical housing policies, including urban renewal projects, exclusionary zoning, public housing siting, and residential redlining, led to racial segregation and areas of concentrated poverty in the United States (Kushner, 1980; Massey, 1990). Individual choices about where to live are necessarily constrained by these larger forces, thereby contributing to and perpetuating high-minority, high-poverty

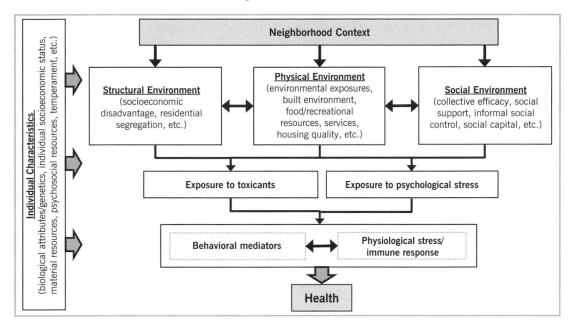

FIGURE 9.2 Conceptual model for the interactions between individuals and their neighborhood social, structural, and physical environments that influence health outcomes.

neighborhoods. Neighborhood social characteristics refer to the social organization and social interactions between members of a community that influence the health and behavior of residents. If residents of a neighborhood have strong social relationships, these may act to modify deleterious health behaviors, exchange information about health behaviors, the causes and consequences of disease, and the resources available for health care, or mobilize community members to change negative aspects of their physical environment. Neighborhoods also have unique physical environments; some are full of trees and green space with good sidewalks, dog parks, and community gardens, while others are not as pleasant, characterized by vacant housing, graffiti, garbage, and a distinct lack of green space. Some neighborhoods have well-stocked grocery stores, gyms, good schools, and community centers that offer programs for children and adults alike, while others have an abundance of fast-food restaurants and no community recreational facilities to speak of. The environmental justice literature also shows that disadvantaged neighborhoods have a disproportionate number of environmental hazards such as landfills, chemical emitting factories, and (more recently) fracking sites. The structural, social, and physical environments interact to create differentials in exposure to both toxic substances and psychological stress. Just living in a neighborhood characterized by high levels of socioeconomic disadvantage, segregation, poor housing, no green space, and poor social support can cause much higher levels of stress! Individuals, of course, have the means to cope with these environmental circumstances and the choices we make can mediate the effect of our environment and reduce (or increase) stress. Perhaps we chose to meditate for an hour each day to reduce our stress levels, or spend a significant amount of our free time studying so we can get into college and move out of the (poor) neighborhood in which we grew up. Finally, the impact of neighborhood conditions is modified by individual-level characteristics. For example, some individuals may have the economic means and social support that allow them to overcome a poor neighborhood environment while others may not.

SOCIAL CONTEXT AND HEALTH

In 1978, Sir Michael Marmot, a British researcher from Oxford University, published the first in a series of articles on the "Whitehall Studies" that would fundamentally change the ways in which people viewed human health. After a decade of collecting data on risk factors for cardiovascular and respiratory disease from British civil servants, Marmot discovered an inverse relationship between employment grade (e.g., how high you are up the job hierarchy) and mortality from a wide variety of illnesses. In fact, those in the highest employment grade had one-third the mortality rate of those in the lowest grade (Marmot, Rose, Shipley, & Hamilton, 1978). These findings are now known as the "social gradient in health" and have been replicated both by Marmot himself and by those conducting numerous other studies around the world. Time and time again, evidence shows that in general, the lower an individual's socioeconomic position, the worse their health. It is a global phenomenon, seen in low-, middle-, and high-income countries, which creates deep health inequalities both within and between countries. Figure 9.3 shows infant

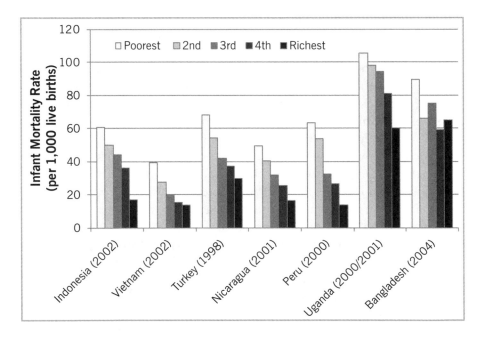

FIGURE 9.3 The social gradient in health. Infant mortality rates for selected countries by wealth quintiles. Created using data from Gwatkin et al. (2007).

mortality rates by wealth quintiles for seven countries around the world. The lowest wealth quintile has the highest rate of infant mortality. In some countries, such as Indonesia, the wealthiest segment of the population has infant mortality rates close to those of the United States. Neighborhood socioeconomic environment shows the same social gradient in health. For example, Bosma, van de Mheen, Borsboom, and Mackenbacket (2001) found that persons living in a neighborhood with a high percentage of unemployed or poor persons had a higher risk of death than did those living in a neighborhood with a low percentage of unemployed or poor persons. Mortality was about 50% higher in the poorest neighborhoods compared to the wealthiest ones. But many researchers who study socioeconomic inequalities in health struggle to explain exactly why disadvantaged communities have poorer health. After all, just because you're poor doesn't mean you are biologically less fit than a person who is wealthy. Thus, there must be something about living in poor communities that affects people's health behaviors, ability and willingness to utilize health care, or stress levels. We now think that health can be shaped by the social environment of neighborhoods; that is to say, by the social relationships and social ties between residents. This is the notion of social capital.

Social Capital

Communities that are close-knit are more likely to work together to achieve common goals, such as creating cleaner public spaces, strengthening schools, and advocating for new community centers. Residents are also more likely to exchange information, including information about health behaviors and the availability of health resources. The benefits derived from the connections that develop between neighbors are often referred to as **social capital**. Formally, social capital is the tangible material and psychosocial "resources embedded in social relationships that are available for community members to access" (Kawachi & Berkman, 2014, p. 291). Think of it this way: if you help a friend move, you have built social capital that allows you to call on him or her for help if and when you yourself ever move. It is often discussed in terms of three separate mechanisms that operate together to influence health: social contagion, collective efficacy, and informal social control.

Social contagion is similar to the concept of diffusion discussed in Chapter 6. In this case, behaviors, behavioral norms, and information spread more quickly through social networks that are stronger and more closely knit. These behaviors can be both health-promoting and deleterious. For example, Christakis and Fowler (2007) constructed social network ties for 12,067 adults enrolled in the Framingham Heart study, one of the largest longitudinal studies conducted in the United States. They found that obesity (using body mass index [BMI]) seemed to spread through social networks. That is, a person's chances of becoming obese increased by 57% if he or she had a friend who became obese. In addition, the risk of becoming obese appears to "trickle down" in social networks; so, if you become obese, this raises your closest friends' risk of becoming obese by 45% (one degree of separation away), your friends' friends' chances by 20% (two degrees away), and your friends' friend's friends' risk by 10% (three degrees away). Christakis and Fowler suggest that "people are connected, and so their health is connected" (Christakis & Fowler, 2007, p. 378).

Collective efficacy, another aspect of social capital that can affect health, refers to the ability of a group to mobilize to undertake collective action. Often, members of a group are connected to each other through a community organization, church, or some other institution and individuals benefit from their connection to this group. In the aftermath of the devastation caused by Hurricane Katrina in 2005, many communities mobilized to begin to clean up after the disaster, to find and provide food and shelter, and to care for the elderly and disabled. Studies show that depression was significantly lower in communities characterized by high levels of collective efficacy, even after controlling for the socioeconomic characteristics of the community (Fullerton et al., 2015). However, in disadvantaged neighborhoods, many residents began to leave their communities, often because recovery efforts were too slow, and community organizations began to splinter and social support became "spatially displaced," which resulted in the dissipation of many of the ties that had supported the community immediately after the hurricane (Elliott, Haney, & Sams-Abiodun, 2010). This finding suggests that collective efficacy is both a spatial and a temporal process; for collective action to take place, people must be members of spatial neighborhoods and the ability to support collective action may weaken over time.

Informal social control is a concept that is often used to understand differences in neighborhood-level health behaviors, especially among adolescents. The term refers to social interactions among members of a community that maintain social order, that is, the willingness of community members to influence and regulate the deviant behavior of others. This concept originated in the criminology literature but has been adapted to examine a range of health behaviors such as underage smoking, drinking, and drug use. Smoking bans are a good example of informal social control. They rely not just on people adhering to rules about smoking in public places, but on the collective action of community members in policing people who chose to ignore the rules. Studies have shown lower smoking rates among adolescent peer networks in which smoking is considered unacceptable (Alexander, Piazza, Mekos, & Valentee, 2001; Hoffman, Monge, Chou, & Valentee, 2007). Among adults, Westmaas, Wild, and Ferrence (2002) found that social control (measured as pressure from family or friends) predicted greater smoking reduction among men and women. A study in Agincourt, South Africa, found a similar result for alcohol use (Leslie et al., 2015). High levels of informal social control, measured using a series of survey questions about how likely residents of a community were to intervene if certain negative events occurred, led to a decrease in heavy alcohol consumption and potential problem drinking in South Africa. Neighborhoods offer important contexts within which social control occurs; children and adolescents often interact with other children in their neighborhood and are less likely to engage in poor behaviors when other adults are willing to step in and stop such behaviors.

The network of social ties people develop that give rise to social capital is not necessarily geographically bounded, though much of the neighborhoods-and-health research makes just this assumption. Decades ago when families were not as mobile as they are now (see the discussion of the mobility transition in Chapter 4), people often lived and worked in the same town in which they were born. They grew up in relatively close-knit communities where children were more collectively parented than they are today. Today, only about 37% of the population lives in the same town in which they were born—an additional 20% live in the same state but not the same town, which leaves a whopping 43% of the population that has lived in two or more states. In fact, an estimated 12% of the population moves in any given year! This kind of mobility translates into a geographic rearrangement of social ties. Your closest friends upon whom you rely for social and emotional support may no longer live in your town; rather, your cell phone allows you to keep in close contact with high school friends even after you've moved away to attend college or find a job. In areas of the United States where people move less—mainly the Midwest—social ties within a community are stronger. Areas with highly mobile populations are less likely to be characterized by strong and lasting social networks. In fact, the obesity study by Christakis and Fowler discussed above found that obesity spread through social networks, which were spread across the United States, but that the weight gain of immediate geographic neighbors did not affect the chance of weight gain. However, in a study of cholera risk and the effect of a cholera vaccine in Bangladesh, Root, Giebultowicz, Ali, Yunus, and Emch (2011) found that higher levels of vaccine coverage in social networks can lead to increasing levels of indirect protection of nonvaccinated individuals and progressively higher levels of total protection of vaccine recipients. Results were nearly identical when cholera risk and vaccine efficacy were examined using geographic proximity rather than social network ties. This suggests that the geographic boundedness of social ties is likely related to a country's phase of development. The Matlab Health and Socioeconomic Survey, conducted in a small rural area in Bangladesh, has shown that in the mid-1990s, extended families typically lived very near to each other, and even women who married were often still quite close to home. A second round of the survey, done in 2014–2015 after Bangladesh had undergone a rapid phase of economic development, showed that nearly 40% of the population of Matlab had migrated to one of the big cities for work (mostly young men and women) and even abroad to places like Singapore, Malaysia, and the United Arab Emirates. This movement has significantly altered the social relationships in the area. This is all to say that in some geographic regions at certain points in time neighborhoods may be a good indicator of social capital, but in other regions they may not.

Social Disorganization Theory

The structural and social environments presented in Figure 9.2 are assumed to be closely related. Structural characteristics, and the historical practices that produced and reinforced residential segregation and poverty, help to create and reinforce the social aspects of a neighborhood such as collective efficacy and informal social control. **Social disorganization theory** identifies specific dimensions of neighborhood structure that are relevant to a wide range of phenomena, including poor health and well-being (Shaw & McKay, 1942; Sampson et al., 2002; Browning & Cagney, 2003). In particular, three structural factors—low economic status, ethnic heterogeneity, and residential instability—lead to the disruption of community social organization, for example, social capital that influences the values, attitudes, aspirations, and motivations of residents. The breakdown of these social processes account for variation in child and adolescent health and behavioral outcomes by affecting social interactions, parental health and well-being, home environments, school and religious settings, and other local institutions.

Empirical evidence supports the idea that higher levels of residential turnover, concentrated disadvantage, and minority populations are associated with lower levels of collective efficacy and higher levels of social disorder (Brody et al., 2001; Rankin & Quane, 2002; Sampson et al., 1997) and a few studies have found direct links with poor mental health (Caughy, O'Campo, & Muntaner, 2003), psychosocial distress (Gary, Stark, & LaViest, 2007), child mistreatment (Couldton et al., 1999), reduced physical activity (Molnar, Gortmaker, Bull, & Buka, 2004), and low birth weight (Schempf, Stabino, & O'Campo, 2009). Neighborhoods with low levels of collective efficacy and high social disorder: (1) fail to effectively monitor and enforce resident behavior, (2) lead to an uncomfortable and high-stress residential environment, (3) create low cohesion among families and community members that may undermine healthy behaviors, and/or (4) create environments where people are less likely to engage in physical activity (Leventhal & Brooks-Gunn, 2000; Molnar et al., 2004; Tempiero et al., 2010). Thus, the social and structural elements of disadvantaged neighborhoods work together to create environments that are disruptive and maladaptive for the health of neighborhood residents.

Social Environment and Stress

While most researchers agree that factors such as social capital, neighborhood SES, and residential stability work in combination to affect health, there is still an open debate about how exactly these sociostructural factors "get under the skin" and influence health. One hypothesis is that poor neighborhood environments elicit the body's stress response. Stress is both a biological and a psychological response to external circumstances we feel we do not have the resources to deal with (Kubzansky, Seeman, & Glymour, 2014). When an individual perceives that external events are overwhelming his or her capacity to cope, a sense of stress results, which in turn triggers a biological stress response. The body responds by releasing "stress hormones" (e.g., cortisol and adrenalin), which produce physiological changes in our body and affect our autonomic, metabolic, and immune systems. High levels of stress may weaken your immune system so that you are more likely to get sick, cause your body to age faster, or decrease your body's ability to regulate important functions like blood pressure and metabolism of glucose. While these stress hormones may produce fairly standard physiological responses, people's perceptions of stressful circumstances are quite different. Living in a high-crime neighborhood may feel extremely stressful for one person, and barely affect his or her neighbor. Differential perceptions of stressful situations could explain why some people appear to be more sensitive to poor neighborhood environments than others. Strong social support systems (and social capital) can act as buffering mechanisms, protecting people from becoming overwhelmed by stressful external events. When asked, individuals that perceive they have a large social network they can activate in times of need report lower levels of stress compared to individuals who perceive that they are alone or without friends, neighbors, and relatives to rely on.

But results from studies examining SES, perceived stress, and hormone response are highly inconsistent (Dowd, Simanek, & Aiello, 2009). For example, the Chicago Community Adult Health Study found that respondents in neighborhoods with high levels of perceived and observed stressors or low levels of social support had lower levels of cortisol. But Cohen, Doyle, and Baum (2006) found that lower income and education were associated with higher levels of cortisol. Studies of psychological (rather than physiological) response to stress are much more consistent. Exposure to stressful neighborhood environments coupled with low levels of social support have been linked to depression and poor mental health in individuals. Neighborhood environments that are stressful, typically characterized by low SES, high crime, and high resident mobility, are related to higher levels of mental health problems. In a study of psychological distress in four large cities in the Netherlands, researchers found that living in neighborhoods with high SES and high social cohesion was associated with lower levels of psychosocial distress (Erdem, Prins, Voorham, van Lenthe, & Burdorf, 2015). Similarly, a study of adults in Chicago found that high levels of social support and lower levels of neighborhood stress translated to lower levels of depressive symptoms (Mair, Roux, & Morenoff, 2010). An individual's response to stressful situations can also directly trigger unhealthy behaviors. People who are stressed are more likely to smoke, drink alcohol, use drugs, overeat, or undereat.

EFFECTS OF THE BUILT ENVIRONMENT ON HEALTH

Where we live and all the places we move through in our daily lives affect our health. Much of Chapters 11 and 12 explore how the natural environment affects our health, but the way we plan and build our neighborhoods, towns, and cities also influences our health. Much of this topic is explored in Chapter 10 on urban health, but this topic has been so thoroughly covered in the neighborhoods and health literature, it deserves some mention here. Neighborhood built environments—including features such as land use, density, street connectivity, transportation systems, and availability of healthy food and recreational options—have been tied to dietary choices and physical activities levels and a variety of related health outcomes such as diabetes, obesity, and hypertension. The concept of how our built environment affects us can be intuitively understood through the example of obesity. Although there may be genetic differences between individuals in whether they are more or less susceptible to becoming overweight or obese, much of the recent obesity epidemic can be attributed to changes in caloric input versus output (see Chapter 8 on nutrition). Clearly people are taking in more calories and burning off fewer than in the past. The built environment influences these patterns at a population level.

Our urban centers have changed dramatically over the past century. "Traditional" neighborhoods are characterized by higher residential density, a mixture of land uses (residential and commercial), and grid-like street patterns with short block lengths. These features of the built environment encourage people to walk or bicycle for transport. In contrast, newer low-density, automobile-dependent neighborhoods with segregated patterns of land use and weak public transport system development (most of these are suburban) often prohibit walking or biking for transport. Studies have shown that the **walkability** of a neighborhood can foster both physical activity and a sedentary lifestyle. Walkability is a measure of how friendly a neighborhood is to walking, which can be measured in a variety of different ways. The proximity of people's homes to other people's homes, measured by calculating the population density, can be important. If there are more people living in an area, then there are more opportunities to interact with neighbors, friends, or schoolmates without driving. A child who lives within walking distance of friends is more likely to walk to that friend's house to play. The proximity of people to commercial enterprises is also important. Mixed-use areas with residential, industrial, and commercial space foster walking because people can walk to their jobs and to shopping centers or to recreation areas. Connectivity also influences walkability. An isolated neighborhood with many dead ends that do not connect to other residential or commercial areas does not encourage walking. Saelens, Sallis, Black, and Chen (2003) interviewed people from two different neighborhoods in San Diego, California: one high-walkability neighborhood with a mixture of single-family and multiple-family residences and higher residential density, and one low-walkability neighborhood with predominantly single-family homes. Respondents were asked to report features of the neighborhood environment including residential density; proximity and ease of access to nonresidential land uses, such as restaurants and retail stores; street connectivity; sidewalks and pedestrian/bike trails; and traffic and crime safety. They also asked about walking behavior and asked participants to wear an accelerometer, a small device that measures physical activity. They found that adult residents living in high-walkability neighborhoods engaged in 70 more minutes of physical activity per week and had lower obesity prevalence than residents of low-walkability neighborhoods. While the walkability of a neighborhood is important for fostering physical activity, the presence of structural resources, such as gyms, parks, community centers, and other recreational settings, is also related to increases in exercise and physical activity. It seems that if you live closer to a gym or park you are more likely to be physically active on a regular basis (Kaczynski & Henderson, 2008). A neighborhood that is more "obesogenic"—where people have low physical activity and therefore higher obesity rates—will have low population density, fewer travel destinations within it, single-use zoning, and low connectivity.

The increasing obesity rates do not just involve caloric output, but also caloric input. Therefore, the local food environment is probably just as important as the walkability of a neighborhood. Access to supermarkets, exposure to fast food, and availability of healthy food options are all important aspects of the local food environment. The nutritional environment, including prices of different types of foods, is important to understand how neighborhood context can influence obesity rates. An aspect of the poverty syndrome discussed in Chapter 3 is the absence of outlets for fresh vegetables and fruits in many inner-city neighborhoods. "Food deserts" are neighborhoods where residents do not have access to an affordable and healthy diet (Cummins, 2002). In general, residents of neighborhoods with better access to supermarkets and other food retail stores that provide access to health-ful foods have healthier diets. They consume more fruits and vegetables, have a lower caloric intake, and are less likely to eat fast food (Larson, Story, & Nelson, 2009). For example, the Multi-Ethnic Study of Atherosclerosis (MESA) measured availability of healthy foods in a variety of different ways—including the density of supermarkets in an area and participant reports of available food options—and found that more healthy food options equated to a better quality diet. People ate less fats and processed meats and had higher intakes of fruits and vegetables, nuts and soy protein, and white meat (Moore, Roux, Nettleton, & Jacobs, 2008). Beyond the availability of food, advertising is also important. Fast-food restaurants adver-tise on billboards in some places more than others, and it's likely that this has an effect on people's usage patterns. The fast-food restaurants themselves have built-in advertising with easily recognizable signs. Often children can recognize these signs at a very early age, which is hardly surprising, given that the mascots of these res-taurants are clowns or cartoon characters.

QUICK REVIEW 1

✓ Individual health can be influenced by neighborhood environments.

✓ The spatial variation in health and behavioral outcomes across neighborhoods is caused by differences in the types of people who live in those neighborhoods and differences in the physical and social environment.

✓ People in lower social classes, and with fewer socioeconomic resources, tend to have poorer health. This is known as the "social gradient in health."

✓ The social gradient extends to neighborhoods—people living in poor neighborhoods also have higher mortality and worse health outcomes.

✓ Health is shaped by the social relationships and social ties between residents residing in shared neighborhood environments.

✓ Many researchers believe that poor neighborhood environments elicit a physiological stress response that adversely affects the metabolic and immune systems and leads to poor health outcomes.

✓ The physical structure of neighborhood environments (the built environment) can affect the activities people engage in and the choices they make about food and healthy behaviors.

✓ Neighborhoods that are walkable with good sidewalks and interesting places to walk to, and with healthy food options such as grocery stores and public gardens, encourage physical activity and healthy eating choices.

OPPORTUNITIES AND CHALLENGES IN NEIGHBORHOOD EFFECTS STUDIES

Modeling Neighborhood Effects

Studying the effects of neighborhood environments on health is not easy. Researchers continually struggle to find the best way to separate the effects of neighborhood from all the other factors that influence health—that is, to differentiate between neighborhood-level and individual-level determinants of health. The goal is to disentangle the context of a neighborhood from the composition of the population. Context is about place, and composition is about people. The following is an example of how one might study a health outcome that involves both compositional and contextual factors. In North Carolina many birth defect rates exhibit a significant amount of spatial variation (Root, Meyer, & Emch, 2009). Environmental health studies suggest that these patterns are due to toxicants in a mother's environment (contextual factors). However, there is also evidence that birth defects are related to both behavioral risk factors (e.g., smoking) and socioeconomic risk factors (compositional factors). Spatial patterns of birth defects may be due to the fact that individuals with similar backgrounds and risk factors live in the same geographic area, producing larger area-level patterns of disease.

A neighborhood health study distinguishes between the environmental and socioeconomic factors that explain the geographic variation. The study in North Carolina found that gastroschisis (an abdominal wall defect) is more common in births to poor white women than in births to other women, and that there are significant clusters of gastroschisis births in the Piedmont region of the state, which are hypothesized to be related to environmental factors (Root et al., 2009). Few studies of birth defects use methods appropriate for determining whether differences in prevalence across areas are due to characteristics of the areas themselves (contextual factors) or to differences between the types of people living in the areas (compositional factors). To address this limitation, multilevel regression models can be used to pull out the independent effects of individual-level and area-level characteristics on birth defect outcomes. When modeling area-level variables, one must avoid committing the ecological fallacy—that is, avoid making inferences about individuals when one is investigating aggregations of individuals. (See Chapter 1; Chapter 5 provides a more comprehensive discussion of the ecological fallacy in a geographic context.)

Nonrandom Sorting and the Problem of Causality

Another problem often encountered with neighborhood studies is related to the question of how people end up living in the neighborhoods they do. This is sometimes referred to as nonrandom sorting, **selection bias,** or the endogeneity problem (Oakes, 2004). Selection bias occurs when individual (or family) characteristics that are related to health are also related to the neighborhood an individual has chosen to live in. For example, children in low SES families are more likely to report poorer health, often because of lack of health insurance, worse nutrition, and other factors related to parental and family resources. But these low SES families are also more likely to live in low-income neighborhoods. If we cannot pull these two things apart (e.g., control for selection bias), we do not know whether poor child health is caused by the neighborhood or by family characteristics. Children with poor health may move to neighborhoods with fewer doctors and clinics, worse school environments, and unsafe outdoor play options because this is the type of community their family can afford. But the fact that there are no doctors, playgrounds, or safe schools in their neighborhood can also lead to poor health. This interplay between parental or family characteristics and neighborhood muddies the relationship between neighborhood environment and health. So what causes poor child health? The neighborhood environment or family and parent characteristics? This is the problem of showing causality.

Propensity scores are a commonly used statistical method for mitigating selection bias (Root & Humphrey, 2014b; Smith, 2011). A *propensity score* is the estimated probability of living in a certain type of neighborhood (e.g., low income) given a set of individual-level characteristics; for example, the probability a child lives in a low-income neighborhood given individual characteristics such as race, and parental/family characteristics such as income, education, and structure (e.g., single- or married-parent family). This technique limits the influence of confounding variables, such as family SES, by using them to estimate the propensity of living in a disadvantaged neighborhood. After doing this, a researcher can model the effect of neighborhood-level variables on health outcomes.

While statistical methods, such as regression modeling, are the most common way to study neighborhood effects on health, many researchers argue that **experimental study designs** are the only way we can truly understand the independent effects of neighborhood on health. The basic idea behind them is that individuals are randomized into a treatment group and a control group. The treatment group is exposed to some type of intervention, which could be anything from a medical intervention like a new vaccine to a behavioral intervention like a smoking cessation program. The control group carries on as usual. At the end of the experimental period outcomes between the two groups are compared.

There are only a few examples of true experimental studies that examined neighborhood effects on a variety of health outcomes. The first, and arguably most famous, is the Moving to Opportunities Study (MTO). Starting in 1994, nearly 4,500 low-income families living in extremely disadvantaged urban neighborhoods in Baltimore, Boston, Chicago, Los Angeles, and New York were given the opportunity to move into a different neighborhood. Families were randomly assigned to one of three groups: a group offered a housing voucher that could only be used to move to a low-poverty neighborhood, a group offered a traditional Section 8 housing voucher, which could be used to move into a different high-poverty community, and a control group that did not move. Families were followed up 10–15 years later and results of the study found that the housing vouchers did indeed improve the quality of neighborhoods people moved into; they lived in lower poverty, lower crime, and less racially segregated neighborhoods and they felt safer and more satisfied with their communities. The health impacts of the MTO were very small for some outcomes and nonexistent for many others. Adults who moved into low-poverty communities showed slightly lower severe obesity and diabetes rates than the control group—about 4% lower (Ludwig et al., 2011)—but there were no differences in hypertension or self-rated health. Children who moved into low-poverty neighborhoods showed little to no differences in health outcomes. MTO did appear to improve mental health outcomes; adults reported lower levels of depression and psychological distress (Kling, Liebman, & Katz, 2007). The MTO was important because it was one of the first experimental studies of neighborhood health and showed some (very modest) benefits to moving from low-income to higher-income communities: in particular, people in the treatment group felt better about where they lived, which was reflected in better mental health. However, given the time and money involved in the MTO study, these initial outcomes were disappointing for many researchers and policymakers and contradicted a large body of social science theory and experimental evidence from educational researchers. The design of the study was subsequently attacked, calling into doubt the ability to draw strong conclusions from the study (Clampet-Lundquist & Massey, 2008). Other researchers suggested that the MTO study didn't truly address neighborhood effects because it moved people to new neighborhoods, rather than improving the neighborhoods they lived in (Sampson, 2008).

A much more encouraging evaluation of MTO occurred during the writing of this book. Chetty, Hendren, and Katz (2015) followed children from the MTO study into their mid-20s and found that children who moved to low-poverty neighborhoods had incomes that were 31% higher than children who did not. The timing of the move seemed to matter, however, as this effect on income was not evident for children who moved after the age of 13. All this suggests that the positive effects of leaving poor neighborhoods as a child cannot be observed until children are old enough to finish college and enter the adult labor market. Longitudinal studies, which we discuss in more detail below, become important tools for understanding the lifetime effects of living in different neighborhood contexts.

Measuring Neighborhood Environments

Nearly all the early neighborhoods and health studies used census data to measure neighborhood socioeconomic environment. The U.S. Census Bureau (as well as census-taking organizations in other countries) divides the country up into many small areas in order to facilitate the enumeration of the population. Most people are familiar with counties, but counties are subdivided into census tracts, which in turn are divided into census block groups and, at the smallest level of geography, census blocks. Every year, the Census Bureau publishes population estimates for demographics such as race, median household income, population living below the poverty level, occupation, and many more socioeconomic and demographic variables. Early on, researchers realized that these data could be used as proxy measures of neighborhood environments. Essentially, if you knew which census block group or tract a person lived in, you could attribute all the census data for that area to the individual. These data could be used to characterize that individual's neighborhood—wealthy, poor, racially diverse, segregated, white, or black. The census is the only large-scale collection of data of this type, so it was convenient and inexpensive to use census data to measure neighborhoods. One major flaw of this technique is that census geographies don't necessarily follow true neighborhood boundaries. They were created to divide the population up into manageable groups so they could be easily and efficiently counted.

Neighborhoods develop through social processes that create a sense of community among residents, which in turn lead to the accumulation of social capital and collective efficacy. It is likely that census geographies do not correspond to the theoretically relevant neighborhood area that affects health outcomes. To further understand this idea, let's go back to the idea of an obesogenic environment discussed earlier in the chapter. Suppose a woman's probability of being obese is influenced by social norms regarding obesity in her community. There is some evidence to suggest that if it is socially acceptable to be overweight, people in the same social group are more likely to be overweight. In order to define the neighborhood environment that might influence this woman's choices around food consumption and physical activity we need to know how she interacts with her environment. Where does she encounter other overweight individuals? Does she see people in her neighborhood exercising? Studies have also shown that obesity rates cluster spatially in low-income communities. There are a number of reasons for this phemonenon, one being that high-calorie, high-fat foods are very inexpensive and affordable for people living in poor communities, leading to higher obesity rates. Much early research used census data on the proportion of the population living in poverty to predict poverty rates. This modeling technique worked, in that associations between poverty and obesity were found, but the use of poverty rate may only tell half the story.

More recent studies of neighborhoods have begun to use direct measures of social and physical environment collected through survey data or systematic observation. Measures of the neighborhood environment created using census data, or simply by counting the presence of neighborhood resources, such as parks, grocery stores, fast-food outlets, bars and liquor stores, may not match up with people's perceptions of their environment or their actual use of neighborhood resources. This mismatch between objective data-driven measures of the environment and subjective perception-based measures of neighborhood is related to the concept of access that is discussed in Chapter 13. While census proxies and business listings can be used to measure the mere presence of specific neighborhood characteristics and resources that are theoretically related to health behaviors and outcomes (access by availability), surveys that ask study participants to report on the physical and social conditions of their neighborhood can measure whether people feel they are affected by those conditions, know what resources exist in their neighborhood, and are willing to use resources and engage in the social life of their community (access by contact and acceptability).

A variety of scales have been developed that measure people's opinions on concepts such as social cohesion, collective efficacy, and social support among neighbors. For example, the Project on Human Development in Chicago Neighborhoods developed an instrument that asks people if their neighborhood is close-knit and whether neighbors help each other, get along, can be trusted, and share the same values (Sampson et al., 1997). Using these data, Morenoff (2003) found that active participation in community organizations, reciprocal social interaction and social support were related to lower levels of low birth weight among mothers in Chicago. He suggests that these social mechanisms reduce maternal stress and lead to healthier birth outcomes. The MESA study used the same instrument and found that neighborhood social cohesion was associated with a higher probability of regular physical activity and consuming the recommended daily fruit and vegetable servings (Samuel et al., 2015) but not with the incidence of type-II diabetes (Christine et al., 2015). The Neighborhood Environment and Health Study (NEHS) in Denver asked a similar series of questions about neighborhood social involvement and again found that individuals who reported higher levels of social involvement reported higher levels of fruit and vegetable intake but found no effect on BMI (Litt et al., 2011). These studies suggest that neighborhood social environments may affect health through constraints on, or enhancements of, health-related behaviors, but not necessarily on the health outcomes themselves.

Systematic observation is another way of collecting objective data about neighborhood environments. This approach was pioneered by the Project on Human Development in Chicago Neighborhoods as a standardized approach for directly observing the physical, social, and economic characteristics of neighborhoods, one block at a time. Early in the study, researchers drove a vehicle down a select set of city blocks to videotape both sides of each block, while two observers recorded characteristics of each block face on observer logs. Back in the lab, research staff watched the videos and examined the observer logs and coded a variety of dimensions of the neighborhood environment. The researchers collected data on land use, drinking establishments, recreational facilities, street conditions, traffic, the physical condition of buildings, cigarettes and cigars on the street or in the gutter, garbage, litter on the street or sidewalk, graffiti, abandoned cars, and condoms, needles, and syringes on the sidewalks. Information was also gathered on adults loitering or congregating, people drinking alcohol on the street, peer groups, gang indicators present, intoxicated people, adults fighting or hostilely arguing, prostitution on the street, and people selling drugs (Sampson & Raudenbush, 1999). A similar effort in New Orleans used systemic social observation to construct a so-called broken windows index that showed an association between poor neighborhood conditions and high gonorrhea rates, all of which cause premature mortality from cancer, diabetes, homicide, and suicide, even after controlling for neighborhood poverty, unemployment, and low education (Cohen et al., 2000; Cohen et al., 2003). This method of directly observing neighborhood conditions is expensive and labor-intensive, to say the least. Recent efforts have focused on using digital resources such as Google Street View to conduct street audits (Clarke, Ailshire, Melendez, Bader, & Morenoff, 2010; Rundle, Bader, Richards, Neckerman, & Teitler, 2011). These efforts have largely met with success, though concern remains about how often street view data are updated.

Longitudinal Studies and the Changing Nature of Neighborhoods

Your neighborhood changes over time. Many people return to their childhood neighborhood only to find that old buildings and parks are gone, new ones have been built, and the people who live in the area now are older, younger, wealthier, poorer, or of a different racial/ethnic group. Neighborhood contexts are not static but rather are changing and dynamic over time. A quick look at data on the same neighborhood using data from the 1980, 1990, 2000, and 2010 Censuses shows that basic neighborhood demographics have changed in many areas of the United States over the course of 30 years. Hispanic populations have grown tremendously, income inequality has increased, minority populations shifted from the urban core to more suburban areas, and natural disasters such as Hurricane Katrina forced mass migrations and changed the demographics of several major U.S. cities. In addition, and related to this last point, people do not stay in one neighborhood their whole lives. In fact, evidence shows that a large proportion of children moved at least once during childhood and many children move several more times before graduating from high school (Root & Humphrey, 2014a). Adults are also highly mobile—moving for college, job opportunities, or because they want to experience a new city. Given such high mobility rates and the dynamic and changing nature of neighborhoods, many researchers suggest that studies that only examine the effect of neighborhoods at one point in time (cross-sectional studies) are insufficient to understand how neighborhoods affect health over time. In fact, life-course theory (Elder, 1998) emphasizes that circumstances early in a person's life influence later circumstances and transitions with long-term consequences on adult health, independent of adult circumstances. A wealth of studies on a variety of health, cognitive, and educational outcomes support this theory.

Studies need to take into account the dynamic nature of neighborhoods, and thus a *longitudinal* approach to the phenomena being studied is necessary. Longitudinal data sets are those that require measurement of phenomena at many different times. Tracking the same people over an extended period of time is expensive and therefore is not often done. An example of a study that follows a longitudinal design is the China Health and Nutrition Survey, which has collected information on 19,000 individuals in nine provinces in 2006, 2009, and 2011 (Popkin, 2008; Popkin & Du, 2003). There are few examples of longitudinal studies that explicitly integrate measures of neighborhood social and built environments. The aforementioned MESA Study is a longitudinal study that evaluates long-term neighborhood exposures as they related to cardiovascular disease, type-II diabetes, and a variety of risk factors related to these conditions. Though still ongoing at the time this textbook was written, the study has followed a cohort of 5,000 individuals over 10 years. Every few years the study collects data on neighborhood health food and physical activity resources, as well as survey-based measures of social cohesion, safety, walkability, and availability of health food options. Thus, the researchers have dynamic and changing measures of neighborhood environment over time. The study has shown that long-term exposure to residential environments with greater resources to support physical activity and healthy diets was associated with lower incidence of type-II diabetes (Christine et al., 2015). Since so few longitudinal neighborhood studies have been done to date, it is difficult to know if these results can be replicated. Root and Humphrey (2014b) used a longitudinal cohort of U.S. children to examine child self-rated health status between kindergarten and 8th grade and found that neighborhood socioeconomic conditions and racial diversity had no effect on overall health.

Neighborhood Definition and Units

A neighborhood is not simply a spatial unit. You can visualize the neighborhood that you grew up in, but your former nextdoor neighbor may have a different area in mind for the boundaries of the neighborhood. In neighborhood health studies, an area specifically defined as a **neighborhood** is the unit of analysis. Such units may represent traditional neighborhood boundaries; for instance, Chicago is divided into 77 community areas that serve as the basis for urban planning (see Figure 9.4). Other neighborhood health studies may use local census units (such as blocks, block groups, or tracts) or alternative areal aggregations (including ZIP codes). When conducting such a study, the investigators need to think about the spatial scale, but unfortunately many do not. Orange County, North Carolina, home of

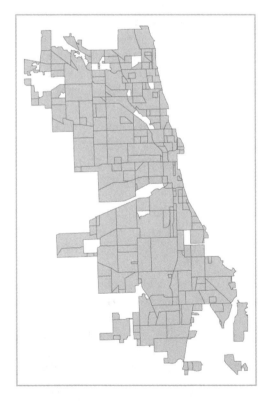

FIGURE 9.4 Chicago community areas.

the University of North Carolina at Chapel Hill (see Figure 9.5), is used below as an example to explain this concept.

There were 133,801 people living in Orange County at the time of the 2010 U.S. Census. When one is studying the context of where people live as it relates to their health, the choice of the unit of analysis could influence the results of the study. For instance, if one were to study the impact of community context (e.g., crime, sidewalks, bike trails, playgrounds) on obesity rates, then dividing Orange County into different-sized census units would determine whether the study was based on a very local context (census blocks) or much coarser units (block groups or tracts). Figure 9.6 shows the various-sized census units that could be used in such a study. They divide the people into from 28 to 2,652 areas. Think about where you live and your choice of whether you will walk or drive to do daily activities, such as shopping, exercise, school, and work. If you consider the small area around your home and a variable such as the number of fast-food restaurants, and then consider the larger block groups and tracts, then that variable would be very different in the larger units. The issue is one of scale. If you have a fast-food restaurant in your block area, then you might go there more often than if it was a mile away. So if you are trying to

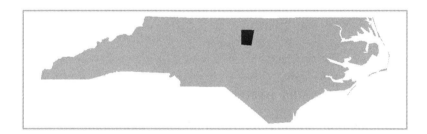

FIGURE 9.5 Location of Orange County, North Carolina.

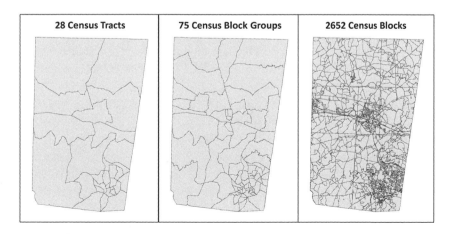

FIGURE 9.6 Orange County, North Carolina: Census geography.

measure the contextual effect of fast-food restaurants in close proximity to people's homes, and the effect is a very local-scale process, then the measurement of the effect must be at a local scale.

Many studies, however, do not use an appropriate spatial scale when measuring neighborhood contextual effects; they simply analyze data that are conveniently available. Many studies use ZIP code areas because they are widely available, but as Figure 9.7 illustrates, they are very large, irregular units that are used to deliver mail efficiently. Some neighborhood health studies have even used counties to study what the authors call "neighborhood effects." Although there is no unified definition of neighborhoods, they usually refer to more local-level areal patterns than would be found by analyzing the county-level or ZIP-code-level context. Geographers also are often concerned about what delimits, or forms, a neighborhood as a functional region of interaction. Where, for example, do children go to school? From how far do people come to use a park? Where do people go for health care, shopping, or church? These social activity areas can define a functional neighborhood relevant to environmental exposures or epidemic diffusion (Bian, 2004), as discussed in other chapters.

FIGURE 9.7 Orange County, North Carolina: ZIP code boundaries.

Neighborhoods do not need to be defined by existing areal units, and sometimes it makes sense simply to consider the area within a given distance of a house. Figure 9.8 shows a hypothetical 1,000-meter buffer (Euclidean distance area) around house number 1, with nine other houses (2 through 9) also in the neighborhood. Because houses 11 and 12 are outside the 1,000-meter buffer area, they are considered to be outside the neighborhood or area of influence. This type of neighborhood area is easy to construct when there is a geographic information system (GIS) database for the study area.

The environmental context of the area being studied can also be used to define a neighborhood. Figure 9.9 shows an area in rural Bangladesh where a neighborhood study of cholera, a water-borne disease, was recently done (Emch, Ali, Root, & Yunus, 2009). In the zoomed-in view of the satellite image on the right, the light gray squares represent two hypothetical households that are connected via a water

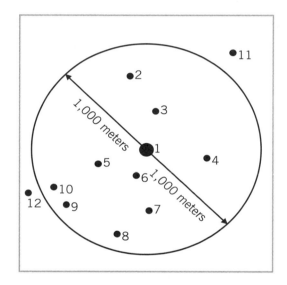

FIGURE 9.8 A neighborhood defined in terms of a 1,000-meter buffer around house 1.

FIGURE 9.9 Households connected via ponds. From Emch, Ali, Root, and Yunus (2009). Copyright ©2009 Elsevier. Reprinted by permission.

body. A neighborhood in this case could be defined as all of the households in close proximity to a common water body.

Figure 9.10 is a GIS database for this study area in rural Bangladesh. The database was created by digitizing ponds from satellite imagery and overlaying them with household points surveyed by using global positioning system (GPS) receivers. In the Emch and colleagues (2009) study, neighborhoods were constructed by modeling connectivity of households via water bodies, and cholera incidence was calculated for the different neighborhoods and compared to ecological variables. This section has shown various ways to define neighborhoods at different scales; all of these methods can be valid, given the right research problem.

Much of the literature on neighborhoods and health, while it is ultimately interested in how health varies by spatial units, has in fact been **aspatial**. For instance, a study that uses census tracts because they are readily available, and uses a traditional multilevel regression modeling approach that does not take into account interaction between areas, is not spatial. The latter approach (using multilevel regression modeling that does not take into account spatial interaction between adjacent areas) can be particularly problematic. This approach probably does not represent reality in most cases because it does not consider the fact that we are more likely to visit the neighborhood next to our own more often than another neighborhood that is farther away. In the birth defects example above, a multilevel

FIGURE 9.10 Household clusters near digitized ponds. From Emch, Ali, Root, and Yunus (2009). Copyright ©2009 Elsevier. Reprinted by permission.

regression model would assume that people are only exposed to toxicants in their own neighborhood. In such a model, neither adjacent neighborhoods nor neighborhoods that are far away would be assumed to have an effect on the health outcome. This assumption is likely to be true for the neighborhoods far away, but the adjacent neighborhoods are likely to have some effect on a person's risk of birth defects. The answer to this problem is something called a "spatial multilevel model," and there are several examples of these which have already been developed.

An aspatial approach may not reveal reasons for spatial differences in the health outcome being studied. Alternatively, a study that starts with a theory about why a health outcome is related to space and proceeds to investigate the spatial variation of that outcome rigorously is more likely to be successful. Also, health outcomes may be multiscalar processes. People living in a particular census block in a certain town are influenced by the small neighborhood around them as well as by their neighbors themselves, from whom they get information, contagion, and help. But they are also influenced by their community decisions about planning and zoning, as well as by state-, regional-, and national-level organizations.

QUICK REVIEW 2

- Neighborhood health studies examine the environmental and social processes that explain geographic variation in health outcomes.

- Statistically modeling neighborhood effects on health is a difficult process because researchers must disentangle the effects of neighborhood from other factors that affect health.

- Some of the methodological challenges to neighborhood effects studies include selection bias, deciding how to conceptualize and measure a neighborhood environment, examining the effect of neighborhood over time, and geographically defining a neighborhood area.

- Much of the time, people choose where to live. Therefore, it is difficult to determine if a neighborhood environment affects health or if people with similar health (due to another characteristic like income) choose to live in the same neighborhood.

- Neighborhood socioeconomic environments can be measured using census data or with direct measures of people's perception of neighborhood from surveys.

- Longitudinal studies of neighborhood effects on health are important because neighborhood environments are dynamic and people move over time.

- The neighborhood as a geographic unit of analysis can be defined in many different ways. Researchers should critically think about what processes they wish to measure and at what geographic scale these processes operate.

CONCLUSION

Studies of neighborhoods and health have become common during the past decade because health is not just about individuals. It is about the context of people's lives, including where they live, work, and perform other activities. The pendulum has swung away from studies assuming that health risk is only related to individual biology to a more holistic approach that looks at complex contextual reasons for health variations in the population. We now know that social contexts are equally important in understanding disparities in health and disease across populations. The existence and quality of social support, social capital, and one's connections with friends, neighbors, and family—all affect your health. Many of these social situations exist within neighborhoods. Although this latter type of research has been termed *neighborhood health studies*, it does not only encompass neighborhoods as traditionally defined. Any areal unit can be considered a neighborhood, including a census unit, a Euclidean distance area around a household, or even a town or county. The areal unit is all about scale, and some processes that may affect a population's health are at a coarser scale and others at a finer scale. Although there are still methodological limitations to our studies of how areal context influences a population's health, this trend in health research is an advance that better reflects the complex reasons why and where people get sick. This approach to disease studies is very much in the spirit of the long tradition of disease ecology described in Chapter 2.

REVIEW QUESTIONS

1. What are some of the factors that contribute to ill health? Of the factors you listed, which are social and which are biological?

2. Consider the problem of infant mortality. List specific factors that might affect health in each level of the spatial hierarchy shown in Figure 9.1.

3. Discuss how neighborhood social networks might encourage the uptake of mammography. What role do social contagion and informal social control play?

4. Describe the major challenges to examining whether neighborhoods actually impact health outcomes.

5. Which U.S. populations do you think live longer, healthier lives? List three reasons why these inequalities exist.

6. Think about the community you grew up in. What social and structural factors promoted good health? What factors encouraged healthy behaviors? What factors contributed to poor health?

7. Stress is hypothesized to contribute to poor health. Some stressors include living in an unsafe neighborhood, being uncertain about where food will come from, and worrying about how you will get care when you get sick. How do exposures to stressors—and neighborhood resources available to manage them—vary with race or class position? Describe the social and structural forces that create and reinforce these stressors.

8. How specifically has a geographic understanding of health contributed to the study of neighborhood effects?

9. Suppose you want to study the effects of neighborhood on exercise behavior. Describe how you would design your study, including the overarching study design, who you would talk to, what types of data you would collect, and how you might analyze your information.

10. How would you design a program to reduce obesity that targets neighborhood populations, rather than individuals?

REFERENCES

Alexander, C., Piazza, M., Mekos, D., & Valente, T. (2001). Peers, schools, and adolescent cigarette smoking. *Journal of Adolescent Health, 29*(1), 22–30.

Barnett, J. R. (2000). Does place of residence matter?: Contextual effects and smoking in Christchurch. *New Zealand Medical Journal, 113,* 433–435.

Barnett, J. R., Moon, G., & Kearns, R. (2004). Social inequality and ethnic differences in smoking in New Zealand. *Social Science and Medicine, 59*(1), 129–143.

Barnett, R., Pearce, J., & Moon, G. (2005). Does social inequality matter?: Changing ethnic socio-economic disparities and Maori smoking in New Zealand, 1981–1996. *Social Science and Medicine, 60*(7), 1515–1526.

Bian, L. (2004). A conceptual framework for an individual-based spatially explicit epidemiological model. *Environment and Planning B, 31*(3), 381–395.

Bosma, H., van de Mheen, H. D., Borsboom, G. J., & Mackenbach, J. P. (2001). Neighborhood socioeconomic status and all-cause mortality. *American Journal of Epidemiology, 153*(4), 363–371.

Brody, G., Ge, X., Conger, R., Gibbons, F., Murry, V., Gerrard, M., et al. (2001). The influence of neighborhood disadvantage, collective socialization, and parenting on African American children's affiliation with deviant peers. *Child Development, 72,* 1231–1246.

Browning, C., & Cagney, K. (2003). Moving beyond poverty: Neighborhood structure, social processes, and health. *Journal of Health and Social Behavior, 44,* 552–571.

Caughy, M., O'Campo, P., & Muntaner, C. (2003). When being alone might be better: Neighborhood poverty, social capital, and child mental health. *Social Science and Medicine, 57,* 227–237.

Chetty, R., Hendren, N., & Katz, L. F. (2015). *The effects of exposure to better neighborhoods on children: New evidence from the Moving to Opportunity experiment* (Working Paper No. 21156). Retrieved from the National Bureau of Economic Research. Available at *www. nber.org/papers/w21156.*

Christakis, N. A., & Fowler, J. H. (2007). The spread of obesity in a large social network over 32 years. *New England Journal of Medicine, 357*(4), 370–379.

Christine, P. J., Auchincloss, A. H., Bertoni, A. G., Carnethon, M. R., Sánchez, B. N., Moore, K., et al. (2015). Longitudinal associations between neighborhood physical and social environments and incident type 2 diabetes mellitus: The Multi-Ethnic Study of Atherosclerosis (MESA). *JAMA Internal Medicine, 175*(8), 1311–1320.

Clampet-Lundquist, S., & Massey, D. S. (2008). Neighborhood effects on economic self-sufficiency: A reconsideration of the Moving to Opportunity Experiment 1. *American Journal of Sociology, 114*(1), 107–143.

Clarke, P., Ailshire, J., Melendez, R., Bader, M., & Morenoff, J. (2010). Using Google Earth to conduct a neighborhood audit: Reliability of a virtual audit instrument. *Health and Place, 16*(6), 1224–1229.

Cohen, D. A., Mason, K., Bedimo, A., Scribner, R., Basolo, V., & Farley, T. A. (2003). Neighborhood physical conditions and health. *American Journal of Public Health, 93*(3), 467–471.

Cohen, D. A., Spear, S., Scribner, R., Kissinger, P., Mason, K., & Wildgen, J. (2000). "Broken windows" and the risk of gonorrhea. *American Journal of Public Health, 90*(2), 230–236.

Cohen, S., Doyle, W. J., & Baum, A. (2006). Socioeconomic status is associated with stress hormones. *Psychosomatic Medicine, 68*(3), 414–420.

Cummins, S. (2002). Food deserts. In C. Trink-Shevrin & R. Park (Eds.), *The Wiley Blackwell Encyclopedia of Health, Illness, Behavior, and Society* (pp. 562–564). New York: Wiley-Blackwell.

Curtis, S. E. (1990). Use of survey data and small area statistics to assess the link between individual morbidity and neighborhood deprivation. *Journal of Epidemiology and Community Health, 44,* 62–68.

Diez-Roux, A. V. (1998). Bringing context back into epidemiology: Variables and fallacies in multilevel analysis. *American Journal of Public Health, 88,* 287–293.

Diez-Roux, A. V. (2001). Investigating neighborhood and area effects on health. *American Journal of Public Health, 91*(11), 1783–1789.

Diez-Roux, A. V., Stein Merkin, S., Arnett, D., Chambless, L., Massing, M., Nieto, F. J., et al. (2001). Neighborhood of residence and incidence of coronary heart disease. *New England Journal of Medicine, 345,* 99–106.

Dowd, J. B., Simanek, A. M., & Aiello, A. E. (2009). Socio-economic status, cortisol and allostatic load: A review of the literature. *International Journal of Epidemiology, 38*(5), 1297–1309.

Dunn, J. R., Veenstra, G., & Ross, N. A. (2006). Relations between self-perceived relative social position and self-rated health: Results from a nationally representative survey of the Canadian population. *Social Science and Medicine, 62*(6), 1465–1473.

Elder, G. H. (1998). The life course as developmental theory. *Child Development, 69,* 1–12.

Elliott, J. R., Haney, T. J., & Sams-Abiodun, P. (2010). Limits to social capital: Comparing network assistance in two New Orleans neighborhoods devastated by Hurricane Katrina. *Sociological Quarterly, 51,* 624–648.

Emch, M., Ali, M., Root, E. D., & Yunus, M. (2009). Spatial and environmental connectivity analysis in vaccine trials. *Social Science and Medicine, 68,* 631–637.

Erdem, Ö., Prins, R. G., Voorham, T. A., van Lenthe, F. J., & Burdorf, A. (2015). Structural neighbourhood conditions, social cohesion and psychological distress in the Netherlands. *European Journal of Public Health,* ckv120.

Fullerton, C. S., Ursano, R. J., Liu, X., McKibben, J. B. A., Wang, L., & Reissman, D. B. (2015). Depressive symptom severity and community collective efficacy following the 2004 Florida hurricanes. *PLoS ONE, 10*(6), e0130863.

Gary, T., Stark, S., & LaVeist, T. (2007). Neighborhood characteristics and mental health among African Americans and whites living in a racially integrated urban community. *Health and Place, 13,* 569–575.

Gordon-Larsen, P., Nelson, M. C., Page, P., & Popkin, B. M. (2006). Inequality in the built environment underlies key health disparities in physical activity and obesity. *Pediatrics, 117*(2), 417–424.

Gordon-Larsen, P., & Reynolds, K. D. (2006). Influence of the built environment on physical activity and obesity in childhood and adolescents. In M. I. Goran & M. Sothern (Eds.), *Handbook of pediatric obesity: Etiology, pathophysiology, and prevention* (pp. 255–270). Boca Raton, FL: CRC/Taylor & Francis.

Gwatkin, D. R., Rutstein, S., Johnson, K., Suliman, E., Wagstaff, A., & Amouzou, A. (2007).

Socio-economic differences in health, nutrition, and population within developing countries: An overview. Washington, DC: World Bank, Human Development Network.

Hoffman, B. R., Monge, P. R., Chou, C. P., & Valente, T. W. (2007). Perceived peer influence and peer selection on adolescent smoking. *Addictive Behaviors, 32*(8), 1546–1554.

Jones, K., & Duncan, C. (1995). Individuals and their ecologies: Analyzing the geography of chronic illness within a multilevel modeling framework. *Health and Place, 1,* 27–30.

Kaczynski, A. T., & Henderson, K. A. (2008). Parks and recreation settings and active living: A review of associations with physical activity function and intensity. *Journal of Physical Activity and Health, 5*(4), 619–632.

Kawachi, I., & Berkman, L. (Eds.). (2003). *Neighborhoods and health.* New York: Oxford University Press.

Kawachi, I., & Berkman, L. (2014). Social capital, social cohesion, and health. In L. F. Berkman, I. Kawachi, & M. Glymour (Eds.), *Social epidemiology* (pp. 290–319). Oxford, UK: Oxford University Press.

Kling, J. R., Liebman, J. B., & Katz, L. F. (2007). Experimental analysis of neighborhood effects. *Econometrica, 75*(1), 83–119.

Kubzansky, L. D., Seeman, T. E., & Glymour, M. M. (2014). Biological pathways linking social conditions and health: Plausible mechanisms and emerging puzzles. In L. F. Berkman, I. Kawachi, & M. Glymour (Eds.), *Social epidemiology* (pp. 512–561). Oxford, UK: Oxford University Press.

Kushner, J. A. (1980). *Apartheid in America: An historical and legal analysis of contemporary racial segregation in the United States.* Arlington, VA: Carrollton Press.

Larson, N. I., Story, M. T., & Nelson, M. C. (2009). Neighborhood environments: Disparities in access to healthy foods in the U.S. *American Journal of Preventive Medicine, 36*(1), 74–81.

Leslie, H. H., Ahern, J., Pettifor, A. E., Twine, R., Kahn, K., Gómez-Olivé, F. X., et al. (2015). Collective efficacy, alcohol outlet density, and young men's alcohol use in rural South Africa. *Health and Place, 34,* 190–198.

Leventhal, T., & Brooks-Gunn, J. (2000). The neighborhoods they live in: The effects of neighborhood residence on child and adolescent outcomes. *Psychological Bulletin, 126,* 309–337.

Litt, J. S., Soobader, M. J., Turbin, M. S., Hale, J. W., Buchenau, M., & Marshall, J. A. (2011). The influence of social involvement, neighborhood aesthetics, and community garden participation on fruit and vegetable consumption. *American Journal of Public Health, 101*(8), 1466–1473.

Ludwig, J., Sanbonmatsu, L., Gennetian, L., Adam, E., Duncan, G. J., Katz, L. F., et al. (2011). Neighborhoods, obesity, and diabetes—A randomized social experiment. *New England Journal of Medicine, 365*(16), 1509–1519.

Macintyre, S., Ellaway, A., & Cubbins, S. (2002). Place effects on health: How can we conceptualize, operationalize, and measure them? *Social Science and Medicine, 55,* 125–139.

Mair, C., Roux, A. V. D., & Morenoff, J. D. (2010). Neighborhood stressors and social support as predictors of depressive symptoms in the Chicago Community Adult Health Study. *Health and Place, 16*(5), 811–819.

Marmot, M. G., Rose, G., Shipley, M., & Hamilton, P. J. (1978). Employment grade and coronary heart disease in British civil servants. *Journal of Epidemiology and Community Health, 32*(4), 244–249.

Massey, D. S. (1990). American apartheid: Segregation and the making of the underclass. *American Journal of Sociology, 96,* 329–357.

Molnar, B., Gortmaker, S., Bull, F., & Buka, S. (2004). Unsafe to play?: Neighborhood disorder and lack of safety predict reduced physical activity among urban children and adolescents. *American Journal of Health Promotion, 18,* 378–386.

Moon, G., & Barnett, J. R. (2003). Spatial scale and the geography of tobacco smoking in New Zealand: A multilevel perspective. *New Zealand Geographer, 59*(2), 6–15.

Moore, L. V., Roux, A. V. D., Nettleton, J. A., & Jacobs, D. R. (2008). Associations of the local food environment with diet quality—A comparison of assessments based on surveys and geographic information systems: The Multi-Ethnic Study of Atherosclerosis. *American Journal of Epidemiology, 167*(8), 917–924r.

Morenoff, J. D. (2003). Neighborhood mechanisms and the spatial dynamics of birth weight. *American Journal of Sociology, 108*(5), 976–1017.

Morenoff, J. D., James, S., Hansen, B. B., Williams, D., Kaplan, G., & Hunte, H. (2007). Understanding social disparities in hypertension prevalence, awareness, treatment, and control: The role of neighborhood context. *Social Science and Medicine, 65,* 1853–1866.

Morenoff, J. D., Sampson, R. J., & Raudenbush, S. W. (2001). Neighborhood inequality, collective efficacy, and the spatial dynamics of urban violence. *Criminology, 39*(3), 517–558.

Oakes, J. M. (2004). The (mis)estimation of neighborhood effects: Causal inference for a practicable social epidemiology. *Social Science and Medicine, 58,* 1929–1952.

Pearce, J., Hiscock, R., Moon, G., & Barnett, J. R. (2009). The neighbourhood effects of geographical access to tobacco retailers on individual smoking behaviour. *Journal of Epidemiology and Community Health, 63*(1), 69–77.

Popkin, B. M. (2008). Will China's nutrition transition overwhelm its health care system and slow economic growth? *Health Affairs, 27*(4), 1064–1076.

Popkin, B. M., & Du, S. (2003). Dynamics of the nutrition transition toward the animal foods sector in China and its implications: A worried perspective. *Journal of Nutrition, 133,* 3898S–3906S.

Popkin, B. M., Duffey, K., & Gordon-Larsen, P. (2005). Environmental influences on food choice, physical activity and energy balance. *Physiology and Behavior, 86*(5), 603–613.

Rankin, B., & Quane, J. (2002). Social contexts and urban adolescent outcomes: The interrelated effects of neighborhoods, families, and peers on African-American youth. *Social Problems, 49,* 79–100.

Root, E. D., Giebultowicz, S., Ali, M., Yunus, M., & Emch, M. (2011). The role of vaccine coverage among social networks in cholera vaccine efficacy. *PLoS ONE, 6*(7), e22971.

Root, E. D., & Humphrey, J. L. (2014a). The impact of childhood mobility on exposure to neighborhood socioeconomic context over time. *American Journal of Public Health, 104*(1), 80–82.

Root, E. D., & Humphrey, J. L. (2014b). Neighborhood racial composition and trajectories of child self-rated health: An application of longitudinal propensity scores. *Social Science and Medicine, 120,* 31–39.

Root, E. D., Meyer, R., & Emch, M. (2009). Evidence of localized clustering of gastroschisis births in North Carolina, 1999–2004. *Social Science and Medicine, 68,* 1361–1367.

Rundle, A. G., Bader, M. D., Richards, C. A., Neckerman, K. M., & Teitler, J. O. (2011). Using Google Street View to audit neighborhood environments. *American Journal of Preventive Medicine, 40*(1), 94–100.

Saelens, B. E., Sallis, J. F., Black, J. B., & Chen, D. (2003). Neighborhood-based differences in physical activity: An environment scale evaluation. *American Journal of Public Health, 93*(9), 1552–1558.

Sampson, R. J. (2008). Moving to inequality: Neighborhood effects and experiments meet structure. *AJS: American Journal of Sociology, 114*(11), 189–231.

Sampson, R. J., Morenoff, J., & Gannon-Rowley, T. (2002). Assessing "neighborhood effects": Social processes and new directions in research. *Annual Review of Sociology, 28,* 443–478.

Sampson, R. J., & Raudenbush, S. W. (1999). Systematic social observation of public spaces: A new look at disorder in urban neighborhoods 1. *American Journal of Sociology, 105*(3), 603–651.

Sampson, R. J., Raudenbush, S. W., & Earls, F. (1997). Neighborhoods and violent crime: A multilevel study of collective efficacy. *Science, 277,* 918–924.

Samuel, L. J., Himmelfarb, C. R. D., Szklo, M., Seeman, T. E., Echeverria, S. E., & Roux, A. V. D. (2015). Social engagement and chronic disease risk behaviors: The Multi-Ethnic Study of Atherosclerosis. *Preventive Medicine, 71,* 61–66.

Schempf, A., Strobino, D., & O'Campo, P. (2009). Neighborhood effects on birthweight: An exploration of psychosocial and behavioral pathways in Baltimore, 1995–1996. *Social Science and Medicine, 68,* 100–110.

Schwartz, S. (1994). The fallacy of the ecological fallacy: The potential misuse of a concept and the consequences. *American Journal of Public Health, 84*(5), 819–824.

Shaw, C., & McKay, H. (1942). *Juvenile delinquency and urban areas.* Chicago: University of Chicago Press.

Smith, R. (2011). *Multilevel modeling of social problems: A causal perspective.* Cambridge, MA: Spring Science + Business Media.

Susser, M. (1994). The logic in ecological: I. The logic of analysis. *American Journal of Public Health, 84*(5), 825–829.

Westmaas, J. L., Wild, T. C., & Ferrence, R. (2002). Effects of gender in social control of smoking cessation. *Health Psychology, 21*(4), 368.

FURATHER READING ·

Balfour, J. L., & Kaplan, G. (2002). Neighborhood environment and loss of physical function in older adults: Evidence from the Alameda County study. *American Journal of Epidemiology, 155*(6), 507–515.

Buka, S. L., Brennan, R. T., Rich-Edwards, J. W., Raudenbush, S. W., & Earls, F. (2003). Neighborhood support and the birth weight of urban infants. *American Journal of Epidemiology, 157*(1), 1–8.

Chaix, B., Merlo, J., Subramanian, S. V., Lynch, J., & Chauvin, P. (2005). Comparison of a spatial perspective with the multilevel analytical approach in neighborhood studies: The case of mental and behavioral disorders due to psychoactive substance use in Malmo, Sweden, 2001. *American Journal of Epidemiology, 162,* 171–182.

Chuang, Y., Cubbin, C., Ahn, D., & Winkleby, M. A. (2005). Effects of neighbourhood socioeconomic status and convenience store concentration on individual level smoking. *Tobacco Control, 14*(5), 337.

Corburn, J., Osleeb, J., & Porter, M. (2006). Urban asthma and the neighbourhood environment in New York City. *Health and Place, 12*(2), 167–179.

Diez-Roux, A. V., Nieto, F. J., Caulfield, L., Tyroler, H. A., Watson, R. L., & Szklo, M. (1999). Neighbourhood differences in diet: The Atherosclerosis Risk in Communities (ARIC) Study. *Journal of Epidemiology and Community Health, 53*(1), 55–63.

Diez-Roux, A. V., Schwartz, S., & Susser, E. (2002). Ecologic studies and ecologic variables in public health research. In R. Detels, J. McEwen, R. Beaglehole, & H. Tanaka (Eds.), *The Oxford textbook of public health* (4th ed., pp. 493–508). London: Oxford University Press.

Diez-Roux, A. V., Stein Merkin, S., Hannan, P., Jacobs, D. R., & Kiefe, C. I. (2002). Area characteristics, individual-level socioeconomic indicators, and smoking in young adults: The CARDIA study. *American Journal of Epidemiology, 157*(4), 315–326.

Ecob, R., & Macintyre, S. (2000). Small area variations in health-related behaviors: Do these depend on the behavior itself, its measurement, or on personal characteristics? *Health and Place, 6,* 261–274.

Emch, M., Ali, M., Park, J. K., Yunus, M., Sack, D., & Clemens, J. D. (2006). Relationship between neighborhood-level killed oral cholera vaccine coverage and protective efficacy: Evidence for herd immunity. *International Journal of Epidemiology, 35,* 1044–1050.

Emch, M., Ali, M., Yunus, M., Sack, D., Acosta, C., & Clemens, J. D. (2007). Efficacy calculation in randomized vaccine trials: Global or local measures? *Health and Place, 13,* 238–248.

Lee, R. E., & Cubbin, C. (2002). Neighborhood context and youth cardiovascular health behaviors. *American Journal of Public Health, 92*(3), 428–436.

Mayer, S. E., & Jencks, C. (1989). Growing up in poor neighborhoods: How much does it matter? *Science, 243,* 1441–1445.

O'Campo, P. (2003). Invited commentary: Advancing theory and methods for multilevel models of residential neighborhoods and health. *American Journal of Epidemiology, 157,* 9–13.

Pickett, K. E., & Pearl, M. (2001). Multilevel analyses of neighbourhood socioeconomic context and health outcomes: A critical review. *Journal of Epidemiology and Community Health, 55*(2), 111–122.

Sampson, R., Morenoff, J., & Gannon-Rowley, T. (2002). Assessing neighborhood effects: Social processes and new directions in research. *Annual Review of Sociology, 28,* 443–478.

Winkleby, M., Cubbin, C., & Ahn, D. (2006). Effect of cross-level interaction between individual and neighborhood socioeconomic status on adult mortality rates. *American Journal of Public Health, 96*(12), 2145–2153.

Yen, I., & Syme, S. L. (1999). The social environment and health: A discussion of the epidemiologic literature. *Annual Review of Public Health, 20,* 287–308.

CHAPTER 10

Urban Health

For the majority of our time as a species we have been rural creatures, first foraging for food and then settling down as sedentary agriculturalists, starting ~10,000 years ago. A radical shift has taken place in the last decade, however, for we are now a predominantly urban-dwelling species as more of us leave the countryside behind. In 2007, half the world's population, some 3.3 billion people, lived in cities, while the other half still lived in rural areas (United Nations, 2014). By 2015, 54% of the world's 7.2 billion people were living in cities. The majority of the growth in the world's urban population is driven by migration and fertility patterns in low- and middle-income countries: the urban dynamics of wealthy countries are quite different (Figure 10.1). While cities are not new, having existed for thousands of years, the scale of today's cities and the numbers of people living within them constitute a profound change in the ecology of humanity.

This chapter focuses on how cities have historically contributed to human health and the special health problems of modern urban areas. Urban areas represent unique ecologies for human health, with massive population densities and interactions with built environments. The chapter examines how cities struggle to provision their populations with food, water, sanitation, and other services, particularly in the face of rapidly growing populations and the urbanization of poverty. Then, the chapter examines traffic in the context of morbidity and mortality due to traffic accidents and air pollution in low- and middle-income countries, where car and motorbike ownership is increasing but traffic safety measures are scarce.

CITIES AND URBANIZATION

What exactly constitutes a city or urban area? Settlements or localities are defined for the United Nations as urban by national statistical agencies according to

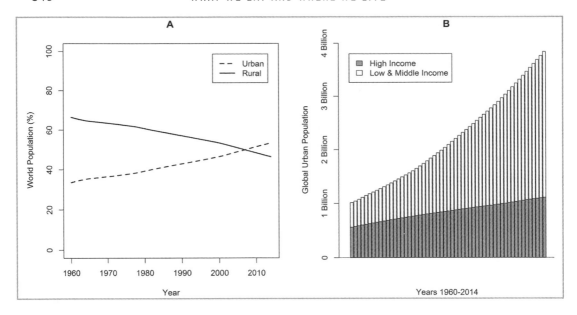

FIGURE 10.1 (A). Percentage of world population living in rural and urban areas, 1960–2014. (B). Total urban population by income level, 1960–2014.

remarkably different criteria. In India, for example, a nuclear settlement can have schools, a hospital, and a population of over 40,000, but may not be defined as urban because its population is more than 80% agriculturally engaged. The U.S. Census Bureau defines urban areas as those consisting of a large central place (with density at least 1,000 people per square mile in core block groups and 500 in surrounding census blocks) and a total population of at least 2,500 for urban clusters, or at least 50,000 for urbanized areas. Urban classification cuts across other hierarchies, and urban areas can be in metropolitan or nonmetropolitan areas.

Urbanization is the process of transition from rural settlements to urban settlements, primarily through rural–urban migration. This destination in the mobility transition became dominant in the now wealthy countries (the United States, Europe, Japan) during the first half of the 20th century. In the United States, for example, those farm boys who had seen Paris during World War I and World War II weren't content to return to their rural isolation. Southern sharecroppers headed for new northern assembly lines. African Americans migrated by the millions from southern rural poverty to northern cities, many to work in manufacturing jobs. At the same time the agricultural frontier, in the Great Plains generally, "hollowed out." America as a population moved to live in cities along the coasts or around the Great Lakes. As the low- and middle-income world goes through the demographic and mobility transitions so much faster with such a larger population base, the "second wave" of urbanization involves human relocation and adjustment at a size and speed that are almost incomprehensible.

Most of Latin America, latest and almost done in the demographic transition, is already highly urban; its great rural–urban migration was a phenomenon of the

1970s and 1980s. Brazil, the population giant, is already more than 80% urban, but it is projected to grow by 2020 to 89%. This means it will add 48,994,000 people to its cities. The population in the South Asian subcontinent (India, Pakistan, and Bangladesh) is still growing from 1.7% to 2.2% a year. That natural increase, whether it is born in the cities or in rural areas, will become almost entirely urban. Tens of millions now underemployed and malnourished in rural areas will migrate to the cities. The subcontinent will add to its cities a phenomenal quarter of a billion people within little more than a decade, and even more people over the following decades as the three countries become majority urban.

China alone will add more than 300 million to its cities by 2020, 200 million of whom are estimated to be the "floating population" living on trains, on boats, and in unofficial urban settlements today (Harrison & Pearce, 2000). In 1980, as economic reform was just beginning in China, only 20% of its population lived in cities, but by 2011 over half of the country's 1.3 billion people were urban residents, a phenomenal rural-to-urban shift in only 30 years. By 2030 it is expected that China will have more than 200 cities with over a million residents and that a billion Chinese will live in cities. To keep up with this pace of urbanization China has used more concrete in three years (2011–2013) than the United States did during the entire 20th century (Smil, 2013). Tower cranes are ubiquitous in Chinese skylines as apartment block after apartment block is constructed (see Figure 10.2).

Sub-Saharan Africa still will more than double in population by natural increase, even as it urbanizes. Already 72% of its urban population lives in squalid slums, and only 80% have sanitation, with some countries far lower (see Chapter 11). As the rural population migrates to the cities, this poorest of regions will add

FIGURE 10.2 Construction of large apartment blocks in Kunming, China.

through natural increase and migration a quarter of a billion people to those cities. How can cities in developing countries possibly provide adequate shelter, sanitation, water, schooling, and health care to that urban population?

Cities grow by three methods. The first of these is through natural increase: urban residents have urban children. Second, cities add or annex surrounding territory to themselves, usually because it too has become urbanized and needs sewerage, transportation for commuting, electricity, and medical and police services. And most of all, cities grow through in-migration of people. As the demographic transition progresses, access in cities to vaccination and health care, basic education, better nutrition, more information, jobs, and economic opportunities for women that redefine traditional family gender roles all lead to lower fertility rates. At the same time infant and child mortality fall and the fertility rate of urban dwellers inches toward replacement level. In country after country, rural–urban immigrants have adjusted to urban life and values, lowering their fertility rates to between the city rates and those of the rural areas from which they came. Since 70% or more of urban populations are migrants from rural areas, early on these populations grow substantially through natural increase. This was the European American experience also. The migration experience in the developing world is different, however.

In the first wave of urbanization, most migration was **step migration**. Few people went straight from the sharecropped, eroded Southern fields directly to Chicago, for example; instead they moved up the Mississippi River cities. People went first to nearby cities, then moved to larger regional centers, and generally moved up the urban hierarchy as they improved their urban living skills and extended their information fields. A city like Durham, North Carolina, could become a financial, insurance, job, and cultural "metropolis" for African Americans driven by poverty from the cotton fields. In the colonialized developing world, by contrast, one or two cities (usually ports) became—by virtue of their infrastructure (railroads, ports, electricity, sewerage, plumbing) and institutions (government, military, educational, hospital, commercial, and [after independence] manufacturing and investment)—**primate cities** that dominated the urban hierarchy. A normal ratio in an urban hierarchy of largest city to second largest is 2:1; for instance, in the state of North Carolina Charlotte is twice the size of Raleigh, while in the United States as a whole New York City is twice the size of Los Angeles. New York City, while reaching 17.9 million people in 2015, will have only 6.9% of the urban U.S. population. In comparison, think of the ratio in the urban hierarchy of Mexico City, Manila, or Kinshasa to the second city in Mexico, the Philippines, or the Democratic Republic of the Congo. This situation is known as "primacy," and in such urban systems the rural–urban migration does not go to places throughout the urban hierarchy. Rural people move to the one primate destination through chain migration. Thus Bangkok, which has been more than 40 times the size of Thailand's second city, grew from 5.9 million in 1990 to 7.4 million in 2000, to 8.9 million in 2010, and to 9.8 million in 2015, while decreasing only from having 59.2% to having 55.9% of all the urban people in the country. In contrast, India, which had a regular subcontinent-wide urban hierarchy before colonization by Europe, has seen Mumbai (formerly Bombay) grow from 12.3 million in 1990 to 16 million in 2000 to a behemoth 21.4 million in 2015, but it will still have "only" 5% of India's urban population.

Most of the global addition of population to urban areas will occur in medium-sized cities with hundreds of thousands or a few million people today. The chain migration into existing primate cities, however, will lead to the growth of **megacities** of more than 15 and even more than 20 million (bigger than almost three-quarters of the countries in the world today). In 2015 there were over 30 world cities with populations greater than 10 million. Table 10.1 shows the top 20 largest cities in the world in 2015. Among cities in currently developed countries, only Tokyo and New York City remain in the top 10, with Osaka among the top 20. The great cities of Europe are not among them. Notice also the primacy of megacities like Lagos and Dhaka in the percentage of population in their countries.

Thus, the majority of urban growth in the coming decades will be in the less-developed regions of the world. The share of total urban residents from Asia and Africa is growing, though these regions also represent the major share of the world's rural residents. The 3.9 billion people currently living in cities are expected to grow to 4.9 billion by 2030, *adding a billion people into cities in less than two decades*. As previously stated, this growth will be taking place not just in megacities but also in smaller cities of a few million. What effects will this have on human health?

TABLE 10.1. Top 20 World Megacities			
2015 rank	City	2015 population (in millions)	% urban share
1	Tokyo, Japan	38.00	32.0
2	Delhi, India	25.70	6.1
3	Shanghai, China	23.74	3.0
4	São Paulo, Brazil	21.07	12.1
5	Mumbai, India	21.04	5.0
6	Mexico City, Mexico	21.00	21.2
7	Beijing, China	20.38	2.6
8	Osaka, Japan	20.24	17.1
9	Cairo, Egypt	18.77	51.4
10	New York, United States	18.59	7.0
11	Dhaka, Bangladesh	17.60	32.0
12	Karachi, Pakistan	16.62	22.8
13	Buenos Aires, Argentina	15.18	39.2
14	Kolkata, India	14.86	3.5
15	Istanbul, Turkey	14.16	25.2
16	Chongqing, China	13.33	1.7
17	Lagos, Nigeria	13.12	15.0
18	Manila, Philippines	12.95	28.7
19	Rio de Janeiro, Brazil	12.90	7.4
20	Guangzhou, China	12.46	1.6

Note. Data from *World Urbanization Prospects*, UNDESA Population Division.

A BRIEF HISTORY OF CITIES

As discussed in Chapter 2, human settlements have typically taken either nuclear, dispersed, or linear forms, each associated with different types of disease ecologies. Major cities of the world typically emerge from linear settlement forms, with rivers and roads acting as natural places for larger and larger numbers of people to accumulate. Our earliest cities, either those that still exist or those that were abandoned long ago, were also located in the agricultural hearths of the world, in Mesoamerica and in the river valleys of China and in the Fertile Crescent of the modern Middle East (see Chapter 8 for more on agricultural hearths). Agriculture enables population growth, as more food can be produced per acre than via hunter–gathering, thus supporting larger populations. Eventually, however, the land reaches a "carrying capacity" and there is an excess of labor, that is, more people exist than are needed to produce food. Individuals are freed from agricultural production and instead specialize in other occupations, and take up residence in gradually larger and larger settlements.

By 750 C.E. the city of Chang'an in central China had close to 1 million people living inside the city walls and a further million living in the surrounding area. It had a complex street and sanitation network, made possible by the domestication of rice and other crops. The city of Ur in Sumer (modern-day Iraq) was located at the mouth of the Euphrates River on the Persian Gulf, at the center of a large-scale irrigation network that supported crops such as wheat and peas. It is believed that Ur was the largest city in the world in 2050 B.C.E. with a population of ~63,000. Tenochtitlan, known today as Mexico City, was a city of 200,000 people before the arrival of the Europeans, with aqueducts to provide drinking and bathing water and a system of latrines and waste removal, all supported by intensive agriculture. Cahokia Mounds in Illinois, just across the Mississippi River from St. Louis, covered nearly 6 square miles and had a population of 10–20,000 people around 1200 C.E.; it was one of the biggest cities in the large and complex culture of the Mississippi River valley. These, and indeed all, cities acted as areas of concentration, drawing in wealth and productivity (in the form of labor and agricultural goods) from the surrounding countryside. Many formerly great cities have disappeared, their collapse caused by agricultural failures. In the case of Ur, the Euphrates changed course and the crop fields that supported the city died, causing the city to die as well. The ancient cities of the Maya, today tourist sites or hidden by jungle growth, fell prey to deforestation, erosion, and drought that so diminished crop yields that cities failed and people dispersed back into the jungle to live in small groups. Even today, as soon as the surpluses of farmers (meaning the ability of farmers to produce more food than is needed to feed themselves) disappear, or as soon as food supplies are cut off to cities, their persistence is threatened. Most major cities have only enough food to support their population for a few (~3) days. When natural or other types of disasters, such as Hurricane Sandy in New York City or Hurricane Katrina in New Orleans, stop the flow of food into cities, food rapidly disappears from the shelves of grocery stores or from restaurant kitchens.

Along with importing food and other raw materials, cities had to constantly import new residents. Cities experienced sharp population fluctuations due to ravaging epidemics. The aggregation of people into towns, villages, and cities provide humanity with their first **crowd diseases**, diseases that have high population thresholds (i.e., a large number of susceptible individuals) necessary for sustained transmission (see Chapter 6 for discussion of diffusion of diseases through populations). Not until the development of agriculture, and the associated emergence of large human settlements, do we enter the Age of Pestilence and Famine in the epidemiological transition (see Chapter 4 and Chapter 7). The freeing of agricultural labor and the rural–urban migration that begins as a consequence of this provides cities with a continuous supply of susceptible individuals who are also experiencing higher probabilities of contact with infected individuals because of the herding of people into high-density settlements. To achieve the persistence of acute community infections (crowd diseases), it was necessary to have large cities (100,000+) connected in an urban network, and the domestication of animals to pass us disease.

LARGE CITIES IN THE MODERN ERA

Starting in the 18th and extending through the early 20th century there were major changes in manufacturing and agriculture in the United Kingdom and the United States that are known as the Industrial Revolution. Population growth driven by lower infant mortality and phase two of the demographic transition fueled a large labor force. Simultaneously, the enclosure movement in the United Kingdom and improvements in agricultural technology meant that less labor was necessary for food production. Technological innovations in manufacturing, such as the steam engine and the spinning jenny (a device that enabled a single worker to spin eight spools of yarn simultaneously), revolutionized the speed and capability of textile and other forms of manufacturing. The expansion of European colonialism also meant an increase in access to both raw natural resources and markets for manufactured goods.

The Industrial Revolution produced major behavioral and environmental changes. As populations left the rural agricultural production of the countryside for mill and factory towns in the United Kingdom and the United States, they brought their diseases with them. Massive migration began from Europe to the United States, acting as a pressure release for the population growth of first northern and then southern Europe. Labor became an indoor activity, as people left the sunshine and heat of the fields for the stifling air of factories. Low wages and being cut off from rural subsistence food production led to widespread malnutrition. Lack of access to educational opportunities combined with the need for small bodies and cheap labor in factories meant an increase in child labor. The widespread use of coal to power factories led to the first real buildup of industrial pollutants.

Problematically, most United States and European cities had little, if any, municipal planning. The population of London in 1800 was one million, by 1900 it had grown to 6.7 million. The infrastructure of London was inadequate to cope with this massive population growth in only a few short decades; sewerage, for instance, was still pumped untreated into the Thames River, the main source of the city's drinking water. Similarly, the population of New York City in 1800 was approximately 60,000 people, by 1900 it had 4.5 million residents, having doubled its population every decade. The city had no systematic street cleaning efforts, flush toilets, or sewage system, and animals such as pigs, horses, and cattle freely roamed the streets.

In both London and New York the new arrivals crowded into **slums**, called "rookeries" in London and "tenements" in New York. In both settings, human congestion combined with a primitive infrastructure to create ideal conditions for a dramatic increase in epidemic disease. A census of New York City indicated that in 1890 there were over 37,000 tenements housing 1.25 million people; close to 200,000 of those were children under 5 years of age. The tenements were dark and crowded, few rooms had exterior windows to let in light, and many lacked electricity or gas lighting and so were lit with candles or kerosene lamps. Rear tenements were buildings located down narrow alleys between the buildings located on the street and were considered much poorer places to live than the front tenements. Typically a tenement would have one outhouse, located in the courtyard between the front and rear tenement, to share between all residents. The water supply of New York was brought in by aqueduct from upstate. Pumps were not powerful enough to carry water all the way up the four or five stories of the tenements, so often a single water pump was located in the courtyard. As one observer put it

> In reference to the deficient water supply, it may be stated, the usual style among tenant-houses is, a hydrant and sink in a central position of the courtyard for the use of all the inhabitants of both the front and rear houses, there being no water in the houses. The water-closets are in the court also. In many instances, the drainage is superficial, by a gutter formed of flagging in the alley obliquely placed for water and slops to run to the street gutters. In some cases it *seemed questionable whether the alley was intended as an entrance way to a rear house, or a sewage ditch for slops, water, garbage, human excrements, and urine.* (Hygiene and Health, 1865)

London's rookeries were much the same. In one house in 1849, 88 men, women, and children were found living in a single five-room house (White, 2008). These settings provided inspiration for Charles Dickens:

> Wretched houses with broken windows patched with rags and paper: every room let out to a different family, and in many instances to two or even three . . . filth everywhere—a gutter before the houses and a drain behind—clothes drying and slops emptying, from the windows; girls of fourteen or fifteen, with matted hair, walking about barefoot, and in white great-coats, almost their only covering; boys of all ages, in coats of all sizes and no coats at all; men and women, in every variety of scanty and dirty apparel, lounging, scolding, drinking, smoking, squabbling, fighting, and swearing. (Dickens, 1833)

Overcrowding, poor or nonexistent sewers, a lack of ventilation, and limited water for sanitation of bodies or clothing combined to create conditions ideal for the spread of infectious diseases among London's poor. Diseases of these slums included cholera, typhus, tuberculosis, and the first pollution disease, rickets. Rickets is the colloquial term for a medical deficiency of vitamin D, a condition most commonly observed in children. Vitamin D is necessary for calcium absorption and metabolism, and the body makes sufficient vitamin D when exposed to sunlight (this is the only vitamin the human body can make). When the body does not have sufficient vitamin D to absorb calcium, the long bones of the legs are too soft to support the weight of the growing body: the characteristic appearance of rickets is bowed legs as the bones bend.

Known as the "English disease," rickets had been observed prior to the Industrial Revolution primarily in wealthy children, those who stayed indoors and experienced little sunlight exposure (Gibbs, 1994). During the Industrial Revolution, however, rickets became widespread in poor children of the rookeries and tenements, children with already inadequate diets who either worked long hours in factories or, even when outside, were not exposed to sunlight because of intense levels of air pollution from coal fires. Rickets began to disappear in children once therapies (exposure to sunlight or ingestion of vitamin D from food sources) were discovered (Rajakumar, 2003). A primary source of vitamin D in the early 20th century was cod liver oil; the livers of several other fish are also high in vitamin D. Later, milk began to be fortified with vitamin D such that even children with minimal sun exposure do not regularly suffer from rickets.

In the early 1800s the residents of London, New York, and other major cities had to contend with a new epidemic disease: cholera. Previously confined to Asia, London's first outbreak in 1832 killed over 6,000 people. A subsequent outbreak in 1840 killed over 14,000. Cholera had diffused across central Asia and then Europe before making its way to the United States, its movement tracked in maps showing city after city falling prey to this deadly new disease. The primary symptom of a cholera infection was exhaustive diarrhea that caused severe dehydration and eventually death. Different theories as to the cause of cholera were put forth and ranged from low altitude to bad air and smells ("miasmas") to poor moral character.

> As with typhus, so with cholera, the first attraction is a vitiated atmosphere. Both these devourers of human existence flourish best in a climate thoroughly impregnated with the odours of decayed vegetable and animal matter; *a pure air is their destruction*; . . . What carrion is to the vulture and the raven, *bad drainage and overcrowded dwellings are to typhus and cholera*. If, then, we would greet these destroyers of our kind, if we would court their presence, we have only to take care that they find multitudes living in lanes and alleys in which there is no drainage, or in which the drainage is inefficient, where open cesspools and accumulated heaps of a filth unnamable abound; *pack these multitudes together in close unventilated rooms,* let the habits of their lives be, as they almost ever will be, in keeping with the atmosphere around them, and you have spread the banquet and prepared the lodging which will . . . insure the visit of cholera. (Letter from Lord Sidney G. Osborne to *The Times,* December 27, 1847 [Osborne, 2014/1890])

Given the lack of sewerage and sufficient clean drinking water inherent in the environments of Industrial Revolution slums, it is unsurprising that cholera was able to sweep like wildfire through poor neighborhoods again and again.

By 1854 London had begun to implement a citywide sewer system, primarily to deal with the smell of human waste that permeated the air. This new system worked in tandem with human labor, the "night soil" men, who would collect human and animal waste from streets and basements and haul it to farms or other places on the edge of the city. That year a small outbreak of cholera took place in SoHo, one of the last neighborhoods to have the new sewers installed. It was during this outbreak that John Snow documented the spatial patterns of cholera deaths and gathered evidence to support his theory that cholera was a waterborne disease (See Figure 5.1, Snow's map of cholera).

The other disease mentioned in the *Times* quote, typhus, also ravaged slum populations during the Industrial Revolution. Caused by *Rickettsia prowazekii,* the disease caused high fever and eventually death when untreated (and no treatment was available until the mid-20th century with the development of antibiotics). Typhus is spread by the human body louse, which feeds on the blood of people. When a louse feeds on an infected individual, *R. prowazekii* grows in the louse's gut and is excreted in its feces. The disease is transmitted to another human host not when the louse bites the new host but rather when the human scratches the louse bites (which itch) and rubs infected feces into the wounds. *R. prowazekii* can remain viable and virulent in the dried louse feces for many days. Human body lice prefer to feed on humans in our normal temperature range around 98.6 degrees. When a human infected with typhus is experiencing high fever, or when he or she has died and his or her body is cooling, the lice living on that person will jump to new hosts to avoid the hot or cold temperatures, taking the disease with them. The extreme crowding of poor into New York, London, and other Industrial Revolution slums allowed typhus to ravage through buildings, as lice jumped from infected to susceptible hosts and as bedding and clothing were rarely, or never, washed.

In New York in 1850 there was a severe epidemic of typhus fever among Irish immigrants fleeing the Great Famine:

> Upon examination of the records of admission of patients, I discovered that from one tenement house upwards of one hundred cases had been received. On visiting the house, I found a veritable "fever nest." The doors and windows were broken, the cellar was filled with sewage, every room was occupied by families of Irish immigrants who had but little furniture and slept on straw scattered on the floor. I learned that the house was the first resort of immigrants, as there was no one in charge and hence no expense. (Dr. Smith, Bellevue Hospital, 1850, quoted in Rosen, 1972)

> The severe and very close and snowy winter so shut up the poorer classes of city inhabitants in their dwellings, that the malignant outbreaks and rapid diffusion of this fever [i.e., typhus] threatened much harm to the tenant population. (Dr. Elisha Bartlett, quoted in Rosen, 1972)

While typhus disappeared from the wealthy world due to improvements in hygiene and housing, and then the introduction of antibiotics, it remains common in the southern hemisphere due to poverty, inadequate clothing hygiene, and poor socioeconomic conditions. Political unrest, civil war, and refugee populations living in camps are factors that contribute to the emergence of typhus epidemics (Ndihokubwayo & Raoult, 1998).

Tuberculosis has a long history in human populations: evidence for tuberculosis infections is observed in Egyptian mummies and was described by the ancient Greeks. Tuberculosis became endemic in the urban poor, however, during the Industrial Revolution. By the early 1800s it was responsible for approximately one in four deaths in England. Caused by the bacteria *Mycobacterium tuberculosis,* tuberculosis was also known as consumption or white plague. Individuals with tuberculosis infections exhibited a constant cough as the bacteria invaded the lungs, blood appeared in their sputum, and they lost weight and turned pale. The environments of urban slums favored the maintenance and transmission of tuberculosis. Sunlight kills tubercle bacteria, but the dark and crowded conditions of cities, with little sunlight and fresh air, were perfect environments for the coughs of tubercular patients to pass the bacteria onward. As people crowded into tenements and were subject to physical stress, suffering from poor diets and physical exhaustion from working in factories and sweatshops, their bodies were incapable of fighting off infection. Low vitamin D, as previously described, also decreases the body's natural defenses against tuberculosis.

Doctors in New York linked tuberculosis in their patients to the living conditions of so many of the city's residents. Overcrowding of already unhealthy individuals into dark rooms with little ventilation was the perfect environment for airborne transmission of tuberculosis bacteria. Thousands of deaths each year took place in New York as a result of tuberculosis. Partly in response to this high disease burden, the Tenement Act of 1901 called for light and ventilation and clean water in all new tenement houses built. Other changes began to decrease the toll that tuberculosis took on urban poor in Europe and the United States: introduction of purification and chlorination of drinking water supplies in 1910, the pasteurization of milk in 1912, the growth of the sanatorium movement (wherein tubercular patients were removed to places with clean fresh air to allow their lungs to heal), and laws against spitting in public places.

Indeed, it is changes in sanitation and hygiene, changes in the behavior of individuals and the places in which they lived, that caused tuberculosis, typhus, cholera, rickets, and the other diseases that plagued the urban slums of the Industrial Revolution to recede by the early decades of the 20th century. Sewage systems, glass windows, water treatment plants, city parks, the provision of hot water indoors, decreases in overcrowded apartments, the implementation of public schooling, and child labor laws—all of these things changed the nature of morbidity and mortality in the urban poor *before* the advent of antibiotics and modern medicine. In the urban slums of low- and middle-income countries, however, the sanitation and hygiene changes of the wealthy world have not taken place, and antibiotics and vaccines are of limited efficacy in the face of water, waste, and nutritional problems.

DEVELOPING WORLD CITIES: DICKENS OR A DREAM?

To many residents of wealthy nations, cities in their own countries or the growing cities of low- and middle-income countries are places to be avoided, places filled with grinding poverty and human suffering. This view is often based on a Dickensian image of cities, one that has its roots in Industrial Revolution London, New York, and Paris. Cities represent disruption, deprivation, disease, and death (Szreter, 1997, 2003). But for the residents of such cities, even for the poorest residents, cities are a symbol of civilization and wealth and opportunity, they are the dream destination for poor rural migrants looking to make their fortune or improve the lives of their family.

The reasons that people move to cities can be divided into **push-and-pull factors**, each operating at a variety of scales. Pull factors are those that draw migrants to cities, to one city over another, or to one neighborhood over another. Push factors are those that tip the scales in favor of a person packing his or her bags and leaving home. Pull factors include job and educational opportunities, family or friend networks, marriage opportunities, better weather, and so on. Push factors include population pressure, land insecurity, conflict or war, natural disasters, lack of employment, and the like. For many rural–urban migrants in low- and middle-income countries, seemingly inescapable poverty and debt from having to provide dowries or weddings for daughters or having to purchase agricultural supplies such as fertilizer and seeds, or a lack of available land now that more children are surviving to adulthood and farms have already been subdivided a number of times, push them off the land and into cities (see Chapter 8). Millions of Bangladeshi women have left rural areas and migrated to cities such as Dhaka and Chittagong to work in garment factories, pushed by debt and pulled both by chances to earn income and to live more freely than at home. Millions of Bangladeshi men have left rural areas and migrated to cities in Saudi Arabia and other countries to labor in the construction industry, remitting money home.

Similar to the lack of urban planning observed in Industrial Revolution cities, there is unrestrained and ill-planned or unplanned growth in the cities of low- and middle-income countries today. City governments are unable to meet the level of demand that growing populations pose. To put urban growth into context, let's consider the case of Dhaka, the capital and primate city of Bangladesh.

Dhaka is an old city: habitation dates to the first century C.E., and it was originally a linear settlement along the Buriganga River. In 2015 there were approximately 14.4 million people living in Dhaka, making it a megacity and the world's 10th largest city. Eighteen million people live in the larger metropolitan area. The UN predicts that by 2025 the population of Dhaka will have risen to over 22 million. This growth is driven both by high fertility among city residents and high in-migration from rural areas. If four million people will be added to the city's population in a decade, this equates to 400,000 new residents each year, ~7,700 new residents each week, and a thousand new residents each day. The speed and scale at which Dhaka, and other major cities such as Lagos, Mumbai, and Jakarta, are growing is unprecedented in human history and is driven by the momentum of population growth from the demographic transition. Any city government, even the wealthiest and most organized city governments in wealthy nations, would have trouble meeting the needs of such a rapidly growing population. Making the issue more complex is the fact that many of these new urban residents arrive with little or no means to support themselves, no concrete job opportunity, and often only a very vague idea of where to live and how to survive in a city.

Many new migrants to cities in low- and middle-income countries will end up in slums, known in other places and other times as rookeries, tenements, and shantytowns. Over one-third of urban households live in slums worldwide. UN-Habitat estimates that the number of people living in slum conditions in 2015 was 863 million, in contrast to 760 million in 2000 and 650 million in 1990. The greatest increase in slum populations is in Asia, already home to the majority of slum dwellers, although increasing rural–urban migration in sub-Saharan Africa is driving slum growth in that region as well. The UN has five defining characteristics of slums: lack of sanitation, lack of clean water supplies, lack of durable housing, lack of sufficient living area, and insecure residential status (UN-Habitat, 2004). There are health risks associated with each of these defining criteria (Unger & Riley, 2007).

The first defining characteristic, a lack of sanitation, affects multiple aspects of life for slum dwellers. When there are no sewers to channel human waste away from dwellings, the viruses and bacteria contained in human waste are transmissible from person to person. When multiple households, sometimes dozens or even hundreds of households, share a single latrine, many individuals will opt to defecate in the open rather than use overfilled latrines. Intimately tied to this first criteria is the second, a lack of clean water supplies. When sanitation is lacking and wastewater is discharged without being treated, local water supplies become contaminated. Multiple families sharing spigots or taps, and the labor involved in collecting drinking water, leads individuals to use less water than necessary for cooking and hygiene. According to the UN Environmental Program's (UNEP) "Sick Water" report, 90% of wastewater discharged in developing countries is untreated (UNEP, 2010). This, they estimate, contributes either directly or indirectly to over 2 million diarrheal deaths per year, primarily in children under the age of 5. The picture below (Figure 10.3) shows children playing in an open area of a Dhaka slum, with garbage and sewerage in the flooded lot behind them.

The UN defines lack of sufficient living area as two or more individuals sharing a room of less than 5 square meters per person. Essentially this describes a family sharing a single room, doing all of their sleeping, cooking, eating, bathing, and, sometimes, their jobs in that space. Overcrowding of individuals into rooms exposes them to health risks such as increased transmission of airborne infectious diseases (e.g., pneumonia, tuberculosis, or pertussis); increased exposure to cooking smoke and fumes, which can cause respiratory illness such as asthma; and increased likelihood of passing on diseases of skin-to-skin contact. Small, unventilated rooms also expose residents to high temperatures, which can exacerbate conditions such as high blood pressure or asthma. There are also psychosocial stresses associated with a family sharing a single room, as there is a lack of privacy for children entering puberty, potential exposure to domestic violence, and so on.

Slums are often densely packed and constructed of substandard materials or built in a rushed and haphazard fashion. This lack of durable housing, with millions of individuals living in dwellings constructed of cardboard, scrap metal, or

FIGURE 10.3 Children playing in an open area of a Dhaka, Bangladesh, slum.

mud on land that is prone to floods or mudslides or fire, physically threatens the health of people living in slums. Slums tend to be located on land that is undesired, at least until urban growth forces slums to relocate. Undesirable land in cities can take the form of low-lying areas that flood seasonally or during a heavy rain, of hillsides that are prone to mudslides (particularly the case in Latin American cities), or locations near trash dumps or factories releasing toxic pollution. The risk of such locations combined with the inferiority of building materials and poor construction means that slum dwellers are more likely to face physical injuries during fires, earthquakes, or floods as their houses collapse around them. Infectious disease risk may also be higher in such locations as standing water provides breeding grounds for mosquitoes and other insect vectors.

As untenable as life in a slum would be to most of us in the wealthy world, to a slum resident holding on to the small piece of land or the small space they have for their family is vitally important. Unfortunately, most slum dwellers have insecure residential status: they do not own the land on which they have built their dwelling or do not have formal rental paperwork with their landlords. Unscrupulous landlords are often squatters themselves, renting space on land that they do not hold the title for. Slums are thus easy targets for city governments or landowners looking to build new, formal housing structures as cities fill in on themselves. Without land titles or any documentation of a rental agreement with a landlord, slum residents are unable to fight for rights to stay, to fight for their houses not to be bulldozed, and are further unable to fight for any improvement in their sanitation/water/environments (Unger & Riley, 2007). In addition to being intimately tied to the health effects associated with the other four characteristics of slums, this criteria for slums is associated with psychosocial stresses of never knowing when your house may disappear, and worrying that any investment you make in your dwelling will be for naught if the bulldozers appear.

Though millions of people are considered to be slum dwellers, many have only one characteristic of a slum. For the 43% of urban households in slums in South Asia, 31% have a single deprivation (Economic, 2008). Of the 62% of urban households in sub-Saharan Africa that live in slums, 31% have a single deprivation. For those with only one deprivation, simple low-cost solutions could work. Providing paperwork that demonstrates secure residential status, for instance, or the provision of more durable housing materials, could move millions of people out of the "slum" category. For millions of others, however, the coincidence of multiple characteristics makes easy or rapid solutions impossible.

As an example of a modern megacity struggling to keep up with population growth, Dhaka's current water supply system has as its foundation a network built by the British during colonization in 1874. While it has been expanded upon since then, only two-thirds of Dhaka's millions of residents are served by the city water supply system. The sewage system serves only a quarter of the population, while 30% are on septic tanks. This leaves close to half of the population, ~7 million people, with no sewerage system whatsoever. The majority of solid wastes, more than 9 million tons, are thus dumped untreated along streets, into low-lying areas, and into water bodies.

From 1996 to 2005 the number of slum residents in Dhaka more than doubled, from 1.5 to 3.4 million, increasing the proportion of Dhaka's population living in slums from 20 to 37% (Islam, Angeles, Mahbub, Lance, & Nazem, 2006). Not only did the number of slum dwellers increase, so too did the number of slum communities: by ~70%, from approximately 3,000 to close to 5,000. Slums in Dhaka are fed by rural migrants streaming into the city looking for employment. In these slums, 65% of residents have no safe latrines and 40% share their water source with more than 11 other families. A survey of Dhaka slum residents in 2013 indicated that the number without safe latrines and who share their water source with many other families remained approximately the same as was observed in 2005/2006, even as other indicators (housing material, housing space, electricity, mobile phone access) improved (National Institute of Population Research and Training [NIPORT], et al., 2015). It is perhaps counterintuitive to think of a slum dweller as having electricity and television and a mobile phone, but these are relatively easier to come by than clean safe drinking water and sanitation. Electricity is often unreliable; in the summer in Dhaka in particular there can be frequent power cuts, but the power grid is relatively easy to hook into. TVs and cell phones are individual purchases, within the power of slum dwellers to access. A slum resident cannot install sewer systems or drill a well for clean water—these are actions that have to be taken by city governments, NGOs, or other organizations.

When considering the expansion of slums and slum populations, it is important to remember that slums themselves, and slum residents, are not the problem! They are the *spatial indicators* of urban poverty, social exclusion, and inappropriate government policies. As Anna Kajumulo Tibaijuka, the executive director of the UN Human Settlements Program, says, "Slums are a manifestation of the two main challenges facing human settlements development at the beginning of the new millennium: rapid urbanization and the urbanization of poverty" (quoted in UN-Habitat, 2004). Slums represent an active, grassroots attempt by the desperately poor to take care of themselves. Usually these are the working poor, as opportunities for new urban residents to secure a well-paying job and move themselves and their children out of poverty, as urbanites of the Industrial Revolution were able to do, are scarce (Konteh, 2009).

TRAFFIC

If you've ever lived in a major urban center and commuted to or from work during rush hour, you know how extensive the problem of traffic has become, especially over the past 30 years. But the increasing use of cars, and the ensuing traffic jams, are not just a problem in high-income countries. Countries in the early stages of the demographic transition, mostly low- and lower-middle-income countries, have many young people, are in the midst of the mobility transition, and have high rates of migration and circulation. Much of this population movement is made possible by the increasing availability of motorized transport. There is also a general increase in SES and an emerging middle class, particularly in Asian countries. The combination of these factors, plus the crowding of more and more people into dense urban spaces, fuels an increase in the importance of traffic and road injuries to global human health.

Between 2005 and 2050 the number of cars in the world is projected to increase by 2.3 billion. The majority of this increase, 1.9 billion, will be in emerging markets in developing countries. Places such as India and China, with their large and newly wealthy middle class, are at a stage in development where car or other vehicle ownership is expected to rapidly increase. The WHO estimates that 1.2 million people are killed on the world's roads each year, and that 50 million more are injured. If these trends continue, the WHO estimates that the number of injuries and deaths due to traffic will increase by more than 60% by 2020. In 2012, road traffic injuries were the ninth leading cause of death worldwide; by 2030 road traffic injuries may be the fifth leading cause of death and mortality may double to 2.4 million people each year.

Speaking about global causes of death masks important variations in the role of traffic and road injuries by age and income. Road traffic injuries were the 14th leading cause of death in the 0–4 year age category in 2004 but the top cause of death in the 15–29 year age category (second for 5–14 and third for 30–44 years). They make up far less of the mortality in people over age 45. The mortality impacts of traffic are thus felt most heavily by young adults, a category that makes up a large share of the population in low- and middle-income countries.

In low- and lower-middle-income countries, the substantial burden of infectious diseases such as HIV, malaria, and tuberculosis, combined with stroke and heart disease, prevent road traffic injuries from being a top-10 cause of death. In upper-middle-income countries, however, where the burden of infectious diseases has waned, it is the seventh leading cause of death. In high-income countries, with efficient provision of emergency medical services and high rates of legislation and compliance for road safety, road injuries are not a leading cause of death. If there were 1,000 people in the year 2012 that were distributed into income brackets the way the world's 7 billion people are distributed, and if these 1,000 people died of the same causes as the world's 7 billion people, then 22 of these hypothetical 1,000 people would have died from being injured on the road. Three of these individuals would be from low-income countries, eight from lower-middle, nine from higher-middle, and two from high-income countries. This distribution highlights the fact that it is the middle-income countries of the world, those with both the highest population sizes and increases in wealth, that bear the burden of road traffic fatalities.

Indeed, current data indicates that while 53% of the world's vehicles are found in low- and middle-income countries, these are home to 91% of the world's fatalities from road traffic. This begs the question, Why is there a mismatch between where vehicles are located and where fatalities due to these vehicles are taking place? There are several reasons, including **mixed traffic patterns** and **vulnerable road users**, legislative and enforcement variation, and emergency service provision.

"Mixed traffic patterns" is a phrase that characterizes a road setting where pedestrians and cyclists are not separated from motorized transport such as cars, trucks, and buses. In the majority of the wealthy world pedestrians are separated from cars and motorcycles on sidewalks, and they can cross roads using pedestrian bridges or at clearly marked crosswalks. Bicyclists sometimes have their own clearly marked lanes. Stop lights and stop signs are in place, and generally observed at intersections, and highways or other high-speed roads have limited access. In low- and middle-income countries there is often little or no separation of pedestrians and bicyclists from the flow of motorized transport. An observer of rush hour traffic at a main intersection in Dhaka, Bangladesh, might see the following modes of transportation types: pedestrian, bicyclist, bicycle rickshaw, auto-rickshaw (three-wheeler), motorbike/motorcycle (two-wheeler), car, truck, and bus. All eight of these types of users share the same space, there is no protection for pedestrians, bicyclists, or bicycle rickshaw drivers or their occupants against being hit by a motorized vehicle. Even the people riding on two- or three-wheelers are exposed to the larger cars, trucks, and buses.

This situation describes what are known as "vulnerable road users." Globally, vulnerable road users (pedestrians, cyclists, three-wheelers) make up 46% of road traffic deaths. In certain countries, however, they may constitute much higher percentages of fatalities. In Thailand, for instance, over 84% of the fatalities in 2010 were from these vulnerable groups. Two-wheelers in particular contribute to large shares of road traffic-related morbidity and mortality, primarily in Asia. Two- and three-wheeled forms of transport are popular for a variety of reasons. They are cheaper to buy and easier to maintain than cars, and they can navigate roads that four-wheelers would have trouble with (steep, narrow, winding, washed-out during rain). While these are all advantages, their lack of enclosure means that the drivers and passengers on two- and three-wheelers (and you can fit a family of five on a two-wheeler) are more vulnerable to injury than if they were enclosed in a car or other four-wheeler. While two- and three-wheeler ownership is already high in Asia, it is also likely to increase dramatically in sub-Saharan Africa in the coming decades for the same reasons that it has in Asia: these types of vehicles do well under difficult road conditions and are a relatively inexpensive first vehicle purchase.

Adding to the morbidity and mortality caused by mixed traffic patterns and by poorly built or maintained infrastructure are issues surrounding road user legislation and enforcement. Driver's education and licensing is absent or unenforced in many countries, meaning that a totally inexperienced driver can drive off the lot with his or her new vehicle while never having been behind the wheel before. Enforcement or adherence to lane markings or stop lights may be nonexistent. Many cities lack speed limit laws, the wearing of seatbelts may not be mandatory or only mandatory for drivers, there may be no legislation relating to the use of child safety seats, and many countries do not have or do not enforce blood alcohol content (BAC) limits.

One of the major drivers of high levels of morbidity and mortality, particularly given the increase in two- and three-wheeled transport, are the effectiveness (presence and enforcement) of helmet laws. Wearing a helmet is the single most effective way of reducing head injuries and fatalities resulting from motorcycle crashes (Liu et al., 2008). In 2005 there were 8 million motorbikes in Vietnam, one for every 10 people. While national legislation in 2000 mandated helmet use on certain roads, and instituted a fine structure for not wearing a helmet in 2003, overall helmet use remained very low and morbidity and mortality remained very high (Hung, Stevenson, & Ivers, 2006). A new law took effect in 2007 that required all riders to wear a helmet on all roads, with large fines (up to 30% of the national per capita income) associated with noncompliance (Enserink, 2014). The picture below (Figure 10.4) was taken on a street in Ho Chi Minh City in 2009: notice that everyone, drivers and passengers, are wearing helmets. The fall in morbidity and mortality in Vietnam since this strict helmet law went into place indicate its success in reducing the negative impacts of traffic on human health.

Once a traffic accident has taken place, this accident resulting in lifelong disability or in death is highly dependent on the speed and completeness of care that an individual receives. Wealthy countries may have a high number of road accidents but they also have high survivability because injuries are tended to quickly and effectively by emergency responders. The WHO estimates that 111 countries have a single nationwide emergency number (e.g., 911 in the United States), but that only 59 countries in the world have ambulance services that serve the majority (75%+) of the population. What in the wealthy world could be a survivable accident may result in death in the absence of emergency medical services in resource-limited contexts, or a relatively minor injury in a wealthy setting may translate into lifelong disability in the absence of care.

Even when accidents do not result in a fatality, morbidity due to traffic accidents has a huge effect on population health. It is estimated that 2 million people in

FIGURE 10.4 Motorbike traffic on a Ho Chi Minh City street.

India per year are disabled due to road injuries. A family may lose their breadwinner, either due to death or injury, after a road accident. Poor families are less able to cope with the direct and indirect costs of traffic accidents, and also experience higher mortality than do wealthy individuals. The increase in road traffic injuries and deaths also places an increased burden on already struggling health care systems. Road traffic injury patients in India can account for 20–50% of emergency department admissions; in Kenya 45–60% of surgical admissions are road traffic injury patients.

Efforts made in Western countries can be effective in developing country contexts. The implementation and enforcement of speed limits, alcohol limits, child restraint and seatbelt laws, and helmet laws are political and bureaucratic changes that can decrease morbidity and mortality. The installation of sidewalks and stop lights and other infrastructural changes to the transportation environment can also decrease injuries and deaths. Finally, improving postcrash care, in terms of a single emergency number in a country and greater availability and training of emergency medical services, can decrease not only mortality but also reduce the disabling of otherwise healthy individuals.

Traffic's other major influence on urban health is via air pollution. Many of the cars, trucks, and other vehicles joining the roads of low- and middle-income cities are highly polluting, releasing particulate matter, ozone, and other pollutants into the air. It is estimated that 1,000 new cars hit the roads in Beijing each day, and that Beijing's traffic levels increased by 130% between 2009 and 2010 (Fan, 2008; Zhu,

2010). This has caused huge increases in air emissions from traffic. In Beijing and other major Asian cities, such as New Delhi and Bangkok, increases in particulate matter, nitrous oxide, and other pollutants are leading to increased rates of chronic obstructive pulmonary disease (COPD), asthma, and other respiratory conditions. For more on this topic, see Chapter 11.

DISAPPEARING CITIES?

Ironically, even as we've spent the previous part of the chapter discussing the phenomenal growth of cities as rural–urban migration continues in many parts of the world, urban growth is actually likely to slow in the coming decades. To be sure, millions upon millions of people will move to cities and millions upon millions of children will be born in cities. While the absolute number of urban residents will increase, their proportional share of urban residents will be smaller, and urban growth will slow. In China, for instance, which has seen such phenomenal rural–urban migration in such a short amount of time, the expansion of cities is expected to stop and the shrinking of the urban population to begin by 2050, due to the impacts of decreased fertility (both via the so-called One Child Policy, which is now ended, and through voluntary fertility reductions associated with higher SES).

Already cities in Europe, the United States, Japan, and South Korea are shrinking. Multiple forces are at play in these shrinking cities. Foremost among them is declining fertility: there are simply fewer people to live in the cities of the late 20th century. Economic breakdown is another cause: major industrial cities of the past no longer need, or employ, a large labor force as industry is transferred overseas or is mechanized. Mill towns in England and car-manufacturing cities in the United States lose population as people are forced to migrate for economic opportunity.

Shrinking cities are problematic for human health just as growing cities are, mainly through difficulties in providing the services necessary to maintain health. Detroit, Michigan, had a population of ~1.8 million people in the 1950 census, but by the 2010 census there were only ~700,000 residents. As the auto industry failed and moved to other areas of the country and world, the city hollowed out, a process that was exacerbated by the economic collapse of 2008. Homes, indeed entire city blocks, were abandoned. The tax base disappeared, and with it the tax revenue necessary to provide emergency services, to maintain streets, to mow parks, to staff schools. Abandoned buildings and lots, as they revert to nature, expose the population left behind to ticks and other insect pests, to rats and other rodents, to crime, and to the psychosocial stresses of living in areas that have been deserted. In 2015, the city of Detroit declared bankruptcy. There are far-reaching implications to this fiscal crisis, but one important impact on human health has been the devaluation of pensions held by municipal workers, pensions that many people rely on during retirement. Detroit and other cities, such as Pittsburgh in the United States, have begun to proactively shrink their cities in order to better serve the needs of the remaining population. Shrinking cities need as much urban planning as do growing cities.

CONCLUSION

While cities occupy only 2% of the Earth's surface, their true footprint is enormous: they host more than half of the world's population, account for over 75% of the world's energy consumption, and are responsible for at least 80% of the world's carbon dioxide emissions (Harrison & Pearce, 2000). How efficiently we run our cities today and in the future, how effectively we decrease air pollution and energy consumption and handle the waste generated by all that humanity, will determine the future impacts of global climate change. At the same time, the large footprint of cities is made possible only by continued influx of food and other items into cities from rural areas. As we have seen from cities lost to history, cities cannot survive without their hinterlands.

All urban residents desire and deserve the same provision of services: education for their children, clean water and sanitation systems, emergency services (fire, ambulance, police), voting and political participation, infrastructure maintenance, and access to health care and employment. The provision of these services requires efficient and effective city management along with a sufficient source of funds (i.e., taxes). Many cities in low- and middle-income countries are struggling to keep up with the pace of in-migration and urban growth that they experience. When too few of their current residents are beneficiaries of the services listed above, how can they hope to extend these services to millions of new residents? At the same time, cities in high-income countries are facing a struggle to provide services caused by a different form of population change: the shrinking of urban populations and the hollowing out of urban cores. Shrinking urban populations will also face Chinese cities and cities in other countries with lowering fertility rates.

Revolutionary changes in ecology, in this case humans becoming an urban species rather than a rural species, bring with them revolutionary changes in human health and the need for adaptation. The crowd diseases of our earliest cities are still with us, and are now compounded by the noninfectious diseases of lifestyle and pollution exposure as well as threats to health from motorized transport. Finding ways to mitigate morbidity and mortality due to traffic injuries, air pollution, lack of sanitation, and precarious living circumstances for the urban poor are all challenges our species will be forced to face in the coming decades.

REVIEW QUESTIONS

1. What are the similarities between the difficulties that urban areas face today and urban areas of the Industrial Revolution faced? What are the differences? How does urban growth in the 21st century repeat that of the 19th?

2. Of the five deprivations that the WHO and the UN use to define a slum, which do you think is the most important? Which do you think is most easily fixed or solved? Do you agree with these five criteria, or is there another that you would include?

3. Watch the video entitled "Geneva Camp in Dhaka, Bangladesh" on YouTube (*www.youtube.com/ watch?v=ZWRp5bNIOmU*). How many of the five deprivations that define a slum do you see in the video?

4. Rickets is a reemerging health issue in the United States. Thinking about the ecology of rickets in the 19th century described above, what form do you think the reemergence of rickets takes? In what populations is rickets being observed, and what behaviors are causing rickets in U.S. children? Use the triangle of human ecology to brainstorm some risk factors for rickets that are behavioral and environmental. Then do a Google search to see if you are correct.

5. Go to the Country Profiles webpage of the WHO's Violence and Injury Prevention program (*www. who.int/violence_injury_prevention/road_safety_status/2013/country_profiles/en*).   Download three different country profiles, one for a country in each of the high-, middle-, and low-income categories (e.g., Australia, Thailand, Mozambique). Look at the data compiled by the WHO on the number of road users in each country, how those road users break down by category (car, three-wheeler, two-wheeler), and then how road deaths break down by category and trends over time. Can you see the phenomena described in this chapter playing out in these three income categories? How does the distribution and number of road deaths in the countries vary by income? Which country has the highest number of road deaths given the total number of registered vehicles? How does national legislation (in the upper left of the profile) on BALs and seatbelt and helmet use interact with the patterns that you see?

6. Visit YouTube and search for "Dhaka traffic." Watch one of the videos. How many types of road users can you see? Which of them are considered vulnerable road users? How does the infra-structure of the city's streets (e,g., sidewalks, lane markers, crosswalks, stoplights) influence the interactions between people and vehicles? Now search for videos of "Beijing traffic" or "Bangkok traffic" or "Los Angeles traffic" and see how they compare to rush hour in Dhaka.

7. How might a city shrink itself in an effective way? What cities around the world are projected to shrink in the coming decades? What types of problems does this cause for residents who are left behind?

8. How does the double burden of disease that accompanies the epidemiological transition play out differently in an urban area versus a rural area? Are urban residents better or worse off than their rural counterparts?

9. Do you agree with the premises of this chapter that urban environments represent a new ecology for us as a species and that how we manage our cities will determine our future?

REFERENCES

Corcoran, E., Nellemann, C., Baker, E., Bos, R., Osborn, D., & Savelli, H. (2010). *Sick water?: The central role of wastewater management in sustainable development* [A rapid response assessment, United Nations Environment Programme, UN-HABITAT, GRID-Arendal]. Arendal, Norway: UNEP/GRID-Arendal.

Dickens, C. (1833). Gin shops. Republished in *The Dent Uniform Edition of Dickens' Journalism, Sketches by Boz and Other Early Papers, 1835.*

Enserink, M. (2014). Hats off to Vietnam's helmet law. *Science, 345*(6202), 1261.

Fan, M. (2008). Creating a car culture in China. *Washington Post.* Retrieved from *www.washingtonpost.com/wp-dyn/content/article/2008/01/20/AR2008012002388.html.*

Gibbs, D. (1994). Rickets and the crippled child: An historical perspective. *Journal of the Royal Society of Medicine, 87*(12), 729–732.

Harrison, P., & Pearce, F. (2000). *AAAS atlas of population and environment.* Berkeley and Los Angeles: University of California Press.

Hung, D. V., Stevenson, M. R., & Ivers, R. Q. (2006). Prevalence of helmet use among motorcycle riders in Vietnam. *Injury Prevention, 12*(6), 409–413.

Hygiene and Health. (1865). *Report by the Council of Hygiene and Public Health of the Citizens' Association of New York Upon Epidemic Cholera and Preventive Measures: New York, November 1865.* Sanford, NY: Harroun.

Islam, N., Angeles, G., Mahbub, A., Lance, P., & Nazem, N. (2006). Slums of urban Bangladesh: Mapping and census 2005. Dhaka, Bangladesh: Centre for Urban Studies.

Konteh, F. H. (2009). Urban sanitation and health in the developing world: Reminiscing the nineteenth century industrial nations. *Health and Place, 15*(1), 69–78.

Liu, B. C., Ivers, R., Norton, R., Boufous, S., Blows, S., & Lo, S. K. (2008). Helmets for preventing injury in motorcycle riders. *Cochrane Database of Systematic Reviews, 1,* CD004333.

National Institute of Population Research and Training (NIPORT); Measure Evaluation, University of North Carolina at Chapel Hill, North Carolina; International Centre for Diarrhoeal Disease Research, Bangladesh (icddr,b). (2015). *Bangladesh Urban Health Survey 2013 Final Report.* Chapel Hill: University of North Carolina.

Ndihokubwayo, J., & Raoult, D. (1998). Typhus épidémique en Afrique [Epidemic typhus in Africa]. *Medecine tropicale: Revue du Corps de sante colonial, 59*(2), 181–192.

Osborne, S. G., Lord. (2014). The letters of S. G. O.: A series of letters on public affairs written by the Rev. Lord Sidney Godolphin Osborne and published in 'The Times' 1844–1888 (A. White, Ed.). Charleston, SC: Nabu Press. (Reprinted from 1890, London: Griffith, Farran, Okeden, & Welsh).

Rajakumar, K. (2003). Vitamin D, cod-liver oil, sunlight, and rickets: A historical perspective. *Pediatrics, 112*(2), e132–e135.

Rosen, G. (1972). Tenements and typhus in New York City, 1840–1875. *American Journal of Public Health, 62*(4), 590–593.

Smil, V. (2013). *Making the modern world: Materials and dematerialization.* Chichester, UK: Wiley.

Szreter, S. (1997). Economic growth, disruption, deprivation, disease, and death: On the importance of the politics of public health for development. *Population and Development Review, 23*(4), 693–728.

Szreter, S. (2003). The population health approach in historical perspective. *American Journal of Public Health, 93*(3), 421–431.

Unger, A., & Riley, L. W. (2007). Slum health: From understanding to action. *PLoS Medical, 4*(10), 1561–1566.

UN-Habitat. (2004). The challenge of slums: Global report on human settlements 2003. *Management of Environmental Quality: An International Journal, 15*(3), 337–338.

United Nations, Department of Economic and Social Affairs. (2008). *The Millennium Development Goals Report 2008.* New York: United Nations Publications.

United Nations, Department of Economic and Social Affairs, Population Division. (2014). World urbanization prospects: The 2014 revision, highlights (ST/ESA/SER.A/352). New York: United Nations.

White, J. (2008). *London in the nineteenth century: A human awful wonder of God.* New York: Random House.

Zhu, K. (2010). China's Great Wall of traffic jam: 11 days, 74.5 miles. *ABC News.* Retrieved from *http://abcnews.go.com/International/chinas-traffic-jam-lasts-11-days-reaches-74/story?id=11550037.*

🖉 FURTHER READING ∙∙∙

Boyden, S. V. (Ed.). (1970). *The impact of civilisation on the biology of man.* Toronto: University of Toronto Press.

Fabos, J. G. (2004). Greenway planning in the United States: Its origins and recent case studies. *Landscape and Urban Planning, 8*(2–3), 321–342.

Harpham, T. (1994). Urbanization and mental health in developing countries: A research role for social scientists, public health professionals, and social psychiatrists. *Social Science and Medicine, 39,* 233–245.

Keiser, J., Utzinger, J., Caldas de Castro, M., Smither, T. A., Tanner, M., & Singer, B. H. (2004). Urbanization in sub-Saharan Africa and implication for malaria control. *American Journal of Tropical Medicine and Hygiene, 71*(Suppl. 2), 118–127.

McGee, T., Ginsburg, N., & Koppel, B. (Eds.). (1991). *The extended metropolis: Settlement transition in Asia.* Honolulu: University of Hawaii Press.

Mutatkar, R. K. (1995). Public health problems of urbanization. *Social Science and Medicine, 41,* 977–981.

United Nations Human Settlements Programme (UNHSP). (2003). *Global report on human settlements 2003: The challenge of slums.* Sterling, VA: UN-HABITAT/Earthscan.

Williams, B. (1990). Assessing the health impact of urbanization. *World Health Statistics Quarterly, 43,* 145–152.

World Health Organization (WHO). (1993). *The urban health crisis: Strategies for health for all in the face of rapid urbanization: Report of the technical discussions.* Geneva, Switzerland: Author.

PART IV

ENVIRONMENTS and CLIMATES

The biophysical environment can affect our health in many ways. The chemicals we release into the ground, water, and air eventually wind up in our bodies and can have subtle and lasting effects on health. At the same time, the human population has developed protective cultural buffers which, at times, limit exposure to unsafe environments. This section focuses on the environmental, weather, and climate impacts on human health.

Chapter 11 introduces a variety of environmental health exposures including toxic hazards, indoor and outdoor air pollution, and water quality. The negative health impacts of lead exposure are described, as are cultural and legislative shifts that have decreased childhood lead exposure in the last decades. Outdoor air pollution is one of the world's most common causes of premature death. As the world rapidly urbanizes, especially in countries with large populations like China and India, air pollution has the potential to affect huge portions of the world's population. Indoor air pollution exposure is driven by the types of cooking fuel people use and the small size and poor ventilation of housing, and is often far worse than outdoor pollution. It is most severe in the same countries with the worst outdoor air pollution. This chapter discusses water quality including the great progress that has been made in bringing improved drinking water and sanitation to most of the world during the past 15 years. The chapter ends with discussions of environmental justice and the production of healthy environments.

Chapter 12 focuses on how weather and climate affect human health. Influences of weather and climate include heat-related illness, changes in disease vector habitats, and vulnerability of coastal populations in many developing countries. Physical geographic variation leads to variability in population and disease distributions. As the global climate changes, we expect to see shifts in these distributions. This chapter provides a diverse set of examples including depression, influenza, rickets, births and deaths. The chapter also offers a short discussion of climate change science and reviews how future climate change scenarios are being used to predict how climate change will impact health and disease distributions.

CHAPTER 11

Environment and Health

Since the beginning of the Industrial Revolution there has been an exponential increase in the kinds and amounts of metals, gases, and chemicals that have been added to the air, water, and soil in Earth's industrial regions. Many of the chemicals have never existed in the environment before and some chemicals persist for long periods. A major source of pressure on the environment today is the desire by newly industrializing countries (NICs), such as China, for a lifestyle approximating that of wealthy nations. New transportation and economic activities are quickly consuming what remains of Earth's fossil fuels. The purpose of this chapter is to survey these environmental processes and their spatial context as a background for research on health and disease. Dubos (1965) pointed out some time ago that humankind is now adapting, genetically and culturally, to the environments that humans have built. We spend most of our time inside, in our homes and workplaces. In industrialized countries most people live in cities, but even in rural areas cultivated land and settlements create the environmental stimuli that surround people. Coping with these new stimuli is largely under the control of culture because genetically we humans as a species change very slowly. Dubos noted, for example, that there is no reason to think that our eardrums are any more able to withstand vibration than those of cave dwellers 200 generations ago. Yet noise levels have increased dramatically.

Few nations currently face more environmental degradation due to industrial development than China (Chen et al., 2012; Economy, 2004; World Resources Institute, 1994). China has changed more rapidly than any society in history, with an average annual growth rate in excess of 9% during the past two decades. Most of that country's rivers, especially in urban areas, are seriously polluted. Dust and chemicals pollute the air of many cities and cause widespread health problems.

Chronic obstructive pulmonary disease (COPD)—which is linked to exposure to particulate matter, sulfur dioxide (SO_2), and cigarette smoke, among other factors—is responsible for 5% of annual deaths worldwide. Approximately 90% of deaths from COPD occur in low- and middle-income countries. China has a COPD morbidity rate of 8.2% and a mortality rate of 1.6% (Zhong et al., 2007; Fang, Wang, & Bai, 2011). Indoor exposure to emissions from poor-quality coal used for cooking and heating is also a major health risk, increasing the incidence of both cardiopulmonary disease and stroke.

This chapter summarizes some of the main environmental health exposures including toxic hazards, indoor and outdoor air pollution, and water quality, followed by discussions of risk assessment and environmental justice, and ending with a discussion of healthy environments. Chapter 12 focuses on the related area of climate and health.

TOXIC HAZARDS

Toxic hazards to health and life are nothing new. In fact, many occur naturally in the environment. Poisonous and carcinogenic gases are emitted by swamps and volcanoes. Toxic chemicals are produced by parasitic fungi on rye, wheat, peanuts, and other crops, especially in wet years. One such fungal toxin that causes **ergotism**, for example, has been blamed by scientists for everything from the medieval Saint Vitus's dance to the Salem witch trials. Food plants, such as soybeans, cabbage, and wheat, produce an array of chemicals designed to protect themselves from fungi and arthropods; these chemicals can damage the liver, destroy red blood cells, block the absorption of protein or iodine, cause allergies, and generally poison livestock and humans. Nature produces many chlorine-containing chemicals (Abelson, 1994). The smoke of burning wood contains more than 100 organochlorine compounds, including the extremely toxic dioxin. Forest fires, domestic wood burning, and slash-and-burn agriculture all create large amounts of these polychlorinated chemicals. Adding to these and other naturally occurring pollutants, however, humans are exponentially increasing pollutants of their own creation. Table 11.1 lists a few pollutants and health effects of current concern. Of greatest concern, perhaps, are the increasing numbers of chlorinated hydrocarbons and organophosphates, many of which are known to be carcinogenic or teratogenic (causing malformation of the fetus).

Many substances that are dangerous as pollutants are important economically. Polychlorinated biphenyls (PCBs) are nonflammable, have a high plasticizing ability, and have a high dielectric constant. They are therefore widely used in transformers and capacitors; as heat transfer and hydraulic fluids; as plasticizers in adhesives and sealants; and as anticorrosion coats for electric wires, lumber, and concrete. As PCBs leak, leach, and vaporize, however, they become air and water pollutants of great concern (and occasional accidental polluters of food and feed). They are stored in fatty tissue, pass along the food chain, and persist for long periods in the

TABLE 11.1. Some Environmental Pollutants and Health Effects

Pollutant	Affected organs	Health effects	Source
Mercury	Brain; bowels; transplacental transmission	Minamata disease; liver and kidney disease; diarrhea; lack of coordination; numbness; convulsions; death	Chloralkali plants; pulp and paper processing; electrical industries; fungicides in water and food supply
Cadmium	Blood vessels; kidney	Hypertension; bone softening and fractures; kidney disease; cadmium emphysema	Mining many metals; electroplating; stabilizer for polyvinyl chloride; batteries, cigarettes, pigment
Chlorinated hydrocarbons	Fat tissue; liver	Hydrocarbon toxicity	Processing, storage, and transfer of petroleum products; organic solvents; rubber, plastics, paints, lacquer; dry cleaning
Organophosphates	Nerve–muscle synapses	Dizziness; headache; muscular weakness; incoordination; liver and cardiovascular diseases; convulsions; bone-marrow disease; blocks breakdown of acetycholine, which transmits nerve impulses; accumulation leading to convulsions, blurred vision; diarrhea; stillbirth	Insecticide (DDT, lindane, dieldrin, aldrin, chlordane, toxaphene); hexachlorophene (shampoos, deodorants, insecticides, herbicides)
Polychlorinated biphenyls (PCBs)	Fat tissue; liver	Inhibit growth of cells and interfere with enzymes; enhance action of organophosphates; yusho disease; growth disturbance; fatigue; nausea; jaundice; diarrhea; cough; asthma; acne; loss of hair; numbness; nervous system disturbance; joint deformity at birth	Used as heat-transfer media in transformers and capacitors; solvents in adhesives, sealants and anticorrosives; paints and rubber; ink; brake linings
Dusts of quartz, silica, carbon, asbestos, cobalt, and iron oxides	Respiratory interstitial tissue	Pneumonconiosis (scarring of lungs); silicosis; black lung	Mining; sandblasting; quarrying, pottery and ceramics; stone masonry
Hydrogen sulfide	Respiratory center in brain	Paralysis of respiration; consequent edema, hemorrhage, death decay in sewers and mines	Oil wells and refineries; sulfur and protein
Fluoride	Bones; teeth	Binds to magnesium, manganese, and other metals to interfere with endocrine function and enzymes; damage to calcium metabolism and pituitary water balance; dental and skeletal fluorosis and osteomalacia if calcium intake inadequate	Food and water; aluminum and other smelting
Asbestos	Pleura and peritoneum	Mesothelioma, rapidly fatal once symptomatic	Mining; brake linings; fireproofing; talcum powder; cement; ceiling tiles; clothing
Beryllium	Lungs	Sarcoidosis	Metal alloys for heat stress and coal burning

Note. Compiled from National Center for Environmental Health (1996).

environment. In addition to disinfecting water, chlorine and its compounds are used in the manufacture of pharmaceuticals and in their content. The organophosphates have become so useful that the world's food supply relies heavily on them.

OUTDOOR AIR POLLUTION

According to the WHO, approximately 7 million premature deaths annually can be attributed to **air pollution**, which is one in eight deaths worldwide. "Air pollution" may be loosely defined as the contamination of the atmosphere by a harmful level of toxic substances. The key words are "harmful level." The level of risk to any given individual is a product of both a substance's toxicity and the level of the individual's exposure. Virtually all substances are potentially harmful if present in sufficiently large quantities. The exact effects of air pollution on health are a function of the dose delivered to the receptor and the ability of the receptor to cope with the resultant stress. In humans, the stress experienced by a critical organ or receptor tissue from particle inhalation depends on the properties of the particles. The delivered dose is a function of the anatomical features of the receptor, as well as the manner of breathing, breathing rate, and the integrity of the body's defense systems. Some pollution-related diseases may take years or even decades to develop. It has proved difficult to connect specific types of air pollution with health hazards such as bronchitis or lung cancer, in part because of the difficulties of valid sampling and appropriate scale. Researchers have often had to resort to classifications of urban or rural, for example, as surrogate measures for air pollution. These classifications, unfortunately, bring with them many confounding factors.

Atmospheric pollution is caused by a variety of substances, some gaseous and some solid. Figure 11.1 summarizes the many different sources and types of pollutants and the chains of pollution, described in more detail below. We focus on a few pollutants for which the U.S. EPA has established national air quality standards: carbon monoxide (CO), lead, nitrogen dioxide (NO_2), ozone (O_3), particulate matter whose aerodynamic size is less than or equal to 10 micrometers (PM-10), and SO_2. The EPA established limits for atmospheric concentrations of these pollutants in 1970, and these limits have been adjusted over time. Table 11.2 lists these pollutants and their quality standards as of 2015. Forty-seven percent (147.6 million) of people living in the United States live in areas with unhealthy levels of air pollution in the form of high levels of ozone or particle population (American Lung Association, 2014). According to the EPA, certain metropolitan regions in the United States have ambient levels of these criteria pollutants that exceed the established limits and thus may create health hazards where they are found (Figure 11.2). The EPA calculates the air quality index, which is a metric that ranges from 0 to 500. Anything over 100 is considered unhealthy for sensitive groups, and anything over 150 is considered unhealthy for anyone. The EPA maintains an online interactive mapping tool at *airnow.gov* that shows the air quality throughout the United States on a daily basis.

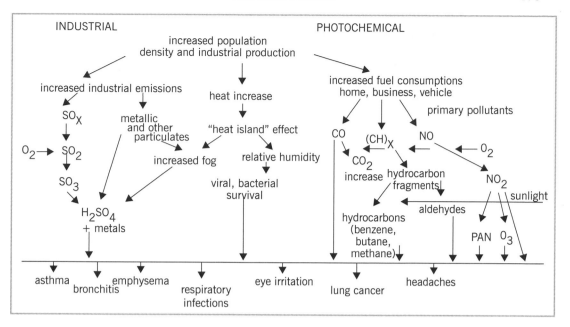

FIGURE 11.1 Chains of air pollution.

The major source of carbon monoxide (CO) is automobile exhaust. Although it can be a regional phenomenon, CO pollution is often localized and occurs predominantly during cold weather. High concentrations may occur near areas of severe traffic congestion. In Tokyo police directing automobiles at street intersections take frequent breaks, during which they inhale pure oxygen to clear their lungs of pollutants. In the developing world the police aren't so lucky: they seldom get breaks and usually work in circumstances where vehicle emission standards are almost nonexistent. CO is the one pollutant that produces a change in human physiology that can be directly related to the concentration to which the subject was exposed. Death occurs in humans exposed to concentrations of about 1,000 parts per million, corresponding to blood levels of 60% blood carboxyhemoglobin. Because CO blocks the transport of oxygen in the bloodstream, and those with certain heart diseases require a high oxygen supply, it is not unreasonable to draw the same parallel as has been made with cigarette smoking and heart disease and to suggest that CO is a contributing factor (Koskela, Mutanen, Sorsa, & Klockars, 2000; Stern, Halperin, Hornung, Ringenberg, & McCammon, 1988; Strauss & Mainwaring, 1984).

One of the most prevalent forms of air pollution is commonly known as **smog** (smoke–fog), which has recently become a pervasive problem in the large cities of China and India (Shao, Tang, Zhang, & Li, 2006; Chan & Yao, 2008; Gurjar et al., 2010; Guttikunda & Calori, 2013). In discussing the policy and health implications of air pollution, one needs to distinguish between two quite different forms with different locational characteristics. The most serious air pollution episodes involve the classical or London smog closely associated with England and the Appalachian

Pollutant	Major sources	Primary (health-related) standards	Health effects
Carbon monoxide (CO)	Two-thirds of emissions from transportation sources	8 hour: 9 ppm (10 mg/m³); 1 hour: 35 ppm (40 mg/m³)	Reduces oxygen delivery to body's organs and tissues; major threat to sufferers of cardiovascular disease
Lead	Lead gasoline additives, nonferrous smelters, and battery plants	Rolling monthly average: 0.15 mg/m³	May cause neurological impairment; fetuses and children may suffer from central nervous system damage
Nitrogen dioxide (NO₂)	Two main sources: transportation and stationary fuel combustion from electric utilities and industrial boilers	1 hour: 100 ppb; annual: 53 ppm	Lowers resistance to respiratory infections, especially in children
Ozone (O₃)	Transportation and industry	8 hour: 0.075 ppm	Impairs lung function, especially among those with impaired respiratory systems
Particulates (diameter < 10 micrometers)	Industry, power plants, transportation, construction, fires, agriculture, mining	24-hour average: 150 mg/m³	Impairs breathing; aggravates existing respiratory and cardiovascular diseases; may damage lung tissue
Particulates (diameter < 2.5 micrometers)		Annual arithmetic mean: 12 μg/m³; 24-hour average: 35 mg/m³	
Sulfur dioxide (SO₂)	Coal and oil combustion; steel mills, refineries, pulp and paper mills; largest source: coal-burning electric power plants	1 hour: 75 ppb	Impairs breathing; aggravates respiratory and cardiovascular disease

TABLE 11.2. U.S. Environmental Protection Agency Criteria for Air Pollutants and Health Effects

Note. ppm, parts per million; ppb, parts per billion; mg/m³, milligrams per cubic meter; μg/m³, micrograms per cubic meter. Data from EPA (2008, updated 2015) (*www.epa.gov/criteria-air-pollutants/naaqs-table*).

region of the United States in the 19th and early 20th centuries. These episodes were characterized by high levels of SO_2 and smoke particulates that built up under stagnant weather conditions lasting 3 days or more. In many industrial processes, sulfur is released during the burning of large quantities of coal and other fossil fuels. Combining with oxygen and eventually with water vapor, sulfur can become dilute sulfuric acid and damage buildings and cloth as well as human lung tissue.

Photochemical smog was first recognized as a problem in Los Angeles during World War II and has since been observed in and around many cities of the world (Stephens, 1987). In those early times in southern California, there was widespread awareness of an eye-burning haze, cracks in the sidewalls of rubber tires, and "bronzing" of sensitive vegetation. At first, this phenomenon was blamed on industry, then on the outdoor burning of trash. Only after exhaustive research did scientists become aware that the major source of these gas emissions was motor vehicles. Tropospheric (ground-level) O_3 (ozone) is the main constituent of Los Angeles smog. It is actually a secondary pollutant formed when hydrocarbons, volatile organic compounds, and oxides of nitrogen (NO_x), known as primary pollutants, react in the presence of sunlight. Fuel combustion sources such as electric

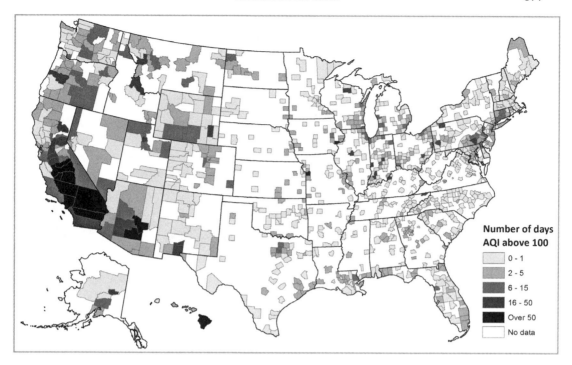

FIGURE 11.2 County air quality index (AQI) map. The legend shows the number of days the AQI was above 100 in 2014 (100 is the threshold for the national air quality standard). From U.S. EPA (*airnow. gov*).

utilities and industrial boilers, as well as transportation vehicles, contribute to atmospheric O_3. Unlike CO, O_3 is worse during warm, sunny weather. Respiratory irritation and breathing problems can occur with chronic exposures to NO_2, particularly in people with asthma and bronchitis, at exposures as low as 100 parts per billion. Children also appear to be susceptible to bronchitis and breathing difficulties in the presence of low levels of NO_x.

Both forms of smog can be common health hazards in places characterized by frequent atmospheric inversions. Normally the temperature decreases with altitude up to 40,000 feet (12 kilometers). The regular rate of decrease is 3.5°F per 1,000 feet (6.38°C per kilometer) and is known as the "adiabatic lapse rate." Because warm air rises, the polluted and heated air from the surface rises and is dispersed through atmospheric mixing. There are several ways, however, in which the air at lower altitude can become cooler than the air above it. Most frequently this can occur near a large body of water such as a large lake or the ocean. When the sun goes down in the evening and the land cools off, a light breeze can direct cool air off the lake or ocean onshore. This forms a cold layer under the warm one, which rises to cover the cold layer. This is called an inversion layer and will trap the body of air below it (Figure 11.3). In valleys between mountains with steep sides, inversions occur as the morning or afternoon sun warms the upper layers of the air, while at the bottom the air is sheltered and remains cold or is cooled by a river

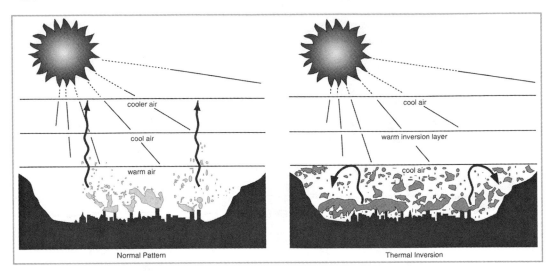

FIGURE 11.3 The effect of temperature inversion. The normal lapse rate may be reversed either near the ground or at high elevations.

flowing through the valley. An inversion at higher elevation can occur when there is atmospheric subsidence and divergence. Such upper-air inversions can occur under stagnating anticyclones, and they tend to occur in certain latitudinal zones where subsidence is common at least seasonally. Los Angeles became notorious for its smog not only because it has so many automobiles, but because its location is characterized by onshore breezes; downhill air drainage; radiational cooling on clear, cloudless nights; lots of sunshine; and summertime atmospheric subsidence. Of all these factors, only the automobile emissions are subject to control.

Frequencies of inversions and the associated health hazards of smog vary greatly not only from place to place but seasonally. Industrial emissions and automobile emissions vary over the course of a week, as well as diurnally. Some pollutants persist for days, and others for seconds or less. The various cycling times of years, seasons, weeks, days, hours, and minutes form a sampling nightmare for the researcher. Some particulate matter, such as lead, has a steep distance decay that may move from dangerous to inconsequential within 200 yards of an intersection; some gases, such as SO_2, may persist for thousands of miles. The several ways of modeling point sources of pollutants usually focus on mathematical models of plumes including using numerical air pollution models to simulate air pollution levels. Trigonometry can help predict where the pollutants will settle by calculating from the height of the ejection point (smokestack), wind speed, thermal stability, topography, and so forth. As numerous point sources merge into the areal base of a large metropolitan area, the dynamics of atmospheric mixing and movement become more relevant. The rate of emission is a major concern, of course, but so are wind speed and mixing height.

According to studies of U.S. metropolitan areas by Dockery et al. (1993), the finest particles (smaller than 2.5 micrometers, known as PM2.5) of dust, soot, smoke, or tiny droplets of acid are believed to cost tens of thousands of lives each year in the United States. The Dockery and colleagues research found statistically significant and robust associations between fine-particulate pollutants, on the one hand, and lung cancer and cardiopulmonary disease, on the other. Fine particles may be more dangerous than large ones because people can breathe them deeply into the lungs, where tissue damage can lead to breathing difficulties or worse. This research was important because the research design controlled for cigarette smoking and other risk factors. Figure 11.2 demonstrates that in general, the worst fine-particle pollution is located in southern California and a belt stretching from the East Coast to the Great Lakes.

SO_2 is emitted largely from stationary coal and oil combustion sources, such as steel mills, refineries, pulp and paper mills, and coal-burning electric power plants. Human exposure to concentrations of SO_2 exceeding the national standard can affect breathing and aggravate existing respiratory and cardiovascular disease. Sulfur is especially hazardous to individuals with asthma, bronchitis, or emphysema, as well as to children and the elderly. Smelting of ores and other industrial processes may emit small particles of heavy metals such as lead, iron, manganese, and titanium, but most particulates are commonly carbon-based (i.e., soot). Mortality and respiratory illness increase significantly during heavy smog episodes. In London during a week in early December 1952, there were 3,500–4,000 more deaths than the expected average for that time of the year. The industrial valleys of Pennsylvania suffered a similar fate in 1948. Such dramatic events tend to raise public and official consciousness about air pollution. It has been estimated that air pollution in China kills 400,000 of its citizens each year (Watts, 2005). China has some of the highest levels of small particulate matter (PM 10 & PM 2.5) in the world and has been shown to be responsible for large increases in mortality due to respiratory and cardiovascular diseases (Chen et al., 2012). The growth of urban areas in China (and India) has resulted in millions of people exposed to hazardous pollutants from both industry and traffic.

There is another family of hazardous air pollutants, commonly referred to as air toxics; these have been shown to cause cancer, poisoning, and immediate illness in humans (U.S. EPA, 1996). Other less measurable effects include respiratory, immunological, neurological, reproductive, and developmental ones. Examples of air toxics include (but are not restricted to) dioxins, benzene, arsenic, beryllium, mercury, and vinyl chloride. In fact, U.S. clean air legislation lists 189 pollutants as hazardous air pollutants (HAPs) and targets them for regulation. Air toxics emanate from both stationary and mobile sources. Toxins may be inhaled directly or deposited onto soil or into water bodies, thereby indirectly affecting human health through ecological chains. Top consumers in the food web, usually consumers of fish, may accumulate chemical concentrations many millions of times greater than the concentrations present in the environment. As a result, fish consumption advisories have been issued for hundreds of water bodies in the United States, including the Great Lakes. The distribution of HAP emissions by state appears to correlate positively with the distribution of hazardous waste sites (U.S. Census Bureau, 1996, p. 238). The top 17 ranked states in terms of waste sites include 12 of the states in the highest HAP emissions category (more than 90,000 tons per year). The good news is that air pollutant emissions have on average decreased throughout the United States since 1970, despite growth in population, energy use, and vehicle activity (National Research Council [NRC], 2004, 2006). The bad news is that, as previously mentioned, NICs including China and India have shown the opposite trend (Guttikunda & Calori, 2013).

The distribution within the United States of the release of toxic substances, including air toxics, is not uniform. Figure 11.4a is a map of **Toxic Release Inventory** (TRI) sites. The TRI is a publicly available database that contains information on toxic chemical releases reported annually. Figure 11.4b is a map of National Priorities List (NPL) sites, which are what the EPA considers the most serious uncontrolled or abandoned hazardous waste sites in the United States. As you can see from these maps, the entire eastern half of the United States is filled with TRI sites, and the Northeast, Midwest, and South have the most NPL sites.

Cutter and Solecki (1989) and Cutter and Tiefenbacher (1991) examined the spatial distribution of air-borne releases of acutely toxic materials in the metropolitan United States throughout the 1980s, and demonstrated through cartographic and statistical analyses that frequency of chemical firms and rail miles per state were good predictors of the number of incidents. Their study dispelled the notion that air-borne releases of acutely toxic chemicals occur only at fixed sites or from spectacular rail accidents. The highest likelihood of toxic chemical incidents was associated with high concentrations of chemical manufacture and distribution facilities. In India, Guttikunda and Calori (2013) found different sources for the various types of air pollutants, with transportation the largest contributor for volatile organic compounds, and NO_X, SO_2, CO, and for particulate matter they found multiple sources including industries such as brick kilns.

According to the EPA, the aggregate emissions of six common pollutants dropped 69% between the enactment of the Clean Air Act in 1970 and 2014. With

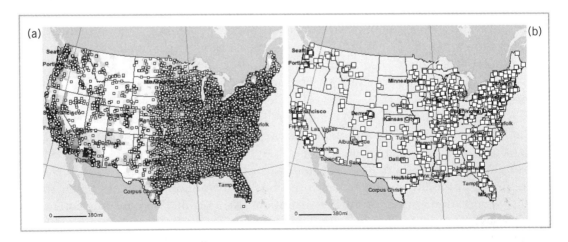

FIGURE 11.4 (a) Toxic Release Inventory (TRI). (b) National Priorities List (NPL) sites. From U.S. National Library of Medicine (*toxmap.nlm.nih.gov/toxmap/superfund/mapControls.do*).

the reduction of air pollution there has been a reduction in exposure to ambient fine-particulate matter. Reduction in that exposure was found to be associated with improvements in life expectancy in the United States. Pope, Ezzati, and Dockery (2009) found that from 1980 to 2000 life expectancy increased in part because of decreased exposure to fine particulate matter at a concentration of 10 µg per cubic meter. In developing countries, including China and India, air pollution regulations are far behind those of the United States, but if they were enacted and enforced they would likely have a large impact on increasing life expectancy.

INDOOR AIR POLLUTION

Recent research has added a new dimension to the air pollution picture (Smith, 2012). A startling finding is that exposures to toxic chemicals are sometimes 70 times higher indoors than outdoors. This is partly because indoor air carries the double burden of outdoor pollutants brought inside and pollutants that are generated indoors. The CDC (1994) has classified indoor pollution as a factor of high environmental risk. A large burden of total indoor air pollutants, at least in the developed world, is probably attributable to environmental tobacco smoke in both homes and workplaces. The U.S. Department of Health and Human Services (2014) estimates that since 1964, approximately 2,500,000 nonsmokers have died from health problems caused by exposure to secondhand smoke. As public awareness of the hazards of exposure to environmental tobacco smoke increases, businesses and communities throughout the United States are taking actions to prevent involuntary exposure. These actions range from prohibiting smoking indoors to limiting smoking to specially designated, separately ventilated smoking rooms.

Indoor air pollution in the developing world is largely caused by cooking and heating (Smith & Mehta, 2003). For example, nearly all of China's rural residents use solid fuels (biomass and coal) for household cooking and heating, and such fuel use is responsible for approximately 420,000 premature deaths annually (Zhang & Smith, 2007). A meta-analysis showed that the risk of pneumonia in young children is increased by exposure to unprocessed solid fuels by a factor of 1.8 (Dherani et al., 2008). In places such as Guatemala where open fires are used for cooking, one solution is the introduction of chimney woodstoves, which have been shown to reduce CO exposure and wheezing (Smith-Sivertsen et al., 2009).

A landmark EPA study (see Shabecoff, 1985, for a report) found it probable that such consumer products as paints, cleansers, propellants, plastics, and cosmetics, as well as adhesives, fixers, resins, and building insulation materials, were major sources of indoor pollution. This study, which measured the chemicals in the participants' bodies on a daily basis, found significant correlations between blood levels of these chemicals and visits to gas stations and dry cleaners, use of paint or solvents at home, and smoking. Important studies of air pollution in domestic space are sure to follow, but the habitat dimension of pollution hazards has been interlocked again with the behavioral dimensions of house type, clothing, hobbies, and customs of personal and domestic hygiene (see Chapter 2).

Modifications to the home environment in industrial countries have been linked with an increase in the prevalence of asthma (Jones, 1998). Home weatherization programs, which aim to "tighten" houses in order to conserve energy, also reduce air exchange between indoor and outdoor environments, in effect trapping pollutants inside the house. Modern homes are much better insulated than was previously the case. Many more homes now have central heating and sealed-unit double glazing. Fitted carpets have generally replaced loose rugs, and advances in construction technology have led to a greater use of synthetic building materials. While these housing modifications prevent heat loss in the winter and make air conditioning more efficient in the summer, they may also create toxic indoor environments. Air-borne contaminants are readily produced and build up to higher concentrations than are encountered outdoors. Various indoor exposures have been related to asthma, including house dust mites, molds and fungus, NO_2 from gas appliances, volatile organic compounds, and formaldehyde.

According to the CDC, in 2012 18.7 million American adults (8% of all adults) and 6.8 million children (9.3% of all children) had asthma. During the past three decades there has been a large increase in the number of people with asthma, predominantly in children under 5. In 2012, there were almost 14.2 million physician office visits due to asthma and 1.8 million emergency department visits. Although the asthma epidemic is clearly related to indoor air pollution, it is simultaneously related to the air quality of the community in which a person lives and works. For instance, Sahsuvaroglu and colleagues (2009) found that traffic-related pollutants, such as NO_2, were associated with childhood asthma in Hamilton, Ontario. In New York, Maantay (2007) found that people living near noxious land uses were 66% more likely to be hospitalized for asthma. This latter finding is also an issue of **environmental justice** (EJ), since the people in New York who were hospitalized with asthma were 30% more likely to be poor and 13% more likely to be minority group members. EJ is discussed later in this chapter.

Researchers have also been paying increased attention to the radioactive gas known as **radon**. Radon is an odorless, colorless gas that arises from the natural decay of radioactive elements commonly found in rocks and many types of soil. In tightly sealed masonry structures, radon gas can build up to hazardous levels. According to EPA surveys, the average radon level in U.S. homes is 1.25 picocuries per liter of air. Almost all scientists agree that prolonged exposure to high levels (more than about 8 picocuries) of radon poses a genuine cancer hazard. The U.S. National Academy of Sciences estimates that radon is the cause of thousands of lung cancer deaths nationwide each year (Committee on Health Risks of Exposure to Radon et al., 1999). A more recent meta-analysis confirmed that radon in homes is carcinogenic (Samet, 2006). Most of these deaths occur among people who smoke cigarettes. Other studies, however, show little evidence linking household exposure to cancer. In Winnipeg, for instance, researchers found that mean household exposures of 3.5 picocuries appeared to produce no increased risk for lung cancer. This raises questions about how much risk radon poses at low levels (Leary, 1994). In the United States, known radon concentrations are highest in the Western mountain states, the Midwest, and Appalachia. The EPA provides county-level radon averages in an online mapping and data visualization tool: *www.epa.gov/radon*. In areas with radon problems, radon testing is mandatory when houses are sold and if high levels are detected remediation by the installation of ventilation systems, usually in basements is required. Radon is a problem not only in the United States but throughout the world including China, England, France, South Africa, and Brazil. While the problem has long been known about, there are no doubt many more places where it is a serious problem, places that have not been studied and where no remediation or prevention is occurring that we know about.

WATER POLLUTION

Several episodes have been responsible for attracting research attention and public interest to the hazards of water pollution. One of the first occurred in Minamata, Japan, during the 1950s, when more than 200 people died or suffered severe brain and nerve damage. The cause was methyl mercury polluting the water and being passed through the food chain in fish to the population.

In the 1970s the Love Canal incident, near Niagara Falls, New York, focused national attention on the problems of hazardous waste disposal and the dangers of groundwater contamination. Hooker Chemicals and Plastics Corporation (now Occidental Chemical) purchased an uncompleted canal site, and between 1942 and 1953 disposed of about 22,000 tons of mixed chemical wastes into the ditch. Shortly after Hooker ceased use of the site, the land was sold to the Niagara Falls School Board for a price of $1. In the late 1950s, residential development of the area attracted 900 families, and an elementary school was constructed on the property. Unusually heavy rain and snowfalls in 1975 and 1976 provided high groundwater levels in the Love Canal area. Portions of the Hooker landfill subsided and chemical residue surfaced, contaminating ponds and other surface waters. Basements began to ooze an oily residue; noxious chemical odors permeated the area; and chemical corrosion of subterranean pumps became apparent. In 1978 the New York State Department of Health declared the Love Canal area a threat to human health and ordered the fencing of the landfill site. Later the school was closed, and pregnant women and young children were evacuated from the site. Eventually, all the properties in Love Canal were purchased, and a Federal Disaster Assistance Agency was created to assist the city of Niagara Falls to remediate the Love Canal site. Since then, gasoline leaks from underground tanks, PCB dumps into rivers and drains, and leakage from municipal landfills have poisoned the water supplies of scores of communities nationwide and forced Congress to create a Superfund program for toxic waste cleanup (Zeigler, Johnson, & Brunn, 1983).

Ninety-five percent of all fresh water available on Earth (exclusive of ice caps) comes from beneath the surface. Although most people in the world get their domestic water supply from surface water or shallow wells, about half the population in the United States and large cities elsewhere tap groundwater sources. The purity of groundwater has long been held as sacrosanct; however, untreated contaminated water can carry many infectious agents. The U.S. EPA (1994) reports the leading sources of water pollution and types of water pollutants in the nation's rivers, lakes, ponds, reservoirs, estuaries, and the Great Lakes. Agriculture is the most widespread source of pollution in the surveyed rivers. Bacteria from fecal contamination pollutes 34% of impaired river miles; excess nutrients, especially nitrogen and phosphorus compounds from agriculture, pollute 43% of lake acreage; and urban runoff and storm sewers are the most widespread source of pollution (nutrients and bacteria) in estuaries. The Great Lakes contain one-fifth of the world's fresh surface water and are polluted from both the air and surface runoff. In addition to nutrients from agriculture, toxic organic chemicals (primarily PCBs) have been found in 98% of the samples taken. Agricultural chemicals have led to feminization in cane toads in Florida because of high hormone levels (McCoy et al., 2008). Air pollution from industry and pesticides from agriculture have been implicated. The U.S. National Primary Drinking Water Regulations limit contaminants to public water systems including microorganisms, disinfectants and their by-products, organic and inorganic chemicals, and radionuclides. The sources for the contaminants vary and include human and animal fecal waste, discharge from industry, additives to water for disinfection, agricultural runoff, mining, and some are naturally occurring.

There are very few microorganisms in U.S. public drinking water supplies, but in the developing world access to clean water is a major problem. Given the number of illnesses and deaths attributed globally to diarrhea, it seems fair to say that human feces remain among the world's most hazardous water pollutants (Fewtrell et al., 2005). Poor sanitation and a lack of access to clean water still constitute two of the world's most serious health problems. In the developed world, these problems have been mitigated by providing indoor piped water and flush toilets to virtually all urban residents. The same is true for the wealthy in developing countries. For most developing nations, however, financially strapped governments can ill afford anything like complete coverage with indoor plumbing.

The United Nations **Millennium Development Goals** (MDG) were developed to try to tackle some of the greatest challenges for the poorest people in the world during a 15-year period starting in 1990. MDG 7c was to halve, by 2015, the proportion of the population without sustainable access to safe drinking water and basic sanitation. The target of halving the proportion of people without access to improved sources of water was met 5 years ahead of schedule. Between 1990 and 2012, 2.3 billion people gained access to improved drinking water sources, but while great progress was made 748 million people remained without access to an improved source of drinking water. Not as much progress was made during this time period in bringing improved sanitation to the poorest people in the world. Over a quarter of the world's population has gained access to improved sanitation since 1990, yet 2.5 billion in developing countries still lack access to improved sanitation facilities. Table 11.3 lists 2012 data for several developing nations that lagged behind in terms of access to clean water and sanitation in the early 1990s. The table also includes the United States as a comparison. As you review the data in the table, note that there are large disparities between countries and in some countries there are large disparities between urban and rural areas. In almost all countries access

TABLE 11.3. Access to Safe Drinking Water and Improved Sanitation, Selected Low-Income Nations, 2012

Country	Access to safe drinking water (%)		Access to improved sanitation (%)	
	Urban	Rural	Urban	Rural
Bangladesh	86	84	55	58
Burkina Faso	97	76	50	7
Central African Republic	90	54	44	7
China	98	85	74	56
Ethiopia	97	42	27	23
Guinea-Bissau	96	56	34	8
India	97	91	60	25
Indonesia	93	76	71	46
Mali	91	54	35	15
Mozambique	80	35	44	11
Myanmar	95	81	84	74
United States	100	100	100	100

Note. Data from Joint Monitoring Programme for Water Supply and Sanitation (*www.wssinfo.org/en/welcome.html*).

to improved water is greater than access to improved sanitation. The Joint Monitoring Programme for Water Supply and Sanitation (2006), which is maintained by both UNICEF and WHO, defines minimum standards of water and sanitation as follows. Safe drinking water includes a household connection to a water supply, a public standpipe, a borehole, a protected dug well, a protected spring, or rainwater collection. Unsafe drinking water includes surface water (river, dam, lake, pond, stream, canal, irrigation channel), an unprotected well, an unprotected spring, vendor-provided water, bottled water, or tanker truck water. Improved sanitation includes connection to a public sewer, connection to a septic system, a pour-flush latrine, a simple pit latrine, or a ventilated improved pit latrine. Unimproved sanitation includes a public or shared latrine, an open pit latrine, or a bucket latrine.

In cities of the developing world, households without indoor piping often obtain their water from one of many sources, such as overcrowded or distant communal standpipes; expensive private water vendors; or heavily polluted, shallow, hand-dug wells. Increasingly, developing countries are adopting borehole technology, where deep tube wells with engine-powered pumps are becoming important sources of uncontaminated domestic water (Emch, 1999; Fewtrell et al., 2005). In countries where vector-borne disease is prevalent, for example, tube wells are providing parasite-free village water sources for bathing and laundry. Those without flush toilets may end up using pit latrines or latrines located over ponds, streams, drains, or open sewers—all of which demand far more rigorous hygiene behavior than is required for the standard technologies of the wealthy.

Figure 11.5 illustrates the strata of an artesian water source. Surface water percolates into the soil, eventually reaching the water table, or a saturated zone of ground above an impervious stratum of rock. This groundwater may also intersect depressions and form springs or help fill lakes. Below the stratum of impervious rock there is often another layer of pervious rocks, such as sandstone, and a layer of impervious rock below that. Most rock strata are tilted. Where the pervious layer has contact with the surface, precipitation percolates into it to form an aquifer. The source area of this precipitation may be many hundreds of miles from a desired well site, so aquifer water may be under pressure if the rock strata are even slightly tilted. An artesian well drilled through the rock cap into the aquifer may, if the location is right, flow under its own pressure. Although aquifers are filled by precipitation, in some places the water that originally filled the aquifer fell in other climatic periods and other continental locations.

There may be several layers of aquifers. Often the lowest one is saline. Injection wells are sometimes used to dispose of toxic wastes in saline aquifers. Under great pressure, and through wells constructed in layers and sealed to prevent leaching into higher strata, liquid wastes are injected into the saline aquifer. Overall, U.S. groundwater quality is considered "good," but some regions have experienced "significant" groundwater contamination. The most frequent sources of this contamination are leaking underground storage tanks, agriculture (pesticides and herbicides), Superfund sites, and domestic septic tanks (U.S. EPA, 1994). Water pollution in the large urban areas of China mirrors the story of air pollution in that country described above (Shao et al., 2006). Most of the rivers

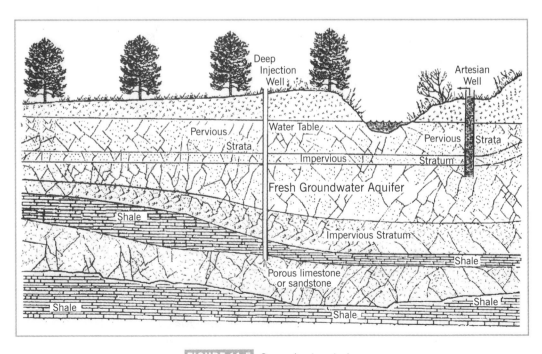

FIGURE 11.5 Groundwater strata.

and lakes in China's urban areas have seen water quality dramatically decline with economic growth. Sources of contaminants include industry and the inability to deal with dramatic increases in sewage treatment as cities have grown. Recently in several parts of the world, including Bangladesh and Vietnam, groundwater has been found to be contaminated with naturally occurring arsenic, which causes cancers and other health problems (Berg et al., 2001; Smith, Lingas, & Rahman, 2000). While the shallow aquifer has very high arsenic levels in some areas, the deep aquifer over 500 feet deep is almost devoid of arsenic. The installation of deep tube wells in rural Bangladesh has not only been an effective mitigation approach for avoiding high levels of groundwater arsenic, but has been found to have the added effect of reducing diarrheal diseases (Winston et al., 2013; Escamilla et al., 2011).

One of the most comprehensive studies of water pollution and human health effects was carried out in Puerto Rico by Hunter and Arboña (1995). Because of its historical linkages to the West and subsequent industrialization, the Commonwealth of Puerto Rico has pushed consumerism to levels comparable to those found in the rest of the United States. The resultant outpouring of sewage and solid waste imperils this small, overcrowded tropical island that is dependent on groundwater. The sudden death of large numbers of fish has provided a sensitive biological indicator of environmental quality. The usual suspects include agricultural runoff and discharges from industry and municipal sewers. Here, as elsewhere in the United States, water supplies are becoming increasingly polluted with toxins; yet water treatment techniques still primarily reflect concern with infectious diseases of the past, rather than the carcinogenic pollutants of today's environment. It has been several decades since the passage of the Safe Drinking Water Act, yet the gamut of toxic chemicals in the public water systems is broadening, and threats to human health have increased rather than diminished.

Concern about pollution of groundwater is growing. In the mid-1980s, the California Department of Health Service reported that one-fifth of the state's large drinking water wells fed by groundwater exceeded the state's pollution limits (Sun, 1986). Studies in two heavily populated counties in that state found high levels of trichloroethylene and dibromochloropropane, which pose reproductive and acute health risks, including stomach cancer and lymphatic leukemia (Kloos, 1995). Other studies, however, have found no relationship between contaminants and disease. This is often the case when there is a failure to control for confounding factors or when conclusions about a population are based on samples obtained in small areas. This can lead to ecological fallacies, which have been mentioned elsewhere in this book. Pesticides have been detected in half of Iowa's city wells. More than 1,000 wells in Florida have been shut down because of contamination with EDB, a chemical used to kill soil nematodes. The common beliefs that pesticides would decompose in the soil and that nature would cleanse itself through percolation can no longer be accepted. While the National Primary Drinking Water Regulations help protect the public water supplies used by most people in the United States, they don't pertain to people using well water in rural and periurban areas of the country.

SOURCES AND HEALTH EFFECTS OF LEAD

Although lead is also toxic to adults and affects virtually all organ systems, adverse effects on cognitive development and behavior in children up to the age of 6 are of special concern. The CDC estimated that at least 4 million households have children living in them who are being exposed to high levels of lead exceeding 10 micrograms per deciliter of blood. This level is considered by the CDC as being high enough to trigger primary prevention campaigns in the local community. There are many sources of lead in the environment. Sewage sludge is commonly used as fertilizer in irrigation systems. Domestic sewage has almost everywhere been contaminated by industrial effluents, with the result that heavy metals such as lead and cadmium are present at relatively high levels. With repeated application of sewage, lead can quickly build up in the soil and thus pass into the human food chain.

A main source of lead exposure among lead-poisoned children in urban areas is lead-based paint in homes built before 1978. Although the use of lead-based paint in exterior and interior home surfaces was banned in 1978, the U.S. Department of Housing and Urban Development (2011) estimates that there are 23.2 million homes, which is 21.9% of all homes in the United States, that have lead-based paint hazards and are predominantly low-income households. Children may ingest lead directly from paint chips (lead makes the paint taste sweet), but an important route of exposure is the normal mouthing of hands or objects such as toys, resulting in the ingestion of small amounts of lead-paint-contaminated house dust and soil.

Automobile emissions became another focus of urban lead poisoning in the 1970s. Lead was used as a gasoline additive to improve the fuel's burning characteristics. It is emitted as a fine particulate in exhaust gases and settles in urban soils. There it is taken up by vegetation (including food crops in gardens), or children playing in or near streets, who are likely to ingest it as a result of breathing or handling food with soiled hands. The vertical (from the ground) and horizontal (from the road) gradients of lead levels in the air, as well as isotopic matching of local aerosol lead and local gasoline lead, provide irrefutable evidence of the importance of automobile emissions. The transportation network constitutes the broad geographical pattern of lead pollution. Levels of lead in the dirt of agricultural fields and in the dust of urban apartments decay sharply away from traffic. Since 1973, a reduction in tetraethyl lead in gasoline from over a gram per gallon to 0.1 grams per gallon was mandated in the United States and was banned in 1986. However, lead does remain in soils for long periods of time and is therefore still an environmental health hazard in many areas of the United States (Mielke et al., 1983, 2013).

Hunter's (1977) research demonstrated that childhood lead poisoning was a "summer disease" due to the seasonal exposure to gutter dirt or automobile emissions. Earickson and Billick (1988) described elevated lead levels in children in Louisville and Detroit in the 1970s, demonstrating a statistical association between airborne lead and other pollutants and poor neighborhoods. Bailey, Sargent, Goodman, Freeman, and Brown (1994) investigated the geographic variation in lead poisoning among children living in Massachusetts in 1990. Their statistical analysis corroborated the CDC's (1992) report that poor minority children living in female-headed households in deteriorating housing built before 1950 are at disproportionately high risk for excessive lead exposure. Geographer Howard W. Mielke showed that soil lead from automobile exhaust is a more important source than paint (Mielke et al., 1998) and he recently found spatial and temporal relationships between urban lead levels in air and violent crime in several American cities (Mielke & Zahran, 2012). Some have suggested that the switch to unleaded gasoline is largely responsible for the overall decrease in lead poisoning and, through improvements in cognition, the decrease in crime rates in the United States in the past several decades (Nevin, 2000; Wright et al., 2008).

Lead poisoning has recently been found to be associated with municipal water supplies in Washington, DC, and Flint, Michigan. Water pipes in distribution systems were constructed out of lead and corrosion of those pipes can cause lead to leach into drinking water. EPA standards require cities to monitor lead levels and provide the best corrosion control systems possible. In late 2015, the Flint drinking water crisis brought this problem to national attention. The problem stemmed from a switch in drinking water sources from the less corrosive water source of Lake Huron to the more corrosive water source of the Flint River. In a study that investigated changes in childhood lead levels before and after the water source change, Hanna-Attisha, LaChance, Sadler, and Schnepp (2016) found that incidence of lead increased from 2.4 to 4.9%; using a GIS they found the highest elevated lead levels in the poorest neighborhoods.

RISK ASSESSMENT AND PREVENTION

With more than 3 million chemicals registered, 70,000 chemicals in general use, and new ones appearing at a rate of more than 1,000 a year, it is not surprising that the environment in the industrialized countries and NICs has become contaminated. Exposure to pollutants can be either acute or chronic. A person may be exposed once for a few minutes at work to a leakage or spill, or may consume over a lifetime undetectable levels of contaminant in food or water. Toxic waste from a dump may contaminate a nearby residential water supply to very high levels, or low levels of benzene may be breathed throughout a lifetime in the ambient air of an urban area. Even for most of the chemicals thought to be hazardous, little is known about their threshold levels or dosage effects. Threshold limit values for major chemicals have been announced by government agencies responsible for protecting workers. These values establish time-weighted average concentrations that are acceptable for occupational exposure, in the same way that standards have been set for exposure to radiation, noise, and air-borne dust. Few standards have been estimated for nonoccupational exposures. Animal studies indicating risk or toxicity examine a single pollutant in isolation, but exposure to pollutants occurs in complex groups and a negative outcome may be the result of a mixture of pollutants experienced simultaneously or in succession. As each individual swims in his or her own personal toxic soup of chemicals, and since exposing people to potentially harmful chemicals in short or long duration would be prohibited under research ethics, rigorous and "gold standard" studies of pollutant risks are nearly impossible.

Research is further complicated because reliable, sub-national-scale data do not exist for etiological study. There are, for example, no microarea data for industrial exposures, cigarette or alcohol consumption, or dietary composition. One must work between small local surveys of specific points and broad socioeconomic classifications of national associations and trends. It is not surprising that point and area data are often improperly mixed, that researchers generalize across scale, or that spatial autocorrelation is ignored. The personal monitors, currently being tested, that identified indoor pollution may in the future provide reliable data for researchers interested in studying microscale exposure through daily and weekly activity patterns. For now, most studies rely on surrogate but measurable variables, such as substituting ethnicity for diet.

One of the clearest and most concise explanations of the measurement and uncertainties associated with risk estimation is that by Talcott (1992), who claims that sizable uncertainties are associated with quantitative environmental risk assessments. Risk estimates incorporate the following questions: What is the probability that release of a substance will occur? What quantity of this substance will be released? How will the concentration of this substance change as it disperses from the point of release? How many people or other organisms in the environment will be exposed to this substance? How much of the delivered dose of this substance will be taken up by organisms? And what will be the relationship between a particular dose of this substance and an organism's response; that is, how will the toxic effects of this substance increase with increased dose? From the point of view of environmental justice, this last question has deadlocked many a jury. The toxicity of a particular dose of a substance varies not only across species, but among individuals of the same species. Sex, age, size, diet, and the geography of exposure to a substance, among other confounders, affect how toxic increased doses of a substance are to organisms.

One way the EPA assesses populations at risk is by using the tools of GIS to identify the buffer zones around hazardous sources such as dumps or nuclear power plants. Complicating factors include prevailing wind direction; changes in humidity; local turbulence and mixing; air drainage; frequency of inversion; and differences in soil, rock, and topography. Underground movement of water causes a population situated upflow to be at higher risk from the hazard location than a population situated downflow a little farther away. The transportation and routing system, labor circulation, or seasonal changes in population density or industrial production also affect risk. Studies by political scientists, sociologists, and geographers have shown that often the most critical locational determinant for siting a hazardous facility is the political weakness of the local population (Greenberg & Schneider, 1996).

Geographers can help in the detection of old waste disposal sites, the planning of new sites, and the etiological analysis of diseases in developed economies. Processes of siting, transporting, storing, processing, concentrating, or dispersing occur in a differentiated environment of short-term micrometeorology, local geomorphology, transportation systems, settlement patterns, and economic and social activity. Risk assessment of population exposure to known and unknown hazards is a multiscale and complicated business of intrinsic geographic interest, but there has been limited geographic contribution thus far.

GLOBALIZATION AND THE PERCEPTION OF HEALTH HAZARDS

Pollution flows without regard to international boundaries. Acid rain, the greenhouse effect of increasing levels of carbon dioxide, and the fallout of strontium-90 as a result of atmospheric nuclear testing are problems shared by humans in general, regardless of the sources. There has been a less generally recognized but concomitant diffusion of useful technology that produces hazardous by-products. Among the most sought-after industrial capacities are petroleum refineries, the manufacture of plastics and synthetic materials, electronics, pharmaceuticals, and the production of fertilizers and insecticides. Mining and smelting are being extended into previously undeveloped areas. Agriculture is being intensified through application of more chemicals. Research is progressing into such useful areas as recombinant DNA and the capacity to produce strains of food crops resistant to fungi and insects—without consideration of what effects the "naturally occurring" plant chemicals that produce resistance will have on human health.

Chlorofluorocarbons (CFCs) emitted into the atmosphere, because of their extraordinary chemical stability, reach the stratosphere where they release chlorine by photolysis. This free chlorine scavenges O_3 and destroys it (Lindley, 1988). Release of CFCs occurs through industrial activity, equipment leakage, or the disposal of old refrigeration and air-conditioning units, as well as by use of aerosol cans using CFCs as propellants. Even some members of the health care professions use CFCs as a diluent for ethylene oxide in cold sterilization procedures in hospitals, clinics, and the manufacture of health-related devices. Whereas CFCs are the most voluminous O_3 depleters, there are several other O_3-depleting substances in use, including carbon tetrachloride, hydrochlorofluorocarbons, halons, methyl bromide, and methyl chloroform.

The O_3 layer was observed to be thinning over Antarctica in the late 1970s. Repeated observations have confirmed the attenuation and charted its progress (Stolarsky, 1988). There is some attenuation in the northern hemisphere as well, but at about 60° north, not over the pole. On average, the global stratospheric O_3 layer declined by 2–4% between 1980 and 1988 (Last & Guidotti, 1991). Between 1988 and 1995, global production of CFCs dropped 76%, due mainly to a reduction in use of O_3-depleting substances in industrial nations.

There has been noticeably increased human exposure to ultraviolet (UV) light in recent years. O_3 in the stratosphere absorbs much of the UV light that would otherwise reach the Earth's surface. It has been estimated that a 1% reduction in the O_3 shield would result in a 2% increase in the UV light reaching the surface. The human health effects of increased UV irradiation due to O_3 depletion include higher risks of nonmelanoma skin cancer and malignant melanoma; cataract and retinal degeneration; and possible impaired immunological responses that increase risks of other conditions, including systemic malignancies. Even at high northern latitudes, exposure to sunlight is the principal determining factor for nonmelanoma skin cancer (MacKie & Rycroft, 1988). Oddly, the amplification factor appears to decrease with increasing latitude, which means that residents in the very highest latitudes of North America may be at less increased risk for the carcinogenic effect of increased UV penetration than those at the latitudes where most of the Canadian and some of the U.S. population lives. There have been successful attempts to curb CFC generation and release (French, 1997). On January 1, 1996, production of CFCs was to have ceased in developed countries under the provisions of the Montreal Protocol of the Vienna Framework Convention on Ozone-Degrading Substances. However, given that there was not a freeze of CFC production until 2013, and given the long half-life of CFCs (75 years or more), O_3-depleting activity is expected to persist at significant levels well into the 21st century.

The experience gained by industrial nations as they passed through their technological transition should have been of great preventive value to poor developing nations, but there is little evidence of its application. Modern health hazards are diffusing rapidly. Most developing countries perceive the immediate benefits of technology more acutely than they see the distant, nebulous hazards. Their national budgets are stretched thinly to cover the needs of education, health, and infrastructural development. Civil servants with the education and skills required to inspect and regulate plants that use advanced chemical technology are in limited supply. The general populace is usually uninformed and naive about modern chemicals. Farmers, for example, are at risk of severe poisoning from handling insecticides and herbicides, and they sometimes rinse their equipment in streams that provide water for drinking and washing. If they are illiterate, they cannot read labels. Farmers in Southeast Asia have been known to dump insecticide in streams to kill fish for harvest, as they used to do with native plant toxins. Chemical and drug manufacturers have been known to export to developing nations poisons and other substances that have been banned for use in developed countries. Herbicides similar to Agent Orange, used in the Vietnam War, have been marketed to countries such as Colombia, resulting in reports of miscarriages and birth defects. The attitude in developed countries is often that other countries must regulate themselves, but the expert knowledge and resources are seldom available in the developing world. As is described above, some of the largest environmental problems are now being experienced in the two most populated countries in the world, China and India. They have experienced unprecedented economic growth during the last 30 years and the standard of living for most people in those countries has improved. However, the air and water pollution problems are also unprecedented.

HAZARDS, POWER, POLICY, AND ENVIRONMENTAL JUSTICE

There are many spatial dimensions to environmental health hazards. Atmospheric and water pollution constitute areal hazards; mines, manufacturing sites, refineries, and waste and storage dumps constitute point hazards; and the roads and sea lanes of transportation constitute linear hazards. Each is of concern at a different scale, and each involves different kinds of regulatory and preventive policies and different levels of government. Technology control and alternatives must be carefully considered in terms of social justice and compensation, private investment, and tax costs. Geographically, the costs and benefits may or may not coincide in one place. In the case of automobile accidents or pharmaceutical side effects, for example, the zone of maximum benefits from the technology and the zone of maximum risk of its hazards coincide.

Zones at risk from atmospheric or water pollution usually contain, but extend well beyond, the place that economically benefits from production. In contrast, the hazards of agricultural technology are usually felt most by farm workers, but the benefits extend to a much wider area. Toxic wastes, radioactive wastes of all kinds, and their transportation form a class by themselves because the maximum risk is usually borne by a place far removed from the zone of maximum benefit. For example, malfunctions at the Chernobyl nuclear power plant endangered the local populations as far north as Scandinavia, and most of the chemical industry that fills South Carolina's major toxic waste dump is located in and benefits other states.

Whether the noxious facility at issue is a permanent, national, high-level radioactive waste dump or a mental hospital, such facilities must be located somewhere (Elliott & Taylor, 1996). One research question is this: How do people perceive the risk of noxious facilities and react to them? A team of geographers and others at McMaster University in Canada (Elliott et al., 1993; Eyles, Taylor, Johnson, & Baxter, 1993) discovered that Canadians residing in southern Ontario took for granted proximate solid waste disposal facilities ("just a part of the landscape") because these sites had existed for an extended time without incident. It is not that people were unconcerned about risks to safety and property value. Indeed, across the three sites studied by this research team, levels of reported concern ranged from a low of 28% (existing solid waste incinerator) to 67% (existing solid waste landfill) to 74% (proposed solid waste landfill). However, those interviewed perceived that risks were beyond their control: "uncertain, perhaps invisible, sometimes stigmatizing impacts of social and environmental change" (Eyles et al., 1993, p. 811).

A longitudinal analysis of one of the original sites studied by the Canadian research team confirmed these earlier findings (Elliott et al., 1997). Although levels of reported concern about a proposed solid waste disposal facility were relatively high (74% of residents within 4.5 kilometers of the site), levels of reported concern decreased significantly over time (measured at the time of landfill construction and 2.5 years after operation began), while levels of unsolicited "dislike" increased significantly. It would appear, therefore, that residents in close proximity to the site have learned how to cope with the existing landfill in their midst. This in no way implies that area residents are happy about the situation. As one respondent put it, "It's not because you like it any better; when there's nothing you can do about it, you just accept it and go on" (quoted in Elliott et al., 1997, p. 236).

At what scale should democratic control of technology be exercised? Noxious facilities are usually located where people have the least political power. For urban facilities, this usually means areas where ethnic minorities or the poor live, rather than upper-class suburbs. It makes sense to locate hazardous facilities in areas of sparse population, if the geology is compatible. The sparseness of population, however, also means that there are fewer people to oppose the will of the numerous people elsewhere who have reaped the benefits.

Eyles (1997) sets out a critical and theoretical environmental health research agenda, which includes the language and perception of environmental risk in everyday life; concepts of structure and power; consensus and conflict in environmental health policy; and questions about definitions of environmental quality. This agenda has a sociopolitical dimension because it pits property owners, workers, consumers, and others concerned about health risk against economic entities and government agencies. It also has an ethical dimension because it involves balancing individual desires to eliminate involuntarily imposed risks (however small) with society's need to have reliable technology at an affordable price. Geographic research cannot afford to ignore these sociopolitical and ethical concerns.

Environmental justice (EJ) is an area of inquiry that investigates the inequitable environmental burden borne by minorities and people living in economically disadvantaged areas. For example, asthma has been a common condition used in EJ studies because the disease burden disproportionately affects disadvantaged groups. EJ is both a political movement and an academic field. Some consider the formal beginning of the EJ political movement to have occurred in 1982 in Warren County, North Carolina, where the state sited a PCB toxic waste facility in a predominantly African American area. The political history of the movement cannot be understood without considering the context of the 1964 Civil Rights Act, which prohibits the use of federal funds to discriminate based on race, color, and national origin. In 1994 President Bill Clinton signed Executive Order 12898, which instructed federal government agencies to explicitly consider human-health and environmental conditions in minority and low-income communities in their activities. It also ordered the formation of a federal government working group on environmental justice. The EPA administrators under the next two presidents, George W. Bush and Barack Obama, sent formal memos reaffirming the original EJ executive order. The EPA is the main federal agency that implements environmental policies including those related to EJ. The agency defines EJ as "the fair treatment and meaningful involvement of all people regardless of race, color, national origin, or income with respect to the development, implementation, and enforcement of environmental laws, regulations, and policies" (*www.epa.gov/environmentaljustice*). In 2014 the EPA implemented a plan, called EJ 2014, to integrate environmental justice into the Agency's programs, policies, and activities (*www.epa.gov/environmentaljustice/plan-ej-2014*).

While federal government now considers EJ when it implements policies and practices, much of the EJ political action in the country is grassroots. There are countless local political EJ movements around the United States and the rest of the world that focus their energies on particular environmental problems in local areas.

EJ is also an academic pursuit and it is a theoretical underpinning for many geographic studies focusing on disparities in environmental risk and/or in health disparities. Health geography is an ideal field from which to tackle these complex issues because they are inherently spatial. An example is the aforementioned lead exposure study in Flint, Michigan, where children with high lead levels were more likely to live in more deprived communities. There have been many geographic studies that have measured spatial disparities in environmental burden. One study by Jerrett et al. (2001) found that air-borne particulate matter is more common in low-income areas of Hamilton, Ontario. Interestingly, more recent work found reductions in the disparities in environmental burden of air pollution over time in that industrial city where the economy has dramatically changed during the past few decades including factories closing (Buzzelli, Jerrett, Burnett, & Finklestein, 2003). Not all studies that measure environmental equity find the inequity that has been hypothesized. One U.S.-based study found no racial and ethnic disparities of the siting of treatment, storage, and disposal (TSD) of hazardous waste by census tract (Anderton, Anderson, Oakes, & Fraser, 1994). But at a different spatial scale Pulido (2000) found significant racial and ethnic disparities in the neighborhood siting of TSD and TRI facilities, and that study offers a historical and theoretical context that describes the complex reasons for the spatial patterns. Health geographers have built on EJ work and measured not just that there are social and ethnic inequities in the environmental burden but also whether the inequities are related to adverse health outcomes. In a national-level study in New Zealand, Pearce, Richardson, Mitchell, and Shortt (2011) investigated whether environmental deprivation scores were related to all-cause mortality, cardiovascular disease, respiratory disease, and breast cancer. They found that environmental deprivation had modest effects on health but not on all outcomes including breast cancer. Collins, Frineski, Chakrabarty, Montgomery, and Hernandez (2015) offered a case study analysis of air pollution in Houston and suggested that much of the EJ research has relied on ecological analysis, and they called for household-level studies.

QUICK REVIEW

✓ Air pollution is responsible for 7 million premature deaths each year, which is one of the main causes of death globally.

✓ Air pollution sources include industrial and transportation emissions as well as indoor sources such as exhaust from solid fuels used in developing countries.

✓ Sources of water pollution include chemical contamination from industrial sources, microbiological contamination usually from poor sanitation services, and natural sources such as groundwater arsenic in well water that originate in certain soils.

✓ Access to improved drinking water has increased dramatically in much of the developing world since 1990, whereas access to improved sanitation has lagged behind.

✓ Lead exposure causes developmental delays in children and has several sources including inhaling dust from paint and ingestion of paint chips produced before 1978, soils near roadways contaminated from leaded gasoline before 1986, and municipal water supplies where older distribution pipes were made of lead.

✓ Environmental justice is the area of inquiry that investigates the inequitable environmental burden borne by people living in economically disadvantaged areas, for example, lead contamination from the municipal water supply in more deprived communities in Flint, Michigan.

HEALTHY ENVIRONMENTS

This chapter has thus far focused on the deleterious aspects of the environment on health, but some environments can be good for your health. Chapter 8 described how food environments vary; when there is a dearth of healthy food options some scholars have referred to these areas as food deserts. At the other extreme the food environment can have many healthy food options and very few unhealthy ones. If a community has farmers' markets and no fast-food restaurants, then that would be a healthy food environment. Chapter 8 also uses the term *obesogenic* environments for built environments that do not encourage physical activity. As described in Chapter 9, the built environments of neighborhoods can also foster physical activity. A neighborhood that is safe, has nice parks, bike trails, and walking access to schools and stores and other services would promote physical activity. As described in Chapters 8 and 9, these neighborhood contextual factors ultimately have an effect on the health outcomes of people living in them.

As described in the introductory chapters of this book, the field of health geography for more than half a century has and continues to consider diverse determinants of health and disease in a holistic way. The concept of **therapeutic landscapes** focuses on the positive or therapeutic characteristics of places. Wil Gesler introduced the concept in his 1991 book called *The Cultural Geography of Health Care.* He described how geographers from the humanist perspective include the concept of social space in their scholarship and that some social spaces can promote health. One example he gave was how in Renaissance Europe mental health problems were often treated by having patients bathe or drink water from mineral springs that were thought to have healing powers. This concept was also present in the postcolonial United States where health spas and resorts were built around mineral springs because of their mysterious healing powers. One example is Poland Springs, Maine, where beginning in the mid-19th century a healing place centered on the spring was thought to cure people of anything from heartburn to mental illness. This is the same place that Nestlé now bottles water for mass consumption for the U.S. market.

The concept of therapeutic landscapes has evolved during the past 25 years. Gesler differentiated between different types of therapeutic landscapes, describing some as being based on beliefs and social relations and others related to natural settings, the built environment, and symbolic landscapes (Gesler, 1992). Williams (2010) offers a more specific taxonomy of types of therapeutic landscapes that include (1) physical places known for health, (2) health care sites/services, (3) sites specific to marginalized and special populations, and (4) everyday sites of varied therapeutic value. Williams (1998) further theorized Gesler's concept of therapeutic landscapes by describing how positive feelings about a place are psychologically rooted and thus facilitate the formation of a sense of place. She expanded the notion of sense of place to include health care facilities such as alternative birthing centers. Anyone who has ever been in an old hospital room with its institutional design will know they are the opposite of a caring environment. Williams (2003) explored cultural components of the link between health and place in everyday life in a case study of indigenous communities and their understanding of land and health in Canada. Conradson (2005) further theorized the concept by interrogating people's therapeutic encounters with the healing landscapes using the example of a respite care center focused on treatment of people with physical impairment as a case study.

CONCLUSION

The hazards faced by urban and rural areas, or by old industrial countries and NICs, are converging as technologies and pollution spread. As many industries decentralize and spread to other countries, so do certain hazardous occupational exposures. As trucks and automobiles, insecticides and herbicides, radios and air conditioners become universal, so do the products of their manufacture. Less than a decade after the 1978 declaration in the United States of the first national emergency due to technological rather than natural causes (at Love Canal) came the disasters of thousands of deaths from the release of poisonous gas by a pesticide plant in Bhopal, India, in 1985, and the nuclear catastrophe at Chernobyl in 1986. Agricultural intensification; proliferation of industrial occupations; production of toxic and radioactive wastes; and consumption of petrochemicals, pharmaceuticals, synthetic textiles, electronics, glues, and paints are now almost universal. The scale of environmental change does not match the scale of contemporary regulation and control.

The ultimate causes of our environmental problems are those of population numbers, inequalities, and diminishing natural resources. As humans multiply, they seem to be oblivious to the fact that resources are finite, and that everything on Earth has not been placed here explicitly for their benefit. Human gains have been achieved at great costs: depletion of material resources and extinction of many other forms of life. The greatest "gains" have accrued disproportionately to the wealthy nations and industrial organizations, who have been best able to marshal technology. Developing world populations, though not wealthy by Western standards, have also exacted their cost on the environment. Over the past few decades, intensive logging and burning of forests worldwide, partly to open up land for agriculture, have contributed to air pollution and microclimatic change. Unfortunately, technology has not been employed as effectively as it might have been to reduce pollution of the environment. To effectively reduce emissions, nations must adopt policy measures such as energy or fuel efficiency standards for appliances and transport vehicles; they must also eliminate market distortions such as low, subsidized energy prices that reduce the economic benefits to consumers of saving energy (World Resources Institute, 1996, pp. 315–338). Chapter 12 extends this discussion to another aspect of our environment, our climate, and describes how the field of health geography has contributed to understanding how global climate change can affect our health.

REVIEW QUESTIONS

1. Table 11.1 lists many different environmental pollutants, their sources, and health effects. Choose one and do a Google Scholar search (scholar.google.com) for the pollutant and the words "health effects," for example, "cadmium health effects." What are the main sources of the pollutants in the papers you found with the search? What are the health effects?

2. There are many different types and causes of outdoor pollution and the impacts are spatially variable. Where are the following types of air pollution worst in the United States and world and why: photochemical smog, PM, SO_2, and air toxics?

3. Log onto the *airnow.gov* website and type in your Zip code at the top of the page next to Local Air Quality Conditions. What is the air quality index where you live? Are the pollutants PM2.5 and ozone levels high? Is the air quality good where you live? Choose another location on the map on the front page where you think the air quality is very different from where you live. Click on the location, and the air quality values will be listed. Compare the air quality index, PM2.5, and ozone levels with where you live. Why do you think they are different? What do you think are the sources of the differences?

4. What are the main sources of indoor air pollution in different parts of the world? What part of the world does indoor air pollution lead to the most health problems? What are some of the interventions that can be used to limit the health effects of indoor air pollution? Are the interventions different in different parts of the world?

5. How successful has the world been in achieving the goals of MDG 7c? You are tasked with developing the post-2015 MGD goals for the next 15 years. How would you change the focus to reduce consumption of contaminated water throughout the world?

6. Lead is no longer in paint or gasoline in the United States, yet risk of environmental lead still exists. The risk is spatially variable as well. Explain why the problem persists and summarize the spatial patterns of environmental lead risk in the United States.

7. EJ is a political endeavor and an academic pursuit. Its practitioners are interested in the inequitable environmental burden of populations. In the area that you live do you think that some groups of people have a higher environmental burden than others? What are the environmental burdens and what groups live near them (e.g., factories, dumps, water treatment plants)?

8. Healthy environments have been described as therapeutic landscapes by some scholars. Think about a therapeutic landscape where you live or have visited. What are the characteristics that make it a therapeutic landscape? Based on Williams's taxonomy, what type of therapeutic landscape is it and why?

REFERENCES

Abelson, P. (1994). Chlorine and organochlorine compounds. *Science, 265,* 1155.

American Lung Association. (2014). *State of the air.* Chicago: Author.

Anderton, D., Anderson, A., Oakes, J., & Fraser, M. (1994). Environmental equity: The demographics of dumping. *Demography, 31*(2), 229–248.

Bailey, A., Sargent, J., Goodman, D., Freeman, J., & Brown, M. (1994). Poisoned landscapes: The epidemiology of environmental lead exposure in Massachusetts children 1990–1991. *Social Science and Medicine, 39,* 757–766.

Berg, M., Tran, H., Nguyen, T., Viet Pham, H., Schertenlieb, R., & Giger, W. (2001). Arsenic contamination of groundwater and drinking water in Vietnam: A human health threat. *Environmental Science and Technology, 35*(3), 2621–2626.

Buzzelli, M., Jerrett, M., Burnett, R., & Finklestein, N. (2003). Spatiotemporal perspectives on air pollution and environmental justice in Hamilton, Canada, 1985–1996. *Annals of the Association of American Geographers, 93*(3), 557–573.

Centers for Disease Control and Prevention (CDC). (1992). Blood lead levels among children in high-risk areas—California, 1987–1990. *Morbidity and Mortality Weekly Report, 41*(17), 291–294.

Centers for Disease Control and Prevention (CDC). (1994). *National Health and Nutrition Examination Survey (NHANES III)*. Atlanta, GA: Author.

Chan, C., & Yao, X. (2008). Air pollution in mega cities in China. *Atmospheric Environment, 42*(1), 1–42.

Chen, R., Kan, H., Chen, B., Huang, W., Bai, Z., Song, G., et al. (2012). Association of particulate air pollution with daily mortality: The China Air Pollution and Health Effects Study. *American Journal of Epidemiology, 175*(11), 1173–1181.

Collins, T., Frineski, S., Chakraborty, J., Montgomery, M., & Hernandez, M. (2015). Downscaling environmental justice analysis: Determinants of household-level hazardous air pollutant exposure in Greater Houston. *Annals of the Association of American Geographers, 105*(4), 684–703.

Committee on Health Risks of Exposure to Radon (BEIR VI); Commission on Life Sciences; Division on Earth and Life Studies; National Research Council. (1999). *Health effects of exposure to radon*. Washington, DC: National Academy Press.

Conradson, D. (2005). Landscape, care, and the relational self: Therapeutic encounters in rural England. *Health and Place, 11,* 337–348.

Cutter, S., & Solecki, W. (1989). The national pattern of airborne toxic releases. *Professional Geographer, 41,* 149–161.

Cutter, S., & Tiefenbacher, J. (1991). Chemical hazards in urban America. *Urban Geography, 12,* 417–430.

Dherani, D., Pope, D., Mascarenhas, M., Smith, K., Weber, M., & Bruce, N. (2008). Indoor air pollution from unprocessed solid fuel use and pneumonia risk in children aged under five years: A systematic review and meta-analysis. *Bulletin of the World Health Organization, 86,* 390–398.

Dockery, D., Arden Pope, C., Xu, X., Spengler, J., Ware, J., Fay, M., et al. (1993). An association between air pollution and mortality in six U.S. cities. *New England Journal of Medicine, 329*(24), 1753–1759.

Dubos, R. (1965). *Man adapting.* New Haven, CT: Yale University Press.

Earickson, R., & Billick, I. (1988). The areal association of urban air pollutants and residential characteristics: Louisville and Detroit. *Applied Geography, 8,* 5–23.

Economy, E. (2004). *The river runs black: The environmental challenge to China's future.* Ithaca, NY: Cornell University Press.

Elliott, S., & Taylor, S. (1996). Worrying about waste: Diagnosis and prescription. In D. Munton (Ed.), *Hazardous waste siting and democratic choice: The NIMBY phenomenon and approaches to facility siting* (pp. 290–318). Washington, DC: Georgetown University Press.

Elliot, S., Taylor, S., Hampson, C., Dunn, J., Eyles, J., Walter, S., & Streiner, D. (1997). "It's not because you like it any better . . ." Residents' reappraisal of a landfill site. *Journal of Environmental Psychology, 17*(3), 229–241.

Elliot, S., Taylor, S., Walter, S., Stieb, D., Frank, D., & Eyles, J. (1993). Modeling psychosocial effects of exposure to solid waste facilities. *Social Science and Medicine, 37,* 791–804.

Emch, M. (1999). Diarrheal disease risk in Matlab, Bangladesh. *Social Science and Medicine, 49,* 519–530.

Escamilla, V., Wagner, B., Yunus, M., Streatfield, P., van Geen, A., & Emch, M. (2011). Impact of deep tubewells on childhood diarrhea in Bangladesh. *Bulletin of the World Health Organization 89*(7), 521–527.

Eyles, J. (1997). Environmental health research: Setting an agenda by spinning our wheels or climbing the mountain. *Health and Place, 3*(1), 1–13.

Eyles, J., Taylor, S., Johnson, N., & Baxter, J. (1993). Worrying about waste: Living close to solid waste disposal facilities in southern Ontario. *Social Science and Medicine, 37*(6), 805–812.

Fang, X., Wang, X., & Bai, C. (2011). Burden and importance of proper management about chronic obstructive pulmonary disease in China. *International Journal of Respiratory Care, 31*(7), 493–497.

Fewtrell, L., Kaufmann, R., Kay, D., Enanoria, W., Haller, J., & Colford, J. (2005). Water, sanitation, and hygiene interventions to reduce diarrhoea in less developed countries: A systematic review and meta-analysis. *The Lancet Infectious Diseases, 5*(1), 42–52.

French, H. (1997). Learning from the ozone experience. In L. Brown (Ed.), *State of the world* (pp. 151–171). New York: Norton.

Gesler, W. (1991). *The cultural geography of health care.* Pittsburg, PA: University of Pittsburgh Press.

Gesler, W. (1992). Therapeutic landscapes: Medical issues in light of the new cultural geography. *Social Science and Medicine, 34,* 735–745.

Greenberg, M., & Schneider, D. (1996). *Environmentally devastated neighborhoods.* New Brunswick, NJ: Rutgers University Press.

Gurjar, B., Jain, A., Sharma, A., Agarwal, A., Gupta, P., Nagpure, A., et al. (2010). Human health risks in megacities due to air pollution. *Atmospheric Environment, 44*(36), 4606–4613.

Guttikunda, S., & Calori, G. (2013). A GIS based emissions inventory at 1 km × 1 km spatial resolution for air pollution analysis in Delhi, India. *Atmospheric Environment, 67,* 101–111.

Hanna-Attisha, M., LaChance, J., Sadler, R. C., & Schnepp, A. C. (2016). Elevated blood lead levels in children associated with the Flint drinking water crisis: A spatial analysis of risk and public health response. *American Journal of Public Health, 106*(2), 283–290.

Hunter, J. (1977). The summer disease: An integrative model of the seasonality aspects of childhood lead poisoning. *Social Science and Medicine, 11,* 691–703.

Hunter, J., & Arboña, S. (1995). Paradise lost: An introduction to the geography of water pollution in Puerto Rico. *Social Science and Medicine, 40,* 1331–1335.

Jerrett, M., Burnett, P., Kanaroglou, J., Eyles, N., Finklestein, N., Giovis, C., et al. (2001). A GIS–environmental justice analysis of particulate air pollution in Hamilton, Canada. *Environment and Planning A, 33*(6), 955–973.

Joint Monitoring Programme for Water Supply and Sanitation. (2006). *Meeting the MDG drinking water and sanitation targets: The urban and rural challenge of the decade.* Geneva, Switzerland: World Health Organization/UNICEF.

Jones, A. (1998). Asthma and domestic air quality. *Social Science and Medicine, 47,* 755–764.

Kloos, H. (1995). Chemical contaminants in public drinking water wells in California. In S. Majumdar, E. Miller, & F. Brenner (Eds.), *Environmental contaminants, ecosystems and human health* (pp. 30–43). University Park: Pennsylvania Academy of Science.

Koskela, R., Mutanen, P., Sorsa, J., & Klockars, M. (2000). Factors predictive of ischemic heart disease mortality in foundry workers exposed to carbon monoxide. *American Journal of Epidemiology, 152*(7), 628–632.

Last, J., & Guidotti, T. (1991). Implications for human health of global ecological changes. *Public Health Review, 18,* 49–67.

Leary, W. (1994, September 6). Studies raise doubts about need to lower home radon levels. *New York Times.*

Lindley, D. (1988). CFCs cause part of global ozone decline. *Nature, 323,* 293.

Maantay, J. (2007). Asthma and air pollution in the Bronx: Methodological and data considerations in using GIS for environmental justice and health research. *Health and Place, 13,* 32–56.

MacKie, R., & Rycroft, M. (1988). Health and the ozone layer. *British Medical Journal, 297,* 369–370.

McCoy, K., Bortnick, L., Campbell, C., Hamlin, H., Guillette, L., & St. Mary, C. (2008). Agriculture alters gonadal form and function in the toad *Bufo marinus. Environmental Health Perspectives, 116,* 1526–1532.

Mielke, H., Anderson, J., Berry, K., Mielke, P., Chaney, R., & Leech, M. (1983). Lead concentrations in inner-city soils as a factor in the child lead problem. *American Journal of Public Health, 73*(12), 1366–1369.

Mielke, H., Gonzales, C., Powell, E., & Mielke, P. (2013). Environmental and health disparities in residential communities of New Orleans: The need for soil lead intervention to advance primary prevention. *Environment International, 51,* 73–81.

Mielke, H., & Reagan, P. (1998). Soil is an important pathway of human lead exposure. *Environmental Health Perspectives, 106*(Suppl. 1), 217–229.

Mielke, H., & Zahran, S. (2012). The urban rise and fall of air lead (Pb) and the latent surge and retreat of societal violence. *Environment International, 43,* 48–55.

National Research Council (NRC). (2004). *Air quality management in the United States.* Washington, DC: National Academies Press.

National Research Council (NRC). (2006). *State and federal standards for mobile-source emissions.* Washington, DC: National Academies Press.

Nevin, R. (2000). How lead exposure relates to temporal changes in IQ, violent crime, and unwed pregnancy. *Environmental Research, 83*(1), 1–22.

Pearce, J., Richardson, E., Mitchell, R., & Shortt, N. (2011). "Environmental justice and health: A study of multiple environmental deprivation and geographical inequalities in health in New Zealand. *Social Science and Medicine, 73*(3), 410–420.

Pope, C. A., III, Ezzati, M., & Dockery, D. W. (2009). Fine-particulate air pollution and life expectancy in the United States. *New England Journal of Medicine, 360,* 376–386.

Pulido, L. (2000). Rethinking environmental racism: White privilege and urban development in southern California. *Annals of the Association of American Geographers, 90*(1), 12–40.

Sahsuvaroglu, T., Jerrett, M., Sears, M., McConnell, R., Finklestein, N., Arain, A., et al. (2009). Spatial analysis of air pollution and childhood asthma in Hamilton, Canada: Comparing exposure methods in sensitive subgroups. *Environmental Health, 8*(1).

Samet, J. (2006). Residential radon and lung cancer: End of the story? *Journal of Toxicology and Environmental Health, Part A: Current Issues, 69*(7–8), 527–531.

Shabecoff, P. (1985, June 11). U.S. calls eleven toxic air pollutants bigger threat indoors than out. *New York Times.*

Shao, M., Tang, X., Zhang, Y., & Li, W. (2006). City clusters in China: Air and surface water pollution. *Frontiers in Ecology and the Environment, 47*(7), 353–361.

Smith, A., Lingas, E., & Rahman, M. (2000). Contamination of drinking-water by arsenic in Bangladesh: A public health emergency. *Bulletin of the World Health Organization, 78*(9), 1093–1103.

Smith, K. (2012). *Biofuels, air pollution, and health: A global review*: New York: Springer.

Smith, K., & Mehta, S. (2003). The burden of disease from indoor air pollution in developing countries: Comparison of estimates. *International Journal of Hygiene and Environmental Health, 206*(4–5), 279–289.

Smith-Sivertsen, T., Diaz, E., Pope, D., Lie, R., Diaz, A., McCracken, J., et al. (2009). Effect of reducing indoor air pollution on women's respiratory symptoms and lung function: The RESPIRE randomized trial, Guatemala. *American Journal of Epidemiology, 170*(2), 211–220.

Stephens, E. (1987). Smog studies of the 1950s. *Eos: Transactions of the American Geophysical Union, 68*(7), 1–5.

Stern, F., Halperin, W., Hornung, R., Ringenburg, V., & McCammon, C. (1988). Heart disease mortality among bridge and tunnel officers exposed to carbon monoxide. *American Journal of Epidemiology, 128*(6), 1276–1288.

Stolarsky, R. (1988). The Antarctic ozone hole. *Scientific American, 258,* 30–36.

Strauss, W., & Mainwaring, S. (1984). *Air pollution*. London: Arnold.

Sun, M. (1986). Ground water ills: Many diagnoses, few remedies. *Science, 232,* 1490–1493.

Talcott, F. (1992). How certain is that environmental risk estimate? *Resources, 107,* 10–15.

U.S. Census Bureau. (1996). *Statistical abstract of the United States: 1996*. Washington, DC: U.S. Government Printing Office.

U.S. Department of Health and Human Services. (2014). *The health consequences of smoking–50 years of progress: A report of the surgeon general*. Atlanta: U.S. Department of Health and Human Services, Centers for Disease Control and Prevention, National Center for Chronic Disease Prevention and Health Promotion, Office on Smoking and Health.

U.S. Department of Housing and Urban Development. (2011). *American Healthy Homes Survey: Lead and arsenic findings*. Washington, DC: Office of Healthy Homes and Lead Hazard Control.

U.S. Environmental Protection Agency (EPA). (1994). *National water quality inventory, 1994 report to Congress*. Research Triangle Park, NC: Author.

U.S. Environmental Protection Agency (EPA). (1996). *National air quality and emissions trends report, 1995*. Research Triangle Park, NC: Author.

Watts, J. (2005). China: The air pollution capital of the world. *The Lancet, 366*(9499), 1761–1762.

Williams, A. (1998). Therapeutic landscapes in holistic medicine. *Social Science and Medicine, 46*(9), 1193–1203.

Williams, A. (2003). Therapeutic landscapes and First Nations Peoples: An exploration of culture, place, and health. *Health and Place, 9,* 83–93.

Williams, A. (2010). Therapeutic landscapes as health promoting places. In T. Brown, S. McLafferty, & G. Moon (Eds.), *A companion to health and medical geography* (pp. 207–223). West Sussex, UK: Wiley-Blackwell.

Winston, J., Escamilla, V., Perez-Heydrich, C., Carrel, M., Yunus, M., Streatfield, P., et al. (2013). Protective benefits of deep tube wells against childhood diarrhea in Matlab, Bangladesh. *American Journal of Public Health, 103*(7), 1287–1291.

World Resources Institute. (1996). *World resources 1994–95*. New York: Oxford University Press.

Wright, J., Dietrich, K., Ris, M., Hornung, R., Wessel, S., Lanphear, B., et al. (2008).

Association of prenatal and childhood blood lead concentrations with criminal arrests in early adulthood. *PLoS Medicine, 5*(5).

Zeigler, D., Johnson, J., & Brunn, S. (1983). *Technological hazards* (Resource Publications in Geography). Washington, DC: Association of American Geographers.

Zhang, J., & Smith, K. (2007). Household air pollution from coal and biomass fuels in China: Measurements, health impacts, and interventions. *Environmental Health Perspectives, 115,* 848–855.

Zhong, N., Wang, C., Yao, W., Chen, P., Kang, J., Huang, S., et al. (2007). Prevalence of chronic obstructive pulmonary disease in China: A large, population-based survey. *American Journal of Respiratory and Critical Care Medicine, 176,* 753–760.

FURTHER READING

American Lung Association. (1996). *Health effects of outdoor air pollution.* Washington, DC: Author.

Blumenthal, D., & Ruttenber, A. (1995). *Introduction to environmental health* (2nd ed.). New York: Springer.

Briggs, D., & Elliot, P. (1995). The use of geographical information systems in studies on environment and health. *World Health Statistics Quarterly, 48,* 85–94.

Burton, I., Kates, R., & White, G. (1993). *The environment as hazard* (2nd ed.). New York: Guilford Press.

Croner, C., Sperling, J., & Broome, F. (1996). Geographic information systems: New perspectives in understanding human health and environmental relationships. *Statistics in Medicine, 15,* 1961–1977.

Gould, P. 1990. *Fire in the rain.* Baltimore: Johns Hopkins University Press.

Hall, B., & Kerr, M. (1992). *1991–1992 green index: A state-by-state guide to the nation's environmental health.* Washington, DC: Island Press.

Hester, R., & Harrison, R. (1998). *Air pollution and health.* London: Royal Society of Chemistry.

Lopez, A., & Murray, C. (1998). The global burden of disease. *Nature Medicine, 4,* 1241–1243.

Monmonier, M. (1997). *Cartographies of danger: Mapping hazards in America.* Chicago: University of Chicago Press.

Platts-Mills, T. (Ed.). (1999). *Asthma: Causes and mechanisms of this epidemic inflammatory disease.* Boca Raton, FL: CRC Press.

Waller, L. (1996). Geographic information systems and environmental health. *Health and Environment Digest, 9,* 85–88.

World Bank. (1994). *World development report.* New York: Oxford University Press.

CHAPTER 12

Climate and Health

Climate, weather, and biometeorological conditions affect the health status of people both directly and indirectly. In terms of the triangle of population–habitat–behavior interactions, this chapter is especially concerned with the influences of the natural environment. All living things on earth have evolved to cope with either the presence or absence of light. They have to cope with cold and/or heat. They have to get water. None of these essentials are uniform over the earth. Because of the tilt of the Earth's axis, and the time period of the rotation on that axis and of that axis around the sun, predictable patterns of climate have been created; these are associated with predictable patterns of vegetation, or *biomes*. Realms of evolution and biomes of vegetation are each partially characterized by the species and life cycles of arthropods, amphibians, reptiles, birds, primates, and other mammals living there, and of their fungi, bacteria, viruses, and **ectoparasites**. Culture, of course, has a lot to do with how this natural environment affects people, how they are protected, and how people alter that environment. Climatology is one of the oldest studies in geography, but too often the physical geographers who write about correlations of heat waves and deaths, or flood and mosquito hazards, completely ignore cultural buffers, adjustments, and active responses. Some, however, do not ignore the complex and integrated social, cultural, and climatological complexities that sometimes lead to human health problems.

Climate is the large term referring to weather conditions averaged and prevailing over long periods, many decades or centuries, of time. Such broad descriptive average parameters as how cold the winters are or how hot the summers get, average last or first days of frost, onset of the monsoon rains, general humidity levels, and direction and strength of winds—these parameters describe climate. Climate determines what crops can be grown where, and how many blood meals mosquitoes usually take before they die and disease transmission is broken (or not). The daily and weekly conditions of temperature, humidity, precipitation, wind, and cloud cover that are experienced constitute the **weather**. The fluctuations that always occur in heat, rain, cold, drought, snow cover, wind strength, and storms from week to week and year to year are weather. Fluctuations in the climate—extended periods of severe droughts, several years of mild or severe winters—have always occurred. Cycles of hurricane activity—for example, decades of more or fewer storms occurring and hitting the mainland United States—have been known for more than a century. One of the great questions of our times has been when extremes of weather and concurrence of cycles become enough to change the average (i.e., to change climate). Such a climate change, to a warming earth, is upon us. It will affect temperatures, frost dates, patterns of precipitation (as well as its amounts and forms), wind strength, humidity, surface UV light, floods, and droughts. It will thereby affect crops; insect production; the synchronization in timing of birds, bees, and mosquitoes; water availability for irrigation, sanitation, and hygiene; and the potability of drinking water.

Climate is composed of long-term averages over large areas; weather consists of daily and weekly local experience; and **biometeorology** consists of the variation and change in the physical and chemical characteristics of the atmosphere that affect variation and change in the physicochemical systems of living organisms. The human mammal is sensitive to a far wider range of atmospheric characteristics (Audy's physical insults, or stimuli) than is commonly realized. These not only include temperature, humidity, air movement, atmospheric pressure, solar radiation, sound, and gaseous pollution, but also infrasound, magnetism, and electrical charge. In addition, a host of indirect influences derive from the biometeorology of each location on Earth.

Elements of the built environment, such as microwaves or buildings' central heating, can also be important influences on the physiological status of the population—that is, on the chemical–electrical systems of people's bodies. Physiological status is influenced by the time of year and the time of day; it responds to the light, temperature, and altitude of a place. A person's susceptibility to disease, toxins, and pharmaceutical drugs can be altered by biometeorological changes. Although there is a tendency in medicine to treat all people as a homogeneous population, in fact human physiological characteristics differ from place to place as well as from individual to individual. The temperature at which water boils varies with altitude around the Earth, so it shouldn't be surprising that the potency of a drug or infectious dose of a pathogen does too. However, some of the most talked-about influences of the weather (such as the short tempers and increased violence associated with a long, hot summer) may be more constructs of social rhythms in employment and school vacation, sports, and what might be called the political economy of air conditioning than they are biometeorological effects of the weather itself.

This chapter briefly examines how humans receive and biologically respond to such stimuli, and surveys the most direct effects on human health. Biological rhythms and acclimatization to temperature and altitude are examined. The physiological basis of climatic influence is described in some detail because sound geographic hypotheses and choice of relevant variables depend on an understanding of the disease process. The influence of heat waves, air masses, and their passage is described. Seasonality in the basic vital events of birth and death is analyzed. Finally, attention is turned to the impact of changing climate upon all these, and upon the water-related and vectored infectious diseases. With global warming, disaster preparation and response will undoubtedly become important applications of health geography.

DIRECT BIOMETEOROLOGICAL INFLUENCES

The Radiation Spectrum

Figure 12.1 illustrates the spectrum of electromagnetic radiation. Electromagnetic radiation is propagated through space in the form of packets of energy called *photons,* which travel at the speed of light. Each photon has a frequency, wavelength, and proportional energy. Short wavelengths of radiation are known as ionizing radiation because they can detach electrons and damage atomic structure. These photons of the higher-energy ranges—cosmic, gamma, and X-rays—can penetrate into genes and disrupt the DNA sequences of our design. X-rays are principally a human-made health hazard; the atmosphere mostly shields us from their natural sources.

Ultraviolet (UV) light is best known for causing sunburn. Because the damage it does depends on the intensity and duration of exposure, it creates some important occupational hazards. Welding, for example, produces such intense **ultraviolet radiation** that an instant's unshielded exposure can cause severe inflammation of the eye's membrane. Only a small band within the range of UV radiation can penetrate the atmosphere to any extent because it is blocked by a stratospheric layer of ozone. As described in Chapter 11, this ozone layer was partially destroyed—unintentionally, of course—by industrial production of halogenated chemicals such as the chlorofluorocarbons (CFCs, used in refrigeration and spray-can propellants). CFCs, inert and seemingly harmless at normal surface temperatures, react with the ozone when catalyzed by the extreme cold. International treaties now banning CFCs

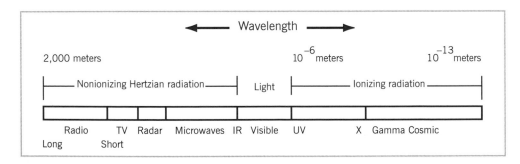

FIGURE 12.1 The spectrum of electromagnetic transmission. Radiation, from cosmic to radio, is a continuum of wavelength. IR, infrared; UV, ultraviolet.

and related chemicals seem to have stopped enlarging the hole in the ozone layer over the Antarctic. We now know, however, that other chemicals are involved in this process, and that when the greenhouse gases trap heat in the lower atmosphere, they will accelerate the destruction of ozone in the then colder stratosphere. The weakened shield of the ozone layer against the harmful effects of UV radiation for human and other life on land is expected during the first part of this 21st century to let the highest amount of ground-level radiation through in the midlatitudes.

Humans evolved in the presence of UV radiation and are dependent upon it. It activates a human enzyme to produce vitamin D. The human body is dependent on UV radiation to catalyze the body's synthesis of vitamin D; when it is not synthesized it leads to a disease called rickets, described in Chapter 10. Visible light is biologically very powerful. It stimulates photoreceptive tissue, such as the retinas in human eyes. It also changes the electrical charge of protoplasm (the physical base of all living activities) and the viscosity, permeability, and colloidal behavior of proteins. It thereby causes living things to turn toward light, and it energizes photosynthesis. Light seems to be the most important factor influencing biological rhythms. Only these photons in the intermediate range of electromagnetic radiation can be detected by human senses.

Infrared and longer-wave radiations are important for heating the Earth's surface and its cold-blooded plants and animals, either directly or through reradiation by the Earth. The health hazards of these long-wave radiations result from their function as penetrating heat sources. Infrared radiation used to be a cause of blindness in such occupations as the manufacture of glass, but careful shielding and new industrial processes have removed most such occupational hazards. Today microwaves are of some concern as the machinery that produces them proliferates. The waves can penetrate the body at considerable distance and cause internal heating that the body's sensors do not detect. The eyes are at greatest risk because the lens of the eye has no blood circulation to remove the heat. At the end of the spectrum are the longer radio waves. They are all around us, but are not known to have ever done anyone harm, not even people intensely exposed by occupation.

All of these types of electromagnetic radiation vary over the surface of the Earth. The Earth's magnetism affects the path and concentration of much radiation. The tilt of the Earth results in seasonal and latitudinal differences described in this radiation. High elevations that project through thousands of feet of atmosphere— and the people who live at those elevations—experience different amounts and compositions of radiation. Human-made sources of several kinds of radiation are becoming important, and they have a very irregular distribution. Consequences of radiation have been much studied in terms of cell chemistry, microbiology, and physiology, but the spatial distribution of hazards and health effects has received little attention, except for risk of skin cancer.

Biological Rhythms

The presence or absence of light is one of the oldest and most universal selective pressures to which all living things have had to adapt. An important part of that adaptation has been the development of daily and seasonal rhythms of many body processes. It is clearly beneficial to an organism to be able to anticipate when cold, night, rain, high tide, or spring flowers and grass are going to come, so that the slow physiological processes of growing hair, losing leaves, and storing food can be initiated in good time. The precise synchronicity of breeding swarms of many sea creatures and insects, the beaching and egg laying of turtles and crabs that travel great distances to appear simultaneously at a specific place, and the migration of birds and bats are phenomena that have long fascinated scientists and testified to the existence of biological clocks. The equivalent body rhythms of human beings have been recognized much more recently. Some social scientists have resisted recognizing any environmental influences on people, and some religious belief systems resist any suggestions that humans are animals living in the biosphere.

The existence of endogenous clocks is now well established for a wide variety of animals, including humans. Even in constant temperature and total darkness, some biological processes continue to oscillate rhythmically. The rhythms with a span of about 24 hours are known as *circadian*. In humans they include sleep and wakefulness; body temperature; cognitive performance; serum hormone levels; urinary cycles and excretion of ketosteroids, chloride, sodium, and urea; and ionization of blood calcium and phosphate, which affects hormone regulation. It is clear that the physiological state of individuals is constantly changing. A tenfold difference in the susceptibility of mice to bacterial toxins at different times of night has been demonstrated, and it is likely that human circadian rhythms also influence susceptibility to infection and toxins.

The study of circadian and other rhythms can help us understand the ways in which environment directly affects and alters the status of a population. Reception and interpretation of environmental stimuli influence virtually all the body's major organs and endocrine systems (Figure 12.2). There have been major advances in understanding the nature of the body's clock (Barinaga, 1999; Somers, Devlin, & Kay, 1998; Thresher et al., 1998). A cluster of nerve cells in the brain, the suprachiasmatic nucleus (SCN), is now believed to be the organ responsible for synchronizing body rhythms with environmental light (Bernard, Gonze, Cajaves, Herzel, & Kramer, 2007). Pigments called cryptochromes have been found in the eye, skin, and part of the brain. They are believed to drive the body's clock. The two forms of cryptochromes located in a part of the retina absorb blue light and transfer the light signal to the SCN—a process for which the individual does not have to be awake or conscious. In individual neurons there, two proteins alternately build up and then turn on and off each other's gene for production, thus forming an accurate internal clock in each of tens of thousands of cells in the SCN. This oscillation also turns on and off a gene for production of vasopressin and maybe other endocrines. The photoreception of the cryptochromes sets the body's elegant master clock of oscillating proteins. Individual light sensitivity can result in *seasonal affective disorder* (SAD; winter depression) and sleeping problems among the elderly, as well as difficulties adjusting to changes in longitude.

The main body mediator, the hypothalamus, controls the pineal gland in mammals, which acts as an endocrine transducer (i.e., it converts one form of energy to

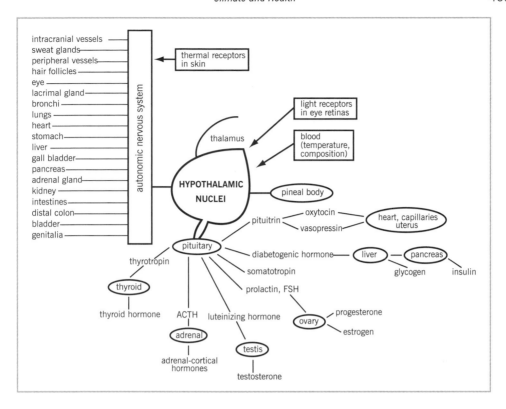

FIGURE 12.2 Hypothalamic environmental mediation. Sensing the environment, the hypothalamus acclimatizes the body through its direct neural connections to many organs and through the hormones it induces in the pituitary gland. ACTH, adrenocorticotropic hormone; FSH, follicle-stimulating hormone.

another). It synthesizes and secretes the hormones melatonin and serotonin. These are two of the most powerful chemical transmitters in the brain and can penetrate everywhere in the body. During the day, the pineal gland synthesizes serotonin. As darkness falls, the pineal gland produces a surge of melatonin. The cycle in the pineal gland is implicated as the clock that controls the development of puberty (Fellman, 1985). A sharp decline in melatonin production at puberty seems a trigger. Middle-aged adults secrete only half as much melatonin as children do. Besides causing sleepiness, melatonin dampens release of estrogen and so may be related to timing of conception. The blue light absorbed by cryptochromes, as described above, suppresses the production and release of melatonin. This is why screen usage, in the form of smartphones, tablets, and computers, with the blueish light that is emitted, in the hours before bed is associated with restless or poor sleep.

The hypothalamus also controls the pituitary, the body's master gland, and so it can adjust most of the body's metabolism, endocrine system, blood vessels, and other organs as appropriate for environmental conditions. As we shall see, this is important for acclimatization. In the 1980s SAD was recognized—the "winter blues," with depression, excessive sleep, weight gain, and loss of interest in daily activities. The major therapy that has been developed is *phototherapy*, which consists simply of bright lighting for a short period in the morning, to communicate with the pineal gland.

Most people become aware of circadian rhythms when they travel across longitudes. The physical and mental effects of desynchronizing the body's rhythms from those of the environment are familiar as "jet lag." It is not a matter of fatigue; the effects are the same even if one slept well on the airplane. Rather, body temperature and kidney/hormone cycles are no longer appropriate for the new schedule of social activities and daylight. Each biological clock is reset, advancing or slowing an hour or two each cycle, until all cycles are again synchronized with the environment. During the period of jet lag, however, the body may be more susceptible to infectious agents and (through impaired judgment and coordination) to accidents. There is growing use of melatonin to help reset the clock.

The large volume of international travel has made circadian rhythms and their adjustment a matter of concern to the business, sports, and tourist industries, as well as to diplomats and the military. One topic of interest is the importance of circadian rhythms for work shifts. When people continuously work at night, first the sleep cycle, then the body temperature cycle, and eventually all the other cycles will be appropriately set to the social activity pattern. In some professions, however, shifts are regularly changed around or even rotated. This means that nurses, police officers, and other 24-hour-service personnel may frequently be involved in critical situations while effectively in a state of jet lag.

Most research on circadian rhythms has been medical. Are people who live in the long winter nights of high latitudes, for example, more susceptible to infectious agents than those in equatorial lands? Light is used today to induce chickens to lay more eggs and to bring cattle and sheep back into estrus after a miscarriage or failed fertilization. Does nighttime electric lighting affect human fecundity at all? Geographically designed studies of circadian rhythms would include contrasts between places and cultures (and occupations/activities by gender and class) and could be very illuminating.

Acclimatization

Acclimatization is the genetically given capacity of humans to adapt their physiological systems over long periods to heat, cold, and altitude. People are adapted to living in an extraordinary range of temperatures and other environmental conditions. They live and reproduce in hot deserts and Arctic tundra, as well as in the high Andes and the Himalayas.

Behavioral and genetic adaptations are of great importance to acclimatization. Cultural geographers have studied adaptive differences in house types, designs, and construction; the materials and design of clothing; and the cultural rhythms of daily activity patterns. Genetic adaptations often involve the morphology of the body. In many animal populations, for example, long limbs and slimness serve to dissipate heat efficiently, while squat bodies with a low surface-to-volume ratio serve to conserve heat. Physical anthropologists have identified some such tendencies among human populations, but extensive migration has created as much exception to the rules (tall Norwegians in the Arctic and compact Chinese in tropical Asia) as conformity (compact Inuit in the Arctic and tall, slender Tutsi in tropical Africa). As important as genetic and behavioral adaptations are, this chapter looks at the physical processes of acclimatization and the implications for health that result from variations in physiological status of the population.

The physiological processes of thermoregulation are known in some detail. The human body must maintain a temperature of about 98.6°F (37°C), regardless of the environmental temperature. Over short periods, people react by shivering to generate heat from muscle movement and sweating to cool the body by evaporation. People everywhere sweat and shiver at the same skin temperature. Acclimatized people, however, use other physiological means to control body temperature. Their skin temperatures may not reach the threshold values for sweating or shivering.

The hypothalamus learns about environmental temperature from nerve channels to thermoreceptors in the skin, and about body temperature from nerves and the hypothalamic blood vessels. The hypothalamus causes the peripheral blood vessels to dilate (vasodilation) under heat stress, so that more blood can be brought to the surface for cooling, and to constrict (vasoconstriction) in protection against heat loss, so that more blood is held in the core of the body. Over time, the density of the capillary network near the skin will change. Most animals sweat from a few sites (e.g., armpits in humans) controlled by the sympathetic nervous system. Humans have evolved an unusual capacity for eccrine (excretory) sweating by glands located all over the body. This involves excreting dilute solutions of water, salts, urea, sugar, and lactic acid for purposes of cooling the surface of the body through evaporation. Under conditions of prolonged heat, the efficiency of this sweating may be greatly increased. An acclimatized person sweats from the body's entire surface; the flux of water can reach several liters in an hour. The consequent drain of blood electrolytes (such as magnesium and sodium) is enormous, even though urinary output is considerably diminished.

By far the most complex adaptations to temperature involve the endocrine system (see Figure 12.2). Endocrines alter the basal metabolic rate to generate more or less heat from food. Because different kinds of food, such as fats or carbohydrates, have different capacities for heat generation, dietary needs change. Levels of thyroxin (which regulate carbohydrate metabolism) change, as do levels of iodine required to make thyroxin. Thus someone living at a cold, high elevation would need more iodine to maintain a high basal metabolic rate and would be more at risk of goiter than someone in a warm place, even if iodine were equally available in the environment. In fact, iodine is less available in glaciated mountains, which have soils deficient in iodine and are usually remote from the transportation systems that would facilitate trade and the import of iodine-rich seafood, so goiter is often a serious problem in such areas.

Another endocrine system, the adrenocortical system, stimulates the pancreas to produce more or less insulin and controls the metabolism of the liver. It also affects capillary resistance and the removal and excretion of waste products from the blood. Even the relative amounts of types of blood cells change from winter to summer. Blood volume also increases under heat stress and decreases under cold stress. An important adaptation to cold is deposition of a layer of fat under the skin. The work of the heart, fat deposition in the blood vessels, blood-clotting time, blood sugar levels, and innumerable other physiological characteristics also differ in people acclimatized to different degrees of heat or cold, and even within an individual in different seasons.

Acclimatization to high altitude affects far fewer people, but in some ways is more dramatic than adaptations to heat or cold. The greatest adaptive stress is from low oxygen tension in the rarified atmosphere: more red blood cells must be produced, and hemoglobin (and iron nutrition) must be increased. The body must change its endocrine balance to prevent acidosis resulting from changed levels of carbon dioxide in the blood. Permeability of cell and capillary walls, and the facility with which red blood cells give up oxygen to body tissue, must be increased. Lung size, surfaces, and permeability are altered. When fully adjusted to elevations of over 10,000 feet (3,000 meters), the physiology of people alters sufficiently to create much more susceptible, and often hazardous, reactions to therapeutic doses of drugs. Fecundity also seems to be lower at high altitudes. When people acclimatized to high altitudes descend to sea level to seek work, they may be susceptible to respiratory infections.

THE INFLUENCES OF THE WEATHER

Weather is experienced by people on an immediate basis. It changes hour by hour and mile by mile. The frequency with which, and the amount that, it changes— the variability of weather—are its most important characteristics for health. The extremes of conditions reached are also significant.

Air masses—huge chunks of relatively homogeneous air flowing over the Earth's surface, with a depth of up to a few kilometers—are as familiar to television weather report watchers as the fronts that separate air masses. Although tremendous energy exchanges occur at the fronts, the air masses do not generally mix. They retain their own characteristic temperature, humidity, atmospheric (barometric) pressure, and other properties. The fronts that separate them are accompanied by winds and, especially if associated with squalls and thunderstorms, by the generation of ionization, extremely low frequency (ELF) waves, and infrasound (too low in frequency for humans to hear). Let us review these individual elements and their health implications before considering what happens when fronts pass or extreme conditions develop.

Temperature, Humidity, and Air Movement

Temperature, humidity, and movement of air together determine how readily the human body loses its heat. Such heat is generated by metabolic activity, even when the body is at rest. We have considered the major physiological processes by which the body controls the generation, conservation, and dissipation of heat during acclimatization. The ability of the body to radiate heat depends on the surrounding temperature. The effectiveness of evaporative cooling depends on the surrounding humidity and air movement. The heat index (a combination of temperature and humidity) and wind chill (a combination of temperature and wind velocity) measures portray environmental stress much more accurately than a simple measurement of temperature, which is commonly used by medical scientists.

The most obvious health impact of the weather involves extremes of heat and cold. Death from exposure to cold involves no mysterious mechanisms: When the body's core temperature falls below a certain point, its chemical reactions slow down and finally cease. Heat causes increased peripheral circulation, which can be stressful in its own right, and copious sweating can drain the body of electrolytes. In heatstroke, the sweating mechanism shuts down, allowing body temperatures to rise rapidly. Certain proteins in the brain can permanently change, even if the body is cooled before death results.

Heat waves have long been associated with increased mortality. This association seems to be discovered *de novo* after every event. Increased heat-related mortality usually occurs in the first 2–3 days following the heat wave's peak. Greenberg, Bromberg, Read, Gustafson, and Beauchamp (1983) studied the 1980 Texas heat wave and compared it with the heat wave of 1950 and the period 1970–1979. Temperatures of over 100°F (37.7°C) existed for 61 of 71 days in 1980, and 107 deaths were attributed to heat. Not surprisingly, the risk of death was highest among those engaged in heavy labor and among the elderly. Death rates were higher during heat waves of former years; this was presumed to be because of the increased prevalence of air conditioning over time. In terms of our triangle, the lack of access to the (behavioral) technological intervention of cooling in the built habitat is a killer. Yet climatologists, who are geographers, can focus so intently on the natural habitat that they can fail to consider cultural buffers or even the built environment. Kalkstein and Davis (1989), for example, conclude that "regional acclimatization" appears to be especially important in summer, since the strongest associations of heat with mortality occur in parts of the United States where hot weather is uncommon (the Pacific Northwest and the continental North), and the weakest associations occur where heat is most common (the Southeast and continental South). This they took to be evidence that people respond to weather in a relative rather than an absolute fashion because the population responds more dramatically where hot weather occurs rarely. In social sciences, addressing such associations, one speaks instead of the vulnerability of a population, usually due to socioeconomic circumstances. Today the social risks involved in vulnerability include the increased numbers of elderly persons (whose physiology is less efficient at cooling) living alone; the risk of living alone, compounded when people decline to open their windows for improved circulation because of fear of crime; and, of course, lack of money to buy cooling technologies. Chicago, St. Louis, and many other U.S. cities have responded to these risks by organizing the provision of fans and warning family members and neighbors to check on the elderly.

There were almost 700 deaths attributed to a heat wave in Chicago during a 5-day period in 1995; most who died were poor, elderly, and socially isolated people

(Semenza et al., 1996; Johnson et al., 2012). Most heat-related index research has focused on mortality in particular heat waves in cities. Kovach et al. (2015) investigated heat-related morbidity throughout North Carolina and found that rural areas had greater morbidity than urban areas. They found that the highest rates were in poor areas with labor-intensive agriculture with poor housing stock including mobile homes.

There are several population predisposing factors for suffering heatstroke. These include the presence of such degenerative diseases as cardiovascular disease, renal (kidney) disease, and hypertension; preexisting acute diarrheal or febrile disease; salt deficiency and dehydration; and the use of drugs that affect the thermoregulatory system (alcohol, amphetamines, and such therapeutic drugs as diuretics and anticholinergics). People who undertake sustained activity in the heat are at risk, as are people whose lack of fitness or acclimatization makes their cardiovascular systems less able to cope with the necessary heat dissipation.

In August 2003 in Europe, almost 15,000 more people died than expected for the average August; in Italy alone there were 4,000 excess deaths; altogether at least 70,000 lives were lost in the largest European natural disaster in 50 years (Robine et al., 2008). The heat wave was caused by a high-atmospheric-pressure cell that sat unmoving and blocked any instability, winds, or rain. Switzerland and other places had the hottest summer in 250 years. In addition to the heat, however, it was important for France that during August most of the population goes on vacation, including doctors and hospital staffers. Elderly persons at home without air conditioning in this unexpected weather, with few cultural buffers in place and few social networks, were found dead (sometimes days later) or taken to hospitals that were not prepared to receive them. Shocked and now forewarned, French social welfare and medical care systems responded and planned, so that in the heat of the summer of 2008 there was no notable excess of deaths. A 2010 heat wave in Russia was responsible for about 55,000 excess deaths, likely because the built environmental buffers could not withstand temperatures above 100 degrees Fahrenheit. Buffers to such extreme temperatures such as air conditioning will not be available to the poor. An example of this is in central and south India during 2015 when temperatures reached 118 degrees Fahrenheit in several cities and there were approximately 2,500 excess deaths during a 10-day period beginning at the end of May. While most who died were poor, there were massive power outages so some of the affected were not.

Heat waves have also been associated with increased crime, especially homicide, but the pattern is not systematic or consistent. A plausible biological pathway is that heat interferes with sleep and therefore with dream cycles. This topic cries out for some behavioral-vertex research. As noted above, an increase in the number of deaths following heat waves has been repeatedly observed. It is not clear, however, how much of the increase in mortality occurs in those about to die from degenerative disease anyway. Several studies have reported less-than-normal mortality shortly after the peak associated with a heat wave, but the decrease does not usually offset the previous increase. There is some concern that people surrounded by air conditioning may be losing their acclimatization to heat, so that they are less able to cope with extreme heat stress. Mortality, however, is certainly higher among those socioeconomic groups least likely to enjoy air-conditioned homes.

Air-Borne Life

The air is charged with living things. The transport and survival of bacteria, viruses, fungi, and allergens such as pollen depend on certain conditions of atmospheric temperature, humidity, condensation, and movement. Agents of human disease are injected into the air by coughing and sneezing; by the shedding of hair and dead skin; and by the spray of cooling towers, air conditioners, and irrigation systems. Soil bacteria, fungi, and pollen are picked up by the wind. Dispersal depends on atmospheric turbulence. To all such life forms, UV radiation ultimately is lethal.

Temperature and humidity are the limiting factors for survival in air; they act together with different effects on different organisms. Bacteria often can withstand extremes of temperature, but may be seriously affected by humidity. The ubiquitous intestinal bacterium *Escherichia coli* survives for less than 3 hours at 68°F (20°C) and 50% relative humidity, but *Mycoplasma* of human pulmonary origin can survive up to 5 hours at 82°F (28°C) and 50% humidity. Viruses can multiply only in living organisms, but they can survive in the air anywhere from a few seconds to a few hours. Influenza viruses survive longest in cool but nonfreezing air, and stay suspended longer when there is low humidity (Lowen, Mubareka, Steel, & Palese, 2007). Polio viruses have been shown to be progressively desiccated in the air, but to have sufficient longevity for dissemination over several miles. Tolerance and preference for environmental conditions are highly specific to the type of organism. In one study, for example, a polio virus survived best at 80% relative humidity, but a vaccinia virus survived best at 20% (Hyslopo, 1978).

Air-borne bacteria and viruses have caused epidemics miles from their sources. An outbreak of Q fever in San Francisco, for example, was traced to the fumes from the fat-rendering plant of a slaughterhouse. The combined effects of temperature and humidity are thought to result in different survival times for air-borne microorganisms in summer and winter, and thus to be critical factors in the seasonal incidence of disease (described later in this chapter). Both fungi and pollen are suspended in the air. A few fungal infections of the lungs are serious diseases, especially histoplasmosis and coccidioidomycosis. Their distribution is clearly related to temperature, humidity, rainfall, and local winds, which pick up the fungi from disturbed soil. The distribution of human disease may be limited by prevailing winds as well as soil temperature and moisture. Ragweed, the most common pollen source for the hay fever that afflicts more than 6% of the U.S. population, grows best on cultivated land. Again, rain, wind, and humidity affect dispersal.

Atmospheric Pressure

The weight of the atmosphere presses on the surface of the Earth. Subsiding air exerts more pressure than rising air. One atmosphere of pressure, usually defined as 1,013 millibars (mb), is felt as 14.7 pounds per square inch (1,034 grams per square centimeter) at sea level. Atmospheric (barometric) pressure changes are slight in equatorial zones. Over North America normal pressure changes between air masses are on the order of 25 mb. At high latitudes pressure changes may reach 120 mb. The most extreme low and high pressures ever recorded on Earth amount to a difference of about 3 pounds per square inch (211 grams per square centimeter). (Altitudinal changes, such as those encountered in balloons or depressurized aircraft, can cause much greater pressure differences.)

The major mechanism postulated for why normal changes in atmospheric pressure affect the body's biochemistry is that body volume expands slightly, leading to retention of water and therefore to alteration of electrolyte balance. Eventually levels of disequilibrium are reached that trigger intervention by higher homeostatic control systems to restore proper fluid levels. These changes would result in water storage in certain parts of the body—in joints, eyes, and so forth—and could cause joint pain, glaucoma pain, increased blood pressure, blood clotting, and general irritability.

Winds

Winds blow between the different atmospheric pressures that characterize air masses, and between zones of subsiding and rising air, as part of the heat balancing of atmospheric circulation. The strength, frequency, and duration of winds vary greatly from place to place. Winds transport dust, fungi, bacteria, and insects long distances. Early approaches to eliminate river blindness in local areas of West Africa, for example, were stymied by seasonal winds carrying in new infected adult blackflies.

Winds have another effect. The movement of air, especially air with low humidity, promotes the ionization of atmospheric gases. Electrons are stripped from their atoms, producing positive and negative charges. In the extreme form associated with convective buildup of cumulus clouds, segregation of electric charges may result in lightning. The degree of atmospheric electrical charge is one characteristic of an air mass. Dry winds, such as the Santa Ana, *foehn, sirocco,* and *harmattan,* are associated across cultures with irritation, bad temper, accidents, and violence. Despite an extraordinary richness of folklore, however, there has been little in the way of controlled studies.

SEASONALITY OF DEATH AND BIRTH

Death Seasonality

It has been known for at least 2,500 years that some causes of death occur more often in one season than in another. A century ago it was common knowledge in the United States that in summer people died of malaria, yellow fever, cholera, typhoid, gastroenteritis, and tuberculosis, whereas in winter people died of influenza, stroke, and cold-related causes. The seasons are not much changed, but many of the former mortality patterns have been obliterated. How have people caused this?

The late-20th-century literature on the seasonality of mortality was dominated by one person, Masako Sakamoto-Momiyama (1977) of the Meteorological
Research Institute in Tokyo. Her work over 25 years discovered, explained, and
popularized the importance of seasonal patterns and how they are changing. After
analyzing the seasonal pattern of occurrence of scores of diseases at various latitudes; in various climates; and among different age, socioeconomic, race, and ethnic groups, Sakamoto-Momiyama developed general models of variation (Figure
12.3). Mortality patterns in most midlatitude countries were bimodal; that is, they
had both winter and summer peaks. With economic and social development, the
summer peak disappeared, and mortality now had its highest incidence in winter.
Sakamoto-Momiyama differentiated several types of shifts in seasonality. In the
transitory type (cancer in Figure 12.3), the summer peak of incidence gradually
shifted into autumn as it decreased, perhaps moving toward a winter peak. The
reversing type reversed from a summer to a winter peak. It had two subcategories:
diseases for which the former summer peak became a trough and a new winter peak
was created (reversing type A), and diseases that formerly had both a winter and
a summer peak but lost the summer one (reversing type B). Sakamoto-Momiyama
classified gastroenteritis and tuberculosis as type A, and many of the degenerative
diseases associated with aging as type B.

As Sakamoto-Momiyama (1977, p. xiv) related, while doing research in 1965–
1966 in New York, she was surprised to find the winter peak disappearing for many
causes of death in the United States. Her detailed studies of infant mortality in

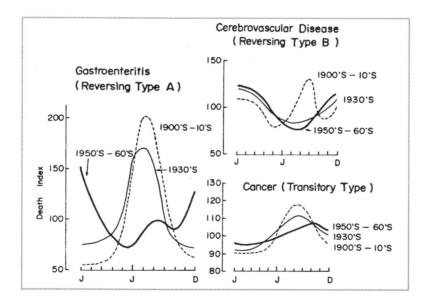

FIGURE 12.3 Models of seasonal shifts in mortality. When the death index by month is plotted for
several time periods, the peak of mortality for some diseases reverses from summer to winter; for
others the summer peak gradually disappears, leaving a previous winter peak as the only seasonality;
for still others the summer peak decreases and shifts toward fall. From Sakamoto-Momiyama (1977,
p. 33). Copyright ©1977 University of Tokyo Press. Reprinted by permission.

particular led her to revise her previous idea that technological progress would lead to concentration of mortality in winter, as she had concluded from her experiences in Japan. Instead, she hypothesized that a "deseasonalization" of mortality would occur. Her later work noted that Japan and Western Europe were following the United States and Scandinavia into this deseasonalized pattern of mortality (Sakamoto-Momiyama, 1977).

Why does this matter? It is amazing that people in the mass, society, could do this. The relevant question is how human behavior could alter such a strong and stable pattern of mortality association. The seasons are constant, but obviously human–environmental interaction has been altered fundamentally. There can be many reasons in cultural ecology why mortality incidence would lose seasonality under conditions of economic and technological development. Heavy labor and exposure to the elements decrease. Diet changes, and a variety of nutritious foods become available all year. Medical technology advances. Health care becomes more accessible in winter. For some causes of death, one can point to specific behavioral developments: refrigeration and the chlorination of water supplies control summer gastroenteritis, for example; vaccination for polio means that swimming pools no longer start summer epidemics; changes in house type, screening, occupation, and other factors have led to the eradication of such diseases as malaria and yellow fever in some countries. Sakamoto-Momiyama (1977) advanced the hypothesis that much of the deseasonalization of mortality was due to central heating and air conditioning. She pointed to the relationships between temperature and cerebrovascular disease (Figure 12.4). Sweden and New York City, which had widespread central heating by 1960, showed consistent linear relationships without changes of slope

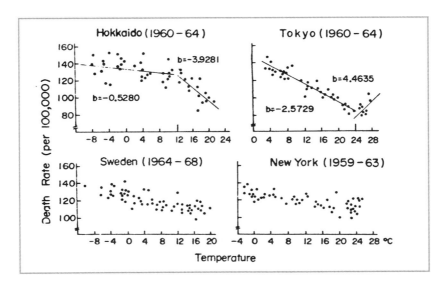

FIGURE 12.4 Association between cardiovascular mortality and temperature. In Hokkaido (northern Japan), Sweden, and New York, deseasonalization of mortality seemed to be associated with use of central heating. From Sakamoto-Momiyama (1977, p. 165). Copyright ©1977 University of Tokyo Press. Reprinted by permission.

at high temperatures, whereas Western Europe and Honshu in Japan, where many houses were still not heated in 1960, had seasonal changes of slope. In Hokkaido, which is cold in winter, people used heating systems that were turned off when the temperature reached 54°F (12–13°C); the winter slope was small until that point and afterward resembled Honshu's (represented by Tokyo in Figure 12.4).

Kevan and Chapman (1980), for example, confirmed the deseasonalization of many diseases in Canada, but found great seasonality continuing for bronchitis, pneumonia, influenza, and circulatory diseases (winter) and for violence, accidents, and poisoning (summer).

Influenza continues to be a winter disease. The association of cooling winter temperatures with increased influenza mortality, after a certain lag, is unquestionable. The association persists over widespread locations and different climates.

Speculation has blamed several factors:

1. Biometeorological conditions cause the nasopharynx and trachea to be dry, and membranes to become more susceptible to virus penetration. (This is also a problem often associated with air travel.)

2. The virus can survive in the air between hosts more easily when the air is relatively dry and cold than when it is hot and humid.

3. The lower solar radiation, and hence UV radiation, promote virus survival.

4. The body's seasonal metabolic changes make it more susceptible.

5. People gather indoors in closer proximity for longer periods in winter; schools are in session; and even recreation tends to be indoors—points emphasized by the importance of room density as a correlate in many influenza epidemic studies. However, prisons are crowded all the time, and influenza is a winter disease there too. Any explanation of influenza's seasonality must account for its winter incidence in Florida and Hawaii, on the Great Plains, and in New England, and in perpetually crowded institutions such as prisons.

Not only infectious diseases are involved in seasonality. Mortality from heart disease is negatively correlated with temperature in several midlatitude countries. Between 30° and 70°F (–1° and 21°C), correlation coefficients of –.95 have been found. A 2.5% increase in mortality has been estimated for each degree Celsius decrease in mean monthly temperature, although the exact relationship differs by climate area. Sometimes mortality from ischemic heart disease but not cerebrovascular disease, and sometimes the reverse, have been correlated with temperature change.

In general, respiratory disease shows the strongest association with winter, and cancer shows the least seasonality—which is hardly surprising, given its long incubation period. The meteorological conditions and physiological stresses of acclimatization are undoubtedly involved in seasonality, but so are culture rhythms and technology. The etiology of continuing seasonal patterns of mortality remains puzzling.

Birth Seasonality

Seasonality of birth, found everywhere, has been much less studied than seasonality of death. The seasonal incidence of birth in Japan is quite different from that in European countries, and that of the United States is again distinct. The index of birth in Figure 12.5 illustrates the bimodal pattern that has persisted in the United States for decades but has shown recent signs of change. There have been a few good descriptions of the seasonal patterns by region, race, and other population characteristics in the United States and Europe, including at least one study in which no seasonality was found in a population (Arcury, Williams, & Kryscio, 1990). Despite minor differences in the amplitude of seasonality among population subgroups in the United States, however, the pattern of seasonality has been remarkably similar. The relative peaks vary in magnitude and timing across the country; they have changed over time, but the manner of this change has not yet been studied.

One possible reason for birth seasonality is that high temperatures may have an effect on spermatogenesis (the creation of sperm). High temperatures may also injure existing sperm. As with other biometeorological effects, threshold and range are presently unknown. How high a temperature over how long is necessary to

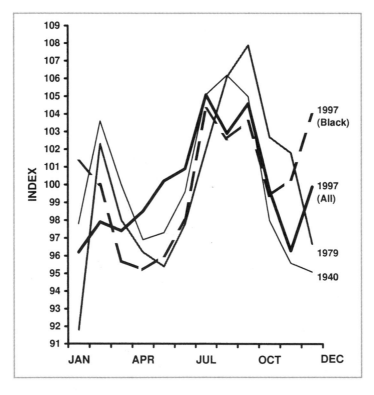

FIGURE 12.5 Seasonality of birth in the United States. The bimodal pattern has remained remarkably constant for decades. In recent years, the April trough (summer conceptions) and February minipeak have been lost overall, but African Americans still retain the southern pattern.

affect fertility? Is a single episode, such as playing tennis in tight athletic clothing on a hot day, enough exposure? Or are weeks of hot, humid weather needed? Is birth, like mortality, also becoming deseasonalized?

The seasonal patterns of birth may have nothing to do with temperature. They may be a result of agricultural patterns; school calendars; the timing of holidays and vacations, Christmas–New Year celebrations, and leisure time in the United States; and the timing of equivalent events in other cultures. When people migrate, does their seasonality of birth pattern change relatively suddenly under the new environmental conditions or gradually, over a generation or more, through cultural adaptation?

Birth seasonality has innumerable consequences for health and disease. The most obvious is the variable need for health services. Obstetrical wards built for peak periods will have many idle beds during the trough. Birthing classes and maternity leaves from work are affected. More significant effects, however, may involve congenital birth defects, premature births, and neonatal mortality (Keller & Nugent, 1983). The point often made by scientists who study congenital disease is that if there is seasonality to it, the condition is not the result of genetics alone. Something in the environment after conception—whether in the womb, at birth, or after birth—must actually cause the expression of the disease. This is often referred to as gene–environment interaction. As bones, nerves, palates, and endocrine systems form at different times in the development of the fetus, biometeorological influences that may be relevant to congenital disease may not occur at the time of conception at all. This is especially true for one mental disease, schizophrenia. A strong peak of birth for individuals with schizophrenia seems to occur during late winter and early spring in many countries. These individuals would have been fetuses in their third month, when the central nervous system is forming, during the late summer heat that is associated (in the United States) with the minimum of conception. This, of course, is also (especially at high latitudes) the time of maximum sunlight and effect on serotonin production and other hormone cycles.

The presence of seasonal patterns in the occurrence of human birth is ubiquitous on Earth. The patterns of northern Europe are quite different from those of tropical Africa. In the United States the pattern has been consistent throughout the 20th century: a minor peak of births in February–March (spring conception), a great concentration of births in August–September (early winter conception), and a deep trough of births in April–May (hot summer conception) (see Figure 12.5). There has been spatial variation in the exact timing and sharpness of this pattern between the North and the South, or between depression and growth decades, but the overall pattern has held for at least half a century—despite the facts that the U.S. population has become overwhelmingly contraceptive-using, has seen massive migration from South to North, has become three-quarters metropolitan instead of rural, and has seen the origins of immigration shift from eastern and southern Europe to Asia and Middle America. Only during the end of the 20th century did the pattern of seasonality begin to alter (Figure 12.5).

The fact that there is a seasonality to human birth that has a spatial pattern; varies in associations by scale of analysis; and reflects complex interactions of biology, environment, socioeconomic and demographic structure, and behavior poses a challenge to interpretation and explanation in human ecology and health geography. Those researching and publishing in biometeorology, anthropology, biological reproduction and fecundity, demographic sociology, or epidemiology often do not read each other; they lack geographic integration.

Since the efforts of geographer Ellsworth Huntington (1938), there has been remarkably little research on the etiology of seasonality of birth. Although his statistical methodology would not pass in an undergraduate term paper today, he asked some profound questions about the seasonal patterns he documented with data collected from all over the world. He ran latitudinal transects (straight lines along which data are collected) through the United States and Japan to examine climatic change in different cultures. In the former Soviet Union he compared the same ethnic groups in different locales and different ethnic groups in the same locale. He was interested in the effect of climate on people, and was the most famous and accomplished proponent of a school of thought known today as *environmental determinism*. Probably because of the subsequent discrediting of many aspects of the school of environmental determinism, geographers have ever since ignored the questions he raised.

PHYSICAL ZONATION OF CLIMATES AND BIOMES

The diurnal and annual variations in intensity of solar radiation, and the amount and timing of precipitation, temperature, and humidity, have been described, classified, and analyzed in dozens of ways. The variations are complicated by the tilt of the Earth's axis, the physiography of mountains and depressions, the contrasts of land and sea, and the size of continents. The broad zones of vegetation, or biomes, that result are also zones of agricultural cropping patterns and of arthropod habitat. The schema in Figure 12.6 of a prototype continent is simplified but provides a general framework for considering the study of health through the physical geography of the earth.

Seasons occur as the tilt of the Earth's axis alternately exposes the northern and southern hemispheres to more direct rays from the sun. The vertical rays are limited to the latitudes between the Tropic of Cancer and the Tropic of Capricorn, with the highest annual amounts being received at the equator and the least at the poles. Because hot air rises and cold air sinks, the equatorial area is a zone of low pressure as the air rises, and the poles are zones of cold, subsiding, dry air. Because cold air can contain less water vapor than an equal volume of warm air, water condenses as the air rises at the equator and cools at higher elevations, producing continually high levels of rainfall over the year. Similarly, the subsiding air at the poles is warmed as it approaches the Earth's surface, and because its capacity to hold water vapor is increased, there is little precipitation. Due to processes of atmospheric circulation, air also subsides at about 30° north and south latitudes, forming a zone of little precipitation at each latitude. Because of the lack of cloud

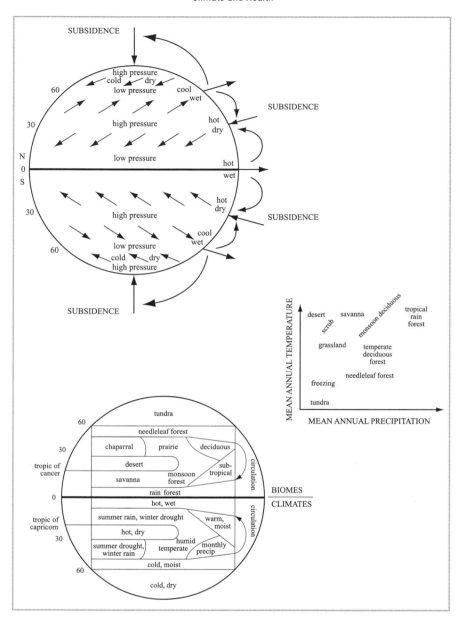

FIGURE 12.6 Physical zonation of climates and biomes.

cover, this zone receives more radiation at the surface than does the equatorial zone. As subsiding air at each 30° latitude and subsiding air at the poles move over the Earth's surface and converge, air rises along a broad frontal zone in the middle latitudes and creates a wet but cooler zone. There is less precipitation in this midlatitude zone than in the equatorial zone, but because there are also less evaporation and less transpiration from plants, there is high availability of water.

Different vegetational zones, known as biomes, extend over large areas. The biomes are adapted to differences in temperature and precipitation, as well as to seasonal variation. The complex and luxuriant rain forest needs continually wet conditions and high solar radiation, whereas lichen and other tundra vegetation can grow in cold, dry conditions with a short season of long days of solar radiation. As Figure 12.6 indicates, from the poles to the equator there is a transition of forest types: from needleleaf (coniferous, taiga, boreal) to temperate deciduous forest (which loses its leaves in the cold season), to subtropical forests (which include evergreen plants such as magnolias and palms), to the tropical rain forest. Desert may be hot or cold, but it is dry; types of desert vegetation such as creosote bush and cactus merge into types of grassland or evergreen scrub as water increases. The savanna biome is grassland (with scattered monsoon deciduous trees) that has adapted to conditions of summer rain and winter drought. At its dry margins the trees are sparse and the grass short, but as it extends toward the equator the trees increase in density until savanna merges into forest. "Chaparral" is the American name for the biome of evergreen brush that has adapted to the severe conditions of summer drought and winter rain. Grassland, known variously as prairie, steppe, pampas, and veld, extends from short, bunched grass on the arid margins to tall, dense grass on the woodland border.

A last complication is introduced by altitude. The atmosphere cools with increasing altitude. Mountains, furthermore, form barriers so that surface air must go up on the windward side, causing precipitation, and come down on the leeward side, forming a "rain shadow," or dry area. An altitudinal zonation of vegetation results that is very similar to the latitudinal zonation that was presented earlier. A tall mountain on the equator (as occurs in Kenya and Ecuador) has a vegetation zonation that proceeds from rain forest at its base to subtropical, deciduous, needleleaf, tundra, and ice on its windward side and from rain forest to savanna, thornbush scrub, needleleaf, and tundra on its leeward side. No matter in what climatic zone people are living, therefore, a great variety of vegetation, agricultural, and living conditions exist for human settlements.

The distribution of diseases transmitted by arthropods—ticks, mosquitoes, flies, and so forth—in different parts of the world can be partly understood from biome distribution. Certain nutritional deficiencies are associated with specific staple crops, and cultivation of these crops often follows environmental constraints of biome distribution. Conditions of hazard from air pollution (see Chapter 11) or solar radiation (this chapter) also largely coincide with biome distribution patterns. Understanding the location of major earth biomes can provide a useful framework for understanding the spatial patterns of the distribution of many health hazards and it is necessary to understand the spatial patterns of the Earth's biomes in order to understand how climate change will influence health in different parts of the world in the future.

CLIMATE CHANGE AND HEALTH

This section begins with a short discussion of **climate change** then reviews how models of climate change have been used to predict how it will impact disease distributions. Many thousands of land and ocean temperatures from around the world have been used to develop a record of the global average temperature. As can be seen in Figure 12.7, there is a long-term **global warming** trend (Wise, 2010, 2012; National Oceanic and Atmospheric Administration [NOAA], 2015). Human activity is the main driver of global warming with the increase in greenhouse gases especially CO_2 (Intergovernmental Panel on Climate Change [IPCC], 2013). Figure 12.7 shows the CO_2 concentration of the Earth's atmosphere since 1880, which has risen dramatically and is projected to rise into the future, thus leading to increases in global warming. Because of global warming, sea levels are rising and precipitation patterns are changing and these effects will vary spatially in different areas of the world. Plate 14 is a global map of the variation in surface temperature changes and Plate 15 is a global map of precipitation changes.

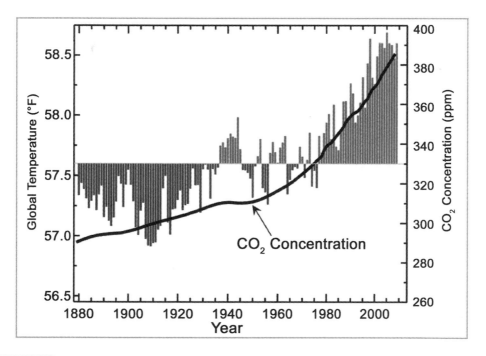

FIGURE 12.7 Global annual average temperature measured over land and oceans and atmospheric CO_2 concentration from 1880 to 2015 (NOAA, 2015).

Climate change is predicted to have many negative effects on human health including extreme heat, natural disasters, variable rainfall patterns, and increased water-borne diseases (World Health Organization [WHO], 2014; IPCC, 2014). According to the WHO, rising sea levels and extreme weather events will lead to more flooding, which will destroy homes, and variable rainfall patterns are likely to affect water supplies and increase the risk of diarrheal and other water-borne diseases. Changes in climate could lengthen the transmission seasons of several vector-borne diseases and change their geographic ranges. Malaria and cholera are two diseases that are strongly influenced by climate and thus more changes would be expected.

While climate change will have impacts on human health, presently there is limited research showing what the impacts will be and where they will occur. The media is not bothered by the limited research. News outlets spread their impressions of information, facts still being gathered, and half-baked hypotheses in the scientific studies addressing the momentous process that is upon us. The climate is changing, and the biosphere of Earth is warming. The media reports, however, leave out many aspects of scale and proportion; and, like those who discovered the effects of heat waves described earlier in this chapter, they almost always leave out the cultural buffers and social organization we first discussed in Chapter 2.

The issue of health consequences from global warming needs to be addressed. The field of health geography is well positioned to conduct studies that measure the likely impacts on health. There are also many misunderstandings about what climate change means for a particular place, and whether events there right now prove or disprove that warming is happening. There are now many efforts to understand the effects of climate change on health. Government agencies and scientific organizations are in the agenda-setting period (McMichael et al., 2003; Confalonieri et al., 2007; Ebi et al., 2008; United States Global Change Research Program [USGCRP], 2009; National Institute of Environmental Health Science [IEHS], 2010) and individual scientists and groups of scientists have begun conducting studies to tackle the agendas, several of which are cited below.

One common misperception is that global warming will happen wherever one happens to be, and at the amount forecast globally. The questions then become whether Minnesota, Texas, or France is warming 1° or 3° and by what decade. One may then be tempted to respond, "So what? That's not so much hotter than it is now." Energy comes mostly from the vertical rays of the sun moving over the equator from one tropic to the other. This tropical heat has to be circulated and dispersed around the globe. Climatic zonation and associated biomes were described earlier in this chapter. The process of global warming will cause some climatic zones to expand latitudinally and altitudinally and may alter seasonal timing in a particular place. The exchange of energy will result in warmer seas, causing stronger hurricane winds, shifting ocean currents that transport heat, and (probably most important for human health) shifts in the pattern and amounts of precipitation. Precipitation that falls as rain instead of mountain snow will run off and not be available to sustain vegetation and agriculture through gradual snow melt in summer. Monsoon extremes and unpredictability will increase. There may or may not be more storms, but the hurricanes and tornadoes that occur will frequently have stronger winds. The initial forecast of the 1970s climatic models was for "growing extremes"—the blizzard of the century again, the 100-year flood repeated, the "worst drought" recurring. Using the classifications that we have used so far in this text, Table 12.1 summarizes these and other concerns. It incorporates health-related ideas from the Centers for Disease Control and Prevention's Policy on Climate Change (CDC, 2014); Committee on Climate, Ecosystems, Infectious Diseases, and Human Health (2001) of the National Research Council; and the International Panel on Climate Change's Fourth Assessment Report of impacts on human health (Confalonieri et al., 2007).

Biometeorology

Fortunately, the tilt of the Earth's axis and the Earth's rotation around the sun cannot be changed by greenhouse gases. Solar intensity, sun angles, and seasonal shifts in the vertical rays remain constant. There may be some shift in visible radiation

TABLE 12.1. Climate Change and Warming: Health Impacts

Potentially impacted	Weather event	Health effects
Biometeorological	Heat waves	Heat stress; heat stroke
	Cold; seasonal change	Respiratory distress
	Fronts, air masses	Arthritic attacks, CVD effects
Air	More heat, humidity, atmospheric inversions	Allergies to increased pollen, fungi; aerosol contagion exposure to pollutants
Water-borne pathogens	Drought	Shortage of water for hygiene and sanitation increases disease transmission, virulence, compromises buffers
	Flood	Groundwash of enteric pathogens, chemicals into water supply
	Warming	Greater and longer survival of pathogen transmission
Water-based pathogens	Drought	Reduction and elimination of vector breeding except for domiciliated vectors/reservoirs which increase contact
	Flood	Increase in mosquito & other vector breeding
	Warming	Greater winter survival & altitudinal extent of vectors
Nutrition	Drought, flood, warming causing salinization, failed pollination, and other ecosystem disruptions and economic consequences; repetitive crop failure	Regional: severe malnutrition; local: human and livestock starvation; forced migration to cities and consequences of crowding in inadequate sanitation
Hazards	Number and severity of storms, flooding increase; coastal land loss to sea rise; mass population displacement	Increased trauma (drowning, injury); exposure to elements (PTSD, depression)

Note. CVD, cardiovascular disease; PTSD, posttraumatic stress disorder.

(see the earlier discussion of the radiation spectrum). Greater differences between air masses may affect arthritis or the timing of heart attacks as the fronts pass. Cold in severe winter storms, "dropping" from the Canadian Arctic deep into the U.S. South, may affect winter mortality. These are minor concerns; the main effect will be through heat waves (Robine et al., 2008). As you have read, however, we humans can erect strong cultural buffers and use our cultural adaptability to cope. France was much more prepared, with community and medical support, for the heat in the summer of 2008 than for its historic heat wave in 2003.

Air

Most of the attention about nongaseous changes to the air we breathe has been paid to the particles (spores of fungi, pollen, etc.) suspended in it and to carbon monoxide, ozone, and other pollutant gases (see Chapter 11). Mold allergies and hay fever will probably be adversely affected. Concerns that the warmer, more humid air will raise the risk of aerosol contagion, however, overlook the fact that viruses such as influenza like it cool and dry so that they can stay suspended in the air and maximize contagion. In some places, warming may make for less influenza! Stronger winds over dryer land will also pick up more dust. This has been a historic health problem for northern China, as well as for West Africa, where the latitudinal band of dust along the grassland biome (Figure 12.6) that has been associated with linear expansion of meningococcal meningitis epidemics may worsen.

Precipitation and Hydrological Effects

The central importance of water for the transmission of human disease has been developed in Chapters 2 and 11. Water-borne pathogens do not have a simple relationship with warmer weather, but decreased precipitation is sure to make health conditions worse. In water that is warmer, or that is warm for longer periods of time, some organisms like *Vibrio cholerae* may survive longer. The predators of amoebas, helminth eggs, bacteria, and mold may also thrive, however. During drought—as water sources shrink in area and as wells become almost airless masses of algae—not only can pathogens become concentrated, but humans and other animals lose the option of drinking elsewhere. In wet–dry climatic areas like northeastern Thailand, the cycle has always been that when the monsoon rains break, they wash the chicken, pig, bovine, and human manure on the ground into sheets of water that contaminate the wells; and during the dry season, when other surface water dries up and even fewer wells have water, people get sick from using the contaminated water. Something like this threatens to occur on a much larger scale. The cultural buffers of chlorinating and otherwise treating water, pumping deep water, or finding renewable fuels to boil water cheaply will become more important. Such buffers will need improved social and economic organization as well as new physical infrastructure to support them. This impact of warming works against the United Nations' millennial goal of greatly decreasing the proportion of the world's population without safe drinking water. Arnell (2004) estimated that with climate change global water resources will become severely stressed. He estimates that the number of people living in water-stressed watersheds will be 3.4–5.6 billion in 2055 with climate change as opposed to 2.9–3.3 billion in the absence of climate change and that the number depends on how water resources are managed in the future.

Perhaps the most attention has been given to the vectored diseases, especially those classified as water-based. Remember that effects of climate warming are going to be species-specific and not simple generalities. Greater drought and less water on the land after rain should mean fewer mosquitoes in general and less malaria and snails for schistosomiasis. But on the other hand, the need for greater container storage can easily result in more *Aedes aegypti* mosquitoes and greater risk of dengue fever, Zika, and other arboviruses. Hales, de Wet, Maindonald, and Woodward (2002) estimated that with projected climate change about 5–6 billion people would be at risk of dengue fever compared with 3.5 billion without climate change. With climate change, more reservoirs of water will also need to be constructed, and these will include drainage and irrigation channels that will support more snails. Zhou et al. (2008) predicted that with global warming there will be an expansion of schistosomiasis in China into currently nonendemic areas in the north putting an additional 8.1% of the land area of that country at risk of the disease. Without human intervention to make it worse, however, drought should be bad for more than water-based vectors, as tsetse flies, sand flies, and even reduviid bugs need humidity.

Climate warming does not simply mean greater drying, however. It means greater storms, more flooding, erratic monsoons, and displacement of downpours from where they have historically occurred to other places entirely. In recent years the U.S. Southeast has repeatedly had prolonged drought as a ridge of high pressure remained stationary over it for months, blocking passage to other air masses and the frontal rain these would bring, as well as suppressing any instability from ground convection that could produce thunderstorms. The Midwest and Pennsylvania got the Southeast's rain as well as their own normal precipitation, resulting in devastating floods in those regions. Is this just an example of normal weather fluctuations, or is it the beginning of a new climate pattern? It can be difficult to understand what one is in the middle of! The consensus, nevertheless, is that government should build more reservoirs and that citizens should learn and implement better water conservation.

Satellite imagery of water on the land in East Africa and after hurricanes in the United States is now used to predict mosquito-borne outbreaks of Rift Valley fever, West Nile virus, Eastern equine encephalitis (in the United States), and of course malaria. More subtle is the effect of warming on the survival of mosquitoes over winter. La Crosse encephalitis in the Midwest flyway is a virus that has been shown to pass transovarially under favorable conditions, such as a warm winter. *Aedes albopictus* mosquitoes have a range extending from India to Japan in Asia; in warm winters it can expand in the United States. The eggs and nymphs of ticks usually winter under the leaf debris. When the disease agent is passed transovarially, as is the rickettsia that causes Rocky Mountain spotted fever, more infection will survive the start of the spring cycle.

The scariest journalism reports that malaria will be spreading and malaria-causing mosquitoes will be breeding even in the United States. Most of the United States was malarious in the past and it still has *Anopheles* mosquitoes everywhere; the mosquitoes have not been eradicated. Malaria is frequently reintroduced to the United States when students and other travelers come back infected and are treated. What prevents the possibility of new malaria here, however, is not that the temperature is 1° lower than it might become, but that strong cultural buffers have been established—ranging from laws on water impoundment to air-conditioned, glass- and brick-walled houses with indoor plumbing. Might malaria and other mosquito-borne diseases occur over a wider area, and especially at higher elevations? There are already reports of malaria being transmitted in schools and villages higher in the East African highlands than occurred even a decade ago. However, these are marginal lands for the mosquitoes. The best defenses against expansion of the disease range are public policy, economic development, and better education.

Nutrition and Hazards

For all the attention given to mosquitoes and hay fever, the risk of all the possibilities described above is trivial compared to the much more likely impacts on human health of global warming: destruction of good agricultural land and crop cycles; impoverishment of regional populations; ecological disruption to birds, bees, and many other seasonal cycles; coastal flooding; and mass displacement in forced migration of hundreds of millions of people from such areas as Bangladesh, the Pacific islands, and the Mediterranean coast. In hurricanes (typhoons, cyclones) and other extraordinary rain events, people drown as their houses and the cars they're driving are washed away. Crops are destroyed; rice fields and other agricultural land are spoiled with salt ocean water.

The displacement of populations because of rising sea levels is the extreme, even if realistic, scenario. It is limited to specific, identifiable areas. The impact of global warming and climate change on food insecurity is much more pervasive and insidious. Not only will changed rain patterns cause droughts, floods, and storms, but frost and thaw will become erratic and difficult to predict. Warm winters will draw fruit trees out in February, when there are no insects to pollinate them. Insects and caterpillars will hatch out before bird migrations arrive for the nestlings to eat them. The biological rhythms of millennia with which agriculture has evolved will be disrupted. Not only will there be local crop failures of many kinds and widespread malnutrition, but systems of food trade, marketing, and transportation will need to raise their flexibility to barely imaginable levels.

Whereas increases in these nutritional, epidemic, and hazardous risks to human health are clearly based in the biology of climatic effects, their impact is a matter of human agency and organization—the behavior vertex of the triangle. Changes will not occur uniformly around the globe or to all people in any place. Low coastal areas are especially at risk. The poor are everywhere the most vulnerable. They have the least education and knowledge (human capital), the fewest social contacts and political power (social capital), and the poorest current health. Where will new physical, economic, and health care infrastructure go, and who will be served by it? Whether *Plasmodium falciparum* expands its territory, as has so often been predicted, depends more on success in distribution of insecticide-treated bed nets than on 1° of warming (Chaves, Kaneko, Taleo, Pascual, & Wilson, 2008). As Fothergill (1998) points out, men and women have different degrees of exposure in all phases of disaster, including risk, warning, preparation, response, and recovery. Some relief efforts have effectively targeted women with relief kits, building grants, and control over resources (e.g., after the 1999 cyclone in Orissa, India, and the 2004 tsunami in Thailand). The elderly, the very young, the poor, and those with disabilities are everywhere the most vulnerable to the health effects of climate change.

Remember the fundamental premise of Dubos and Audy: health is adaptability. This is true not only of individuals, but of community, social, and economic systems; political organizations; and world organs of cooperation and support (the strengthening of which was the target of one of the United Nations' millennial goals). Minimizing the deleterious health impacts of climate warming needs to be a project in political ecology.

QUICK REVIEW

✓ Climate, weather, and biometeorology make up the physical environmental conditions that affect people's health. *Climate* refers to weather conditions averaged and prevailing over long periods, many decades or centuries, of time. *Weather* is the daily and weekly conditions of temperature, humidity, precipitation, wind, and cloud cover. *Biometeorology* is the variation and change in the physical and chemical characteristics of the atmosphere that affect variation and change in the physicochemical systems of living organisms.

✓ Weather can have direct influences on human health. Extreme heat and cold can directly affect health, as can hazardous weather such as tornados and hurricanes. Cultural and social factors modify people's vulnerability to those hazards.

✓ The Earth can be divided into different vegetational zones called *biomes* that are associated with different weather and climate patterns. These biomes form the biophysical basis for disease risk in different areas of the world. The risks, however, can be overcome through cultural buffers such as the alterations to housing and the aquatic environments for malaria eradication in the southeastern United States.

✓ Global warming is predicted to be associated with rising sea levels and extreme weather events that will lead to more flooding and destroy homes and infrastructure. Variable rainfall patterns are likely to affect water supplies and increase the risk of diarrheal and other water-borne diseases.

✓ Climate change is predicted to indirectly affect human health and could be associated with lengthening of the transmission seasons of several vector-borne diseases and change their geographic ranges. Malaria and cholera are two diseases that are strongly influenced by climate and therefore could be affected by climate change.

CONCLUSION

Climate change will affect human health through both grand, sweeping patterns and displacements, and in local and particular ways. For example, vectored diseases (especially those vectored by mosquitoes) on a warmer Earth will extend their distribution a few degrees further in latitude and a couple thousand feet higher in altitude. Across large areas where they already exist, however, they do no harm because of local standards of living, education, health care systems, and vigorous cultural buffers. Expansion of vectors' extent even by miles in such areas will not matter. Since impacts are species-specific, broad patterns of impact may not reflect the local. Some mosquitoes (including some species of *Anopheles*), for example, prefer to breed in brackish water, although most prefer fresh water. Increased rainfall and runoff will freshen some brackish areas, even estuaries. Raised sea levels, including storm surge overwash, will salinate deltaic lakes, groundwater, and wells. Changes in local vector species will be locally affected to an extent little appreciated.

Climate change, of course, will have an impact on more than disease ecology. It will affect all the biometeorological factors described earlier in this chapter as well. There is little proof, however, of the direct effects of weather on health, aside from temperature extremes. Conditions and data vary with age, season, SES, and cultural buffering (e.g., through the use of air heating and cooling). The present status of research is such that little is known of the relevant periodicities. Variations in temperature, pressure, or other weather elements and their rates of change occur over seconds, minutes, hours, days, months, and years. Finding the pathways for influences is difficult because of the problem of specifying the relevant periodicity (Should we correlate heart attacks with temperature change in a day or over a number of days? With passage of a front or with the number of fronts that passed within a limited time?) and our ignorance of threshold or relevant range (Is a temperature change from 80° to 60°F the same in effect as one from 40° to 20°F? Can winter in countries whose mean daily temperatures never get below freezing be compared with that in countries whose temperatures stay below freezing for weeks?). The general scientific law of initial value, applied to biometeorological studies, would state that the magnitude of any weather influence is going to depend on both the level from which it started and the level of the physiology it is affecting.

The impacts of climate change on health will be overwhelmingly upon food security, nutrition, and the availability of usable water (for drinking, sanitation, and hygiene), for both cities and rural areas. We already know who the vulnerable are: the poorest, the children, the elderly, those living in low-lying areas. We can already develop GIS applications that map their location block by block, mile by mile, pretty much anywhere in the world. Planning for disaster relief—actually coping with the aftermath—is proving the integrative powers of GIS. What new technologies, organizations, and policies can develop and deliver which new cultural buffers? How can these be optimally located and designed in space? How can local populations, not satellites, be involved in information gathering, processing, construction, and implementation of effective helpful policies? These are research questions and applications for geography as social science.

REVIEW QUESTIONS

1. Climate can affect people's health in various ways because it forms part of the human habitat, which is one of the three components of Meade's triangle of human ecology. Create a diagram of a climate-related disease using Meade's framework of human ecology. Identify specific climate-related variables involved in the triangle for that disease and categorize each by type of habitat (i.e., natural, built, social).

2. In Chapter 2 you learned that Audy defined health as a continuing property that can be measured by a person's ability to rally from physical insults. Using a climate-related health problem (e.g., malaria, cholera, heat-related illness, SAD), diagram the life cycle of two different individuals using Audy's framework as shown in Figure 1.3. One individual should be from a very poor country with few societal protections from the climate-related health problem and another from a rich country.

3. Humans evolved in different parts of the world to be able to receive UV radiation from the Sun and synthesize vitamin D. Also, people have migrated around the world for millennia but the pace has increased dramatically during the past 200 years. Explain how humans differentially evolved and how rickets has become a modern problem because of migration. Where are your ancestors from and did they migrate from somewhere else in the world? Does this put you at an increased risk of rickets? Does it put you at risk of other health problems related to climate such as cancer?

4. Biological rhythms are related to problems such as jetlag and SAD. Explain the biological basis for these problems and how geography is also part of their etiological pathways.

5. Heat is an important biometeorological cause of morbidity and mortality seasonally in some parts of the world. Characteristics of places and individuals are important determining factors to where heat-related illness occurs. Using the neighborhoods and health framework (see Chapter 9), differentiate between the neighborhood- and individual-level factors that are related to heat-related illness.

6. In this chapter you learned that mortality is seasonal but in some countries including the United States and Japan, it has become less seasonal over time. Explain why in these countries mortality has changed and in many other poorer countries the seasonality of mortality remains.

7. What biome do you live in now and what are its climate characteristics? What health problems occur in your biome that do not occur in some of the Earth's other biomes? List some health problems that do not occur in your biome.

8. There is growing scientific evidence that human impacts have led to increases in atmospheric CO_2, which is causing a gradual increase in the annual global temperature of the Earth and changes in precipitation patterns. Table 12.1 lists some of the potential health impacts of climate change. What are some that you think could affect you during your lifetime either directly or indirectly and how?

9. You learned about cholera in Chapter 11 and the maps in Plates 14 and 15 show some of the spatial patterns of global temperature and precipitation change. Based on your understanding of how cholera is related to the environment and therefore could be affected by environmental change, where do you think there will be changes (increases and decreases) in the burden of cholera in the future? What can we do to minimize the effects of climate change on cholera?

10. Malaria is another disease that is related to environment because it has a mosquito vector and the spatial extent of their habitat could change. Based on the maps in Plates 14 and 15, how do you think the spatial distribution of malaria will change in the future? What cultural buffers might minimize the effects of climate change on malaria?

REFERENCES

Arcury, T., Williams, B., & Kryscio, K. (1990). Birth seasonality in a rural U.S. county, 1911–1979. *American Journal of Human Biology, 2,* 675–689.

Arnell, N. (2004). Climate change and global water resources: SRES emissions and socio-economic scenarios. *Gobal Environmental Change: Human and Policy Dimensions, 14*(1), 31–52.

Barinaga, M. (1999). The clock plot thickens. *Science, 284,* 421–422.

Bernard, S., Gonze, D., Cajavec, B., Herzel, H., & Kramer, A. (2007). Synchronization-induced rhythmicity of circadian oscillators in the suprachiasmatic nucleus. *PLoS Computer Biology, 3*(4).

Centers for Disease Control and Prevention (CDC). (2014). CDC policy on climate and health. Retrieved from *www.cdc.gov/climateandhealth/policy.htm.*

Chaves, L., Kaneko, A., Taleo, G., Pascual, M., & Wilson, M. (2008). Malaria transmission pattern resilience to climatic variability is mediated by insecticide-treated nets. *Malaria Journal, 7,* article 100.

Committee on Climate, Ecosystems, Infectious Diseases, and Human Health. (2001). *Under the weather: Climate, ecosystems, and infectious disease.* Washington, DC: National Research Council.

Confalonieri, U., Menne, B., Akhtar, R., Ebi, R., Hauengue, M., Kovats, R., et al. (2007). Human health. In M. Parry, J. Canziani, P. Palutikof, J. van der Linden, & C. Hanson (Eds.), *Climate change 2007: Impacts, adaptation and vulnerability* (pp. 391–431). Cambridge, UK: Cambridge University Press.

Ebi, K., Sussman, F., & Wilbanks, T. (2008). Analyses of the effects of global change on human health and welfare and human systems. In J. Gamble (Ed.), *A report by the U.S. Climate Change Science Program and the Subcommittee on Global Change Research* (p. 220). Washington, DC: U.S. Environmental Protection Agency, Climate Change Science Program.

Fellman, B. (1985). A clockwork gland. *Science, 85,* 76–81.

Fothergill, A. (1998). The neglect of gender in disaster work: An overview of the literature. In E. Enarson & B. Morrow (Eds.), *The gendered terrain of disaster: Through women's eyes* (pp. 9–25). Westport, CT: Praeger.

Greenberg, J., Bromberg, J., Reed, C., Gustafson, T., & Beauchamp, R. (1983). The epidemiology of heat-related deaths in Texas—1950, 1970–79, and 1980. *American Journal of Public Health, 73,* 805–807.

Hales, S., de Wet, N., Maindonald, J., & Woodward, A. (2002). Potential effect of population and climate changes on global distribution of dengue fever: An empirical model. *The Lancet, 360*(9336), 830–834.

Huntington, E. (1938). *Season of birth.* New York: Wiley.

Hyslopo, N. (1978). Observations on the survival of pathogens in water and air at ambient temperatures and relative humidity. In M. Loutit & J. Miles (Eds.), *Microbial ecology* (pp. 197–205). Berlin: Springer-Verlag.

Intergovernmental Panel on Climate Change (IPCC). (2013). Climate change 2013: The physical science basis. In T. Stocker, D. Qin, G.-K. Plattner, M. Tignor, S. Allen, J. Boschung, et al. (Eds.), *Contribution of Working Group I to the Fifth Assessment Report of the Intergovernmental Panel on Climate Change* (pp. 15–35). Cambridge, UK: Cambridge University Press.

Intergovernmental Panel on Climate Change (IPCC). (2014). Climate change 2014: Impacts, adaptation, and vulnerability: Part A. Global and sectoral aspects. In C. Field,

V. Barros, K. Dokken, M. Mach, T. Mastrandrea, M. Bilir, et al. (Eds.), *Contribution of Working Group II to the Fifth Assessment Report of the Intergovernmental Panel on Climate Change.* Cambridge, UK: Cambridge University Press.

Johnson, D., Stanforth, A., Lulla, V., & Luber, G. (2012). Developing an applied extreme heat vulnerability index utilizing socioeconomic and environmental data. *Applied Geography, 35*(1–2), 23–31.

Kalkstein, L., & Davis, R. (1989). Weather and human mortality: An evaluation of demographic and interregional responses in the United States. *Annals of the Association of American Geographers, 79*, 44–64.

Keller, D., & Nugent, R. (1983). Seasonal patterns in perinatal mortality and preterm delivery. *American Journal of Epidemiology, 118*, 689–698.

Kevan, S., & Chapman, R. (1980). Variations in monthly death rates in Canada. In F. Barrett (Ed.), *Canadian studies in medical geography* (pp. 66–77). Downsview, ON, Canada: York University, Department of Geography.

Kovach, M., Konrad, C., II, & Fuhrmann, C. (2015). Area-level risk factors for heat-related illness in rural and urban locations across North Carolina, USA. *Applied Geography, 60*, 175–183.

Lowen, A., Mubareka, S., Steel, J., & Palese, P. (2007). Influenza virus transmission is dependent on relative humidity and temperature. *PLoS Pathogens 3*(10).

McMichael, A., Campbell-Lendrum, D., Corvalan, C., Ebi, K., Githeko, A., Scheraga, J., et al. (2003). *Climate change and human health: Risks and responses.* Geneva, Switzerland: World Health Organization.

National Institute of Environmental Health Sciences (NIEHS). (2010). *A human health perspective on climate change: A report outlining the research needs on the human health effects of climate change.* Research Triangle Park, NC: Interagency Working Group on Climate Change and Health.

National Oceanic and Atmospheric Administration (NOAA). (2015). Global change indicators. Retrieved from *www.ncdc.noaa.gov/indicators.*

Robine, J., Cheung, S., Le Roy, S., van Oyen, H., Griffiths, C., Michal, J., et al. (2008). Death toll exceeded 70,000 in Europe during the summer of 2003. *Comptes rendus biologies, 331*(2), 171–178.

Sakamoto-Momiyama, M. (1977). *Seasonality in human mortality.* Tokyo: University of Tokyo Press.

Semenza, J., Rubin, C., Falter, K., Selanikio, J., Flanders, W., Howe, H., et al. (1996). Heat-related deaths during the July 1995 heat wave in Chicago. *New England Journal of Medicine, 335*(2), 84–90.

Somers, D., Devlin, P., & Kay, K. (1998). Phytochromes and cryptochromes in the entrainment of the Arabidopsis circadian clock. *Science, 282*, 1488–1490.

Thresher, R., Vitaterna, M., Miyamoto, Y., Kazantsev, A., Hsu, D., Petit, C., et al. (1998). Role of mouse cryptochrome blue-light photoreceptor in circadian photoresponses. *Science, 282*, 1490–1494.

United States Global Change Research Program (USGCRP). (2009). In T. Karl, J. Melillo, & T. Peterson (Eds.), *Climate change impacts in the United States* (p. 196). New York: Author.

Wise, E. (2010). Tree ring record of stream flow and drought in the upper Snake River. *Water Resources Research, 46*(11).

Wise, E. (2012). Hydroclimatology of the U.S. Intermountain West. *Progress in Physical Geography, 36*(4), 458–479.

World Health Organization (WHO). (2014). *Climate and health.* Retrieved from *www.who. int/mediacentre/factsheets/fs266/en.*

Zhou, X., Yang, G., Yang, K., Wang, X., Hong, Q., Sun, L., et al. (2008). Potential impact of climate change on schistosomiasis transmission in China. *American Journal of Tropical Medicine and Hygiene, 78,* 188–194.

📖 FURTHER READING •

Aron, J., & Patz, J. (Eds.). (2001). *Ecosystem change and public health: A global perspective.* Baltimore: Johns Hopkins University Press.

Bovallius, A., Roffey, R., & Henningson, E. (1978). Long range air transmission of bacteria. *Applied and Environmental Microbiology, 35,* 1231–1232.

Buor, D. (2004). Water needs and women's health in the Kumasi metropolitan area, Ghana. *Health and Place, 10,* 85–103.

Campbell, D., & Beets, J. (1979). The relationship of climatological variables to selected vital statistics. *International Journal of Biometeorology, 23,* 107–114.

Cech, I., Youngs, K., Smolensky, M., & Sargent, F. (1972). Day-to-day and seasonal fluctuations of urban mortality in Houston, Texas. *Biometeorology, 23,* 77–87.

Coleman, J. (2008). Atmospheric circulation patterns associated with mortality changes during transitional seasons. *Professional Geographer, 60*(2), 190–206.

Cowgill, U. (1966). Season of birth in man: Contemporary situation with special reference to Europe and the southern hemisphere. *Ecology, 47,* 614–623.

Dalen, P. (1975). *Season of birth: A study of schizophrenia and other mental disorders.* Amsterdam, The Netherlands: Elsevier.

Driscoll, D. (1971). The relationship between weather and mortality in ten major metropolitan areas in the United States, 1962–1965. *International Journal of Biometeorology, 15,* 23–39.

Driscoll, D. (1983). Human biometeorology in the 1970's. *International Journal of Environmental Studies, 20,* 137–147.

Emch, M., Feldacker, C., Islam, M., & Ali, M. (2008). Seasonality of cholera from 1974 to 2005: A review of global patterns. *International Journal of Health Geographics, 7*(31), 1–33.

Epstein, P. (1997). Climate, ecology, and human health. *Consequences, 3,* 3–19.

Folk, G. (1974). *Textbook of environmental physiology.* Philadelphia: Lea & Febiger.

Hay, S., Randolph, S., & Rogers, D. (2000). *Remote sensing and geographic information systems in epidemiology.* London: Academic Press.

Jusatz, H. (1966). The importance of biometeorological and geomedical aspects in human ecology. *International Journal of Biometeorology, 10,* 323–334.

Kavaler, L. (1981). *A matter of degree: Heat, life, and death.* New York: Harper & Row.

McMichael, A. (2004). Climate change. In M. Ezzati, A. Lopez, A. Rodgers, & C. Murray (Eds.), *Comparative quantification of health risks: Global and regional burden of disease due to selected major risk factors* (pp. 1543–1649). Geneva, Switzerland: World Health Organization.

McMichael, A., Powels, J., Butler, C., & Uauy, R. (2007). Food, livestock production, energy, climate change, and health. *The Lancet, 370*(9594), 1253–1263.

Moeller, D. (1997). *Environmental health* (rev. ed.). Cambridge, MA: Harvard University Press.

Perkins, S. (2004). Dead heat: The health consequences of global warming could be many. *Science News, 166,* 10–12.

Rodgers, J., & Udry, J. (1988). The season-of-birth paradox. *Social Biology, 35,* 171–185.

Sakamoto-Momiyama, M., & Katayama, K. (1967). A medical–climatological study in the seasonal variations of mortality in the United States of America. *Papers in Meteorology and Geophysics, 18,* 209–232.

Sakamoto-Momiyama, M., & Katayama, K. (1971). Statistical analysis of seasonal variation in mortality. *Journal of the Meteorological Society of Japan, 49,* 494–509.

States, S. (1977). Weather and deaths in Pittsburgh, Pennsylvania: A comparison with Birmingham, Alabama. *International Journal of Biometeorology, 21,* 7–15.

Terjung, W. (1966). Physiologic climates of the coterminous United States: A bioclimatic classification based on man. *Annals of the Association of American Geographers, 56,* 141–179.

Tromp, S. (1980). *Biometeorology.* London: Heyden.

United Nations Development Program (UNDP). (2008). *Fighting climate change: Human solidarity in a divided world* (Human Development Report 2007/2008). New York: Author.

Warren, J., Berner, J., & Curtis, T. (2005). Climate change and human health: Infrastructure impacts to small remote communities in the north. *International Journal of Circumpolar Health, 64*(5), 487–497.

West, R., & Lowe, C. (1976). Mortality for ischaemic heart disease: Inter-town variation and its association with climate in England and Wales. *International Journal of Epidemiology, 5,* 195–201.

Yang, G., Vounatsou, P., Zhou, X., Tanner, M., & Utzinger, J. (2005). A potential impact of climate change and water resource development on the transmission of Schistosoma japonicum in China. *Parasitologia, 47*(1), 127–134.

PART V

HEALTH CARE
and FINAL THOUGHTS

The final section of this book focuses on geographic understandings of the provision of health services and access to health care. Some of the earliest studies by health geographers explored the ways in which health systems were spatially organized and how this organization affected people's access to care. Effective provision of health services is essential for keeping a population healthy, and is an important factor in understanding the ecology of disease.

Chapter 13 explores access to care as a multidimensional concept that spans both geographic accessibility (Are health resources located in the best place to provide care?) and sociocultural accessibility (Is care provided in an appropriate manner such that people feel comfortable seeking care?). Provision of care is context-specific and the barriers people face in getting good care differ by location, socioeconomic position, and a host of other structural and social factors. We explore many of these barriers, how health systems have been structured in different settings (including the legislative framework in the United States), and the ways in which geographers have examined access across diverse settings. Many countries around the world do not rely entirely on Western medicine, and so we also explore how complementary and alternative systems (such as ayurvedic and Chinese medicine) have been integrated into modern health systems. This chapter ends by considering how the current health systems will change in order to manage the changing demographic and disease landscape that has come with development and globalization.

Chapter 14 concludes with our thoughts on the field of health geography as a whole. The field is diverse and complex and integrates the best ideas from geography with contributions from other fields including sociology, anthropology, public health, political science, genetics, and ecology. Health geographers use a variety of methods to study health and disease—some quantitative and some qualitative in nature. But we are all concerned with a deeper understanding of how the many factors that structure the world around us, the choices we make, and the behaviors we engage in coincide in space and time to promote or prevent health and disease. With such a strong set of tools, there are many opportunities for health geographers, and the field will continue to thrive.

CHAPTER 13

Health Services and Access to Care

The Dartmouth Atlas of Health Care (Dartmouth Medical School, 1998) proclaimed over a decade ago in its opening overview (p. 2) that in health care "geography is destiny." The amount of care consumed by Americans "is highly dependent on where they live—on the capacity of the health care system where they live, and on the practice styles of local physicians." The results of the research for this atlas are discussed later in this chapter, after the concept of access is discussed and the role of distance, central place efficiencies and scale issues of equity, and spatial rationalization by region are presented. The spatial patterns identified by the Dartmouth medical faculty and research staff have exposed differences of enormous importance not only to improving health outcomes, but to controlling health care costs.

The first American geographic studies on locational analysis and regionalization of health services began with studies by Pyle; Morrill and Earickson; and Shannon and Dever in the 1970s. This bloomed in the 1980s into health service research that received the majority of all grants and dominated the annual medical geography paper sessions. It is difficult to appreciate today how little was known about America's "nonsystem," which had just grown, seemingly without a plan. When Pyle began his studies of patient reallocation in Chicago hospitals (discussed shortly), and when researchers in disaster emergency medicine at UCLA Medical School first began to think about distribution of blood supply and accessibility if overpasses should collapse in an earthquake, they both found that before they could address their research question they first had to locate the hospitals. Not only did no one know where patients came from, where blood was collected or stored, or where doctors were located (with or without specialties); there was not even a map of where hospitals and clinics were located or what their bed capacities were. There were no spatial data whatsoever.

Forty years later, locational analysis in health service delivery per se has almost disappeared from the geographic research menu. Most of its research questions have been solved and turned into applications for specific problems; its methodological struggles and innovations have been turned into computer programs that planners can use. The remaining needs and questions have mostly become problems of regionalization. The availability of digital data, and the capacity of a geographic information system (GIS) to manage them, analyze them, and present them for visualization, has changed everything. Using a GIS, a geographer can do in a blink what took a professor months to accomplish. Now, at Dartmouth, medical doctors, statisticians, and epidemiologists can map the country by hospital referral areas; can use hospital and Medicare data to examine medical specializations, distributions of specialists, uses of procedures, outcomes, and expenditures; and can address the big questions: Which rate of use is right? Why do some regions spend more than others? Where are there areas of mismatch in the supply and demand for health services? What is the value of procedures, given differential outcomes in hospital areas? Other questions, however, have morphed into higher policy issues, have been focused on different social ends, racial/ethnic/gender inequalities, and/ or have moved off into the area of health promotion. In addition, as health services research has moved toward examining problems of global health equity, how should we view access to care given that it means so many different things depending on context and setting?

This chapter proceeds with four major sections. The first is concerned with building an understanding of the concept of access to care—what does access mean and how does this understanding differ in different (developmental) settings. The second section summarizes the questions geographers have asked and what they have learned about health care provision. This is not a recitation (or referencing) of many hundreds of articles, but a general identification of some of the key researchers, some of the issues they raised for analysis, and what they and others found. It especially looks at the role of distance on provision of care. The third section looks at concerns of cultural geography about both health and health care, as well as at alternative views of these. The final section addresses (too briefly) the social issues of health promotion and health status—in other words, the human behaviors that create health, maintain it, and prevent disease to begin with.

WHAT IS ACCESS?

In a small village in rural Bangladesh, a young woman sits by the bed of her child. The child has been sick for several days with a fever and a cough, but late last night he began to have trouble breathing. It is the middle of the rice harvest season and the woman spent the morning convincing her husband to leave the fields and take the child to the hospital, located many kilometers to the south. It will take much of the day to travel there by rickshaw and it is likely they will have to wait for hours for the child to be seen by a health professional. Thousands of miles away in Detroit, Michigan, a similar situation exists. A young mother is concerned by her daughter's cold, which started as a cough but rapidly progressed to a high fever and difficulty breathing. The woman lives only a few miles from the closest hospital, which is easily accessible by car and bus, but she has not taken her daughter in because she doesn't have health insurance. The cost of a hospital visit would mean she couldn't pay rent at the end of the month. There is a free clinic located nearby, but the woman doesn't like taking her child there because wait times are long and she needs to get to work soon; she also doesn't feel like the doctors and nurses at the free clinic treat her daughter well because they are African American. Though the two women live in very different contexts, on different sides of the world, they both struggle with accessing health services for their child. But because of their circumstances, their struggles are very different.

Access is a multidimensional concept determined by a variety of geographic, social, and economic factors. It is influenced by the structure of the health system, people's attitudes and behaviors about seeking health care, and financial considerations. In general, *access* is defined as the timely use of services according to need (Peters et al., 2008). Populations living in low-income countries tend to have less access to health services than those in high-income ones. According to the World Bank, low- and middle-income countries account for approximately 12% of global spending on health, but for 90% of the global burden of disease. The average per capita spending on health in high-income countries is around US $3,000—nearly 100 times more than the average of all low-income countries, which was approximately $30 (Gottret & Schieber, 2006).

Access is a continuum that has many levels—from availability of services to actual "appropriate" delivery and use of those services (Ricketts, Savitz, & Gesler, 1994). This continuum is conceptualized in Figure 13.1. Historically, geographers have contributed the most to our understanding of the availability of health services (Where are services located?) and geographic accessibility (Can individuals get to them?), as we will see below. Fewer studies have focused on affordability (Can the patient pay for it?), acceptability (Does it meet patient expectations?), and actual use (Do patients actually receive care?); these topics have fallen more squarely on the shoulders of health economists and other public health researchers.

But geography plays a very important role in understanding access; the relative importance of each dimension of access is context-specific, meaning that individuals in one geographic setting may encounter barriers along one set of dimensions while individuals in another setting may not. In Bangladesh, access is severely limited by geographic distance—how far the family lives from a hospital—because travel in the rural areas is difficult and hospitals and clinics are usually far away. Women's mobility is severely limited in Bangladesh for cultural reasons (e.g., Muslims practicing purdah), so a child's father often must take him or her to the hospital. This has significant economic implications for the family, since he will not be available to work the family farm for several days. It is doubtful this Bangladeshi family will ever be asked if the care they received met their expectations! In Detroit, access is determined by affordability; while the hospital is relatively easy to reach, the family has no health insurance and cannot use it without serious financial hardship. Employment by one or both parents may be part-time or hourly, without sick leave,

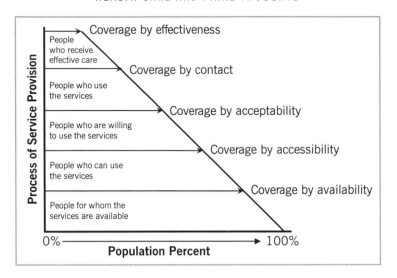

FIGURE 13.1 Model of access to health care. Adapted from Ricketts, Savitz, Gesler, and Osborne (1994).

and with the possibility of being fired for not arriving on time. Racial inequalities also make care unacceptable to certain populations who would rather forego care than cope with perceived discrimination. Access to health services also differs by geographic scale; population-level indicators of access often do not translate to the reality of accessing care experienced by smaller localized groups within that population. In the United States, African Americans and other minority groups are much more likely to be uninsured and less willing to seek care when needed. To truly understand access to health services across geographic scales, it is important to examine the multiple dimensions of access both within and across geographic settings.

THE PROVISION OF MEDICAL CARE

It is not hard to identify many of the components of a health care system. There are the providers of care (types of physicians, nurses, dentists, pharmacists, paramedics, and all their assistants); there are the hospitals, clinics, pharmacies, medical offices, and all their testing equipment and technicians for analyzing everything from the activity of brain structures to urine composition; there are home health aides, social workers, nursing homes, and (today) hospices. Most of these components are shared in some form by alternative medical systems, as we shall see. Not actually part of the health care system, but forming it, surrounding it, and struggling to adapt and preserve it, are the political, economic, and social structures of society and the beliefs of its people. The way that health care is provided is very different in low-income countries than in high-income countries such as the United States. At the same time, most countries agree that a basic set of health services should be provided to all citizens, and this requires that the health sector be structured in an efficient way.

Structure of the Health System

The WHO's (2008) report *Primary Health Care, Now More Than Ever* reintroduced the notion that all people deserve access to health care. This concept has its roots in an international conference that took place in Alma-Ata, Kazakhstan, in 1978. Thousands of delegates from 134 countries and 67 international organizations attended and rallied behind the goal of universal primary health care for all and committed to making this a reality by the year 2000. While this target has clearly not been achieved, despite an investment of billions of dollars in development aid, many of the concepts discussed at the conference are relevant to today's health systems. Delegates outlined nine "essential components" of primary health care: (1) health education; (2) environmental sanitation; (3) a strong system of health workers; (4) maternal and child health programs, including immunization and family planning; (5) prevention of local endemic diseases; (6) appropriate treatment of common diseases and injuries; (7) provision of essential drugs; (8) promotion of nutrition; and (9) traditional medicine. Many of these components can be provided by primary care doctors, nurses, and midwives in small clinics and health posts. They do not require specialists armed with high-tech solutions in large hospitals.

Central place and urban hierarchies have been explained in Chapter 10. These concepts are referred to again here because they can help us understand how goods and services are distributed within a well-structured health system. Contagion may cascade down the urban hierarchy, but we also need to think of health services as being gathered up in the hierarchy. Low-level services (e.g., general practitioners, pediatricians, obstetricians, pharmacists, dentists), which need to be used by many people frequently, should theoretically be widely dispersed. High-order services, such as medical specializations and regional facilities (e.g., those for severe burns), are used rarely and so need a centralized location that can draw upon more potential patients to provide the demand. Central place organization in the ideal sense has been adopted as the "best practice" model for developing countries in the process of improving their national health systems. Strengthening national health systems involves getting basic primary and preventive care to villages and rural areas, even if only by mobile clinic. Therefore most countries have tried to build a hierarchy in which more specialized care is accessible to as many people as possible, while facing the practical need of serving enough people at each level to fully utilize their precious and too-rare medical resources. The hierarchy can be portrayed as a pyramidal organizational system ascending from rural health centers or clinics, to district health centers, to regional/provincial health centers, and finally the national health, picking up specialized personnel and equipment and medical beds as use rises (Figure 13.2). In many developing countries, community health workers (CHWs) or village health workers (VHWs) are an integral part of the primary health care system. CHWs are selected, trained, and work in the communities from which they come; they provide a wide variety of primary health services such as

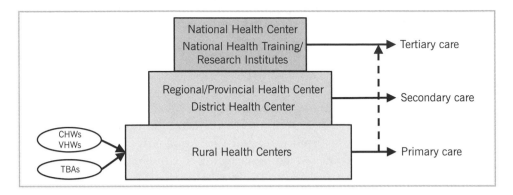

FIGURE 13.2 Pyramidal organizational structure of the health care system.

immunizations, well-child visits, prenatal and postnatal care, nutrition screening, contraception, and oral rehydration therapy. They are trusted members of the community and serve as a liaison or intermediary between higher-level health services and the community. This facilitates access to services and improves the quality and cultural competence of service delivery. Traditional birth attendants (TBAs) serve a similar role but focus on maternal care during and after birth; they are trained to assist women during childbirth and provide referrals for complicated deliveries and education for infant care and breastfeeding.

Geographically, a pyramidal health system looks somewhat like Figure 13.3. Different types of hospitals have different "demand cones": local (satellite), suburban, city, and major research-university hospitals. Think of the center of the largest cone as 0 on a graph of distance decay (see the explanation of distance decay below) for the *Y*-axis, and the curves of demand that form the cones are then the distance decay curves of usage. The *X*-axis represents the distances people travel. The cones overlap as people go to different facilities for different needs, taking

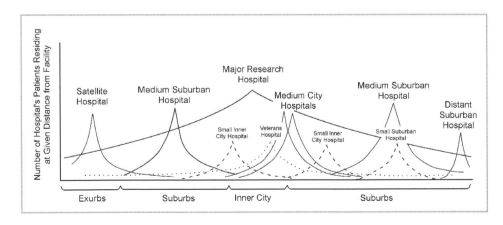

FIGURE 13.3 Distance demand cones for different types of hospitals in the typical eastern American city. Data from Pyle (1979, p. 220).

rare conditions or problems needing special equipment or specialized physicians to higher-order places. The central major research hospital in Figure 13.3 draws people from across the state, perhaps providing chemoprophylaxis and radiation for cancer; however, the satellite and suburban hospitals are more cost-effective and more accessible for ordinary births (premature births are transported up the hierarchy), heart attacks, ruptured appendixes, and broken bones. As can be seen, with the array of hospitals and clinics from small and local (sometimes multiple), to regional and research hospitals drawing from an entire region or the whole nation, it is easy to link central place theory to the functional and spatial organization of an idealized health care system. Actually applying this theory, however, must take into account many factors that distort the ideal. These factors include physical barriers, transport networks, and income distributions. The result is a system that usually retains its functional nature, but whose spatial configuration rarely resembles the nested hexagons of theory. An ideal spatial organization is harder to realize in practice than is a functional hierarchy.

Figure 13.4 shows the plan for development of health care made by the Malaysian government after independence. At the bottom, in every village there was to be at least one government-trained (i.e., not traditional, although she might have been that once) midwife who understood the importance of sterilization of equipment, the danger of germs, and the symptoms of difficulty that required medical monitoring or intervention. Every five midwives were to be part of a subclinic located in the largest of the villages. Here there would be a nurse practitioner who could treat skin infections such as scabies, infected wounds, and children's earaches; take blood and sputum, and send it to be checked for malaria or tuberculosis; vaccinate babies; and provide continuing medicine for tuberculosis or leprosy. The subclinic also included a drug dispenser, as well as a public health field worker who could chlorinate wells, promote sanitation, and provide mosquito nets. The nurse practitioner acted as a screener, telling patients whom the nurse was unable to treat or who weren't responding well to return on Friday (1 day a week), when there would be a doctor present from the clinic—the district health center in the largest nearby town. At the district-clinic level there were a doctor and dentist in residence (beside the ones who traveled "on circuit"), a pharmacist, the district malaria control office, and in the town a small hospital with its medical staff. The town of 10,000 thus had, by plan, five midwives; a subclinic for quick treatment and screening of the town population; a dispenser and a pharmacist; and (with hospital, resident, and traveling personnel) 10 physicians, including surgeons and specialists in infectious disease and heart disease. For treatments and illness beyond the district-clinic level, patients would be referred to the medical center in the state capital, which for Selangor State was also at the time the national capital, Kuala Lumpur, with its medical school and medical research institute.

This was operationalized reality, not just the idealized plan. Of course, not everything developed so smoothly. Changes were needed. Chapter 5 has described research on a land scheme (a settlement) of the Federal Land Development Authority, and also a traditional village that was studied as a control. People on the scheme had a midwife, as did the control village, which was also host to the subclinic. To

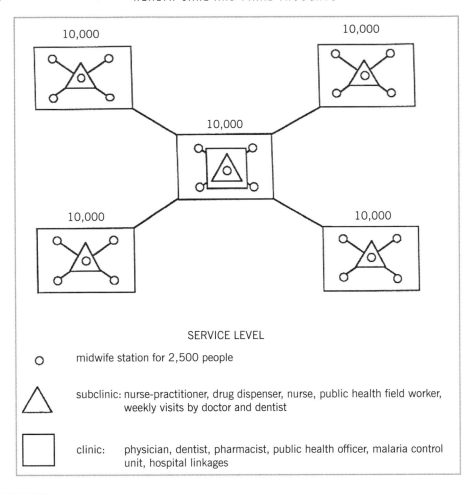

FIGURE 13.4 The ideal distribution of health care, according to the central place hierarchy planned in Malaysia.

travel to the subclinic with a sick child, or one who needed vaccination, a person had to travel 7 miles along a winding, paved road through forest—and then perhaps come back another day for the doctor. It was a long walk, unless someone with a motorcycle or an official with a car could be convinced to miss work time and transport the person and the small child (or children). Beyond the village subclinic, it was another 12 miles to the town district clinic, but there was also a hospital there. For the first 15 years after independence, the ministry of health had thought the hard parts of the plan would be finding and training enough personnel for the subclinics and midwives, and then fully staffing the plan through villages in settled and well-populated agricultural areas as well as the new settlements being pushed into the rain forest. Over time, ministry officials realized that there was a great demand for primary care in the villages: community health workers needed to be "trained-up" to provide vaccination, take care of ear and skin infections, and make sure women were healthy during pregnancy. People often preferred to bypass the subclinic, however. If they were able to get transport out of their settlement, they preferred to go all the way to the clinic in town where there were doctors and direct referral to a hospital was possible. In other words, there was a barrier to getting on the road at all, but once that barrier was overcome by need, higher-level services, expertise, less time waiting to be seen, and perceptibly better help easily offset the greater distance traveled. Other research has found the same pattern of bypassing the middle to seek higher-quality care in India, and it has since become a generally recognized adjustment of the system's morphology.

If these were the distortions in the "ideal" resulting from rather minor distance and time delays in higher-quality service, could a central-place system be envisioned to work at a scale of hundreds of miles and more? Moreover, the market efficiency approach of planning hierarchies of medical care in a central place framework can be challenged as unjust and immoral. Access to more than primary-level health care is a need and a right; Why should a rural person's life and pain be less important than an urban person's? Doesn't everyone need access to X-ray radiology and mammography for diagnosis and screening? Shouldn't C-sections and appendectomies be available to all, and clearing of clogged arteries as well? Perhaps not heart transplants or treatment for first-degree burns, but laser surgery for the eyes? Surely incubators for prematurely delivered babies should be immediately accessible. Often this issue is phrased in terms of money. For the British National Health Service, there was the "inverse care" argument that those with the greatest medical needs received the lowest spending, while the well-off middle class received attention and investment to meet its perceived needs.

There is a tremendous geographic dimension to this dilemma of justice. In many parts of the world there are very large, quite sparsely populated areas with people who live hundreds of miles away from any city. Think of the tundra country in Canada, Alaska, Russia, and Scandinavia; the Australian Outback; the Great Plains and western mountains of the United States; the rain forest of the Congo; and the thousands of islands of Indonesia or other archipelagic nations in the Pacific Ocean. On the one hand, there is the problem of transporting people over vast areas of very difficult terrain to medical care—the broken bones to X-ray, the obstructed births to surgery, the delirious fevers to diagnosis and treatment. On the other hand, however, there is the problem of maintaining highly skilled, specialized personnel and equipment in small, remote centers of health care under difficult conditions for themselves and their families. These personnel are then underemployed and treat fewer than 1% of those they could serve elsewhere. This dilemma is not simply a matter of market greed, but of the spatial function of a central-place hierarchy.

The Health Care System of the United States

Most of the following discussion of health service research revolves around the unique conditions of the U.S. system, so let us examine it and its history first. Health care delivery in the United States is unique among large industrialized nations in that it is provided through an entrepreneurial rather than a governmental system.

The main source of health care for most citizens is in a private market of thousands of independent medical practitioners, pharmacies, clinics, and so on, subject to significantly less government regulation than in other developed countries. Policies developed in other industrialized countries to provide efficient and equitable health care have entailed either direct government provision of health services or significant government involvement in various schemes of national health insurance, planning, and regulation. In the United States, however, there has always been a reliance on private markets as the favored instruments for allocating health care in all but a few categorical programs. One result of this political and economic philosophy has been fragmentation—a collection of systems and subsystems that are based on local and institutional initiatives. There have been attempts at national, state, and local planning—hence all that geographic modeling—but for the most part each segment (e.g., hospital services, community health services, emergency medical services, or pharmaceutical supplies) has gone its own way. As in other sectors of life, in the United States a wide range of special-interest groups has had an impact on health care policy. These groups include major industrial interests; organized labor; federal, state, and local health care planning agencies; hospital organizations; professional physician groups; and the insurance industry. Attempts to bring order out of this chaos have met with limited success.

Several federal legislative acts have influenced the organization of the U.S. health care delivery system. The Hill–Burton Act of 1946, today universally agreed to have been "landmark legislation," was the first of these. One of President Truman's initiatives, the main objective of Hill–Burton was to provide federal funding for hospital construction in "needy" areas. Need was defined simply as a bed-to-population ratio; specifically, the bill mandated that there should be 4.5 beds for every 1,000 people, without discrimination (but with a separate-but-equal provision that wasn't struck down until 1963). The federal money had to be matched by states and municipalities. It also came with conditions: states were obliged to identify general hospital service areas of one or more counties, to assume responsibility for needs assessment and planning, and to provide a "reasonable volume" of uncompensated care for those who could not pay to use a facility. Data were obtained at the county level and within metropolitan areas at the census tract level. This led to the establishment of many (usually voluntary) state health planning agencies. There was much variation among states and cities regarding adherence to the provisions of the act, given the lack of any regulatory or oversight provisions. After being extended a couple of times, in 1975 the Hill–Burton Act was transformed into Title XVI of the Public Health Service Act, with the addition of a few regulatory mechanisms.

In the 1960s the major emphasis of national health policy turned toward increased expenditures for the poor and aged. Medicaid was established to provide medical care to the poor who received government payments, and to others unable to pay basic medical costs. Medicare made a similar provision for the elderly. Both programs were and are administered by the states, with financial support from the federal government. Both pay for outpatient treatment, physician fees, hospitalization, surgery, and equipment such as walkers and pacemakers. Hospitals were originally paid their costs (without restriction) plus 2%, and lower-cost substitutes to hospitals were not covered. Over the years, health care costs rose dramatically. Some analysts thought that insufficient controls on these programs were to blame.

Another landmark in U.S. health care legislation came in 1974, with passage of the National Health Planning and Resources Development Act. This act was an attempt to update and revise existing programs and to foster cooperation among the national, state, and local health care planning agencies. The new program called for each state to create clearly defined health systems agencies. These agencies' populations were to number between 500,000 and 3 million. Their boundaries were to coincide with other administrative units as much as possible (i.e., to follow county or other boundaries). The health systems agencies became the basic units for determining health care needs, administering federal funds for approved programs, and reviewing and evaluating programs. Each state formed a board of consumers and providers, all appointed by the state's governor for the purpose of health services planning.

The 1974 act by no means overcame the problems presented by earlier programs. Although consumers were supposed to constitute a majority on state planning boards, many felt that providers had the dominant voice in planning decisions. Providers continued to resent government interference and to complain about the bureaucratization of health care delivery, the costs of maintaining health service agencies, and the long review process. Nevertheless, these agencies have continued to exist, as shown by their use to analyze data collected for the Dartmouth atlas series introduced above. Health care costs have escalated dramatically in the United States over the last several decades, from $12.7 billion in 1950 to almost $1 trillion today. The U.S. population has aged; technology and medicines have become enormously more powerful and complex; and cost-based payments have provided little incentive for efficiency. Uniform payments to all hospitals since the 1980s have been based on predetermined rates for each of the 468 diagnosis-related groups. This change was a response to findings that there were wide differences (spatially) among hospitals in treatment charges, types of treatment, lengths of stay for similar problems, and uses of ancillary resources.

Perhaps the most dramatic development in the medical care field in the United States occurred in the 1990s with the emergence of **managed care**. Managed care refers to the efforts of organizations such as insurance companies and large health maintenance organizations to hold down medical spending by "managing" the financial aspects of the practice of medicine. This translates into companies and organizations specifying how much and what form of health care will actually be allowed to be dispensed by providers. The majority of people in the United States depend on third-party payers (insurance) to cover most of their medical care costs. Supporters of managed care claim that "unnecessary" drugs, tests, procedures, and welfare programs are what drive up costs. Opponents claim that the managers are interfering with the proper practice of medicine and endangering the lives and health of patients. Supporters claim that keeping people well through early detection and promoting healthy behaviors in exercise and eating will keep costs down by preventing serious illness from developing. (That was the great hope in the beginning, at least.) Detractors claim that the way to maximum profit is not to let people who are sick or at risk into a managed care organization in the first place—to "cherry-pick" healthy clients who will demand little care, and leave the others to be without care. The practice of managing health care is still relatively new and is heavily influenced by politics and corporations that have a great economic stake in the outcome of the system. Under these conditions, it is inevitable that inequities have arisen and will continue to arise (Anderson & Poullier, 1999).

Quite possibly the most important change to the United States health system occurred in 2010 with the passage of the Patient Protection and Affordable Care Act (ACA), also commonly called "Obamacare" after President Barack Obama, who championed the legislation throughout much of his presidency. The basic provisions of the law ensure a near-universal guarantee of access to affordable health insurance coverage, from birth through retirement. It is the first time in American history that health care has been a right for all American citizens.

The ACA is a complex and far-reaching law that has two main goals: to provide access to quality, affordable health insurance and to curb the growth of health care spending in the United States (Rosenbaum, 2011). The law is much too complex to discuss in detail in this text (indeed, entire textbooks have been written about it!) but the ACA addressed many of the dimensions of access discussed earlier in this chapter. First and foremost, the ACA requires individuals to have health insurance (and imposes tax penalties for those that don't) and makes affordable insurance plans available through a series of state-run, government-supported health insurance marketplaces. It also stops insurance companies from refusing to cover people, and from revoking coverage when people become seriously ill. People also can't be forced to pay extra for insurance because of preexisting conditions. But the law contains so much more than insurance reform. Changes in health care delivery, how we address chronic disease, and how we beef up our primary care workforce are included in the law. These changes are important to the fundamental infrastructure of the U.S. health care system and were sorely needed.

At the time this book was written, the ACA had been in effect for only a short period of time, so little was known about the outcomes of the program. It certainly reduced the number of uninsured individuals during the first few years of the program. Between 7 and 16 million individuals gained health insurance due to the ACA, and groups that have historically lacked health insurance—blacks, Hispanics, young adults, and the poor—have seen the greatest gains. This has reduced inequalities across the country. For example, a county-level study by Sommers, Chua, Kenney, Long, and McMorrow (2015) in California found that the state's 2011 insurance expansion program produced significant increases in insurance coverage for low-income individuals, particularly Latinos. We have also seen an unprecedented slowdown in the rate of increase of national health care spending in the United States since the ACA was implemented in 2010. While researchers disagree about whether the ACA was the cause of this slowdown, evidence shows that from 2010 through 2013, per capita U.S. health care expenditures increased at the historically low rate of 3.2% annually, as compared with 5.6% annually over the previous

10 years (Blumenthal, Abrams, & Nuzum, 2015) and Medicaid spending actually decreased over that time.

The ACA gave each state significant latitude in how to set up and structure their health insurance marketplace, which has led to geographical differences in access across the United States. Early geographic analyses of insurance premiums (the amount a person pays each month to have insurance) showed a substantial penalty for living in rural areas—premiums were higher for smaller populations living in less densely populated areas with fewer health care providers (Barker, McBride, Kemper, & Mueller, 2014). You'll pay more for insurance if you live in the Midwest! Premiums are paid out of pocket, so another way of looking at inequalities is the proportion of your income that you pay toward health insurance. In large portions of Wyoming, Colorado, North Dakota, and South Dakota you could pay 13% or more of your monthly salary. In many other states, this percentage is closer to 8–10% (Graetz, Kaplan, Kaplan, Bailey, & Waters, 2014). While it is too early to truly understand the effects of Obamacare, it has fundamentally altered access to health care in the United States.

Distribution and Issues of Equity

One of the first attempts to examine the distribution of health resources in the United States was Shannon and Dever's (1974) book *Health Care Delivery: Spatial Perspectives*. The book contained a series of crude grey-scale maps that were some of the first generated using computers. For the first time geographers (and public health officials) saw state-level maps of the distribution of general practitioners, black doctors, female doctors, pediatricians, cardiologists, midwives, and a dozen other categorizations. Midwives and general practitioners were concentrated in southern states; black physicians were mainly in coastal states; female physicians were few and primarily in West Coast states; and specialists were in northeastern and West Coast cities. Any health practitioner who saw those maps had to ask: *Why* are they distributed like that? Economic/urban/social geographers thought they had theories and methodologies that could explain; some began to study health service delivery.

Medical personnel, clinics, beds, and the like are regarded as **health resources**. There are two main questions about resource distribution: Are there enough? And are distributions fair? The first great question, though, is simply this: Where are the resources? As explained above, nobody knew the answer to this question, by county or city or street. Then came more queries: How many of what? At what scale? All those cardiologists mapped at state scale might have been in the biggest city, with no one at all available in the rest of the state; or all the black physicians, so few, might have been in one university hospital. Once some information about distribution was established, there came a further question: Is this distribution of resources supply- or need-driven? Why are they located where they are?

Resources like registered nurses, pharmacies, pediatricians, and hospital beds were mapped "per 1,000" or "per 100,000" at various levels: state, county, city, and minor civil divisions. Ratios of physicians to populations were calculated for specialties and populations at risk for health service areas and other regionalizations. Distributions were examined for equity by mapping ratios called **locational quotients** and values of the **index of segregation**. Location quotients are a valuable way of quantifying how concentrated a particular industry, or aspects of an industry such as physicians or pharmacies, is in a region. Indices of segregation are similar but focus on concentrations of racial and ethnic groups. They were summarized for ethnic and economic groups, using **Gini coefficients** (all of these are standard distributional measures and basic spatial statistics). They were compared with standardized mortality ratios and age-adjusted death rates. It was gradually realized that subgroups of the population (ethnic/racial groups, age categories, families vs. singles, etc.) were not distributed uniformly, and thus that the specialists who served them shouldn't be either. Normative measures of distribution were developed. Some places had many times more physicians than others; small towns and rural areas were losing doctors rapidly.

Why did doctors locate where they did? Many researchers started with the assumption that money was the main cause, but apparently it wasn't at the top of the list because physicians had a good living and community respect everywhere. Money differentials mattered mostly at the state and regional scales. Racial composition was not a factor for location at the state or city scale, but sometimes influenced location of offices within cities. The ability to pay and the needs of patients were both poor predictors of location. What mattered most were the social needs of the physicians: proximity to other professionals, place of training (which helped establish an information field), and amenities for relaxation from stress during leisure time. Physicians were concerned about the quality of the schools their children would attend, about their ability to stay up to date on medical developments, and (for those attracted to group practices) about the ability to obtain relief and share the 24-hour demands with at least one other physician. The best advice given to those small towns that wanted doctors was to build a golf course!

At a different scale, distance seemed important. Physicians seemed to cluster around hospitals where they had residence and practice privileges. In some rapidly growing suburban areas, location of offices depended on where professional office space could be built or was available in the urban morphology. Midwives and dentists in general practice might be found in small towns, but rarely a medical specialist. Clearly other locational factors were at work. Before an optimal distribution could be promoted, the reasons for the existing distribution had to be understood better.

There are two dimensions of study for this subject: the **macroscale** and the **microscale**. Macroscale studies, because they deal with aggregates, should be used to describe rather than to explain general patterns of resource distributions. Attempting to extrapolate from the macroscale to smaller areas, and vice versa, can easily lead to committing the ecological fallacy. Also, some variables associated with resource locations may be correlated among themselves, and it is often difficult to ascertain which variables are the most important ones. Furthermore, the relative importance of these variables may change over time. In some developing countries the data for macroscale analysis are often simply not available. Microscale studies are the preferred approach to analyzing resource location factors; at the micro-level the actual mechanisms behind resource distribution decisions are clearer. A microscale study, for example, might focus on practitioners' motives for selecting a location.

Locational Analysis

Given the existence of inequalities in health care resource distributions, no matter how one chooses to measure that inequality, the next logical step is to try to devise means of redressing imbalances. If they're not all in the best place, where should they be located? Having a medical resource present abundantly at some scale does not make it available at another. As discussed earlier in the chapter, accessibility depends on many things besides physical presence. Accessibility can depend on ability to pay; on membership (as in certain facilities for union workers, veterans, actors); on race (especially in the segregated South, where in the 1950s even children attacked by polio had to go to different places); and on many other factors. Social science has learned that **social distance** is also important. The language used by doctors, the ability to understand and be understood, and the attitude of medical staff toward patients matters a lot. People want to avoid being humiliated or treated as commodities. Accessibility is sometimes measured in waiting time, which can be hours in a medical reception room for some people, and even days for people in the developing world who have already walked with sick children for many miles. In 2008 in the United States, a woman with a history of mental health problems went into convulsions and died on the floor of a hospital reception room where she had been waiting to be seen for 30 hours. This event at least made the national news. People without health insurance have so overloaded the emergency rooms with their advanced problems that teenagers coming in with merely broken legs may have to wait hours to be seen. People without health insurance can be, and are, turned away even from the emergency rooms of private hospitals unless their problems are life-threatening. Certainly these are all issues of accessibility, ones the ACA is trying to address.

The important measure of accessibility for **locational analysis** is distance. Distance, too, can be measured in various ways. The most common way, and the one used below, is linear or "as the crow flies" distance. Today any GIS program measures this instantly as a vector. Linear distance, however, can be of limited usefulness. For example, a low-income woman in Detroit who needs three bus changes to get to a public clinic to get vaccinations for her child may not be far away from the clinic as the private automobile drives; however, the bus trips may take her 2 hours instead of 10 minutes. For rural people without cars, such as elderly women needing to keep medical appointments, arranging a ride from a neighbor or relative who has child care or daytime work obligations can take a long time and be unreliable. Then, too, driving distance in the winter snow on mountain roads is not at all the same thing as the same driving distance on the coastal plain in summer. So distance can be measured as a linear vector, as road travel, and as travel time. Despite such cases and special conditions, it has been repeatedly found that distance measures are so intercollinear that it usually does not matter which measure is used in statistical analysis. Map vector, "as the crow flies" distance is much easier to collect than data on travel time from interviews for large groups.

Factors in Provider–Consumer Contacts: Utilization and the Role of Distance

The most important concept for geographic accessibility and locational analysis is that of **distance decay**, or the concept that "distance has friction." Figure 13.5 illustrates three lines of decay. The *Y*-axis is the frequency of use—how many people visit a hospital, dentist, or optometrist. The *X*-axis is distance from the health resource. Because distance has friction, facilities and services of all kinds show a dropoff in use with greater distance. Distance decay for health care is not quite as linear as that for movie theaters or fast-food outlets, however. Figure 13.5 shows a hospital/clinic facility. Line A shows a steep distance decay, illustrating that only people living (or perhaps working) near the facility use it. This may be true for getting a flu shot, checking out a chronic cough, or getting a mammogram. Line B shows a low gradient: Use of the facility is much less, but people come from long distances. Someone who cuts a toe off with a lawn mower will get to the facility; greater or lesser distance is not then an issue. Line C represents the interesting case of a steep distance decay with convenience mattering, followed at a greater distance by renewed effort to get help as people fear they may not be able to if they wait too long. So someone with food poisoning (e.g., an *E. coli* variant) may seek help quickly if it is conveniently accessible; may use over-the-counter household drugs and hang around the toilet if there are many miles to go; but may fear being unable to get to help if vomiting, dizziness, and high fever develop, and so may choose to drive there while doing so is still possible. Medical records from physician offices, clinics, and hospitals can be classified by condition (giving birth, chest pains, surgery, etc.), and then patients can be matched by address to streets and blocks, or at least ZIP codes. Classes of such conditions have predictable distance decay curves, we now know. Hospitals or other facilities sometimes use such knowledge to "steal" rival facilities' patients, as in taking actions to induce obstetrical patients to go a little farther to them and fill their special beds. Unusually long distance decay curves, by category, indicate underserved populations and sometimes a need for other doctors, dentists, or resources.

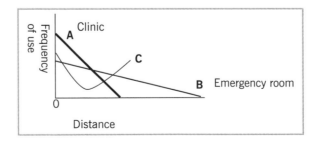

FIGURE 13.5 Three slopes of distance decay.

Hunter, Shannon, and Sambrook (1986) provided an excellent historical example of the concept of distance decay using mental health. They traveled to and copied admissions and address records at most of the pre-Civil War asylums that still existed (e.g., the Dorothea Dix Hospital in Raleigh, which closed only in 2008). Figure 13.6 shows the distance decay curve they calculated for the mental hospitals in 1824–1855, for eight states. Figure 13.7 shows the data mapped out for four states in the Southeast: North Carolina, Virginia, Tennessee, and Kentucky. There clearly appear to be more mentally ill people in the vicinity of the asylums, but we know today that this is only because distance has friction and affects utilization.

Patricia Gober (1997) provided another excellent example that examines the role of access in state abortion rates. The number of clinics available per 100,000 women varies enormously from state to state, of course. She looked further at the impact of a federal law requiring a 24-hour waiting period after first attending a clinic and being advised. In Philadelphia at that time, a young woman would have had a choice of going to any of 14 clinics, going home overnight, and returning the next day. In North Dakota a young woman might have needed to travel over 200 miles to Sioux Falls, South Dakota, the only choice, and probably stay overnight there—a prohibitive expense for her, probably, even without that winter snow.

For help in addressing some of its medical care shortfall areas, the United States has come to rely, even at state level, on the importation of foreign-trained physicians (international medical graduates, or IMGs). Aside from the issue of the quality of their training, are IMGs more likely to be found in underserved areas? Mick, Lee, and Wodchis (2000) analyzed American Medical Association data and

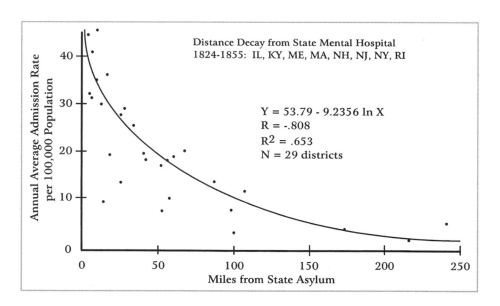

FIGURE 13.6 A distance decay curve of admissions to state mental hospitals in selected states in the 19th century. From Hunter, Shannon, and Sambrook (1986, p. 1046). Copyright ©1986 Elsevier Science. Reprinted by permission.

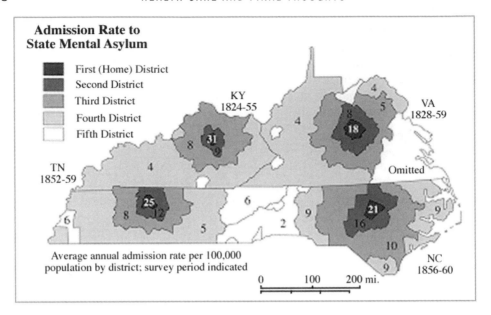

FIGURE 13.7 The effect of distance on state mental hospital admissions in selected southeastern states in the 19th century. From Hunter, Shannon, and Sambrook (1986, p. 1041). Copyright ©1986 by Elsevier Science. Reprinted by permission.

found a positive correlation between disproportionately high numbers of IMGs and low doctor-to-population ratios.

Quite a few ways have been developed to examine accessibility in hopes of improving it. Always measuring distances from the residence does not address how people circulate. People commonly perform chores and access facilities on the way home from work (e.g., a dentist's appointment over lunch hour, a run to the pharmacy on the way home), so workplaces may be the relevant points of origin. People also make multiple-purpose trips. Someone who goes to a pharmacy, a supermarket, an exercise gym, and a dry cleaner on a weekend circulation may seek out a physician or dentist in the same activity area. Gerard Rushton started using **small-area analysis** (subcounty, city block) population data to create **population-sheds** (i.e., population circulation areas, analogous to how the rain falling on many steep hillsides flows down eventually to one river in its basin—a watershed). Such population-sheds, also known as **catchment areas**, have become part of market analysis of many kinds of medical facilities—from dental care in rural Iowa to information centers for inner-city Hispanic teenagers about diabetes prevention.

Location–Allocation

"Whoever said the human race was logical?" That old *Star Trek* line has seldom been applied better than to U.S. health care utilization. The "nonsystem" is a crazy quilt—private and public; preferred provider and walk-in; government Medicare,

Medicaid, veterans' care, and Indian Health Service; insurance paid for by employer, self, the government, or no one at all. In addition, medical technology and digital technology are changing rapidly; ethnic composition and structure are changing; and suburban, exurban, and new urban settlement patterns are changing. Furthermore, apparently Americans were never simply inclined to use the nearest facility. Location–allocation methodologies in health geography were initially premised on a rational system, at least in terms of a rational (economic) person and friction of distance. After first using the methods to plan a logical system, geographers learned to use them instead to determine and measure the other influences upon it. For a decade location–allocation planning employed the most health geographers with master's degrees, dominated the annual professional meetings with days of papers, and filled whole books and courses. Gradually, as its algorithms were solved, computer models took over—and then GIS made it so easy that even nongeographers could do it. It is still important today for geographers concerned with health care to understand the logic behind the planning, however.

The Chicago regional studies of the late 1960s included efforts to plan future health care provision in accordance with the community health areas mentioned earlier in this chapter (Morrill & Earickson, 1968, 1969). The question that Gerald Pyle (1971) tried to address was where future hospital beds and specialized personnel/treatment would be needed 10 and 20 years into the future of Chicago. The population was moving out to the suburbs. Race and ethnicity were shifting territorially, as were age structure, poverty/welfare, and taxpayer money. Could migration and population studies in geography predict where young, middle-aged, and older populations, or richer and poorer populations, would be located in the next decade or two? More to the point, could they predict where the future heart attacks, childbirths, and dialysis needs would be? If millions of dollars were to be spent on new hospitals, clinics, scanners, and medical training of laboratory workers and physician specialists, where should they be located to meet future needs?

This was ambitious research. Whole methodologies had to be developed. As explained above, at first no one knew where the existing hospitals, specialized beds, physicians, or any patients were located. Hospitals had to be mapped, and their types and capacities of beds and personnel determined. Hospital discharge data were used to identify patients by types of problems, but they had to be located, and distance decay calculated and their demography projected. Computer simulation models were developed (at a time, remember, when desktops did not exist and the university mainframe had to be programmed to do procedures that took hours to solve). Pyle plotted where people originated for each hospital, using line maps. Before he could simulate where new capacity would be needed, it was necessary to analyze where capacity was already less than needed, and where there was surplus. Each patient for a hospital (by types such as heart disease, cancer, or stroke) was located and then "allocated" to the nearest bed of that type. Identifying in the models where facilities were filled, and overspilling patients had to be allocated to the next nearest facility, would identify the underserved demand. Future care needs for dialysis and heart attacks were in fact simulated, but what attracted a lot of attention from geographers was the relatively large proportion of the population that did not go to the nearest facility—sometimes even going far away. The difference between ideal location–allocation and what utilization existed helped identify social distance, as well as the use of culture-specific facilities (Catholic and Jewish hospitals, an actors' guild hospital, a union hospital, and of course one public hospital serving the low-income and welfare-receiving public). Morrill and Earickson found that reallocating hospital space could substantially decrease patient travel, but they pointed out that the same objective could be achieved by relaxing existing constraints of income and race. One disadvantage of these early models was that they left out nonspatial aspects of health care delivery, such as disease diagnosis and the complicated rules imposed by Medicare and Medicaid. As such, they were too simplistic to be applied to planning applications today, which require socioeconomic modeling of health service delivery systems.

People or needs can be located and allocated to a facility, once its capacity is known. Think of trying to move patients in nursing homes out of the path of a major hurricane. Where is an empty bed that can be used by someone with hips in traction? Who needs oxygen, or merely careful 24-hour watch, and which facility is a match? These patients are located in this nursing home here and those in that, and where are the nearest available safe beds to which they can be allocated? One of the first uses of location–allocation was in zoning ambulances. If a person having a stroke was picked up from this apartment, or a burn victim from that road crash, where would be the nearest appropriate-for-that-service-level facility to which the patient should be transported? After a few years, the question was reversed. If ambulances should reach all heart attack victims within 20 minutes, where would emergency facilities have to be located in a sprawling urban area to do that without wasting tax money on superfluous centers?

Time went by, and the topics became closing hospitals in cities, or how utilization and unmet needs would shift when chains bought up public hospitals and converted them to use by health care organizations or religious groups. Issues of scale abound. Resource inequality is consistently found within administrative jurisdictions. Ill health related to the poverty common to U.S. inner-city neighborhoods is invisible when national, state, or even county statistics are viewed. In most countries rural dwellers have been at a disadvantage in obtaining basic primary health care, much less specialty services, because medical care resources tend to be concentrated in urban or suburban places—a vexing problem to Canada as well as the United States. As medical students in the 1960s began to shift into medical specialties (which usually are associated with higher-order centers in central place hierarchies), and as rural physicians began to age and retire, there was a great drop in the number of primary care physicians establishing practices in rural and nonmetropolitan areas. By 1988, 111 counties across the United States were without a single physician. As a result, studies of medical personnel distribution were initiated; underserved areas were identified; and federal programs such as the Community Health Centers Initiative (which brought specialists to rural centers 1 day a week or on a similar schedule, while building and maintaining the centers) were developed (Ricketts & Cromartie, 1992). Geographers were active in the work of rural health services research centers in North Carolina and the Pacific Northwest

(Hart, Pirani, & Rosenblatt, 1991), and in studies developing increasingly sophisticated location–allocation models for specific practitioner need and supply.

Regionalization

The dilemma of how to make needed health services accessible to all people, and yet maintain cost efficiencies and quality of personnel and service when people are very few and far apart, was introduced to American geography before there were any who studied health services—before even Shannon and Dever (1974). This dilemma was first examined in graduate seminars in economic geography, where students read Godlund's (1961) account of it as posed by the Swedish government to geographers at the University of Lund. The government was constructing its socialized medical system and wanted the geographers to regionalize the service areas so that everyone would have access to needed care. They were to use the existing medical resources (clinics, hospitals, nursing and medical schools, etc.) and analyze what other facilities needed to be constructed and where to provide these for the nation. Immediately the Swedish geographers confronted the central place dilemma ("on steroids," as we might say today). Their major hospitals, medical schools, and other facilities were located in southern Sweden on the arable plains, along with most of the population. High-order personnel and services, and a large portion of the population, were located around Stockholm. Some of the population, however, lived in the mountains, even north of the Arctic Circle. These people were not just reindeer pastoralists. Fishing vessels sailed out of villages in the fjords. Fishing (with its tackle, lines, and machinery) was dangerous work even without stormy weather and subzero temperatures. Men lost hands and feet. And how could a woman with an obstructed birth, or broken bones, or a heart attack be contacted and transported 500 miles to Stockholm? How could screening for tuberculosis, various cancers, or even dental caries be brought to people there accessibly enough? It seemed obvious that one or more regional centers were needed, but how did one regionalize that? And who had to go live there doing what? Much of the development of "medical geography as health service delivery" as recounted above has involved addressing that dilemma and developing the concepts and methodologies to solve it.

Defining Regions

A **region** may be defined as a bounded area that is internally homogeneous by one or more criteria. Regions may be precisely defined by researchers using numerical measures, or they may be simply mental constructs, such as the perception of the extent of a neighborhood by its residents. Before quantitative, statistical methodology became widely used, regions were mostly formal, defined by nominal characteristics in common (e.g., language, religion, or government administration). In recent decades, thinking about the spatial structure of society has involved examining interrelated activities, usually with core areas and interconnections, which geographers have gradually constructed into spatial systems. Such systems may include the advertising market catchment areas of local television broadcasts, newspaper delivery, and shopping center attraction; they may also include the attendance of children at various schools according to census characteristics of race, median income, welfare, location/capacity of facilities, numbers of buses, and connecting roads. The historic lack of geographic planning in the U.S. health care system is clearly shown by the numerous different definitions of regions. In North Carolina, for example, there have been the following types of multicounty regions: U.S. Public Health Service health service areas, American Medical Association service areas, Department of Veterans Affairs patient service areas (a federal regionalization for hospitals), comprehensive health planning areas, and hospital service areas (a federal regionalization related to Medicare). When all these regions are mapped and overlaid, of course, there is almost no correspondence of boundaries. Nothing can be compared between regions or across the state. Health service areas sometimes incorporate more than one comprehensive health planning area, but at least they share some boundaries. Even the most basic rates are difficult to construct, as census data have until very recently been aggregated and made available only by counties or their subdivisions, not by combinations. Therefore income, education, poverty, housing, or other health-relevant data couldn't be combined to characterize a health service region.

There are both simple and complex methods for constructing service regions. Because the only geographic identifier in patient data is often the postal (ZIP) code of a patient's residence, a popular method of regionalizing medical service areas is to consider a medical facility's primary service area to be those ZIP code areas that contribute the highest proportions of patients to the facility's load. Although this method is based on distance decay, the fact that ZIP codes were never associated with census data or based on any size of population, and that they are not constant but change to suit the postal load, has made them almost useless to associate with any theory or explanatory data. Potential and gravity models (see Chapter 6) have been used, as well as efforts to identify market areas based on those elliptical cells of social activity. There is a large and growing literature on the methodology of regionalization and GIS for emergency use and for health care.

There are at least three ways to rationalize health care delivery spatially: through management of patient flows (e.g., several hospitals funnel their routine outpatient surgery cases to one large free-standing facility); through the allocation of special services, wherein expensive high-tech medical equipment is centered in a few major hospitals, or a hierarchy of clinics and hospitals is established within a region; and through managing the location of health care personnel, which follows the same philosophy as the allocation of facilities (Ricketts, Savitz, Gesler, & Osborne, 1994). There are several issues to consider in creating any regions. Geographic scale is of course important; regions in densely populated areas are necessarily smaller than those in sparsely populated areas. In small-area investigations (county or metropolitan area), data may be collected at the block or neighborhood level, whereas in rural areas only county-level data may be available or meaningful. Theoretically, a region should be geographically contiguous. However, supply and demand in health services do not always conform to spatial contiguity. Patients sometimes bypass many providers to seek care with a distant facility or doctor, as noted earlier;

also, indigent patients do not always conveniently live within the same neighborhood or the same city. Using existing political or postal boundaries as medical service regional boundaries may not be realistic either. Because demographic data are grouped within political/administrative jurisdictions, in which taxes are collected and funds are spent by state and local governments as well, there is a tendency to structure service regions within these same borders. However, people often tend to ignore city, county, and state lines in their daily life transactions (remember activity areas?). Population-sheds, or market catchment areas, and the preconditions for illness certainly are not so constricted. Furthermore, political jurisdictions often lack social and economic homogeneity. For example, one large postal (ZIP) area in the Baltimore, Maryland, metropolitan area extends from within the city into the county. It contains neighborhoods of different racial, economic, and age structures. Chronic disease rates differ significantly in magnitude and kind from one socioeconomic class to another.

The distribution and utilization of emergency medical services have a clear impact on the well-being of society. They are good examples of hierarchical regionalization. Peters and Hall (2000) demonstrated that ambulance resources, which are an important component of emergency health care, are often insufficient, resulting in performances below those required to respond reliably to emergency calls from demand areas. They showed how to use a GIS to assess ambulance response performance and to plan changes in services. The capacity of a GIS to comprehend multiple factors, variously combine them, and select areas for which they are shared and/or different is powerful for regionalization.

As researchers began to understand better how those variables identified earlier (education, race, income, occupation, health insurance status, etc.) worked to "enable" utilization of health care services, and how variables such as knowledge of existing services, health behavior practices, and family size acted to "predispose" people to using them, they increasingly found location–allocation modeling simplistic. More attention turned toward spatial interaction modeling, as in those activity areas of old, only now equipped with network nodes for prevention sites. (This too was paralleled by developments in modeling disease diffusion; see the discussion of Ling Bian's work in Chapter 6.) Recently Yasenovskiy and Hodgson (2007) have combined location–allocation modeling and hierarchical spatial system modeling (i.e., central place) with spatial interaction modeling to create "hierarchical location–allocation with spatial choice interaction modeling." Location–allocation models applied to three levels of health service can optimize objective functions and still respect constraints on the system (such as health insurance requirements). The bypass of "optimal" locations noted above in the description of the Malaysian system has been frequent enough to seriously discount the least-distance factor, while adding attractiveness of size of facility or adjustments for multipurpose trips. Yasenovskiy and Hodgson address the problem by incorporating a spatial interaction model based on competing destinations' attractiveness and distance, but accommodating multiple destinations and other complexities of health care utilization behavior. Thus they powerfully synthesize hierarchical, interaction, and location–allocation modeling.

GIS Analyses and Simulations: Regionalizing and Rationalizing

The availability of GIS and the ability to use it to run many kinds of simulations have both put the previous years of research and conceptualization to good use and made many of the most laborious operations routine. One such operation developed a pair of models—one access-based (i.e., distance to care) and one demand-based (i.e., efficiencies of central place)—to evaluate and address inadequacies of areas of community hospital access (Messina, Shortridge, Groop, Varnakovida, & Finn, 2006). Like the Swedish government as described at the beginning of this section on regionalization, the state of Michigan's Department of Community Health approached geographers at Michigan State University "with questions about how spatial analyses might be employed to develop a revised community hospital approval procedure" (p. 1). The three objectives were (1) visualizing the spatial patterns of population in relation to 136 community hospitals; (2) developing a "clear, defensible" methodology to quantify access to existing hospitals (assessing distance, travel time, road networks, etc.); and (3) contrasting the existing spatial distribution with a theoretical optimum configuration. Plate 16a shows the results of the access-based model. The areas with limited access—by travel time, most importantly—covered much of the Upper Peninsula (but notice the mostly forested and swampy center of the peninsula, with community hospitals in settlements on the two lake shores). When criteria of contiguity and minimum population density were applied, the map (not shown) changed. Only the southeastern area of limited access (north of Detroit), and the smaller area across the state to the west (around Grand Rapids), remained. The Upper Peninsula was too sparsely populated, and areas of limited access were too fragmented.

Plate 16b shows one of the two maps of locations that would maximize coverage, based on the demand model of allocation of existing patient bed-days. The three new locations, each of which would potentially service 25–137 beds per day, were all on the northern edge of metropolitan Detroit when this 16.1-kilometer maximum distance was used. When the maximum distance was increased to 32.2 kilometers, the resulting three new locations shifted northward (map not shown). Then they were in the northeastern corner of the Michigan "mitten," southwest of that in the center, and in the southern protuberance of the Upper Peninsula near the Wisconsin border—all near villages too small to find in an atlas. The main limitation described by Messina et al. for the access-based models was that the road network was not complete, since in Michigan all "M" class roads have been digitized, but the state still ignores federal and private roads. The main, but important, limitation for the demand-based model was that it had to be based on aggregating demand by ZIP code. This made disaggregating analysis impossible, and made the model sensitive to aggregation error. Despite the casual sophistication of the GIS modeling, the outcome is still striking: remote land areas will continue to have lack of accessibility to a hospital, and population demand for facilities and services is still aggregated near a large (the largest) central place, which already has a large cluster of hospitals. It is especially striking to us, however, that the state is now seeking spatial analyses, with its hospitals mapped and its population access modeled.

Hospital Privatization and Closure

This basic inequality examined above between rural and urban areas—the rural areas' vast territory, remoteness, low population density, small or only periodic central places, poverty, and distance to health resources that those in urban areas, with transportation networks and accessible shopping facilities appropriate to their density and affluence, take for granted and demand more of—is only exaggerated at the international scale. Centrally planned, socialist political systems have sometimes promoted equal accessibility to medical resources in all areas or regions— "territorial justice"—at considerable cost to the central government, and generally a lowering of the best quality achieved. When market forces enter into these systems, the rationality of efficiency may improve the quality and accessibility of care for those in the population centers, but tends to devastate the basic-care level once available in peripheral or rural areas. In the United States in the late 1980s/early 1990s, there was a surge of hospital privatization as high-cost public hospitals were bought by health management organizations (HMOs) of various kinds (mostly insurance companies, some church-related groups) for the use of their memberships, or by investors simply for profit. A spate of studies followed analyzing where displaced patients went, whether the region's health care overall became more or less rational, and who got left out of or was denied what services.

The effects of privatization internationally were much more severe. The health service system of the Soviet Union collapsed almost totally with the rebirth of Russia. Epidemics of diphtheria, tuberculosis, and HIV, as well as sharp increases in alcoholism, fetal alcohol syndrome, and heart attacks, increased mortality and decreased life expectancy for the first time in an industrialized country. As mortality soared, fertility fell even lower. The worst conditions have abated, but the rate of natural increase is still negative. In the cone of South America, the changes in government and market forces in Chile and Argentina drastically affected the whole system of health care, including urban–rural differences in accessibility, as Scarpaci (1989, 1990) and others studied.

In China the changes are still expressing themselves as several hundred million Chinese leave rural areas and move not only to cities (see Chapter 10), but in particular to the cities of Guangdong and other rapidly industrializing areas affected by globalization. Under the Mao Zedong regime, doctors were forced from urban hospitals to rural and sometimes remote commune hospitals (county hospitals) for a year or more of service. The top layer of medical school education was removed so that more "barefoot doctors" and other paramedics could be educated about Western medicine (especially germs, sanitation, sterilization, and vaccination). One of the aphorisms memorized by a billion people was "In matters of health, put the emphasis on rural areas." People everywhere were extremely poor (the per capita gross national product at the time was barely U.S. $200), but people had access to both traditional Chinese and Western medicine as practiced by doctors and in hospitals, paid for by their communes. Later, when they were so strongly encouraged to have only one or two children, they felt secure not only that their child would be likely to live but that they would be supported in their old age by the commune pension system. Under Deng Xiaoping's agricultural responsibility system, the communes were dismantled. As the market influences have increased and migration has left hundreds of millions unconnected to the cities in which they live and work, affluence has become important for both patients and care providers. One picture we authors had no plate space left to include in this book was of a beautiful little well-equipped clinic that the government built. Emch's photo showed that it had no patients and no personnel, since doctors were all making money in private practice, whereas the poor peasants could not afford even the government charge. The growing inequality of health care in China has been documented by Smith (1998; Smith & Fan, 1995), who explains that the urban–rural gap has been reinforced by two other trends: the preference among trained medical personnel to live and work in cities rather than rural areas, and the tendency for individuals who might otherwise have chosen a medical career to choose a more profitable one.

The Dartmouth Atlases and Their Truths

In the early 1990s the medical and statistical scholars of Dartmouth University Medical School (Hanover, New Hampshire) began studying the patterns of national utilization of Medicare funds for procedures and hospitalizations mapped and analyzed by hospital service areas. Their investigation of health care utilization and provision, using 306 hospital referral regions and Medicare expenditures, produced *The Dartmouth Atlas of Health Care 1996*. The atlas was so productive of insights, spatial pattern analyses, and findings important for policy analyses that the atlas project has continued for close to two decades now, with each atlas centering on a specialized topic of interest. The atlases are national, but there are regional and even some state volumes. They are for sale, but can be downloaded as .pdf files. Recently more interactive capacity has been added to the project's website (*www.dartmouthatlas. org*) so that typing in specific ZIP codes can bring up hospitals; search means have been added; files can be downloaded; and a tremendous research data bank has been created.

The atlases have shown, according to the website's description of the 1998 edition, that the amount of care consumed by Americans is highly dependent on where they live—on the capacity of the health care system where they live, and on the practice styles of local physicians. Variations in the intensity of use of hospitals, the striking differences in the way terminal care is delivered, and the idiosyncratic patterns of elective surgery raise significant questions about the outcomes and value of health care. The fundamental questions posed by the atlases are Which rate is right?; How much is enough?; and What is fair?

The atlases have found great variation among regions in medical resources available, but even more in practices and resource use. The national atlas edition tracking the care of patients with severe chronic illness (Wennberg, Fisher, Goodner, & Skinner, 2008) examined especially resource use during the last 2 years of life. Surprisingly, rates of elective surgery showed the least variation in pattern and were nearly random among hospitals and regions. Cardiologists, in contrast, performed nearly twice as many heart procedures in the Hospital Service Area of Miami as in that of Minnesota, and more people spent more time in intensive care in the last months of life in Miami. In Minnesota, there were greatly increased rates of knee and hip surgeries, low rates of heart procedures, and greater use of hospices. Higher-spending regions compared to lower-spending regions had a 32% higher per capita supply of hospital beds; had 75% more internists and 26% fewer family practitioners; and generally had *worse* outcomes. Differences were minor regarding rates of major elective surgery, but higher-spending regions had more hospital stays, physician visits, specialist referrals, imaging, and minor procedures. In outcomes for these regions, however, mortality over a period of up to 5 years was slightly *higher* following acute myocardial infarction, hip fracture, and colorectal cancer diagnosis.

In summation, reports on the spatial patterns found that the most intensive use of hospitals did not produce better outcomes, and urged that the rates of use of the hospitals with the best outcomes be established as normative. There was no difference in functional status as an outcome, despite the differences of intensity in treatment. Physicians' "perceptions of quality" in the most intensively served areas were more likely to include poor communication among physicians and inadequate continuity of patient care. Patients where the per capita ratios were highest were more likely to complain that specialists were inaccessible. Studies of variation "provide good evidence that populations in low cost regions are not sicker or in greater medical need than those in high cost regions. . . . A system that rewards higher costs by paying them punishes areas that use fewer resources more efficiently and are reimbursed less" (Dartmouth Medical School, Center for the Evaluative Clinical Sciences, 1998, p. 4).

The greatest surprise, is that "more care is associated with higher mortality" (Wennberg, Fisher, Goodner, & Skinner, 2008, p. 4). The explanation given is that "all medical care poses some risk, and the more care a patient receives, and the more often he or she is hospitalized, the greater the risks. Where more resources are available, patients receive more care, and receiving more care increases the chances of errors being committed, and of patients being faced with complications" (p. 13). It was also found that "more care does not provide patients with greater satisfaction or improved functional status" (p. 13). These findings were important to the efforts in the Fall of 2009 to reform U.S. health care (Obamacare). Among other things, the findings showed substantial savings might be possible and costs could be contained without sacrificing quality, if a "best-practice" model were identified. However, one can imagine the opposition claiming that declining payment for ineffective and redundant practices in expensive-care regions would be "rationing."

In accessibility to available specialists, importance of distance as an obstacle, use of medical services and expectations of outcomes, and all such measures and judgments, we are really dealing with perceptions—with the subjectivity of people, patients, health service providers, and researchers alike. Yet in the United States there has been some commonality of belief and value from which its citizens can proceed to debate their differences. In contrast, even germ theory, let alone the efficaciousness of the medical procedures and practices analyzed in the Dartmouth Atlas series, has not been comprehended or experienced by many hundreds of millions of people in this world.

Health Promotion

The study of health promotion is a large and rich area of research in health geography with frameworks and analyses unique in the field of public health. More importantly, it is a dimension of health geography that needs its own book for today's developments. Geographers concerned with health promotion have worked to address the space, and place, of persons with the disabilities. Whether disabilities are mental or physical, research on accessibility involves more about space than distance. Place is local, and information coming from interviews, focus groups, and other qualitative methodologies is filling it with meaning. The journal *Health and Place* published a whole issue on the geographies of intellectual disability (Metzel & Philo, 2005). The articles in this issue deal with social poverty and service dependency, entangled geographies of social exclusion, residential instability in deprived areas, and places changing as rural areas get restructured. The few geographic studies on mental health have been mostly concerned with the old question of whether mentally ill persons (especially those with schizophrenia) are clustered in areas of old cities because the blighted conditions and treatments made them ill, or gathered there from elsewhere for the amenities and tolerance or support of such areas. There have been studies of the "not in my backyard" syndrome in regard to halfway houses, zoning laws, and such. Dear and Wolch (1987) made an impact in this area of study with their *Landscapes of Despair*. The present book deliberately has not dealt with mental health issues because of both lack of space and the relative lack of research. Looking toward the future, medical and social geographers can be engaged in studies of health promotion and the many economic, social, demographic, and locational issues and needs of mental health.

QUICK REVIEW 1

✓ Access is a multidimensional concept determined by a variety of geographic, social, and economic factors.

✓ Geography plays an important role in understanding access, in particular the availability of health services and geographic accessibility.

✓ Though health systems are very different in low-income countries compared to high-income countries, everyone agrees that all people deserve access to basic and preventive health care.

✓ Central place organization of the health system, where primary care services are widely available, and specialty care is concentrated in a few areas, has been adopted as the "best practice" model for developing countries in the process of improving their national health systems.

✓ Health care in the United States has undergone many changes over the past century, and is based on a market-based approach that requires individuals to obtain health insurance, rather than a social health system, which provides health care to all citizens.

✓ Geographers have devoted significant energy to mapping health resources and examining why resources are located where they are.

✓ Locational analyses help determine the best location for health services such that resources are equitably distributed and provide for the populations most in need. They also try to understand the historical and economic forces behind current-day spatial patterns of health resources.

✓ Some modern-day health systems are set up within regions such that all people have access to health care within a specific area.

✓ The *Dartmouth Atlas of Health Care* is an excellent example of how geography can be used to study the health system and the relationship between the structure of the health system and morbidity and mortality.

CULTURAL ALTERNATIVES AND PERCEPTIONS

Up to this point in this chapter, the analyses of health services and their locations have all centered on the health care system variously called Western medicine or biomedicine. This system certainly includes aspects of cultural perception, from social distance and what to do about it to differences in usage. Women have higher morbidity at every age, but men have higher mortality at every age. It's long been "known" (or at least perceived to be known) that this is at least partly because women more readily seek advice and help, whereas men often do not want to show such vulnerability. It also used to be "known" that members of one ethnic group would come to work with fever and aching bones, and that members of another would likely call in sick with a cold.

Most of our world's understanding of illness, what to do about it, and how to prevent it was developed before germ theory, however. Most traditional treatments are derived from millennia of empirical observation, although religious views certainly have had their input. For billions of people, home remedies and traditional practitioners with their medicines remain the front line of health care. As the limitations of biomedicine have become obvious and drug resistance has resulted in a renewed search for new drugs, traditional medicine has developed a new attraction. Biomedicine, it is said, is reductionist, is remote in social distance, and speaks an incomprehensible language. It cares about organs, not people. It never listens, and it always hurries. Traditional medicine, in contrast, is seen as holistic: it cares about healing persons, listens to them, talks to their families, works with their life issues, gives them gentle drugs, and lays hands on them. Indeed, traditional medicine has increasingly become not only an alternative, but a complement, to biomedicine. This section examines the beliefs and structures of traditional medical systems that are developing as professional alternatives, especially Chinese medicine. Their use in contemporary Western society seems likely to grow.

Alternative Medicine: A General Description

Although trepanning (drilling holes in the skull of a living person) seems to have been practiced by Neolithic peoples, the first solid evidence of medical practice has been found in the primary agriculture hearths that produced the earliest written records. The medicine of these ancient civilizations had roots in the supernatural; practitioners were essentially priests. There is evidence, however, of empiricism, practical organization, and the beginnings of public health. Egypt was famous in ancient times for its medical techniques. Physicians and dentists practiced there as early as 2700 B.C.E. The earliest known legal code, the code of King Hammurabi of Babylon (c. > 90 B.C.E.), contains laws on malpractice and setting medical fees. Hot and cold running water systems and urban sewerage systems were already in widespread use in the Harappan civilization of the Indus Valley by 2000 B.C.E. Public health and medical practice in the New World, as reported by the European explorers of the 15th and 16th centuries C.E., were on a par with those in the rest of the world. In particular, the Spaniards highly praised the sewerage systems of the Incas.

Traditional medicine exists in various forms and under a range of names, including nonprofessional or indigenous as well as traditional, although the more inclusive term **complementary and alternative medicine** (CAM) is widely used today. Practitioners of traditional medicine are called healers, native doctors, shamans, or medicine men. The study of traditional medical systems has emerged in the discipline of **ethnomedicine**, which involves the study of beliefs and practices stemming from indigenous cultural development (Good, 1987). Four general types of traditional practitioners may be identified: spiritual or magicoreligious healers, herbalists, technical specialists such as bone setters, and traditional birth attendants. The first category, **spiritual healers**, is the most common and includes respected healers and charlatans alike. Religion and magic have always been closely tied to healing in traditional societies. Supernatural beings are believed to affect, among other things, illness and health. The healer mediates among the supernatural, the patient, and the community (Good, 1987). **Herbalists** focus on the use of medicinal plants to cure illness. Knowledge of herbal formulations is sometimes codified, sometimes simply passed down from healers to apprentices. Efforts to identify and research the nature and potency of herbally derived medicines have become common in Western nations. *Bone setters,* as the term implies, have a somewhat narrow focus, but in some places their practice resembles that of chiropractors. **Traditional birth attendants** assist mothers at childbirth. The necessary skills may be acquired by helping experienced attendants deliver babies, or through modern training efforts sponsored within clinics and hospitals in many communities throughout the world.

Most traditional medical systems are confined to limited areas and specific populations, scattered throughout every continent (Gesler, 1984). Various Native Americans of both North and South America, numerous African tribes, and a variety of groups in Asia (from Mongols to Malays in the Indonesian islands) practice traditional medicine. Traditional healers may be the basic providers of health care for up to 90% of the rural population in South Asia and Africa.

Although it is difficult to generalize about the medical practices of all these people, some ideas and techniques are widespread. For example, diagnosis and treatment can be carried out by immediate family members, other kin, and/or group leaders, as well as by healers. The various treatments held in common include the use of medicinal herbs, prayers, the sacrificing of animals, exorcisms, the wearing of sacred objects, and the transfer of disease from one person to another. Such treatments often depend on an intimate knowledge of intracommunity relations to be successful. They seem especially effective in dealing with mental illness.

A healer may be a judge; many diseases are seen as stemming from violations of the morals and mores of society. A sick person may be seen as having broken relations with the supernatural or with other humans; his or her suffering may be viewed as a social sanction, and the healer's diagnosis as a kind of social justice. Treatment in such cases often involves a cathartic confession. Because village life is very close-knit, tensions must be resolved for the group's survival. Thus the healer also plays a role of creating "psychic unity." Treatment often includes having sick people, their relatives, and other people bring out their ill feelings toward each other. The healer's knowledge of community conflicts is important here. In many African societies, much less distinction is made between the material and the spiritual worlds than in Western nations. Illness may be viewed as having a social or spiritual origin, in addition to physiological causes. A psychosocial approach to medicine is therefore applied in African medicine, where a person's body, mind, and soul are conceived as an indivisible whole. A medicine may be a substance administered to treat and prevent illness, anything used to control the spiritual cause of illness, or both. Although biomedicine appears to be eclipsing traditional treatment in sub-Saharan Africa, indigenous remedies have been considered just as effective and much cheaper, more accessible, and more easily understood.

Alternative Professional Systems

Ayurvedic medicine, unani, Galenic medicine, Chinese medicine, and biomedicine are all "professional" systems. They are highly organized and have established arrays of techniques and codes of conduct. Each has a highly developed pharmacopoeia and a long and progressive history, and they all serve very large populations today.

The origins of **ayurvedic** medicine, which means "the science of living to a ripe age," can be traced to the migration of Aryans into the Indus Valley around 2000 B.C.E. By the sixth and fifth centuries B.C.E., the Indian medical system had approached its present form. In this system, disease represents a disequilibrium of four *humors* (wind, bile, mucus, and blood), and cures attempt to reestablish a proper balance. Health is related to karma, or the effect of good and evil deeds, in both one's former and present lives. Practitioners must have a comprehensive knowledge of the pharmacopoeia and must understand well the influences on health of climate and morality. Ayurvedic medicine today is important for both rural and urban Indians. Its core area is north India. The Dravidian culture of south India has many ayurvedic practitioners but has also established the siddha system, which has employed substances that purportedly could transform base metals into gold as well as aid in rejuvenating the human body. In diagnosis and treatment, siddha generally corresponds with ayurvedic medicine. Indian migrants have carried their ancient medical practices to other parts of the globe.

Galenic medicine, unani, and biomedicine all have their roots in Greek medicine. Hippocrates (c. 460–377 B.C.E.) represented the culmination of early Greek medicine. His "On Airs, Waters, and Places" (see the quotation that precedes Chapter 1) was the first great classic of health geography: it associated certain diseases with certain climates and recognized that cultural practices and social institutions can change or temper climatic conditions. Hippocrates denied that disease had supernatural causes, and he stressed careful observation of patients. One of the many statements ascribed to Hippocrates is "persons who are naturally very fat are apt to die earlier than those who are slender." By his time, the idea of the four fundamental elements in nature (hot, dry, wet, and cold) and the four bodily fluids or humors (blood, phlegm, yellow bile, and black bile) had been developed. A healthy person maintained a proper balance of these humors, which originated from four parts of the body and were associated with the four fundamental elements (see Figure 13.8). Disease was a manifestation of improper balances among the humors and elements. Treatment depended on the nature of that imbalance.

After the time of Hippocrates, many humoral and nonhumoral schools of medicine arose in Greece. Only with the coming of Galen of Pergamon (c. 130–201 C.E.) did humoral medicine gain the upper hand. Galen, the "father of experimental physiology," was a Greek residing in Rome; he was physician to the Emperor Marcus Aurelius. His experimental insights became important texts, first for the Islamic Arab civilization and then in the early Renaissance for Western Europe. He modified the Hippocratic idea that an equilibrium of humors could be achieved by teaching that physical, cultural, and demographic factors could cause one of the four humors to dominate and produce a unique "temperament": sanguine, phlegmatic, choleric, or melancholic (familiar concepts to anyone acquainted with English literature). The Galenic body of medical thought held sway in Europe for 1,500 years. Galenic practice began to lose its importance only over the last 300 years. When Europeans began exploring and colonizing the world after C.E. 1500, Galenic medicine diffused along with many other cultural traits. When American physicians in the early 19th century (before germ theory was developed) performed

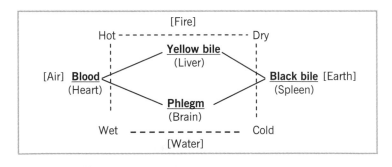

FIGURE 13.8 Hippocratic (Greek) humoral medicine.

bloodletting or used leaches, for example, they were trying to reduce severe fever by releasing blood (hot, heart); and when a person was in a black, melancholic mood (depression) or bilious, Galenic physicians recommended walking outside in the fresh air to get the heart pumping and blood circulating, as well as drinking hot tea, in an effort to rebalance the humors.

The Arabs took over the system of Greek medicine in the seventh and eighth centuries C.E. and called it **unani** ("Greek" in Arabic). In common with other early medical systems, unani had a strong ethical element; it stressed the importance of the doctor–patient relationship in healing, as well as the influence of beliefs on disease and health. Western incursions into the Arab world in the 19th century led to the decline and eventual stagnation of unani, as biomedicine supplanted unani to a large extent. However, it is still important in South Asia, where it was introduced by Muslim conquerors and spread throughout the Mughal Empire. Unani exists alongside systems like ayurvedic, siddha, and biomedicine in India, Pakistan, Sri Lanka, and other South Asian countries. Some of these countries have established state and private pharmaceutical companies that produce unani drugs. Unani is also important in other areas with strong Muslim influence, Southeast Asia in particular.

The foundations of **Chinese medicine** share with Greek medicine such strong associations of elements, organs, influences, and personality states that the roots must go back further than either. They also differ fundamentally. The Yellow Emperor's medical treatise, written between 2600 and 1600 B.C.E., codified the thoughts of Chinese medical practitioners on disease causation and treatment. The world's oldest medical text, it was nevertheless written long after Chinese civilization was expert in both bronze and iron and higher alloy metallurgy. The Chinese added metal to their view of the nature of the cosmos, and so have five elements to the Greek four. As Table 13.1 shows, Chinese cosmology is organized in fives. Besides those in the table, there are also five directions (north, east, south, west, and center), five major planets, and five of many other things. Think of the possible associations that arise from this macrocosm and microcosm. For example, joy is linked with the heart and blood vessels; with fire; with the Chinese color of celebration, red; and with growth and summer. Sadness is associated with lung and nose (congestion) in mature (late middle) age; with drought and sharp metal; and with white, the color of mourning. Anxiety, of course, is associated with "butterflies" or worse in the stomach, tense muscles, and dry mouth; with the uncertainties of change (in the context of Table 13.1, especially puberty); and with yellow, the color of cowardice (even in English). American English even recognizes five seasons: on the East Coast, which has a climate very similar to China's east coast, we know the few weeks of low humidity and clear, sunny, warm days after leaves start to turn as Indian Summer.

Disease is considered in Chinese medicine to be the result of disharmonies within the body, between humans and the environment (the microcosm and the macrocosm), and even throughout the universe. The treatment of disease arises from philosophical concepts such as those of yin and yang, which are the negative and positive principles of the universe. Yin stands for things such as Earth, Moon,

TABLE 13.1. Divisions and Connections of Chinese Medicine

Five divisions of the macrocosm					
Five elements	**Five seasons**	**Five taste qualities**	**Five colors**	**Five atmospheric influences**	**Five stages**
Wood	Spring	Sour	Blue	Wind	Birth
Fire	Summer	Bitter	Red	Heat	Growth
Earth	Late summer	Sweet	Yellow	Humidity	Change
Metal	Autumn	Sharp	White	Drought	Maturity
Water	Winter	Salty	Black	Cold	Storage
Five divisions of the microcosm					
Five elements	**Five sense organs**	**Five structures**	**Five fu's**	**Five tsangs**	**Five emotions**
Wood	Eye	Sinews	Bile	Liver	Anger
Fire	Tongue	Blood vessels	Small intestine	Heart	Joy
Earth	Mouth	Muscles	Stomach	Spleen	Anxiety
Metal	Nose	Hair	Large intestine	Lung	Sadness
Water	Ears	Bones	Bladder	Kidneys	Fear

darkness, and femaleness; yang symbolizes heaven, Sun, light, and maleness. (They are represented by that common symbol of eastern Asia, the circle with the "S" inside it as the yin and yang half perpetually turn into each other.) Therapies to maintain a balance among the elements and between yin and yang were developed over many centuries and are still widely practiced. They include several forms of massage, physiotherapy, a very complex and developed pharmacy, a system of what might be called sanitariums devoted to the therapy of breathing, and the health promotion through balanced yin–yang of the practice of **t'ai chi chuan** (*qidong*, still seen in the parks and balconies of China in early morning). The life energy *ch'i* flows through the "meridians" of the body (these are unblocked by *t'ai chi*). The meridians touch the skin at special points that can be stimulated by heat, pressure, or needles in the healing technique of acupuncture. Of all Chinese therapies, acupuncture is the one that has been most widely accepted in the United States, in good part because its effectiveness can be tested by scientific methods and it can be practiced even in a biomedical setting. It can clearly prevent pain, although many Western doctors are still not certain about that "stimulating the spleen" business. Although China remains the core area of the practice of Chinese medicine, it is practiced wherever Chinese people have migrated. The Nanyang Chinese (Southern Seas, or those in Southeast Asia) have a major school of Chinese traditional medicine and research center on pharmaceuticals in Singapore. Its influence has been greatest in Korea and Japan, part of China's cultural sphere.

Biomedicine arose out of the Galenic medical tradition. Practitioners of galenic medicine in post-Renaissance times were sympathetic toward experimental physiology, which led to many important medical discoveries (such as the circulation of blood). Biomedicine advanced during the development of scientific method and inquiry; in particular, the 19th-century discoveries in bacteriology gave scientific medicine its present high status. These discoveries, collectively known as **germ theory**, focused on the idea that infectious diseases are caused by microorganisms. Two outstanding bacteriologists were Louis Pasteur (1822–1895) of France and Robert Koch (1842–1910) of Germany. Pasteur identified the organisms that cause body decomposition and fermentation of wine. He also identified those that produce anthrax and fowl cholera, and worked on vaccines for these diseases; he even developed an antirabies vaccine, although rabies is caused by a virus and viruses could not yet be seen. Koch identified the bacteria that cause wound infections, developed better techniques to identify bacteria, and discovered the germ that causes tuberculosis. Several other scientists entered the new field of bacteriology, and between 1875 and 1905 about two dozen disease agents were found. During this period there were also discoveries in serology and immunology; disease vectors such as the tsetse fly and *Anopheles* mosquito were identified; surgery and gynecology advanced; and anesthetics and antiseptic conditions became part of surgical procedures.

Specialization was spurred by the rapid accumulation of vast amounts of new information in diverse areas of medicine. Professionalization tightened ethical and educational standards within established medicine, partly in response to the proliferation of nonscientific, nonestablishment health practitioners. For example, midwifery clashed with professional obstetrical and gynecological groups in the 19th century. Such organizations as the British Medical Association (1832) and the American Medical Association (1847) were formed.

Of all the world's major professional medical systems, biomedicine has clearly been the most widely diffused. This medical system has become synonymous with the culture of illness and health in many Western countries, and the Third World has adopted it to varying degrees. Each European power introduced biomedical practices to its colonies. The first European physicians to work in most colonial nations were too few in number to care for any but colonial and military personnel, other Europeans, and a limited number of native persons in the upper social strata (Phillips, 1990). When it was recognized that many diseases are contagious, programs were initiated to treat larger populations who were in contact with Europeans. Despite this, health for most colonial people remained as poor as it had been in precolonial days.

Biomedicine had little real impact in these areas until after World War II, when the colonial governments began to train native assistants in Western medicine, while others went abroad to study medicine. When the colonized countries gained their independence, a native elite had acculturated to Western medicine. This elite perpetuated the Western medical hegemony (Vaughn, 1991). Since then biomedicine has played a major role in the epidemiological transition, one of the primary reasons for today's population growth in developing areas (see Chapter 4). In much of the developing world, Europeans tried for decades to suppress traditional medicine as quackery.

Medical Pluralism

In most parts of the world, people use more than one kind of medicine at the same time. In some countries this is officially tolerated; in others all but biomedicine are suppressed; in still others efforts have been made to integrate multiple systems. This latter approach has worked best with Chinese medicine. The use of more than one kind of medicine or the use of both Western medicine and alternative medicine for health and illness is often called **medical pluralism**.

In Malaysia the indigenous Malay population has its own traditional medicine (which preceded unani but, as the Malays became Muslims, was strongly influenced by it) and *bomoh* to practice it. The bomoh have drugs from the forest for various treatments, and they go into trances and interact even with a powerful tiger spirit in order to treat mental health problems. This land and language, after all, gave us "run amok." The Indian population of Malaysia (about 11%) is mostly Dravidian but also northern Indian, mostly Hindu but also Christian. It widely uses both ayurvedic and siddha medicine, especially for "women's problems." The Chinese of many dialect groups and several religions constitute about 40% of the population. Chinese pharmacies, with practitioners, are found on many streets in most towns. Practitioners have even formed a Chinese traditional medical school. Several of the students there at night work in biomedical laboratories during the day, reading tuberculosis slides and blood slides and analyzing urine and fecal parasites. They have integrated the two systems in their own personal practice, as patients have. People of all the ethnic groups and medical traditions in Malaysia seek help from both Chinese pharmacists and biomedical practitioners. Figure 13.9 shows a sign outside the main hospital in the national capital, where there are parallel waiting rooms, outpatient hours, and pharmacies for Chinese and Western medicine (although the beds and the remainder of the hospital are for biomedicine only).

Such institutionalized pluralism is not common. One difficult problem with more than one health system is that people often use both at the same time. Charles Good (1987) studied traditional medicine in Kenya, Nigeria, and elsewhere. The attraction of Western medicine was especially the power of the needle: antibiotics were magic, vaccines worked. When other medical treatments did not bring quick response, or were not helpful for chronic pains, people repaired to their traditional practitioners. Those herbs were also bioactive, however, and the medical doctors never knew who was taking what. Treatments were not followed up or required drug prescriptions taken; drug interactions were sometimes serious. There was a danger that people would go to traditional practitioners until they were sure that their advice was not working, and then they would present themselves at the Western clinic with sickness or cancer too far advanced for successful treatment. This is the main argument that biomedical doctors continue to raise in opposition to traditional practices, including versions of quackery today in the United States: When people pursue "miracle cures" and other alternative healing, it can cause them to postpone the more efficacious biomedical treatment. Tumors really can be surgically removed and killed with chemotherapy most of the time, but not so easily after they metastasize. Chinese practitioners in Malaysia said that when their second

FIGURE 13.9 Hospital sign in Kuala Lumpur, Malaysia, giving alternative waiting and attendance times for Western and Chinese medicine, in three languages: Malay, Chinese, and English.

treatment for an earache failed to cure it, they sent their patients to a Western doctor because they knew they could not help nasopharyngeal cancer—an important cause of death for the Cantonese.

The holistic health movement in the United States started in California (and North Carolina!). It diffused across the nation rapidly, adding variety as it did so. Holistic therapies include massage of various types, acupuncture (by licensed practitioners), and chiropractic (by licensed practitioners with degrees), as well as things like aromatherapy and advice on healing colors (from Chinese medicine?). The universal complaint against biomedicine was (and still is) that it was reductionist, impersonal, and inhumane. It dealt not with bodies but organs, and increasingly with molecules. People were processed, seldom listened to, and rarely touched. Traditional medical practitioners of whatever kind, it was argued, focused on individuals, took their time, listened, touched, and asked about family and life stresses (Kleinman, 1993).

The professionalization of standards promoted by the American Medical Association in order for individuals to be licensed to practice medicine really did eliminate a lot of "quackery" by get-rich-quick, ignorant, incompetent crooks selling "snake oil." (Today some of them are returning via the Internet.) Practitioners of biomedicine have recognized many of the truths in the arguments for holistic medicine, however, and have started to change some things. The most prominent among these has been the response to the findings of the Dartmouth Atlas, that less can be more, and better. "Preference-sensitive care" is being promoted (especially in choosing procedures and hospital stays), as is listening to the concerns and observations of patients and their relatives. Obstetrical rooms are being redesigned to look more familiar; pain care has become a priority; and simple practices to reduce hospital infections have been renewed, including washing hands with soap and using alcohol wipes instead of antibiotics and detergents. Space has been made to relieve the stress on staff members, and so to improve their own health and the care they render. For example, break areas with water features and plants, and rooms in healing colors, are made available. Hospitals and other medical facilities have also become more open to less traditional therapeutic practices. Many now include lessons in meditation and visualization, or even humor therapy.

One new form of therapy works at several levels in a community. It's been known for more than a decade as pet therapy (see Wilson & Turner, 1998). Petting a dog can lower a person's blood pressure (the dog's, also!), and companionship and the need to care for another living creature provide deep solace for many people. Numerous studies have shown that people who have pets recover faster from surgery and have better survival rates than those without pets (though some of those studies are poorly done—not even considering, for example, how long a person has had a pet, or whether it promotes walking and other healthy activities or just cuddles). Well-done studies in hospitals have shown, by monitoring hormones and stress chemicals, that patients who get visitors are benefited over patients with no visitors, but that visits with a dog produce better results than visits with only a human. Now puppies are taken from shelters out to visit weekly in nursing homes, which helps socialize both the puppies and the residents. People make the effort to train their dogs as "good citizens" and visitors—keeping them under control (no jumping up or dangerous exuberance); teaching them to tolerate wheelchairs, loud hydraulic noises, and crutches; and helping them to ride elevators without fear. They bathe and groom the dogs. Then they spend hours every week volunteering their time, along with the dogs, visiting in children's hospitals, cancer wards, rehabilitation facilities, and other places they're now welcome. Some nursing homes, following the so-called Eden philosophy, have resident dogs that circulate with the medicine dispatchers and hang out in the courtyard. The big problem, it seems, is jealousy among patients over feeding and visiting time!

So it takes a community to visit a patient, with canine help as a tool of our material culture. This requires a joint effort by animal shelters and their workers, dog trainers, and volunteers who use their own valuable time out of care for others. It also requires changed institutional rules and development of new procedures. The result, however, is that people who were previously feeling alone, certainly marginalized, and even outcast are now embraced. It is part of a social change that brings people with various disabilities and infirmities into mainstream activities and places of daily life. It changes *place* to be more supportive, caring, and health-promoting.

There are several directions research can go from here. There have been a few (very few) efforts to look at the spatial dimensions of traditional medicine. There is a hierarchy of expertise and professional reputation among Chinese practitioners in Malaysia, for example, which draws clients from varying distances (catchment areas). In addition, where do practitioners of various forms of traditional medicine get their pharmacopoeia? Where is it grown, how marketed? How are new practitioners recruited and trained? Now that practitioners and clients are all moving to the city, what is happening to all these traditional spatial arrangements? By and large, however, research has not gone in that direction.

QUICK REVIEW 2

✓ Western medical systems are not the only way health care is provided. There are a variety of alternative or traditional systems that have deep cultural and historical roots in many countries around the world.

✓ Traditional systems vary greatly, and are difficult to compare, but many view illness and disease as having a religious or psychosocial aspect.

✓ Ayurvedic medicine, unani, Galenic medicine, Chinese medicine, and biomedicine are all "professional" systems that are highly organized and have established arrays of techniques and codes of conduct.

✓ In Chinese medicine, disease is considered to be the result of disharmonies within the body and are often related to interactions between humans and the environment.

✓ Biomedicine is what is practiced in most Western countries and is based on the germ theory of disease.

✓ In many parts of the world, more than one type of medical system is used at a time, a practice known as "medical pluralism."

CONCLUSION: TRANSFORMING THE HEALTH SERVICE LANDSCAPE

There is a greater, larger movement abroad in developed countries as the processes of cultural ecology work on changing the cultural behavioral interactions with the habitat that have proved so harmful for human health. Our readers may remember that this has happened before. Many infectious diseases had been decreased and even eliminated during the late 19th and early 20th centuries without knowledge of germs by transformations of the built, social, and natural environments. Nutrition and the body's ability to fight infection and survive had been improved by transporting, storing, and (later) refrigerating food and making it available in the cities in the winter, even to the poor (who were then most of the people). Rising incomes and standards of living in the West had improved housing (away with mildew, dampness, chills, and darkness) and clothing and bathing (away with lice, fleas, and encrustations of filth); chimneys and (later) central heating had provided warmth (away with internal wood smoke, cold, and winter peaks of mortality). The "garden city" movement, as governments and industrialists faced the horrendous housing conditions and squalor of the new industrial cities, led to planned gardens (e.g., New York City's Central Park), planting of trees, and regulations on architecture (to lessen crowding and increase light and ventilation). Sanitation systems were created, and chlorinated water was provided for drinking, cooking, and hygiene. The worst of the coal dust was removed from the air. City children were sent to camp in fresh air to exercise and build healthy bodies. In New York City and elsewhere, the upper class mobilized to provide and deliver milk to the most impoverished children of new immigrants. When germs were discovered, vaccines were created as cultural buffers; behaviors that avoided germs were created; practices of hygiene and whole infrastructural systems of public health were developed. Infectious diseases were greatly reduced, controlled, or even eliminated before antibiotics. Remember how the concentrated livers of fish produced relief from bone disease and pelvic malformation even before vitamins were known? The adaptability of humans changed their environment and behavior in ways that created better states of health.

We humans are at it again. Now our main causes of death are noncommunicable, usually degenerative diseases, as well as the violence of automobile accidents, suicide, and murder. Nutrition again has gotten out of balance, now producing obesity and higher risk of disease. We live in a chemical soup, and much of it is consumed with our food. Today a major public health campaign has begun to change our diet—to get the fats, sugars, and chemicals out of it. School lunch programs are being overhauled. Farmers' markets and other local means of purchasing organically grown, fresh produce are increasing exponentially. Genetically modified crops have their own dangers mostly related to a decrease in crop biodiversity, but they promise to almost remove pesticide and fungicide spraying from the environment. Energy change will be removing the particles, heavy metals, and benzene from the air, along with the carbon. Smoking has decreased in most high-income countries, though many low-income countries have yet to tackle the problem. People are urged to exercise, even just to walk. The consequences of all this can be seen in falling rates of heart disease and stroke—for decades the major causes of death.

These shifts in disease burden coupled with the life expectancy gains that accompany economic development will also change the ways in which we provide health care. We need to develop new ways of managing people's health throughout the lifecourse—from infancy to old age—so that the exposures, diseases, dietary problems, and socioeconomic inequalities that occur earlier in life aren't given the opportunity to affect health and quality of life down the road. This is difficult in a disorganized and fragmented health system like that in the United States, but has been achieved in social health systems like those in the United Kingdom and Norway. Investment in and development of health resources will also need to shift in order to address the needs of an aging population. Doctors who focus on geriatrics, cardiac conditions, nephrology (kidney), and pulmonary problems will be in high demand. Longer life expectancy means we as a society have more incentive to ensure high quality of care early in life, so that we do not end up financially responsible (through Medicare and Social Security) for complex degenerative conditions that could have been prevented. We have already seen a shift in the types of drugs pharmaceutical companies are researching; drugs that lower blood pressure or cholesterol, fight off the effects of Alzheimer's and Parkinson's, or treat mental health disorders are more prevalent than ever before. Western health systems have even begun to see the importance of integrating complementary and alternative medicine, including massage, acupuncture, and herbal remedies, to maintain healthy populations and so are becoming more pluralistic. Government health systems in rapidly urbanizing developing countries are struggling to provide quality health care for the rapidly growing population (discussed in more detail in Chapter 10). At the same time, providing care for sparse rural populations cannot be ignored but is resource-intensive and costly.

Changes in the built environment, nutrition, behavior, and cultural buffers can only be a human ecology of building better states of health. The questions of regionalization of care and locational allocation of health resources will need to be revisited, and new systems developed. Our understanding of rural and urban health systems, their similarities and differences, need to be better understood. There is a significant amount of work ahead for intrepid health geographers interested in the geography of health services!

REVIEW QUESTIONS

1. What are the main types of access problems a low-income country has (one that is early in the demographic transition)? As a country moves through the demographic and mobility transitions and begins to urbanize, how do problems accessing care change?

2. You work with a low-income community in Denver and want to know how to improve access to basic health care in this setting. What questions do you ask the community about the problems they encounter getting care? What sorts of resources would you look at in the community?

3. Discuss how you might structure the health system to most adequately provide care to rural populations in the United States. How might this differ for rural areas in low-income settings like Malawi or Uganda?

4. Your company wants to explore the possibility of building a new hospital in a city in the United States. What geographic tools or approaches would you use to decide where to locate the hospital?

5. What is medical pluralism? How does the health system in China exemplify it? Is the health system in the United States pluralistic? If so, how?

6. How would you measure the unequal distribution of health resources in the United States? Would this differ if you wanted to measure resources in a low-income country? If so, how?

7. Why are atlas projects like the *Dartmouth Atlas* series important in the field of public health? What unique information do they provide?

8. Why is a pyramidal structure (like the example in Malaysia) an efficient way of providing care?

9. How has GIS helped with our understanding of access to care? If you were studying access to care in your hometown, how would you use a GIS?

10. Do you think the distance decay concept applies in highly mobile and well-connected countries like the United States? If so, how? If not, why not?

REFERENCES

Access and the Provision of Care

Anderson, G. F., & Poullier, J.-P. (1999). Health spending, access, and outcomes: Trends in industrialized countries. *Health Affairs, 18,* 178–192.

Barker, A., McBride, T. D., Kemper, L. M., & Mueller, K. (2014). Geographic variation in premiums in health insurance marketplaces. *RUPRI Rural Policy Brief, 10,* 1–4.

Dartmouth Medical School, Center for the Evaluative Clinical Sciences. (1996). *The Dartmouth atlas of health care 1996.* Chicago: American Hospital.

Dartmouth Medical School, Center for the Evaluative Clinical Sciences. (1998). *The Dartmouth atlas of health care 1998.* Chicago: American Hospital.

Dartmouth Medical School, Center for the Evaluative Clinical Sciences. (1999). *The Dartmouth atlas of health care 1999: The quality of medical care in the United States.* Chicago: American Hospital.

Dear, M., & Wolch, J. (1987). *Landscapes of despair: From deinstitutionalisation to homelessness.* Cambridge, UK: Polity Press.

Gober, P. (1997). The role of access in explaining state abortion rates. *Social Science and Medicine, 44,* 1003–1016.

Godlund, S. (1961). *Population, regional hospitals, transportation facilities, and regions: Planning the location of regional hospitals in Sweden* (Lund Studies in Geography Series B, Human Geography No. 21). Lund, Sweden: Department of Geography, Royal University of Lund.

Gottret, P., & Schieber, G. (2006). *Health financing revisited: A practitioner's guide.* Washington, DC: The World Bank.

Graetz, I., Kaplan, C. M., Kaplan, E. K., Bailey, J. E., & Waters, T. M. (2014). The U.S. Health insurance marketplace: Are premiums truly affordable? *Annals of Internal Medicine, 161,* 599–604.

Hart, L. G., Pirani, M. J., & Rosenblatt, R. A. (1991). Causes and consequences of rural small hospital closures from the perspectives of mayors. *Journal of Rural Health, 7,* 222–245.

Hunter, J. M., Shannon, G. W., & Sambrook, S. L. (1986). Rings of madness: Service areas of 19th century asylums in North America. *Social Science and Medicine, 23,* 1033–1050.

McLafferty, S. L. (1989). The politics of privatization: State and local politics and the restructuring of hospitals in New York City. In J. L. Scarpaci (Ed.), *Health services privatization in industrial societies* (pp. 130–151). New Brunswick, NJ: Rutgers University Press.

Messina, J. P., Shortridge, A. M., Groop, R. E., Varnakovida, P., & Finn, M. J. (2006). Evaluating Michigan's community hospital access: Spatial methods for decision support. *International Journal of Health Geographics, 5,* 42–60.

Mick, S. S., Lee, S.-Y. D., & Wodchis, W. P. (2000). Variations in geographical distribution of foreign- and domestically-trained physicians in the United States. *Social Science and Medicine, 50,* 185–202.

Morrill, R. L., & Earickson, R. J. (1968). Variation in the character and use of Chicago area hospitals. *Health Services Research, 3,* 224–238.

Morrill, R. L., & Earickson, R. J. (1969). Location efficiency in Chicago hospitals. *Health Services Research, 4,* 127–145.

Peters, D. H., Garg, A., Bloom, G., Walker, D. G., Brieger, W. R., & Rahman, M. H. (2008). Poverty and access to health care in developing countries. *Annals of the New York Academy of Science, 1136,* 161–171.

Peters, J., & Hall, G. B. (2000). Assessment of ambulance response performance using a geographic information system. *Social Science and Medicine, 49,* 1551–1566.

Pyle, G. F. (1971). *Heart disease, cancer, and stroke in Chicago: A geographical analysis with facilities, plans for 1980* (Research Paper No. 134). Chicago: University of Chicago, Department of Geography.

Ricketts, T. C., & Cromartie, E. (1992). Rural primary care programs in the United States. In W. M. Gesler & T. C. Ricketts (Eds.), *Health in rural North America* (pp. 179–205). New Brunswick, NJ: Rutgers University Press.

Ricketts, T. C., Savitz, L. A., Gesler, W. M., & Osborne, D. N. (1994). *Geographic methods for health services research*. Lanham, MD: University Press of America.

Rosenbaum, S. (2011). The Patient Protection and Affordable Care Act: Implications for public health policy and practice. *Public Health Reports, 126*(1), 130–135.

Scarpaci, J. L. (1989). Dismantling public health services in authoritarian Chile. In J. L. Scarpaci (Ed.), *Health services privatization in industrial societies* (pp. 219–244). New Brunswick, NJ: Rutgers University Press.

Scarpaci, J. L. (1990). Medical care, welfare state and deindustrialization in the Southern Cone. *Environment and Planning D, 8,* 191–209.

Shannon, G. W., & Dever, G. E. A. (1974). *Health care delivery: Spatial perspectives.* New York: McGraw-Hill.

Smith, C. J. (1998). Modernization and health care in contemporary China. *Health and Place, 4,* 125–139.

Smith, C. J., & Fan, D. (1995). Health, wealth, and inequality in the Chinese city. *Health and Place, 1,* 167–177.

Sommers, B. D., Chua, K. P., Kenney, G. M., Long, S. K., & McMorrow, S. (2015). California's early coverage expansion under the Affordable Care Act: A county-level analysis. *Health Services Research, 51*(3), 825–845.

Wennberg, J. E., Fisher, E. S., Goodner, D. C., & Skinner, J. S. (2008). The Dartmouth Institute for Health Policy and Clinical Practice, *Tracking the cases of patients with severe chronic illness.* Chicago: American Hospital.

World Health Organization. (2008). *The World Health Report 2008: Primary health care now more than ever.* Geneva, Switzerland: Author.

Yasenovskiy, V., & Hodgson, J. (2007). Hierarchical location–allocation with spatial choice interaction modeling. *Annals of the Association of American Geographers, 97*(3), 496–511.

Cultural Alternatives and Health Promotion

Dear, M. J., & Wolch, J. R. (1987). *Landscapes of despair.* Princeton, NJ: Princeton University Press.

Gesler, W. M. (1984). *Health care in developing countries.* Washington, DC: Association of American Geographers.

Good, C. M. (1987). *Ethnomedical systems in Africa: Patterns of traditional medicine in rural and urban Kenya.* New York: Guilford Press.

Kleinman, A. (1993). What is specific to Western medicine? In W. F. Byrnum & R. Porter (Eds.), *Companion encyclopedia of the history of medicine* (pp. 15–23). New York: Routledge.

Metzel, D. S., & Philo, C. (Eds.). (2005). Geographies of intellectual disability [Special section]. *Health and Place, 11*(2).

Phillips, D. R. (1990). *Health and health care in the Third World.* New York: Longman Scientific & Technical.

Vaughn, M. (1991). *Curing their ills: Colonial power and African illness.* Palo Alto, CA: Stanford University Press.

Wilson, C. C., & Turner, D. C. (Eds.). (1998). *Companion animals in human health.* Thousand Oaks, CA: Sage.

📖 FURTHER READING •

Buor, D. (2003). Analysing the primacy of distance in the utilization of health services in the Ahafo-Ano South district, Ghana. *International Journal of Health Planning and Management, 18,* 293–311.

Buor, D. (2004). Gender and the utilization of health services in the Ashanti Region, Ghana. *Health Policy, 69,* 375–388.

Eyles, J. (1987). *The geography of the National Health.* London: Croom Helm.

Farmer, P., Kim, J. Y., Kleinman, A., & Basilico, M. (2013). *Reimagining global health: An introduction.* Berkeley & Los Angeles, CA: University of California Press.

Fulder, S. (1996). *The handbook of alternative and complementary medicine.* Oxford, UK: Oxford University Press.

Gatrell, A. C. (2002). *Geographies of health: An introduction.* Oxford, UK: Blackwell.

Joseph, A. E., & Phillips, D. R. (1984). *Accessibility and utilization.* London: Harper & Row.

Laditka, J. N. (2004). Physician supply, physician diversity, and outcomes of primary health care for older persons in the United States. *Health and Place, 10,* 231–244.

Loytonen, M., & Gatrell, A. (Eds.). (1998). *GIS and health.* London: Taylor & Francis.

Madge, C. (1998). Therapeutic landscapes of the Jola, the Gambia, West Africa. *Health and Place, 4,* 293–312.

Mayhew, L. D. (1986). *Urban hospital location.* London: Allen & Unwin.

Meade, M. S. (1986). Geographic analysis of disease and care. *Annual Review of Public Health, 7,* 313–335.

North Carolina Rural Health Program website: *www.shepscenter.unc.edu/research_programs/Rural_Program/rhp.html*

Oppong, J. R. (1996). Accommodating the rainy season in Third World location–allocation applications. *Socio-Economic Planning Sciences, 30*(2), 121–137.

Oppong, J. R., & Hodgson, M. J. (1994). Spatial accessibility to health care facilities in Suhum District, Ghana. *Professional Geographer, 46*(2), 199–209.

Scarpaci, J. L. (Ed.). (1989). *Health services privatization in industrial societies.* New Brunswick, NJ: Rutgers University Press.

Smith, C. J. (Ed.). (1988). *Public problems: The management of urban distress.* New York: Guilford Press.

Stribling, D. E. (1983). *Holistic health: A changing paradigm for cultural geography.* Unpublished doctoral dissertation, University of North Carolina.

Techatraisak, B. (1985). *Traditional medical practitioners in Bangkok: A geographical analysis.* Unpublished doctoral dissertation, University of North Carolina.

Thomas, R. (1992). *Geomedical systems.* London: Routledge.

CHAPTER 14

Concluding Words

This textbook describes the geographic approach to studying disease and promoting health. Health-related phenomena have a certain spatial distribution, move in certain directions at varying speeds, and affect people's perceptions of their communities and surrounding environments. Health geographers who examine these spatial processes draw on the concepts and techniques of all the subdisciplines of geography. New diseases emerge from the changed landscape and patterns of settlement of the Earth, and from disease agents' adaptations to our biomedical solutions that have neglected ecological considerations; human population fertility seems to be falling around the world to below replacement levels, and populations are aging and urbanizing; globalization brings democratization of technology and information, but also social and economic change at a pace not seen before, which leads to environmental degradation and health problems of every kind. As these things and more happen over the next several decades, geographers need to bring their focus on scale, space, place, and a holistic approach to understanding health to the public discourse and to the promotion of health.

We have taken you, our readers, on a journey through culture, landscape, and evolution. We have described the power of new ideas, concepts, theories, and methodologies to advance the understanding of health and disease. We have examined why new diseases are emerging, and why old ones have spread or disappeared, as processes of adaptability. Processes of domestication and ecological simplification have resulted in the production of enough food to feed billions of people, but also in many epidemics; we have explained why. Urbanization and older age structures have become worldwide phenomena and will profoundly affect your health in the future; we have explained why. Climate change is upon us and will affect our health in many ways we do not yet understand, some quite local and others more widespread. Movement and variation in space simply are—all around us, and about everything. We have looked at increasingly sophisticated ways to map, analyze, model, integrate, and understand. It is possible, but extremely difficult, to separate the influences of genetics, environment, and group or individual behavior on health; again, we have explained why.

The geography of health care freely incorporates concerns and findings from other disciplines about the social, economic, political, and cultural behavior of individuals and systems. In turn, it contributes its geographic perspective to the emergence of a social science of health care. Similarly, epidemiological design and methodology, parasitology, entomology, microbiology, the anthropology of medical beliefs, and the sociology of groups, along with many other disciplinary insights, are synthesized and used in explaining the spatial distribution of disease occurrence. Health geographers must become better trained in the cognate fields relevant to their specializations and regularly draw on these other disciplines.

We have discussed problems that arise in available medical statistics and in obtaining microlevel data. These issues are changing rapidly as the importance of digitized data and of spatial analysis becomes ever more widely accepted. More important than data limitations to the beginning health geographer is awareness that ignorance of basic and relevant biology can cause research hypotheses even about spatial form to be deficient, and that ignorance of basic sociocultural processes can result in overly simplistic genetic or environmental explanations. Spatial perspective, knowledge about differences of scale and spatial autocorrelation, and general familiarity with physical and social sciences can result in fresh geographic insights, new hypotheses, and sounder planning and policy. Spatial modeling without understanding of the disease system is sterile and perhaps irrelevant.

As we hope this text illustrates, the literature of health geography is far reaching, complex, and exciting. There is a rich diversity to the research—a hybrid vigor and vibrancy derived from the mixing of many new ideas. Individual health geographers may feel most comfortable with some segment of the subdiscipline. Some wish to be more theoretically informed; others explore spatial analysis of diffusion in a positivist manner; and still others focus on microlevel integration and qualitative understanding. Most still pursue the ancient geographic concerns: integrating all the phenomena within space, in order to understand the nature of place and explain the distribution of varied phenomena over space. The subdiscipline is now at a point where most of its practitioners understand that integrating more than one of these areas of expertise will ensure that studies are more complete and relevant. In the age of interdisciplinary scholarship, health geographers of different backgrounds are increasingly working together in research teams.

Employment opportunities for health geographers are abundant, but usually at least a master's degree in the field is needed (although competence in geographic information systems [GIS] can qualify those with a bachelor's degree for some applied geography jobs). Besides holding academic positions, health geographers work for such organizations as the Centers for Disease Control and Prevention, the National Center for Health Statistics, the Research Triangle Institute, Family Health International, the Carter Center's Guinea Worm Eradication Program,

Measure DHS, IFC International, and state health departments, to name just a few. The term "spatial epidemiology" is appearing more often in both the geographic and public health literature. This sometimes translates as "the application of GIS to disease and health problems," but a more sophisticated theoretical and methodological understanding of working with spatial data is necessary to do this type of work well. Students who are interested in this field should pursue training in GIS and mathematics/statistics, as well as geography broadly considered.

Health geography as a well-defined subdiscipline of geography has existed for only a few decades. During that time it has grown from a field with a few individuals producing sometimes excellent work to a corps of researchers and other practitioners producing a substantial and varied body of work. As it has matured, it has become more introspective and critical. The intellectual ties of the geographic perspective link together different projects, data sources, techniques, scales, hypotheses, and general paradigms to contribute a different voice to understanding. Health geography has reached the point as a body of literature that students and health researchers, generally, can expect to draw on before initiating research projects or health program implementation. It has reached the point where a text such as this, drawing from the work of hundreds of individuals and publications, can be written. Its perspectives are basic to addressing broader health questions about the role of social institutions, the consequences of environmental management or misuse, and the impact of sociocultural roles and perceptions—questions that are too frequently ignored. The health impacts of globalization and climate change will still be a matter of "airs, waters, and places," as Hippocrates put it almost 2,500 years ago.

REVIEW QUESTIONS

1. One of the review questions in Chapter 1 asked you to describe "the challenge of health geography" and whether you think that the challenge is being met. Now that you have read the entire book and are likely to finish a university course in health geography, answer the question again by using what you have learned.

2. Chapter 1 emphasizes how the field of health geography involves multiple perspectives within the field of geography and other disciplines and each subsequent chapter of the book described a specific area of the field. Examine the course catalog at your university and list courses that are offered in geography and cognate disciplines that will provide systematic specializations of study within the field of health geography. List the classes that you would like to take at your university that would help prepare you to conduct a health geography study.

3. Choose one chapter of this book and one health outcome that you are interested in studying. Develop an outline of a health geography research study that you would like to implement. What theories or frameworks would underpin the study (e.g., triangle of human ecology, neighborhoods and health) and what data would you use or collect? Explain why the study should be categorized as a health geography study.

Glossary

Acclimatization	Genetically given capacity of humans to adapt their physiological systems over long periods to heat, cold, and altitude.
Acute diseases	Those with symptoms that are severe and whose course is short.
Address matching	The geocoding process that involves taking a street address and converting it to a geographic coordinate pair such as latitude and longitude.
Agent	The causative organism of an infectious disease, variously known as a germ, microbe, pathogen, or parasite.
Agricultural extensification	The expansion of agricultural production into new places, such as forests.
Agricultural hearths	Regions where agriculture developed independently, without the introduction of ideas or technology from elsewhere.
Agriculturalists	People who cultivate the soil, farmers.
Air pollution	The contamination of the atmosphere by a harmful level of toxic substances.
Alleles	Genes that occupy the same locus on a specific pair of chromosomes and control the heredity of a particular characteristic.

Antibiotic paradox	The more antibiotics that are used, the more bacteria will develop resistance to those antibiotics.
Aquifer	An underground layer of water-bearing permeable rock.
Arbovirus	Virus transmitted by mosquitoes, ticks, or other arthropods.
Artifactual	Patterns of health outcomes that may result from errors due to changes in the recognition, classification, or reporting of the disease, or from errors in enumerating the population.
Aspatial	In statistics, models or analyses that ignore or are not associated with a space or place.
Asymptomatic	Infection that does not result in symptoms.
Ayurvedic medicine	An alternative medical system originating in the Indus Valley that views disease as a disequilibrium of four humors (wind, bile, mucus, and blood), and cures attempt to reestablish a proper balance.
Basic reproduction number	The number of cases one case generates on average over the course of its infectious period.
Biomedicine	Medical system based on the application of the principles of the natural sciences and especially biology and biochemistry.
Biomes	Different vegetational zones that extend over large areas of the Earth.
Biometeorology	The variation and change in the physical and chemical characteristics of the atmosphere that affect variation and change in the physicochemical systems of living organisms.
Buffers (cultural)	Cultural characteristic that developed to protect people from disease, such as boiling water for tea or chlorinating water.
Carcinogens	Substances or agents that cause cancer in living tissue.
Cartography	The construction and interpretation of maps.
Case control study	A study that compares behaviors and exposures between people who have a disease (or other outcome) or who died of it, and people who did not have it.
Case fatality rate	The proportion of individuals contracting a disease who die of that disease.
Catchment area	In health care, the region or area from which a medical facility receives most of its patients.

Chinese medicine	A medical system that has been used for thousands of years to prevent, diagnose, and treat disease. It is based on the belief that *Qi* (energy or life force) flows in a regular pattern through a system of channels (meridians) to all parts of the body and keeps a person's spiritual, emotional, mental, and physical health in balance. Traditional Chinese medicine aims to restore the body's balance and harmony between the natural opposing forces of yin and yang, which when out of balance can block *Qi* and cause disease. Chinese medicine includes a variety of techniques including herbal medicine, acupuncture, and tai chi.
Chlorofluorocarbons (CFCs)	Organic compounds used as refrigerants, aerosol propellants, and solvents that have been found to contribute to ozone depletion in the upper atmosphere.
Choropleth maps	Thematic maps symbolizing areas by shades, patterns, or colors.
Chronic diseases	Diseases that are present or recur over a long period of time.
Circulation	Population movement that returns to its place of origin
Climate	Weather conditions averaged and prevailing over long periods, many decades or centuries, of time.
Climate change	A change in global climate patterns that has been attributed largely to the increased levels of atmospheric carbon dioxide produced by the use of fossil fuels.
Clinical	Denoting the appearance of symptoms that can be presented to a physician for observation and treatment.
Cohort study	A study of a whole population. It is divided into those with and without a particular exposure, and the frequency of the disease outcome is noted.
Collective efficacy	The ability of a group to mobilize to undertake collective action.
Columbian exchange	The movement of people, plants, animals, and pathogens between the east and west hemispheres starting in the late 15th century.
Communicable disease	A contagious disease that is transmitted through direct contact with an infected individual or indirectly through a vector.
Complementary and alternative medicine	A variety of treatment approaches that fall outside the realm of Western biomedicine. Complementary medicine refers to healing practices that work in conjunction with Western medicine while alternative medicine is used as a substitute for it.

491

Compositional effects	Occur when individuals with the same set of risk factors live in the same area, giving rise to spatial patterns of disease that are not related to the environment.
Concentrated animal feeding operations (CAFOs)	Agricultural enterprises where animals are kept and raised in confined situations.
Confounding variable	A variable that varies in a systematic way with the hypothesized causal relationship being studied. Although A seems to cause B, in fact another variable, C, is affecting both A and B; the relationship between A and B is therefore spurious.
Congenital diseases	Diseases present at birth.
Contagion	Transmission of infectious disease agents between people, which may be direct through person-to-person contact or indirect through the bites of insect vectors or via fomites (vehicles) such as contaminated blankets, money, or water.
Contagious diffusion	The rapid, widespread diffusion of a feature or trend spreading radially to adjacent places.
Contextual effects	These occur when factors in an individual's social and physical environment have a direct impact on health.
Contingency table	A two-way table used in epidemiological studies that is useful for examining relationships between categorical variables.
Crowd diseases	Diseases that have high population thresholds (i.e., a large number of susceptible individuals) necessary for sustained transmission.
Culture realm	Broad cultural area delimited by the extent of particular cultural practices and beliefs.
Degenerative diseases	Those characterized by the deterioration or impairment of an organ or the structure of cells and the tissues of which they are a part.
Demographic transition	The transition from high birth and death rates to low birth and death rates as a country develops from a preindustrial to an industrialized economic system.
Diffusion	Spread or movement outward from a point or beginning place.
Disability-adjusted life years (DALYs)	The years of life lost due to disability. DALYs for a disease or health condition are calculated as the sum of the Years of Life Lost (YLL) due to premature mortality in the population and the Years Lost due to Disability (YLD) for people living with the health condition or its consequences.

Dispersed settlement	Population is arranged as individual families separated from one another by farmland or forest.
Distance decay	The use of health services decreases or "drops off" with increased distance between a patient and the location of health care.
Domestication	Bringing a wild plant or animal under human management.
Dot map	Maps that use a dot symbol to show the presence of a feature or phenomenon that are used to visualize spatial patterns.
Ecological fallacy	The idea that associations statistically identified at one scale of analysis are valid at or can be generalized to either larger or smaller scales.
Ecological simplification	Creating landscapes that have few species (e.g., one type of corn plant) but many members of that species, and that lack systematic regulation and feedback that comes from having a diversity of species.
Ecology	The study of the interactions between living organisms and their biotic and abiotic environments.
Economies of scale	Cost advantages that enterprises obtain due to size, output, or scale of operation, with cost per unit of output generally decreasing with increasing scale as fixed costs are spread out over more units of output.
Ectoparasites	A parasite, such as a flea, that lives on the outside of its host.
Endemic disease	A disease that is constantly present in an area.
Environmental justice	An area of inquiry that investigates the inequitable environmental burden borne by minorities and people living in economically disadvantaged areas.
Enzootic	Constantly present in an animal population.
Epidemic diseases	Those that occur at levels clearly beyond normal expectation.
Epidemiological transition	A shift from a cause-of-death pattern dominated by infectious diseases with very high mortality, especially at younger ages, to a pattern dominated by chronic diseases and injuries with lower mortality, mostly peaking at older ages.
Epigenetics	Phenotypic trait variations that are caused by external or environmental factors that switch genes on and off and affect how cells read genes instead of being caused by changes in the DNA sequence.

Epistemology	The theory of knowledge.
Epizootic	Disease that is temporarily prevalent and widespread in an animal population.
Ergotism	Poisoning produced by eating food affected by ergot, the Claviceps purpurea fungus, typically resulting in headache, vomiting, diarrhea, and gangrene of the fingers and toes.
Ethnomedicine	The study of beliefs and practices stemming from indigenous cultural development.
Etiology	A cause or set of causes of a disease.
Expansion diffusion	The spread of a feature or trend among people from one area to another in a contagious process.
Experimental study design	A study wherein individuals are randomized into a treatment group and a control group. The treatment group is exposed to some type of intervention, which could be anything from a medical intervention like a new vaccine to a behavioral intervention like a smoking cessation program. The control group carries on as usual. At the end of the experimental period outcomes between the two groups are compared.
Familial cancers	Cancers with a specific gene that has a defined inheritance pattern and so can be "passed down" from parent to child.
Flow maps	Maps that show direction of movement.
Fomite	An object or substance capable of carrying infectious organisms and transferring them from one individual to another.
Food deserts	Areas with restricted access to healthy food, both in terms of financial access and physical access to stores selling fresh fruits and vegetables, whole grains, and lean meats.
Foragers	Individuals who collect food rather than growing it.
Galenic medicine	An alternative medical system originating in ancient Greece that views health as a proper balance of four humors (blood, phlegm, yellow bile, and black bile). Physical, cultural, and demographic factors were thought to cause one of the four humors to dominate and produce ill health.
Geographic information systems (GIS)	Computer systems for the capture, storage, retrieval, analysis, and display of spatial data.
Geophagy	Eating soil to satisfy nutritional needs.
Germ theory	The idea that infectious diseases are caused by microorganisms.

Gini coefficient	A coefficient that measures the extent to which the distribution of income (or some other characteristic) among individuals or households within a region deviates from a perfectly equal distribution.
Global positioning system (GPS)	A system of Earth-orbiting satellites, transmitting signals that enable the locational position of a receiving device on the Earth's surface to be estimated.
Global warming	The gradual increase in the average temperature of the Earth's atmosphere and its oceans.
Globalization	International integration among people, companies, and governments including the interchange of products, ideas, and other aspects of culture.
Gravity model	A model that predicts movement of people, information, and commodities between places based on a modified law of gravitation that takes into account the population size of two places and their distance.
Green Revolution	A set of research, development, and technology transfer initiatives taking place primarily in the 1960s and 1970s that dramatically increased agricultural output, particularly in Asia.
Health	World Health Organization definition: A state of complete physical, mental, and social well-being and not merely the absence of disease or infirmity; Dubos definition: States of health or disease are the expressions of the success or failure experienced by the organism in its efforts to respond adaptively to environmental challenges; Audy definition: Continuing property that can be measured by an individual's ability to rally from a wide range and considerable amplitude of insults, the insults being chemical, physical, infectious, psychological, and social.
Health resources	Aspects of the medical system such as hospitals, doctors, nurses, clinics, beds, and skilled nursing facilities.
Herbalists	Healers who focus on the use of medicinal plants to cure illness.
Herd immunity	Threshold proportion of immune individuals that should lead to a decline in incidence of infection.
Hierarchical diffusion	The spread of a feature from one place to another through the hierarchy of urban places, from large places to smaller.
High fructose corn syrup (HFCS)	A sweetener made from corn starch.
Host	The organism infected by a disease agent.

Human agency — The capacity of a person to act on his or her choices in any given environment.

Human ecology — A descriptive term applied to complex relationships between organisms and their environment.

Human ecology of disease — The ways in which human behavior interacts with environmental conditions to produce or prevent disease among susceptible people.

Hygiene hypothesis — Lack of exposure to infectious disease agents in childhood, such as overuse of antibacterial household products or lack of playing outdoors, does not adequately train the immune system, resulting in overreaction when exposure does take place.

Hyperendemic diseases — Those that occur with intense transmission.

Hypoendemic diseases — Those that occur at low levels, occasionally popping up here or there.

Iatrogenic — Illness or disease caused by doctors or medical treatment.

Immunologically naïve — An individual or population that has never before been exposed to a disease agent; thus the individual or the population is/are totally susceptible to the disease produced by the agent.

Immunosuppression — Decreased functioning of the immune system, either as a result of infection or treatment with drugs.

Incidence — The number of cases of a disease diagnosed or reported for a population during a defined period of time, most commonly a year. It refers to new cases.

Incubation — Time for a disease agent to adapt to its host and multiply and become numerous in its mode of transmission (respiratory droplets, body fluids, etc.).

Index of segregation — A method for measuring how grouped or clustered individuals in the same racial or ethnic group are.

Infectious diseases — Those that result from the activities of living creatures, usually microorganisms, that invade the body.

Informal social control — Social interactions among members of a community that maintain social order (e.g., the willingness of community members to influence and regulate the deviant behavior of others).

Intensification — In reference to agriculture, the increased production of food on the same amount of land. In reference to people and animals, the increased frequency and duration of contact between species.

496

Intermediate hosts	Organisms that are necessary to some stage of an agent's life cycle.
Internally displaced person	Someone who is forced to flee his or her home but who remains within his or her country's borders.
Labor-shed	A region within which people are available and willing to work.
Landscape epidemiology	The study of how disease circulates in human and animal populations and is influenced by environmental conditions.
Latency period	Time between when the infection occurs and the appearance of clinical symptoms.
Linear settlement	Population is arranged along a linear feature, such as a road or river.
Location quotients	A way of quantifying how concentrated a particular industry, or aspects of an industry such as physicians or pharmacies, is in a region.
Locational analysis	Studies the location of health resources in relation to populations in need of care, and attempts to optimize the location of these resources in order to reduce inequalities.
Macroscale studies	These study health by looking at aggregates (e.g., counties, regions, or states) and should be used to describe rather than to explain general patterns of resource distributions.
Managed care	Care refers to the efforts of organizations such as insurance companies and large health maintenance organizations to hold down medical spending by "managing" the financial aspects of the practice of medicine.
Map scale	The amount of area being considered on a map; the ratio between the dimensions of a representation and those of the object being represented, as in "1 inch to a mile."
Maslow's hierarchy of needs	A psychological theory that attempts to understand what motivates humans: physiological needs such as air and water and food must be met before other needs can be satisfied.
Medical pluralism	The use of more than one kind of medicine at the same time.
Megacity	A city with more than 10 million residents.
Metastasis	When cancer spreads from the part of the body where it started (its primary site) to other parts of the body.

Microscale studies	Studies that focus on individuals and the decisions they make. They are the preferred approach to analyzing resource location factors; at the microlevel the actual mechanisms behind resource distribution decisions are clearer.
Microbiome	The world of other species that live inside of each human, which outnumber our own cells by a ratio of at least 10 to one.
Migration	Crossing a political boundary and moving with the intention of permanence.
Millennium Development Goals	Goals developed by the United Nations in 1990 to try to tackle some of the greatest challenges for the poorest people in the world during a 15-year period.
Mixed traffic patterns	Arrangements of motor vehicles and people that do not separate pedestrians or bicyclists from larger vehicles, characterized by lack of sidewalks, crosswalks, lane markers, and other infrastructure.
Modifiable areal unit problem	A problem arising in spatial analysis because the scale at which one chooses to analyze information, or the zones or grouping schemes, can produce different results.
Monocropping	Occurs when an agriculturalist plants the same plant, year after year, without crop rotation.
Monoculture	All plants in a field are the same, as opposed to intercropping or mixed cropping, where there are many species in a single field.
Moran's *I* statistic	A statistic used to quantify spatial autocorrelation.
Mutagen	Any agent (a chemical, radiation, UV light, etc.) that can induce or increase the frequency or extent of mutation.
Natural increase	Population growth due to the difference between births and deaths.
Natural nidus	A microscale region constituted of a living community, among the members of which a disease agent continually circulates, and the habitat conditions necessary to maintain that circulation in the disease system.
Neighborhood	In neighborhood health studies, a unit analysis that represents an area within which people live and interact with their environment and neighbors.
Neighborhood approach	A conceptual framework that is spatial in nature and focuses on the shared environments and the interactions and relationships that occur among individuals who reside in an area, which in turn produce health behaviors and outcomes.

Nosocomial	Infections acquired in hospital or other health care settings.
Nuclear settlement	Population clusters in villages or other settlements, surrounded by forest or farmland.
Nucleotides	Building blocks of DNA, consisting of four types: adenine, thymine, guanine, and cytosine.
Nutrition transition	The gradual transition of populations to a diet that is high in fat and sugar, particularly through the consumption of meats, dairy, and processed foods.
Obesogenic	An environment that is conducive to obesity.
Odds ratio	The number of times an event occurs relative to the number of times it doesn't occur; that is, the ratio of event to no event in exposed individuals divided by unexposed individuals.
Omnivorous	Diet that consists of both plant and animal products.
Oncogene	A gene that has the potential to cause cancer.
Oral rehydration therapy (ORT)	Life saving and simple combination of salt, sugar, and clean water, used to treat diarrheal diseases.
Overlay analysis	A fundamental method in a GIS in which the user is mathematically integrating multiple map layers together, which can be done using both raster and vector operations.
Overnutrition	Increasingly negative health effects, such as diabetes, are associated with an *abundance* of calories and protein.
Paleopathological	Pathological conditions found in ancient human and animal remains.
Pandemic	A disease that is prevalent throughout an entire country, continent, or the whole world; epidemic over a large area.
Pathogens	Agents that cause disease.
Phylogenetic	The ancestral or evolutionary relationships among species or individuals within species.
Political ecology	A field of study that examines the intersection of political, economic, social, and environmental systems that can shape health across spatial and temporal scales. It explores the political and economic realities surrounding decisions to transform the natural environment by examining the links between actors that occur across multiple scales.
Population-sheds	Areas or regions within which a population circulates.

Precautionary principle This principle states that when there is a clear indication of harm (to human life and health), that indication should be a trigger for action even if the scientific evidence of proof is not complete, and especially if delay may cause irreparable damage.

Prevalence The number of people in a population sick with a disease at a particular time, regardless of when the illness began.

Primate city The largest city in its country or region, disproportionately larger than any others in the urban hierarchy, usually the main site of government, culture, education, etc.

Prophylactic A medicine or course of action used to prevent disease.

Prospective study A study that is started before exposure to an agent and in which people are followed forward in time.

Push-and-pull factors Factors that influence where a person moves to (pull) and why they choose to move (push).

Quarantine To separate and restrict the movement of persons to stop the spread of disease.

Race A social construct based upon perceived ethnic differences in skin color, hair type, etc., that are now manifested in socioeconomic, educational, and other differences.

Radon An odorless, colorless gas that arises from the natural decay of radioactive elements commonly found in rocks and many types of soil; radon exposure has been found to be associated with some cancers.

Raster A cell-based GIS data structure.

Realms of evolution Biomes separated by barriers, such as oceans or mountains, that followed separate evolutionary paths.

Refugee A person who has been forced to leave his or her country in order to escape war, persecution, or natural disaster.

Region A bounded area that is internally homogeneous according to one or more criteria.

Relative risk The risk of a health event in those with a risk factor divided by the risk of an event in those without the risk factor.

Relocation diffusion The spread of a feature or trend through movement of people from one place to another.

Remote sensing The science of obtaining information about objects or areas from a distance including from satellites and airplanes.

Replacement-level fertility The number of children a woman must have on average to produce only enough children to replace herself and her partner in the population.

Reservoir	Animal hosts that serve as a continuing source of possible infection for human beings.
Retrospective study	A study design where the disease outcome has already happened and its history is reconstructed backward in time through interviews and records.
Risk	The number of times (frequency) an event occurs relative to the total number in the study population.
Secondary attack rate	The probability that infection occurs among susceptible persons following known contact with an infectious person.
Sedentary agriculturalists	Farmers who utilize the same land year after year, in opposition to nomadic agriculturalists or shifting cultivators.
Selection	The act of choosing certain members of a species to reproduce in subsequent generations, typically based upon some desired quality held by that member, such as larger fruits or earlier ripening.
Selection bias	This bias results when the selection of subjects into a study or their likelihood of being retained in the study leads to a result that is different from what you would have gotten if you had enrolled the entire population. In neighborhood studies, selection bias occurs when individual (or family) characteristics that are related to health are also related to the neighborhood an individual has chosen to live in.
Silent zone	A place where disease is circulating but is unrecognized because no humans are present.
Slum	Living conditions that are defined as lacking the following five things: sanitation, clean water supplies, durable housing, sufficient living area, and secure residential status.
Small-area analysis	Statistical studies done for small geographic areas, such as census tracts or city blocks.
Smog	Air pollution, usually in large cities, caused by combined smoke and fog.
Social capital	The benefits derived from the social connections that develop between individuals or groups of individuals.
Social contagion	The spread of behaviors, behavioral norms, and information through social networks.
Social disorganization theory	This theory identifies three structural factors—low economic status, ethnic heterogeneity, and residential instability—which lead to the disruption of community social organization and, potentially, poor health outcomes.

Social distance	Social situations, such as the attitudes of doctors and nurses, that can put distance between a patient and a doctor such that care is not provided effectively.
Social ecology	The study of the relationships between individuals, social groups, and their environments.
Sociopathological complex	The phenomenon that health outcomes appear consistently more often among the poor than among the affluent are likely the result of stress, lifestyle, diet, housing, pollution, and aging infrastructure.
Spatial autocorrelation	The level of correlation between spatial objects in space because adjacent territories usually influence each other.
Spiritual healers	Practitioners of alternate medicines who mediate among the supernatural, the patient, and the community in order to affect illness and health.
Sporadic cancers	These forms of cancer have no identifiable inherited gene involved. The cancers developed instead as a result of environmental factors (carcinogens such as cigarette smoke) that randomly induce mutations in cells that lead to uncontrolled growth.
Standardization	Conversion of a number to some type of standard score. Age standardization takes two populations with different age distributions and standardizes them to the same population structure.
Staple crops	Food that makes up the dominant portion of a population's diet. Typical staple crops are grains, such as rice or wheat, or starches, such as cassava.
Step migration	Migration to a distant destination that occurs in stages.
Structure	The social, political, and economic contexts that constrain (or enable) an individual's ability to act on their own choices.
Subclinical	When an infectious agent enters the body, multiplies, stimulates the production of antibodies, and is eliminated from the body without the person's being consciously aware of any illness.
Subtherapeutic dose	Consumption of antibiotics for purposes other than treatment of an active infection.
Suitability mapping	A geographic or GIS-based process used to determine the appropriateness of a given area for a particular use.
Syndrome	A set of signs and symptoms that usually occur together, such as acquired immunodeficiency syndrome (AIDS).

t'ai chi chuan	Known also as *quidong,* a practice in traditional Chinese medicine that balances *ch'i* (life energy) flow through the meridians in the body.
Therapeutic	Treatment of a diagnosed condition; in the case of antibiotics, this involves a diagnosed infection.
Therapeutic landscapes	Areas that have a positive effect on a person's health or have a therapeutic effect, as in a natural or spiritual place that promotes health.
TIGER (Topologically Integrated Geographic Encoding and Referencing) files	U.S. Census files used for address matching.
Time–space geography	This field of study examines population movement through space and time using the geospatial "lifeline," a track of what places a person has moved through over time.
Tobler's law of geography	The geographic law that says that things that are close together in space are more likely to have similar values than things that are far apart.
Total fertility rate	The average number of children a woman would have if she went through her life at today's age-specific birth rates.
Toxic Release Inventory	A publicly available database containing information on toxic chemical releases in the United States.
Traditional birth attendants	Women who assist mothers at childbirth and are often trained by helping experienced attendants deliver babies, or through modern programs within clinics and hospitals.
Traditional medicine	The sum total of the knowledge, skills, and practices based on the theories, beliefs, and experiences indigenous to different cultures, used in the maintenance of health as well as in the prevention, diagnosis, improvement, or treatment of physical and mental illness.
Ultraviolet radiation	Radiation in the part of the electromagnetic spectrum that is associated with sunburn.
Unani	A medical system dominant in early Arabic culture.
Urbanization	The process of transition from rural settlements to urban settlements primarily through rural–urban migration.
Vector (disease)	An organism, often an invertebrate arthropod, that transmits a pathogen from reservoir to host.
Vector (spatial data)	GIS data structure associated with spatial objects including points, lines, and polygons.

Virulence	The degree of pathogenicity within a group or species of parasites as indicated by case fatality rates and/or the ability of the organism to invade the tissues of the host.
Vulnerable road users	Individuals using roads who are not protected by the walls of four-wheeled vehicles such as trucks and cars, consisting of pedestrians, bicyclists, motorcyclists, etc.
Walkability	A measure of how friendly a neighborhood is to walking, which can be measured in a variety of different ways.
Water footprint	A way to describe the total amount of water that it takes in all stages of production, from field to factory to store to consumer.
Water-based	A disease dependent on water, as in for vector-breeding habitat.
Water-borne	Diseases transmitted via water.
Water-washed	Diseases preventable by hand/hair/clothes/floor washing and other hygiene practices.
Weather	The daily and weekly conditions of temperature, humidity, precipitation, wind, and cloud cover.
Xeno-zoonoses	An infectious disease transmitted from animal to human by transplantation of an animal tissue or organ into a human body.
Xenograft	Use of tissue from one species in the treatment of another species.
Years of life lost (YLL)	This model measures the burden of disease by counting not only the number of deaths, but the years (and earning power) lost from those lives.
Zoonosis	Infectious disease of animals that can be transmitted to humans.

Index

Note: *f* or *t* following a page number indicates a figure or a table.

About the Authors

Michael Emch, PhD, is Professor and Chair of Geography at the University of North Carolina at Chapel Hill (UNC). He is also Professor of Epidemiology at UNC, a Fellow of the Carolina Population Center, and Director of the Spatial Health Research Group. Dr. Emch has published widely in the subfield of disease ecology, primarily on infectious diseases of the tropical world. He is an associate editor of *Health & Place* and an advisory editor for the international journal *Social Science and Medicine*.

Elisabeth Dowling Root, PhD, is Associate Professor of Geography at The Ohio State University. She is also Associate Professor of Epidemiology in the College of Public Health and a research affiliate at the Institute for Population Research. Dr. Root's work evaluates the short- and long-term impacts of public health interventions—including vaccination campaigns, maternal and child health and family planning programs, and health systems changes—in low-income countries. She is also interested in the long-term effects of neighborhood social and structural environments on child and adolescent health.

Margaret Carrel, PhD, is Assistant Professor of Geographical and Sustainability Sciences at the University of Iowa. She is also Assistant Professor of Epidemiology in the College of Public Health. Dr. Carrel focuses primarily on the geography of infectious disease, with emphasis on how human–environment interactions influence the evolution of pathogens. She is also interested in understanding the impact of food production, particularly of livestock, on human health.